Recent Advances in Atmospheric Science

Recent Advances in Atmospheric Science

Editor: Ronin Massey

Callisto Reference

www.callistoreference.com

Callisto Reference,
118-35 Queens Blvd., Suite 400,
Forest Hills, NY 11375, USA

Visit us on the World Wide Web at:
www.callistoreference.com

ISBN: 978-1-64116-283-8 (Hardback)

Printed in China.

Cataloging-in-Publication Data

Recent advances in atmospheric science / edited by Ronin Massey.
p. cm.
Includes bibliographical references and index.
ISBN 978-1-64116-283-8
1. Atmospheric science. 2. Atmosphere. 3. Earth sciences. I. Massey, Ronin.
QC861.3 .R43 2020
551.5--dc23

Table of Contents

Preface

This book aims to highlight the current researches and provides a platform to further the scope of innovations in this area. This book is a product of the combined efforts of many researchers and scientists, after going through thorough studies and analysis from different parts of the world. The objective of this book is to provide the readers with the latest information of the field.

The study of Earth's atmosphere is known as atmospheric science. It deals with the study of Earth's atmospheric processes, the effects other systems have on the atmosphere and vice versa. Meteorology is the branch of atmospheric science which involves atmospheric physics and chemistry with a focus on weather forecasting. The study of atmospheric changes is known as climatology. It involves the study of atmospheric changes which are both long-term and short-term. It also defines average climates and their changes over time due to both natural and anthropogenic climate variability. The study of upper layers of atmosphere where dissociation and ionization are important, fall under aeronomy. Rocketsondes, radiosondes, satellites, lasers and weather balloons are some of the experimental instruments used in atmospheric science. This book contains some path-breaking studies in the field of atmospheric science. The aim of this book is to present researches that have transformed this discipline and aided its advancement. As this field is emerging at a rapid pace, the contents of this book will help the readers understand the modern concepts and applications of the subject.

I would like to express my sincere thanks to the authors for their dedicated efforts in the completion of this book. I acknowledge the efforts of the publisher for providing constant support. Lastly, I would like to thank my family for their support in all academic endeavors.

Editor

1

30 years of upper air soundings on board of R/V *POLARSTERN*

Amelie Driemel, Bernd Loose, Hannes Grobe, Rainer Sieger, and Gert König-Langlo

Alfred-Wegener-Institut Helmholtz-Zentrum für Polar- und Meeresforschung, Bremerhaven, Germany

Correspondence to: Gert König-Langlo (gert.koenig-langlo@awi.de)

Abstract. The research vessel and supply icebreaker *POLARSTERN* is the flagship of the Alfred-Wegener-Institut in Bremerhaven (Germany) and one of the infrastructural pillars of German Antarctic research. Since its commissioning in 1982, *POLARSTERN* has conducted 30 campaigns to Antarctica (157 legs, mostly austral summer), and 29 to the Arctic (94 legs, northern summer). Usually, *POLARSTERN* is more than 300 days per year in operation and crosses the Atlantic Ocean in a meridional section twice a year. The first radiosonde on *POLARSTERN* was released on the 29 December 1982, 2 days after *POLARSTERN* started on its maiden voyage to the Antarctic. And these daily soundings have continued up to the present. Due to the fact that *POLARSTERN* has reliably and regularly been providing upper air observations from data sparse regions (oceans and polar regions), the radiosonde data are of special value for researchers and weather forecast services alike. In the course of 30 years (29 December 1982 to 25 November 2012) a total of 12 378 radiosonde balloons were started on *POLARSTERN*. All radiosonde data can now be found at König-Langlo (2015, doi:10.1594/PANGAEA.810000). Each data set contains the directly measured parameters air temperature, relative humidity and air pressure, and the derived altitude, wind direction and wind speed. 432 data sets additionally contain ozone measurements.

Although more sophisticated techniques (meteorological satellites, aircraft observation, remote-sensing systems, etc.) have nowadays become increasingly important, the high vertical resolution and quality of radiosonde data remains paramount for weather forecasts and modelling approaches.

1 Introduction

For more than 30 years now, the research vessel and supply icebreaker *POLARSTERN* has been the flagship of the Alfred-Wegener-Institut in Bremerhaven (Germany). The commissioning of *POLARSTERN* on the 9 December 1982, was the result of a political decision to strengthen Germany's role in polar (at that time especially Antarctic) research: in 1978 the Federal Republic of Germany, represented by the Deutsche Forschungsgemeinschaft (DFG), became a member of the Scientific Committee on Antarctic Research (SCAR), and in 1980 the Alfred-Wegener-Institut für Polarforschung was founded (Fütterer and Fahrbach, 2007). R/V *POLARSTERN* was thus bound to become one of the infrastructural pillars of German Antarctic research. Home port of *POLARSTERN* was and still is Bremerhaven, Germany. Since its commissioning, *POLARSTERN* has conducted 30 campaigns to Antarctica (157 legs, mostly austral summer), and 29 to the Arctic (94 legs, northern summer)[1]. Usually, *POLARSTERN* is more than 300 days per year in operation and crosses the Atlantic Ocean in a meridional section twice a year (Fütterer and Fahrbach, 2007). It is therefore the perfect basis for upper air observations in data-sparse regions (i.e. oceans and polar regions).

The first radiosonde on *POLARSTERN* was released on 29 December 1982, 2 days after *POLARSTERN* started on its maiden voyage to the Antarctic. Radiosondes are balloon-borne instruments which record atmospheric ("upper air") profile data, mostly temperature, humidity and pressure. The horizontal wind vector can be estimated when the displacement of the balloon is known. As DuBois et al. (2002) wrote:

"The contributions of this relatively simple device to the late twentieth-century way of life can hardly be exaggerated. No other factor contributed more to the systematization of weather observations, which is beneficial to all who depend upon meteorological prediction. [...]". Radiosonde data are used extensively by weather forecast services, e.g. by the European Centre for Medium-Range Weather Forecasts (ECMWF). There are currently around 600-700 land- and ship-based upper air observation stations routinely in operation worldwide which feed at least daily into the Global Telecommunication System (GTS) of the World Meteorological Organization. These data are used as input for routine weather forecast models such as the one run by the ECMWF. However, Fig. 1 shows that ocean and polar regions are underrepresented (see also Dow, 2004). Regular radiosonde launches on ships therefore are of special value, and icebreakers like *POLARSTERN* can even provide data from the Antarctic coast and the North Pole.

In the following, we describe a data compilation of 30 years of upper air soundings on board of *POLARSTERN* using VAISALA radiosondes. We first provide details on the equipment used, the sampling procedure, and the parameters measured. Then, the data compilation is described in respect to extent, access and quality. A few examples of the data types are also given. In the last part we give a short overview of related radiosonde data in PANGAEA, and finish with a look into the future.

2 Instrumentation, sampling, parameters

The upper air soundings on board of *POLARSTERN* were performed usually once a day at around 12:00 UTC. For this purpose the weather technician in charge released a helium-filled balloon (TOTEX 600 g, 800 g, Japan) equipped with a radiosonde (RS-type, VAISALA, Finland) from the helideck of *POLARSTERN* at a height of 10 m a.s.l. During strong wind conditions ($> 20\,\mathrm{m\,s^{-1}}$), only 350 g balloons could be launched with a reasonable chance of success. All balloons were filled to reach an ascent velocity of $5\,\mathrm{m\,s^{-1}}$; data recordings were made approximately every 25–50 m (until 6 June 1998 the profile data were taken every 10 s = around every 50 m, later the sampling rate was increased to 5 s). The bursting point of the balloon, which set an end to the recordings, mostly was between 25 and 37 km height (König-Langlo et al., 2006). Up to 1996, RS80-15N sondes were used with the Omega wind finding technique. From 1996 on, when GPS came into use, we introduced the RS80-15G (with the change to 15G being a little bumpy resulting in the use of both types in 1996). Both sondes were equipped with RS80A-Humicap radiosonde humidity sensor. In October 2004 the RS80 sondes were replaced by RS90-AG radiosondes. Since 9 April 2005 RS92-SGPW radiosondes have been deployed, see Table 1 and König-Langlo (2015, doi:10.1594/PANGAEA.810000) for details.

Table 1. Overview of the radiosonde types on board *POLARSTERN*.

Sonde type	Deployment date	Campaign
RS80	29 December 1982 to 1 October 2004	Until ARK-XX/3
RS90	15 October 2004 to 4 April 2005	ANT-XXII/1-3
RS92	9 April 2005 to today	From ANT-XXII/4

The radiosondes directly measured air pressure, air temperature and relative humidity. Until 1996, the wind vector was determined with the aid of the OMEGA navigation system, since then a GPS-based windfinding system has been used, leading to a remarkable improvement in the quality of the horizontal wind vector. Altitudinal information was calculated using the hydrostatic approximation.

The recorded data were transmitted from the sondes to *POLARSTERN*. Data reception and evaluation on *POLARSTERN* was carried out by a MicroCora (VAISALA, Finland) until 1996. After that the system was switched to a DigiCORA MW11, in 2003 to a DigiCORA MW21 and since 2012 a DigiCORA MW31 has been used (VAISALA, Finland).

During some cruises ozone profiles through the troposphere and the lower stratosphere were measured by connecting an ozonesonde (ECC-6A, ECC-6B, Science Pump Corporation, USA) to a normal radiosonde using an interface. Ozone concentration was measured by pumping air through a chemical solution and using the principal of iodide redox reaction to release electrons. 1500 g TOTEX balloons were used for these ascents. The DigiCORA was able to handle the data reception and evaluation of both the normal radiosonde and the ozonesonde, simultaneously.

Together with hourly synoptical observations on board of *POLARSTERN*, radiosonde data were fed near real-time into the GTS via the DWD (German weather service) and used routinely by various weather forecast services, such as the ECMWF. In the ECMWF "Monthly Monitoring Report" (available at http://www.ecmwf.int/en/forecasts/quality-our-forecasts/monitoring-observing-system/ecmwf-global-data-monitoring-report-archive), e.g., you can find *POLARSTERN* meteorological data under the "WMO Identifier" DBLK (which is the ship's call sign).

3 Data sets, data access, examples, data quality, related data

3.1 Data sets and data access

In total 12 378 radiosonde balloons were started on *POLARSTERN* in the course of 30 years (from 29 December 1982 to 25 November 2012). All radiosonde data can now be found at König-Langlo (2015, doi:10.1594/PANGAEA.810000). The single radiosonde launches have been grouped into so-called "parents" accord-

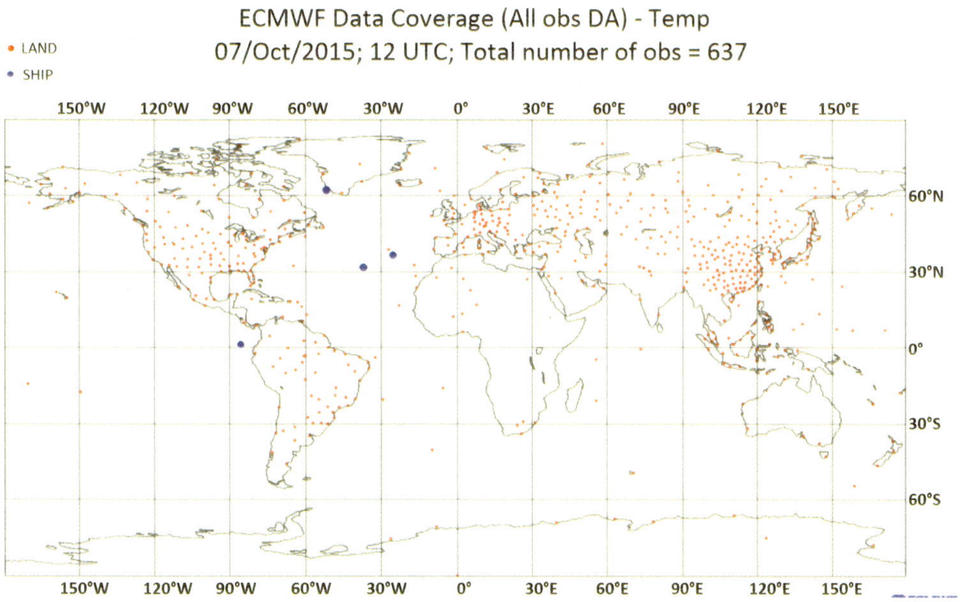

Figure 1. Example of data coverage of upper air soundings feeding into the ECMWF on 7 October 2015 at 12:00 UTC. Ship-based radiosonde launch sites (here $N = 4$) have been enlarged to be able to distinguish them. Source: http://old.ecmwf.int/products/forecasts/d/charts/monitoring/coverage.

ing to the cruise leg they belong to (i.e. a parent contains all data sets of one cruise leg, all in all 210 legs = parents).

At König-Langlo (2015, doi:10.1594/PANGAEA.810000) you will find an overview table with information on the dates of the cruise leg, the number of single observations (i.e. launches), the sonde type used and links to the individual parents and respective cruise reports. Please note that cruise labels for Antarctic cruises always begin with "ANT-", the ones for Arctic cruises with "ARK-". *POLARSTERN* radiosonde data for specific campaigns, areas or dates can also be searched by using the PANGAEA search engine (www.pangaea.de) and adding +PSradio[2] for a search in all parents, or by adding +PSradiosingle[3] for a search in all single radiosonde data sets ("childs"). Apart from the actual data (pressure, temperature, relative humidity, wind direction and wind speed according to altitude), each radiosonde data set contains the information of when and where it was launched. Whenever possible, the height of the tropopause, the precipitable water content and the total ozone are also given. Furthermore, the central parent offers the link to the cruise report of the respective cruise leg (if available).

In Fig. 2 all radiosonde launch sites of 30 years of *POLARSTERN* campaigns are mapped. The location of the homeport (Bremerhaven, North Sea) and the most frequent destinations of *POLARSTERN* (the Arctic and Neumayer Station, Weddell Sea) entail, that the radiosonde launch sites are mostly restricted to Atlantic Ocean regions. In terms of latitudinal coverage of the 12 378 radiosonde launches Fig. 3 confirms that high latitudes are indeed very well represented in our data. For the interested user, Fig. 3 also shows the distribution of launches according to year.

3.2 Data examples for interested users

Of the 12 378 single radiosonde launches, 432 contain ozone data. In Fig. 5 the ozone data of *POLARSTERN* cruise ANT-XVII (launch sites see Fig. 4) are plotted against altitude and latitude, irrespective of the longitude, date or time of measurement (tool used to create Figs. 4–7 is ODV by Schlitzer, 2015). Nevertheless, the distribution of stratospheric ozone, shaped by the Brewer–Dobson circulation, can be seen nicely.

Figure 6 shows an example of a meridional-height section of air temperature (again data from cruise ANT-XVII). Here, the low temperatures of the tropopause near the equator are apparent. With higher latitudes, the tropopause reaches lower altitudes and gets increasingly warmer (Lydolph et al., 1985).

Figure 7 shows an example of the wind velocity of the meridional-height section from cruise ANT-XVII. Pronounced maxima in the tropopause regions are clearly visible. They belong to the four westerly jet streams. In the Northern Hemisphere the polar jet is clearly separated from the subtropical jet while in the Southern Hemisphere both jets coincide to be very close to each other.

[2]Try for example *ozone+PSradio* to obtain all parents (= cruise legs) containing data sets with ozone data

[3]Try for example *ozone+PSradiosingle* to obtain all radiosonde data sets containing ozone data

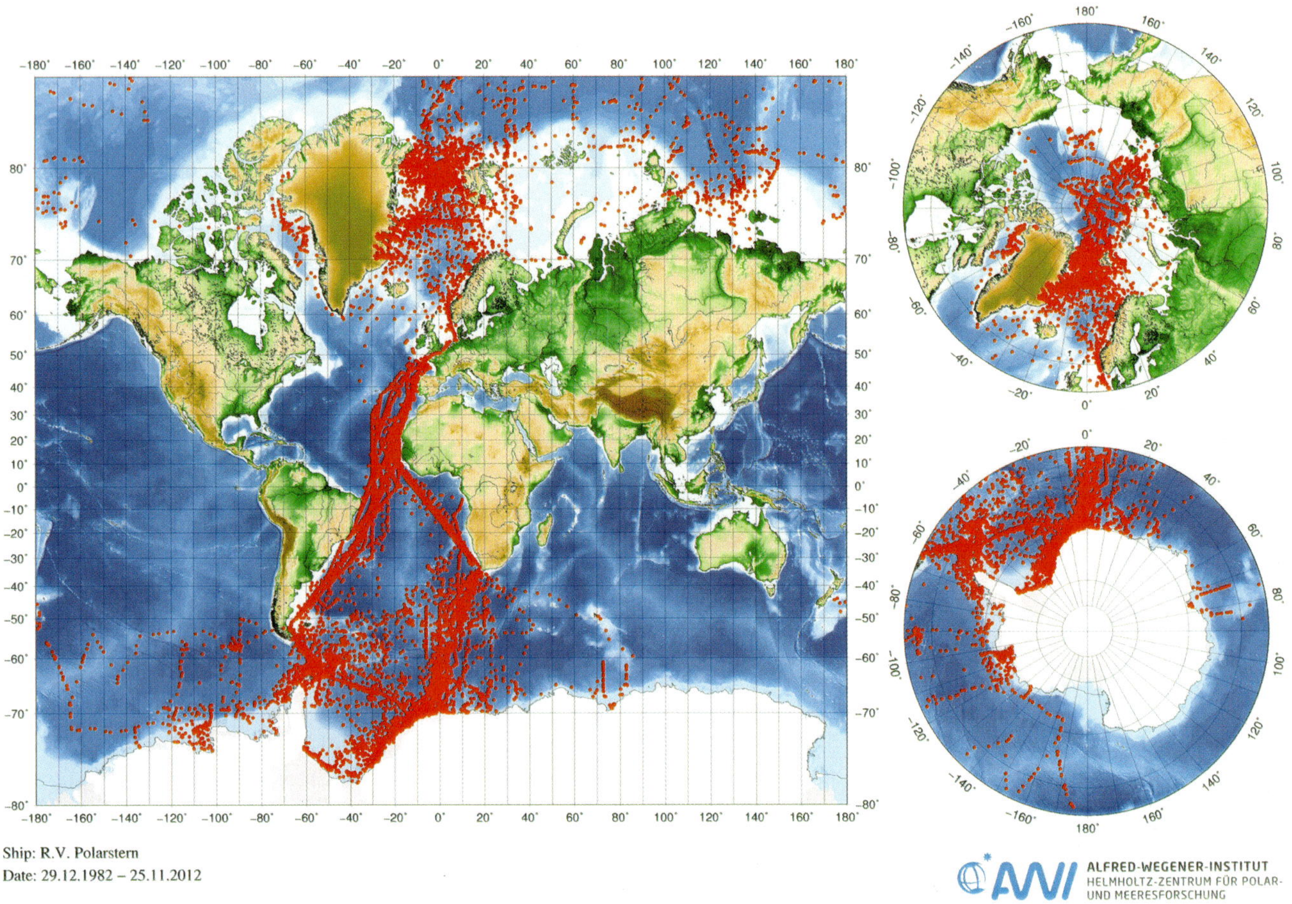

Figure 2. Map showing the locations of 30 years of radiosonde launches performed from *POLARSTERN*. From Antarctic cruise ANT-I/1 on 29 December 1982 until the end of ANT-XXIX/1, on 25 November 2012.

Examples of practical applications of *POLARSTERN* radiosonde data can be found in various scientific articles, for example in Immler and Schrems (2002), in John et al. (2006) or in Yamazaki et al. (2015) to mention just a few.

3.3 Data quality

The radiosonde data presented here have been measured by three different VAISALA sensors (RS80, RS90 and RS92). To optimize the data quality, the changes of the radiosonde types have to be taken into account. The data archived in PANGAEA represent the original values which were provided by the VAISALA reception and evaluation system (i.e. after the intrinsic processing of the reception system). Except for the altitudinal information, which was calculated using the hydrostatic approximation, and a general check of the overall data quality (see below), the data were not corrected for known systematic errors or biases. In the following, we provide an overview of the general checks applied, and we then elucidate the known biases and the respective correction proposals for interested users.

General checks applied:

- Validation of position and date/time via cross-check with other *POLARSTERN* data sets of the respective cruise station
- Removal of outliers exceeding the physically possible limits
- Removal of data points where thermal statification was unstable
- Removal of data points where pressure was not decreasing strictly monotonically
- Removal of data points where relative humidity exceeded 100 % with respect to ice

The general checks led to the exclusion of only very few data points, as the intrinsic data reception systems of the senors described above already removed most of these inconsistencies.

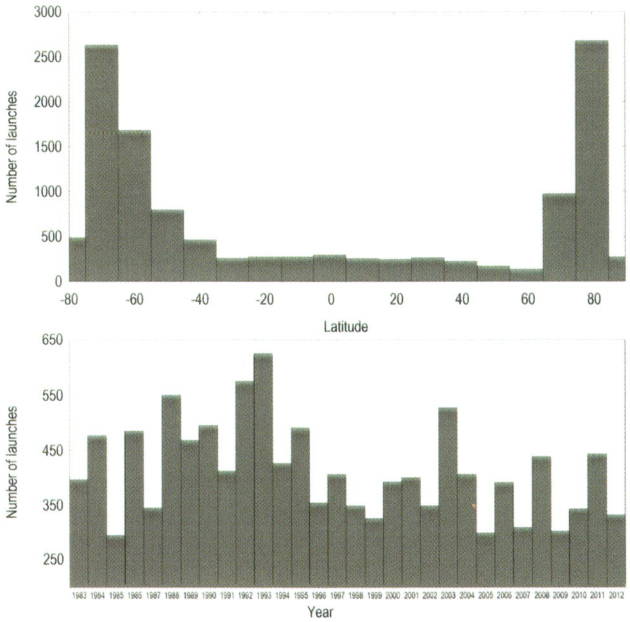

Figure 3. The distribution of *POLARSTERN* radiosonde launches according to latitude (upper part) and year (lower part).

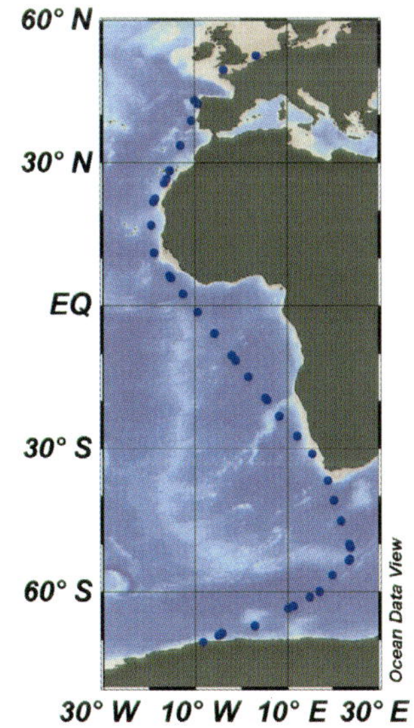

Figure 4. Locations of radiosonde launches performed from *POLARSTERN* during cruise ANT-XVII (14 December 1999 to 20 January 2000).

Known biases and correction proposals:

- RS80

  - The A-type HUMICAP humidity sensor is known to have a considerable dry bias mainly at low temperatures (Miloshevich et al., 2001). Correction methods have been proposed by Wang et al. (2002).

  - Also a time lag has been identified occurring due to the long response time of the humidity sensor at low temperatures. The proposed correction is to apply an algorithm published in Miloshevich et al. (2004).

  - In ice-supersaturated conditions ice-coating of the humidity sensor sometimes causes the measurements to stay near ice-saturation over large parts of the troposphere, not reflecting the conditions of the ambient air.

- RS90:

  The RS90 is equipped with two H-type HUMICAP humidity sensors that are alternately heated to remove condensed water or ice, thus avoiding the effect of sensor icing.

  - A small time lag is still present, even though the H polymer has a faster sensor response time. Again, the according time constant in the correction algorithm of Miloshevich et al. (2004) can be used to correct this.

Humidity data from all three VAISALA radiosondes suffer from the so-called daytime solar radiation dry bias (SRDB). Not being shielded against solar radiation, the sensors can be heated by incoming short-wave radiation during daytime launches (see, e.g. Wang et al., 2003, 2013; Vömel et al., 2007). However, the RS92 operated together with the VAISALA DigiCORA Sounding System MW31 were ranked to be the world's top radiosonde at the 8th WMO Intercomparison of Radiosonde Systems in 2010 and GRUAN compatible (Nash et al., 2011).

Another source of biases can be ship-specific influences such as heating and/or cooling biases discussed in Ciesielski et al. (2004). As already stated in Sect. 2, all soundings on board of *POLARSTERN* were started from the helideck at a height of 10 m a.s.l., resulting in data sets beginning at an altitude of 10 m a.s.l. The values for this first data point of radiosonde launches normally are taken from ground measurements. Due to the fact that no unbiased ground measurements can be taken on a ship, the (largely unbiased) values of the ship's Luv-sensors (29 m a.s.l. for temperature and humidity, 39 m a.s.l. for wind) were used. The air pressure value for this first data point was calculated from the ship's meteorological observatory (16 m a.s.l.). The pressure sensor has an external inlet which ends in a labyrinth in the ship's mast to reduce ship-specific influences. Beginning with the second data point (altitude level), the data originate from the radiosonde itself. In the rare case that the quality check revealed ongoing disturbances for the second data point (e.g. by the ship's plume), the respective data were deleted from the records. For the third altitude level onwards, no ship-based biases were expected anymore.

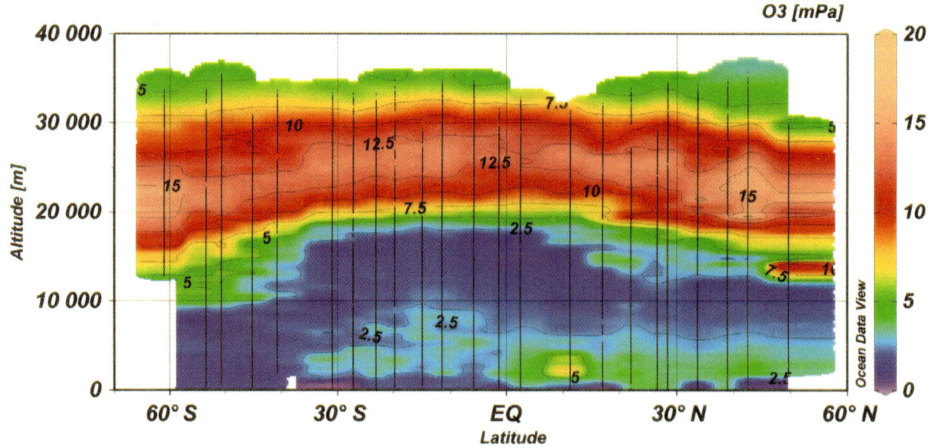

Figure 5. Meridional-height section of ozone (O_3 in mPa) from *POLARSTERN* cruise ANT-XVII (14 December 1999 to 20 January 2000). The lines represent the actual launches, the data were then interpolated to obtain this figure.

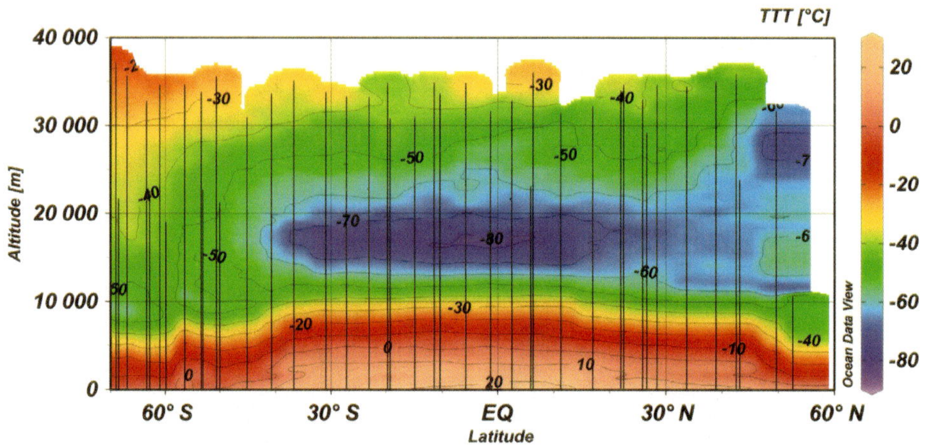

Figure 6. Meridional-height section of air temperature (TTT in °C) from *POLARSTERN* cruise ANT-XVII (14 December 1999 to 20 January 2000). The lines represent the actual launches, the data were then interpolated to obtain this figure.

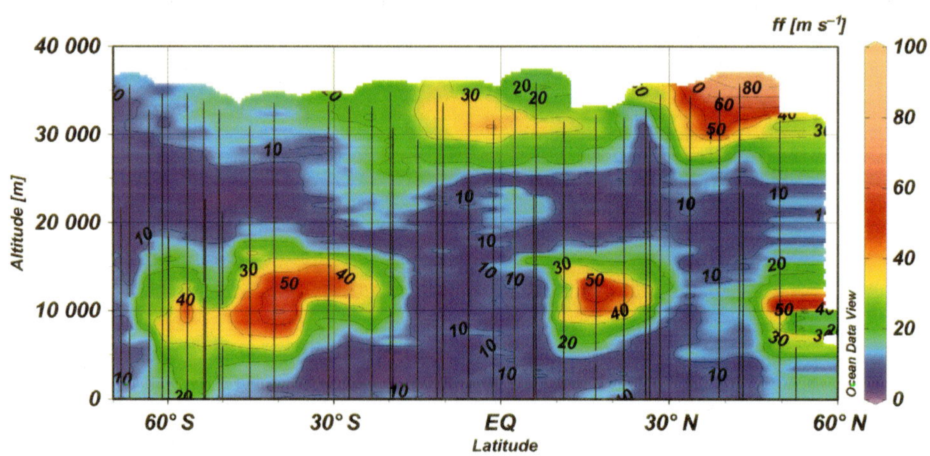

Figure 7. Meridional-height section of wind velocity (ff in $\mathrm{m\,s^{-1}}$) from *POLARSTERN* cruise ANT-XVII (14 December 1999 to 20 January 2000). The lines represent the actual launches, the data were then interpolated to obtain this figure.

3.4 Related upper air data in PANGAEA

Radiosonde measurements on *POLARSTERN* are ongoing. Since 25 November 2012, which is the date of the last radiosonde launch discussed above, data of an additional 1186 *POLARSTERN* radiosonde launches have been archived in PANGAEA (status 15 May 2016). Apart from these, PANGAEA contains 12057 soundings performed at the Antarctic research station Neumayer and 9911 from the Arctic station Ny Ålesund (status 15 May 2016). Both AWI stations are part of the Baseline Surface Radiation Network (BSRN) which uses PANGAEA as its central archive (König-Langlo et al., 2013). For an overview of BSRN radiosonde data (monthly files) see http://www.pangaea.de/PHP/BSRN_Status.php?q=LR1100.

4 Conclusions

Even though more sophisticated techniques (meteorological satellites, aircraft observation, remote sensing systems, etc.) have become increasingly important in weather forecasts, the high vertical resolution and quality of radiosonde data remains crucial for weather forecast quality, for detecting changes in climate, and for the validation of modelling and satellite products (Dow, 2004). The Alfred Wegener Institute is committed to providing high-quality and open-access research data to the scientific community and the interested public, and will thus continue with upper air soundings (and other under-way measurements) on board of *POLARSTERN* until its decommissioning. The new research vessel *POLARSTERN II*, which according to the current schedule will be handed over to the scientific community in 2019, will most likely continue the work of over 30 years of dedicated *POLARSTERN* meteorology.

Acknowledgements. We are very grateful for the time and effort the two reviewers put into improving our manuscript, your help is highly appreciated. We would also like to thank Wolfgang Cohrs for the technical assistance and the creation of Fig. 2. Special thanks to all DWD weather technicians on board of *POLARSTERN* responsible for launching the sondes: Hans Ohlendorf, Dieter Bassek, Herbert Köhler, Wolf-Thilo Ochsenhirt, Juliane Schostak, Hartmut Sonnabend, Roland Menzel, Hans-Peter Lambert, Klaus Buldt, Edmund Knuth, Thorsten Truscheit, and Juliane Hempelt.

Edited by: D. Carlson

References

Ciesielski, P. E., Yu, H., Johnson, R. H., Yoneyama, K., Katsumata, M., Long, C. N., Wang, Junhong., Loehrer, Scot, M., Young, Kathryn., Williams, Steven.F., Brown, W., Braun, J., and Van Hove, T.: Quality-controlled upper-air sounding dataset for DYNAMO/CINDY/AMIE: Development and corrections, J. Atmos. Ocean. Tech., 31, 741–764, doi:10.1175/JTECH-D-13-00165.1, 2014.

Dow, G.: Developments in Observational Requirements for Global Numerical Weather Prediction, University of Reading, Department of Meteorology, Reading, 2004.

DuBois, J. L., Multhauf, R. P., and Ziegler, C. A.: The invention and development of the radiosonde: with a catalog of upper-atmospheric telemetering probes in the National Museum of American History, Smithsonian Institution, Smithsonian studies in history and technology, 53, 88 pp. doi:10.5479/si.00810258.53.1, 2002.

Fütterer, D. and Fahrbach, E.: *POLARSTERN* – 25 Jahre Forschung in Arktis und Antarktis, Bielefeld, Delius Klasing, 293 pp., ISBN: 978-3-7688-2433-0, 2007.

Immler, F. and Schrems, O.: Determination of tropical cirrus properties by simultaneous LIDAR and radiosonde measurements, Geophys. Res. Lett., 29, 2090, doi:10.1029/2002GL015076, 2002.

John, V. O., Buehler, S. A., von Engeln, A., Eriksson, P., Kuhn, T., Brocard, E., and Koenig-langlo, G.: Understanding the variability of clear-sky outgoing long-wave radiation based on ship-based temperature and water vapour measurements, Q. J. Roy. Meteor. Soc., 132, 2675–2691, doi:10.1256/qj.05.70, 2006.

König-Langlo, G.: 30 years of upper air soundings on board of R/V POLARSTERN (1982-12-29 to 2012-11-25), Alfred Wegener Institute, Helmholtz Center for Polar and Marine Research, Bremerhaven, doi:10.1594/PANGAEA.810000, 2015.

König-Langlo, G., Loose, B., and Bräuer, B.: 25 Years of *POLARSTERN* Meteorology, WDC-MARE Reports, 4, CD+booklet, 137 pp., doi:10.2312/wdc-mare.2006.4, 2006.

König-Langlo, G., Sieger, R., Schmithüsen, H., Bücker, A., Richter, F., and Dutton, E. G.: The Baseline Surface Radiation Network and its World Radiation Monitoring Centre at the Alfred Wegener Institute, WCRP Report 24/2013, GCOS – 174, available at: www.wmo.int/pages/prog/gcos/Publications/gcos-174.pdf, 2013.

Lydolph, P., Temple, D., and Temple, D.: The Climate of the Earth, Rowman & Littlefield Publishers, Inc., Lanham, 408 pp., 1985.

Miloshevich, L. M., Vömel, H., Paukkunen, A., Heymsfield, A. J., and Oltmans, S. J.: Characterization and correction of relative humidity measurements from Vaisala RS80-A radiosondes at cold temperatures, J. Atmos. Ocean. Tech., 18, 135–155, 2001.

Miloshevich, M., Paukkunen, A., Vömel, H., and Oltmans, S.. Development and Validation of a Time-Lag Correction for Vaisala Radiosonde Humidity Measurements, J. Atmos. Ocean. Tech., 21, 1305–1327, doi:10.1175/1520-0426(2004)021<1305:DAVOAT>2.0.CO,2, 2004.

Nash, J., Oakley, T., Vömel, H., and Li, Wei.: WMO Intercomparison of high quality radiosonde systems, Yangjiang, China, 12 July–3 August 2010, World Meteorological Organization, Instruments and Observing Methods, Report No. 107, WMO/TD No. 1580, 2011.

Schlitzer, R.: Ocean Data View, available at: http://odv.awi.de (last access:4 January 2016), 2015.

Vömel, H., Selkirk, H., Miloshevich, L., Valverde-Canossa,J., Valdés, J., Kyrö, E., Kivi, R., Stolz, W., Peng, G., and Diaz, J. A.: Radiation Dry Bias of the Vaisala RS92 Humidity Sensor, J. Atmos. Ocean. Tech., 24, 953–963, doi:10.1175/JTECH2019.1, 2007.

Wang, J., Cole, H. L., Carlson, D. J., Miller, E. R., and Beierle, K.: Corrections of humidity measurement errors from the Vaisala RS80 radiosonde – application to TOGA COARE data, J. Atmos. Ocean. Tech., 19, 981–1002, doi:10.1175/1520-0426(2002)019<0981:COHMEF>2.0.CO,2, 2002.

Wang, J., Carlson, D. J., Parsons, D. B., Hock, T. F., Lauritsen, D., Cole, H. L., Beierle, K., and Chamberlain, N.: Performance of operational radiosonde humidity sensors in direct comparison with a chilled mirror dew-point hygrometer and its climate implications, Geophys. Res. Lett., 30, 1860, doi:10.1029/2003GL016985, 2003.

Wang, J., Zhang, L., Dai, A., Immler, F., Sommer, M., and Vömel, H.: Radiation dry bias correction of Vaisala RS92 Humidity data and its impacts on historical radiosonde data, J. Atmos. Ocean. Tech., 30, 197–214, doi:10.1175/JTECH-D-12-00113.1, 2013.

Yamazaki, A., Inoue, J., Dethloff, K., Maturilli, M., and König-Langlo, G.: Impact of radiosonde observations on forecasting summertime Arctic cyclone formation, J. Geophys. Res. Atmos., 120, 3249–3273, doi:10.1002/2014JD022925, 2015.

2

Central-Pacific surface meteorology from the 2016 El Niño Rapid Response (ENRR) field campaign

Leslie M. Hartten[1,2], Christopher J. Cox[1,2], Paul E. Johnston[1,2], Daniel E. Wolfe[1,2], Scott Abbott[2], and H. Alex McColl[1,2,a]

[1]Cooperative Institute for Research in Environmental Science (CIRES), University of Colorado, Boulder, CO 80309-0216, USA

[2]NOAA/Earth System Research Laboratory, Physical Sciences Division, Boulder, Colorado, CO 80305, USA

[a]currently at: Berthoud, CO 80513, USA

Correspondence: Leslie M. Hartten (leslie.m.hartten@noaa.gov)

Abstract. During the early months of the 2015/2016 El Niño event, scientists led by the Earth System Research Laboratory's Physical Sciences Division conducted the National Oceanic and Atmospheric Administration's (NOAA's) El Niño Rapid Response (ENRR) field campaign. One component of ENRR involved in situ observations collected over the near-equatorial eastern–central Pacific Ocean. From 25 January to 28 March 2016, standard surface meteorology observations, including rainfall, were collected at Kiritimati Island (2.0° N, 157.4° E) in support of twice-daily radiosonde launches. From 16 February to 16 March 2016, continuous measurements of surface meteorology, sea surface temperature, and downwelling shortwave radiation were made by NOAA Ship *Ronald H. Brown*. These were largely done in support of the four to eight radiosondes launched each day as the ship travelled from Hawaii to TAO buoy locations along longitudes 140 and 125° W and then back to port in San Diego, California. The rapid nature of these remote field deployments led to some specific challenges in addition to those common to many surface data collection efforts. This paper documents the two deployments as well as the steps taken to evaluate and process the data. The results are two multi-week surface meteorology data products and one accompanying set of surface fluxes, all collected in the core of the eastern–central Pacific's extremely warm waters. These data sets, plus metadata, are archived at the NOAA's National Centers for Environmental Information (NCEI) and are free for public access: surface meteorology from Kiritimati Island (https://doi.org/10.7289/V51Z42H4); surface meteorology and some surface fluxes from NOAA Ship *Ronald H. Brown* (https://doi.org/10.7289/V5SF2T80; https://doi.org/10.7289/V58050VP).

1 Introduction

In June 2015, the weak El Niño conditions that had existed since March 2015 were strengthening and forecasters were confident that they would continue to do so through winter 2015/2016 (Climate Prediction Center (CPC) and International Research Institute for Climate and Society (IRI), 2015). The United States' National Oceanic and Atmospheric Administration (NOAA) wanted to quickly address the unfolding event, which had the potential for large domestic impacts, particularly on the country's Pacific coast, and thus developed the El Niño Rapid Response (ENRR). The Earth System Research Laboratory's (ESRL) Physical Sciences Division (PSD) led the design and implementation of one component, the ENRR field campaign (Dole et al., 2018). Among the many assets put into play were surface meteorological instruments on Kiritimati (pronounced "Christmas") Island and aboard NOAA Ship *Ronald H. Brown*. The primary purpose of the surface meteorological measurements was to provide initialization data for the radiosondes launched twice a day from Kiritimati and four to eight times per day from NOAA Ship *Ronald H. Brown* (Hartten et al., 2018). However, the high temporal resolution of the surface

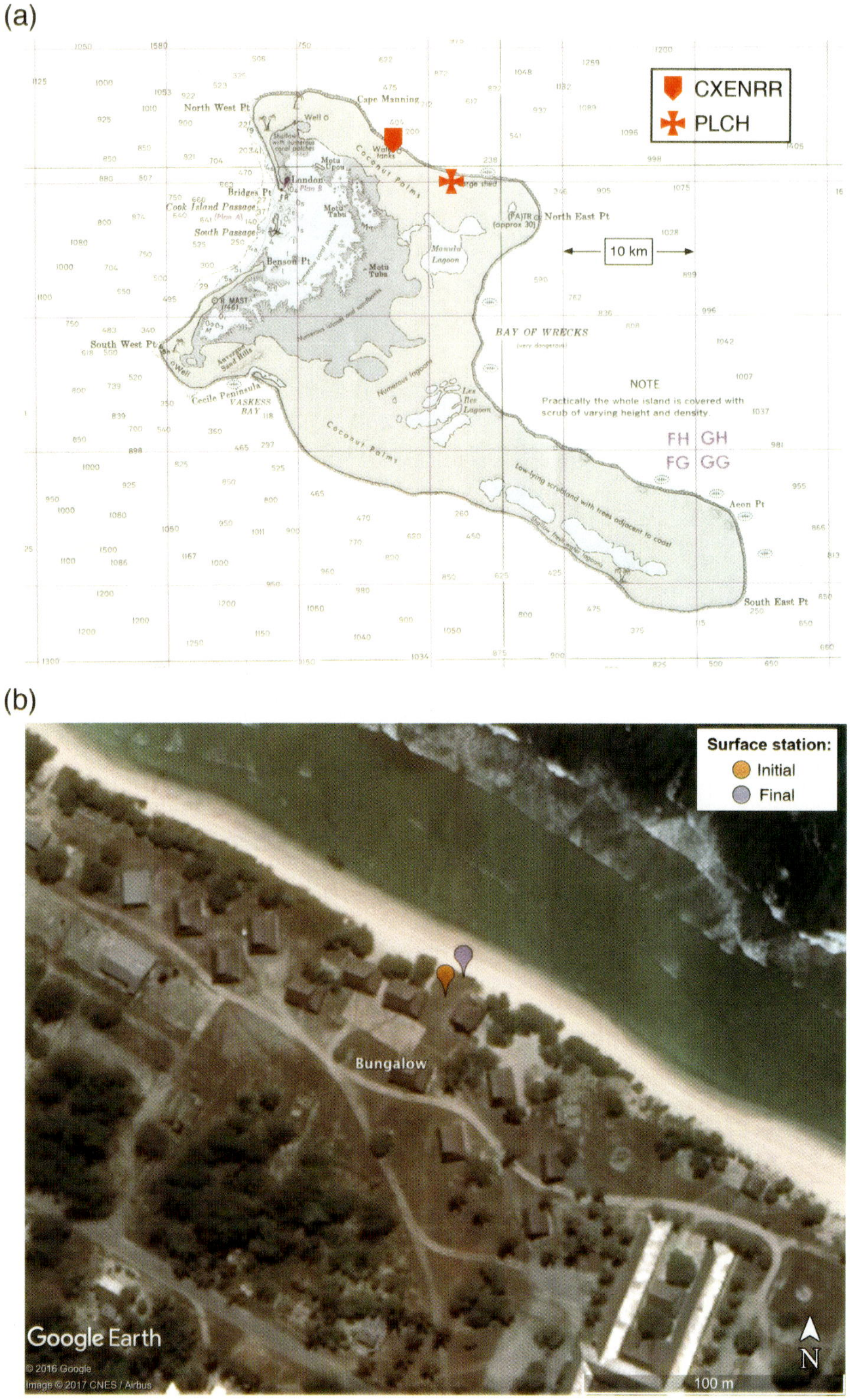

Figure 1. (**a**) Map of Kiritimati showing the locations of Cassidy International Airport (PLCH) and CXENRR. Kiritimati map excerpted from U.S. Defense Mapping Agency (DMA) Stock No. 83BHA83130, courtesy of the University of Texas Libraries, The University of Texas at Austin. Map data from surveys by New Zealand (1938–1941) and the U.S. Navy (to 1962). Heights are in feet above mean high water springs; soundings are in fathoms (or fathoms and feet if less than 11 fathoms). (**b**) Detail of the CXENRR site at the Captain Cook Hotel. The bungalow which served as our base of operations is in the center of the image, with markers to its north indicating the initial and final locations of the surface meteorological instruments.

Table 1. Surface meteorology instruments deployed at Kiritimati Island during the ENRR field campaign.

Instrument	Height (a.s.l.)	Parameter	Collection rate; logged values	Accuracy	Operational notes
Vaisala PTB101B attached to a SPH10 static pressure head	3.6 m	Atmospheric pressure	1 Hz; averaged*	±0.5 hPa at 20 °C ±1.5 hPa at 0–40 °C	
Vaisala HMP45C in an R.M. Young 41003-5 10-plate naturally aspirated solar radiation shield	4.4 m	Air temperature	1 Hz; averaged*	±0.2 °C at 20 °C ±0.25 °C at 30 °C	Sensor bad starting 1 February 2016, 17:59:43 UTC and was replaced 7 February 2016, 03:41:43 UTC
		Relative humidity	1 Hz; averaged*	at 20 °C: ±2 % (0–90 % RH) ±3 % (90–100 % RH)	
R.M. Young 5103 anemometer	6.6 m	Wind speed	1 Hz; vector-averaged*	±0.3 m s^{-1} or 1 % of reading	
		Wind direction		±3°	
		Maximum wind speed	1 Hz; maximum during collection*		
Texas Electronics TE525 tipping bucket	4.4 m	Rain	Accumulation during collection*	±1 % up to 25.4 mm h^{-1} +0, 3 % from 25.4 to 50.8 mm h^{-1} +0, −5 % from 50.8 to 76.2 mm h^{-1}	0.254 mm per tip
Campbell Scientific CR23X datalogger	3.6 m	Battery voltage	1 Hz; averaged*	n/a	Power supply failed; no data logged from 15 March 2016, 11:12:43 UTC to 16 March 2016, 05:04:42 UTC

* The averaging or collection period was 2 min through 11 February 2016, 04:05:43. It changed to 1 min starting 11 February 2016.

measurements in these remote locations makes them useful in their own right. The quick deployment and remote locations led to several data challenges which needed to be overcome both for the sake of creating research-quality radiosonde data sets and to enable the independent value of the surface data to be more fully realized. This article documents the data collection, the problems identified, and corrections applied after the field phase, and the resulting data sets.

2 Instrument specifications and siting

2.1 Kiritimati Island, Republic of Kiribati

Kiritimati is one of the Line Islands, a chain of islands and atolls lying across the Equator south of Hawaii. Its land area of 321 km^2 makes it the world's largest coral atoll (Scott, 1993); its lagoons cover a similar area (Fig. 1a). Much of it lies a few meters above sea level, with its highest point only 13 m a.s.l. Kiritimati's 6456 inhabitants as well as all visitors live in four towns on the northern side of the island (Morate, 2016). The Kiribati[1] Meteorological Service maintains and staffs an office at Cassidy International Airport along the northeast-facing coastline; observers record surface conditions hourly and report them under WMO station ID 91490. Its position in the equatorial central Pacific, coupled with its relatively large land area and advanced infrastructure, have made Kiritimati Island the site of choice for many previous scientific efforts, including the Line Islands Experiment (Zipser, 1970); the multi-year deployment of wind-profiling radars (Gage et al., 1991); climatological studies using coral (Evans et al., 1999); and the Pacific Atmospheric Sulfur Experiment (PASE; Conley et al., 2009).

ENRR operations were based at the Captain Cook Hotel, which is located on the northeast-facing coast about 6 km from the airport (Fig. 1a). This location was chosen because it was familiar to some of our staff; had experience hosting scientific fieldwork; had 60 Hz, 120 V power and Internet access; and had enough driveable paths, staff presence, and public lighting to provide a level of safety during evening radiosonde launches. Field staff lived in a two-unit bungalow

[1]The country's name is pronounced "KEER-eh-bahss".

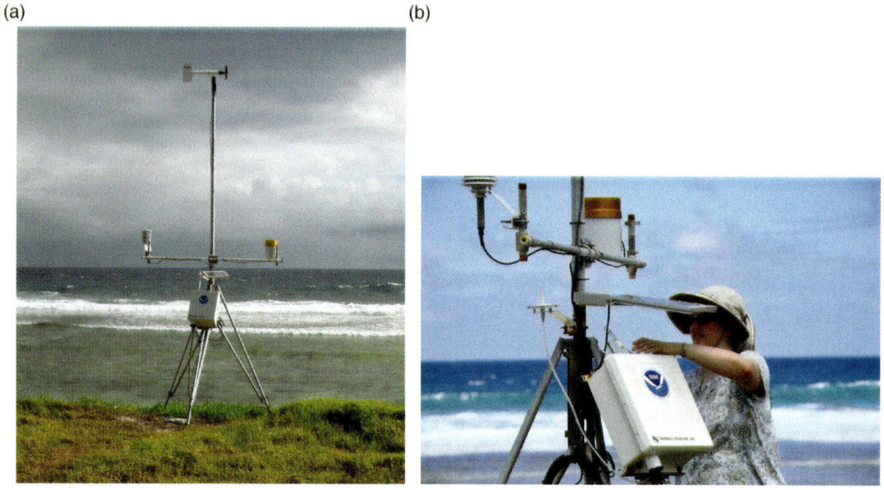

Figure 2. **(a)** The surface met station in its final position above the beach at Kiritimati. The anemometer sits atop the assembly, while the temperature–humidity sensor (on the left) and rain gauge (on the right) are attached to the crossbeam. Just below the crossbeam is the solar panel, and the box containing the Campbell datalogger, the power supply, and the barometer is mounted just below that. A plastic tube runs from the barometer out the tube on the bottom of the box, and can be seen mounted within a white disk just to the left of the solar panel. **(b)** The 26 March 2017, 00:00 UTC radiosonde being put on the Campbell box in the shade of the solar panel prior to launch. The pressure tube egress and mount are clearly visible. Photo courtesy of Gabriele Kerber.

Figure 3. The original position of the surface instruments **(a)**, and their location after being moved back towards the top of the beach on 5 February 2017 **(b)**.

(Fig. 1b). One unit doubled as office space, housing the computers and radiosonde equipment, and for a few weeks supplied power to the outdoor instruments. The bungalow was usually air-conditioned with the temperature set to 24–26 °C. The site was located at (2.01° N, 157.40° W), and upper-air observations were transmitted to the Global Telecommunication System (GTS) using "CXENRR" as a station name.

On 25 January 2016 (UTC), the first day of operations, the surface meteorological instruments were attached to a tripod (Fig. 2) located about 14 m northeast of the concrete pad seen in Fig. 1b. Table 1 contains information about the surface meteorology instruments deployed, the variables measured, and recording details. A solar panel was also attached to the tripod; it was used to charge the battery that ran the Campbell datalogger. This initial location was further from our bungalow than expected, and we had no cable long enough to connect directly to the computers inside the bungalow, so surface conditions could not be monitored in real time and an observer had to go out and read instantaneous surface values directly off the Campbell datalogger during the radiosonde ground check and initialization. Within a few days staff realized that the initial location was somewhat obstructed by the two nearby bungalows (Figs. 1b and 3a), which were approximately 12 m apart from each other. On 5 February the tripod was moved northeast about 10 m, to a location near the top of the beach (Figs. 1b and 3b) that was further from the bungalows, and a newly arrived 30 m cable was run between the station and the computers in our bungalow. This second location was also somewhat obstructed (Fig. 3b), with a tree about 12 m to the west–northwest and the two bungalows about 15 m to the southwest and 6 m to the south. However, it was believed superior because of its more open setting

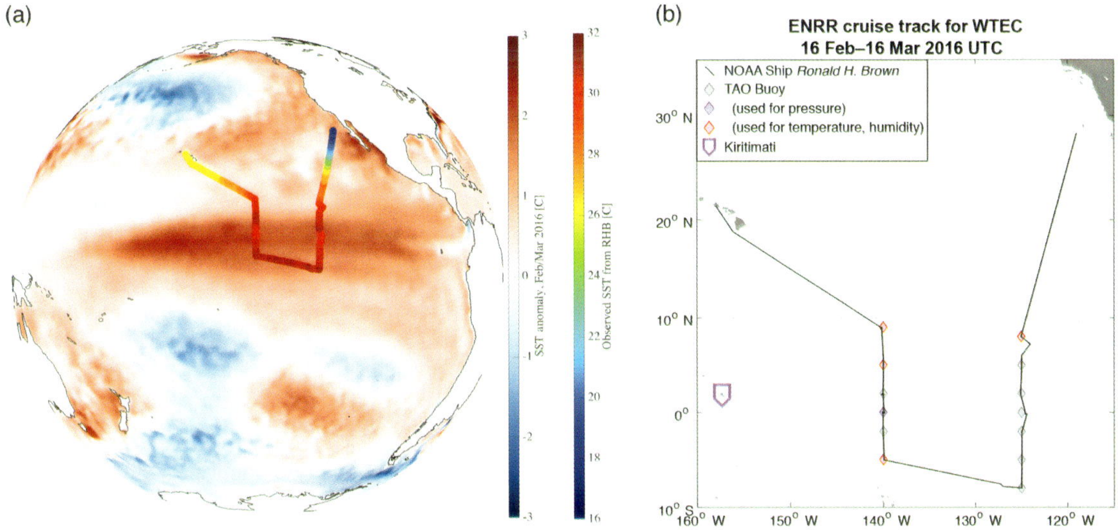

Figure 4. **(a)** SSTs observed by NOAA Ship *Ronald H. Brown* during ENRR, overlaid on SST anomalies. The latter are the departure from the ERA-Interim (Dee et al., 2011) 1979–2016 February through March mean. **(b)** The cruise track of NOAA Ship *Ronald H. Brown* overlaid on a map showing the location of the TAO buoys visited during its cruise and also the location of Kiritimati Island. The buoys used in the post-processing of the ship's surface pressure, temperature, and humidity are highlighted.

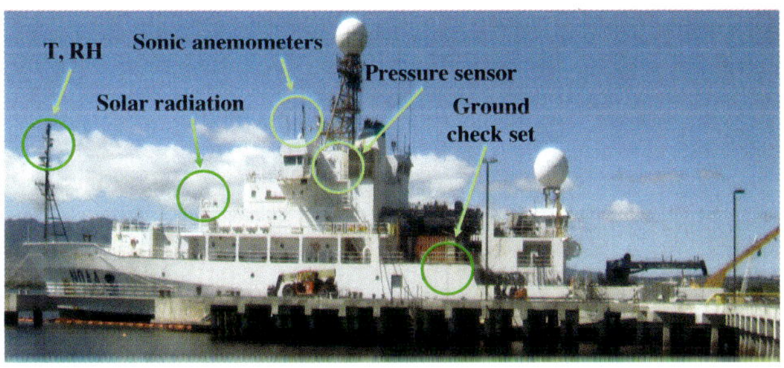

Figure 5. The location of key surface meteorological instruments on NOAA Ship *Ronald H. Brown*.

and long fetch over a 180° sector. This setup was maintained through 28 March 2016, the last day of CXENRR operations.

2.2 NOAA Ship *Ronald H. Brown*

NOAA Ship *Ronald H. Brown* is a Global Class[2] research vessel in service since 1997 as part of the NOAA Marine Operations Center – Atlantic (MOC-A) fleet. The vessel is 83.5 m long with a 16.2 m beam and has a cruising speed of 11 knots. When the ENRR field campaign was being planned, the ship was already scheduled to cruise from Hawaii to San Diego during February and March 2016 to conduct routine maintenance of buoys that are part of the Tropical Atmosphere Ocean (TAO) array along 9° N to 8° S meridional transects at 140 and 125° W (Fig. 4). Because the scheduled timing and path of the ship were well suited for the objectives of the ENRR campaign, three scientists from ENRR were deployed with the ship to acquire soundings for the full duration of the cruise. These were transmitted to the GTS as TEMP MOBILE messages using the station name "WTEC".

The ENRR surface meteorology and oceanographic observations were collected by some of the ship's standard instrumentation suite (Fig. 5). Technical information about the instruments deployed, the variables measured, measurement and recording details, and operation notes are given in Table 2. As at Kiritimati, our primary interest in the surface meteorological data collected during the cruise was for the radiosonde ground check and initialization. The radiosondes were launched four to eight times per day from the main deck of the ship, and the Vaisala GC25 ground-check set was located indoors on the main deck of the ship approximately 3.8 m above the ship's designated water line.

[2]"Global Class ships are the largest and most capable with the ability to work worldwide with large scientific parties and the longest endurance" (Joint Subcommittee on Ocean Science and Technology (JSOST), 2016).

3 Observational issues and post-deployment data correction

There were difficulties with some surface measurements at both Kiritimati Island and aboard NOAA Ship *Ronald H. Brown*. Some problems were almost immediately apparent; others became known as the campaign unfolded or upon post-deployment review. The problems and the methods used to correct them were different for each site and are described below. Only those measured quantities that required detailed

Table 2. Surface meteorology instruments deployed aboard NOAA Ship *Ronald H. Brown* during the ENRR field campaign. The Improved Meteorology (IMET, Hosom et al., 1995) instruments are modified versions of the original manufacturer instruments, as described in the Woods Hole Oceanographic Institution (WHOI, 2010a, 2010b).

Instrument	Height (a.s.l.)	Parameter	Collection rate; logged values	Accuracy	Operational notes
Vaisala PTB330	15.24 m	Atmospheric pressure	1 Hz; averaged*	±0.20 hPa at 20 °C ±0.25 (B) at 0 to 40 °C	Mounted on bridge Calibrated 29 December 2015
IMET Rotronic MP101A	approx. 15 m	Air temperature	1 Hz; averaged*	±0.05 °C	Mounted on bow mast Calibrated 26 July 2014
		Relative humidity		±1 %	
Vaisala WMT700 ultrasonic anemometers	approx. 19 m	Wind speed	1 Hz; vector-averaged*	$\pm 0.1\ \mathrm{m\,s^{-1}}$ or 2 % of reading	Mounted on poles atop the port and starboard sides of the bridge roof
		Wind direction		±2°	
Sea Bird SBE45 thermosalinograph	−5 m	Sea surface temperature	1 Hz; averaged*	±0.002 °C	
IMET Eppley PSP	11.2 m	Downwelling shortwave radiation	1 Hz; averaged*	±2 %	
Furuno GP90	approx. 17 m	Latitude, longitude	1 Hz; averaged*	±10 m	
		Course over ground		±3° at 1–17 kt ±1° at > 17 kt	
		Speed over ground		0.2 kt at ≤ 10 kt 2 % of ship speed	
Sperry MK37	unknown	Heading	1 Hz; averaged*	unknown	

* The averaging or collection period was 1 min.

post-deployment analysis are discussed in subsections; all other information is given in the main site sections.

3.1 Kiritimati Island

Issues in the original Kiritimati Island surface data fall into five categories: instrument calibration issues; instrument failures; non-representativeness of the data caused by poor siting; changes in the data acquisition methods and archival timing; and gaps in the data. Some of these issues were consequences of the "Rapid Response" aspect of the project. Because there is only one international flight to Kiritimati Island each week and the campaign was pulled together in only a few months, there was no time for a site survey prior to the start of fieldwork. This limited the choices of available sites and led to some of the shipped equipment being our "best guess" at what would be needed. It also meant that we often had to make do with what was available on the island, or wait a week or more until something could be shipped.

The "Rapid Response" also affected the exact instruments deployed, which had to be those available on short notice from PSD's instrument pool. Questions about the calibration of some of them arose soon after the campaign started, which led us to develop a practice of placing the initialized Vaisala RS92 radiosonde atop the instrument box for an extended period before launch. Radiosonde initialization, which was done inside the air-conditioned bungalow, includes a ground-check procedure that uses high-precision and high-accuracy temperature and humidity sensors to calibrate each radiosonde (Vaisala Oyj, 2006, 2015). Setting the radiosonde in the shade of the surface station's solar panel (Fig. 2b) allowed the radiosonde to equilibrate to the outside environment; it also provided as much as several minutes of co-located pressure, temperature, and humidity data from the radiosonde and the surface instruments. Details of how we used these coincident measurements to refine the surface observations are discussed in the relevant subsections below, and are in accord with Wang et al.'s (2002) suggestion that "comparisons of prelaunch radiosonde data with the surface

Table 3. Major gaps in the Kiritimati surface data set. Data collection started 25 January 2016, 03:07:43 UTC and ended 28 March 2016, 19:14:42 UTC.

Start of gap (UTC)	End of gap (UTC)	Reason	Measurement(s) affected
25 January 2016, 03:07:43	26 January 2016, 21:49:43	Sensor not aligned	Wind direction
1 February 2016, 17:59:43	7 February 2016, 03:41:43	Sensor failure and replacement	Temperature, relative humidity
5 February 2016, 21:21:43	5 February 2016, 22:01:43	Surface station relocation	All
15 March 2016, 11:12:43	16 March 2016, 05:04:42	Solar panel failure and power supply replacement	All

data from independent surface sensors can always be used to evaluate the accuracy of radiosonde data (and/or surface data), and may provide some guidance on how to correct the data".

There were two instrument failures during the experiment. During the second week of operations it became clear that the temperature–humidity probe had failed. Fortunately, we had brought a second unit with us. We swapped the sensors on 6 February and finalized the replacement sensor's settings on 7 February 2016 at 03:41 UTC; it operated well for the rest of the campaign. On 15 March 2016, at 11:12 UTC, the solar panel providing power for the datalogger failed. It took about 18 h to diagnose the problem, find a replacement power supply and install it, and restart the surface station.

The surface observations were critical for the initialization of the radiosondes, so the surface station had to be located near the radiosonde release location, i.e., close to the bungalow housing the radiosonde data system. The surface station was originally set up on a grassy area northeast of the concrete pad. This location looked good at first, but after a few days of operations the staff realized that the anemometer sometimes spun in circles, indicating swirling winds; that the range of recorded (2 min) wind directions was quite narrow; and that recorded winds were sometimes inconsistent with the values reported by the radiosondes after they had risen above the bungalows and trees. The surface station was moved to a better location, at the edge of the beach and further from all the bungalows, on 5 February 2016 (21:00–22:00 UTC).

On 11 February 2016, at 04:05:43 UTC, four changes were made to the datalogger program. The execution interval, which determines how often each instrument is read, was changed from 2 to 1 s. The internal offset for the barometer was changed from 599.5 to 399.5, so that pressure would be reported with a resolution of 0.01 hPa instead of 0.1 hPa. The averaging time was changed from 2 min to the more standard 1 min. All these changes were made to give better data to help with radiosonde surface initialization and with data evaluations.

The failure of the T/RH sensor, the station move, the change to the datalogger software, and the failure of the datalogger power supply all caused gaps in the data (Table 3). We have not attempted to interpolate across any of the data gaps,

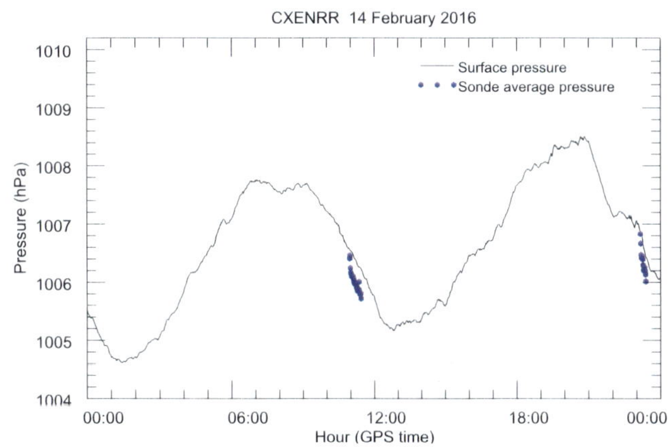

Figure 6. Surface pressure measured at Kiritimati by the PTB101B (1 min resolution) on 14 February 2016 together with pressures measured by the day's two radiosondes prior to launch (1 min averages). The first two points of the averaged radiosonde data are omitted; one is a partial average beginning just after the radiosonde was powered up on the battery, and the second was from inside the room rather than outdoors on the datalogger box.

nor have we tried to interpolate the 2 min data to 1 min values. Instead we have carefully examined data recorded before and after instrument moves, failures, and replacements, and have replaced suspicious data with the flags used to indicate missing data.

There is one final general note about the surface meteorology data. The radiosonde system maintains time very accurately, but it does so by getting the time from GPS satellites. The DigiCORA sounding software (Vaisala Oyj, 2010; version 3.64.1) sets the computer clock to GPS time, which was 17 s ahead of UTC time during ENRR, and the Campbell software sets the datalogger clock to the computer's time. For the final Kiritimati surface data set, all times have been shifted by 17 s, so that all observations are reported in UTC.

3.1.1 Surface pressure

The radiosonde ground-check procedure requires an external pressure reading to provide a calibration check and correction to the radiosonde pressure sensor. This was provided by the Vaisala PTB101B pressure sensor mounted at

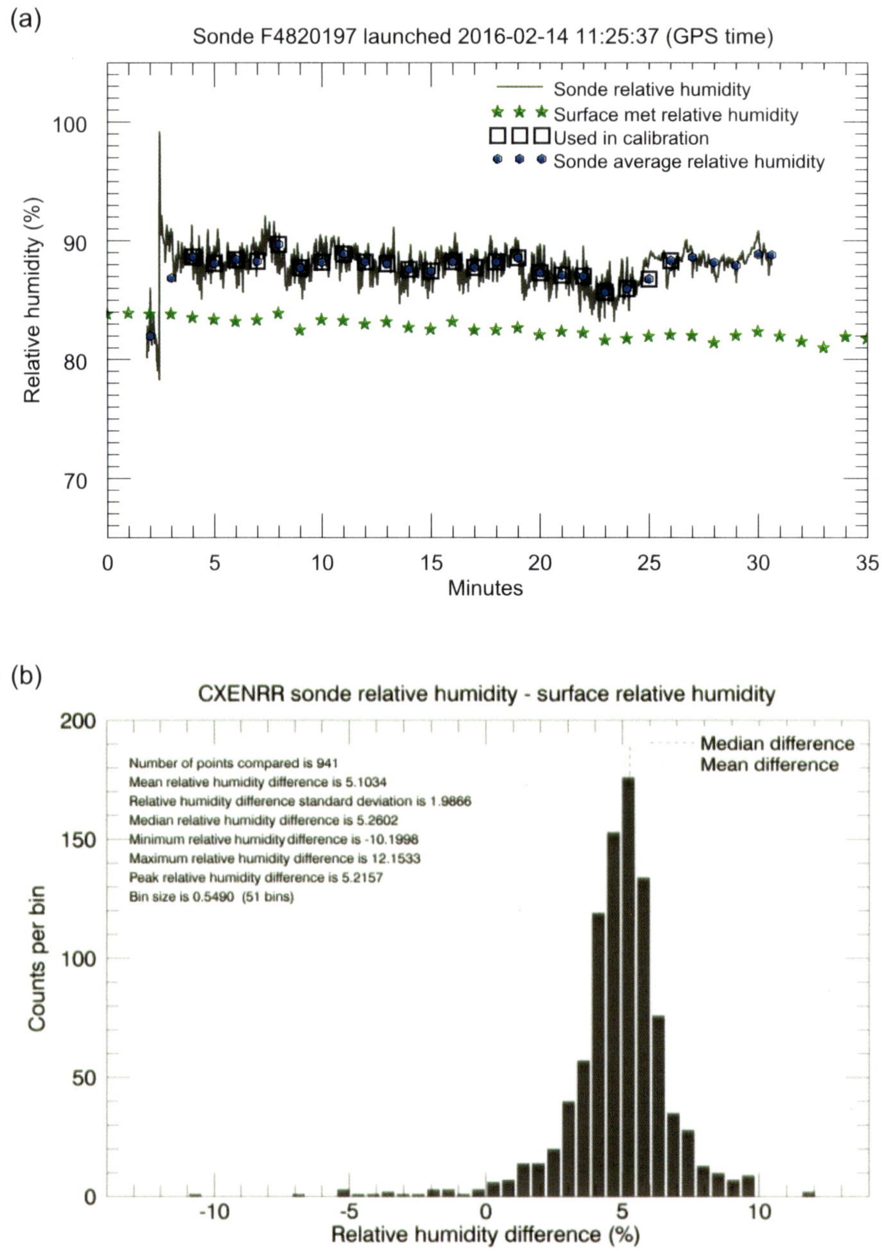

Figure 7. **(a)** 1 s radiosonde relative humidity and 1 min average radiosonde relative humidity prior to the 14 February 2016, 12:00 UTC launch at Kiritimati, together with 1 min values of HMP45C relative humidity. The initial portion of the radiosonde data covers the period when the radiosonde was transported from the Vaisala GC25 ground-check unit inside the bungalow out to the surface station. The averaged values from this radiosonde that were part of the set used to calibrate the surface humidity instrument are indicated. **(b)** Differences between 1 min average radiosonde humidities and 1 min values of relative humidities from the HMP45C when the radiosonde was co-located with the Kiritimati surface instruments prior to launch. All available good surface data and the corresponding radiosonde data (101 radiosondes) are included.

3.6 m a.s.l. on the instrument tripod. The barometer was installed with a Vaisala SPH10 static pressure head to minimize wind-induced errors. During the first month of the experiment, the staff noted pressure differences between the PTB101B and both the radiosondes (Fig. 6) and other barometers. The PTB101B was checked against a high-quality calibrated standard after returning to Colorado. The PTB101B read 0.64 hPa too high, so in the final data set all the surface pressure data were decreased by 0.64 hPa to account for this calibration offset.

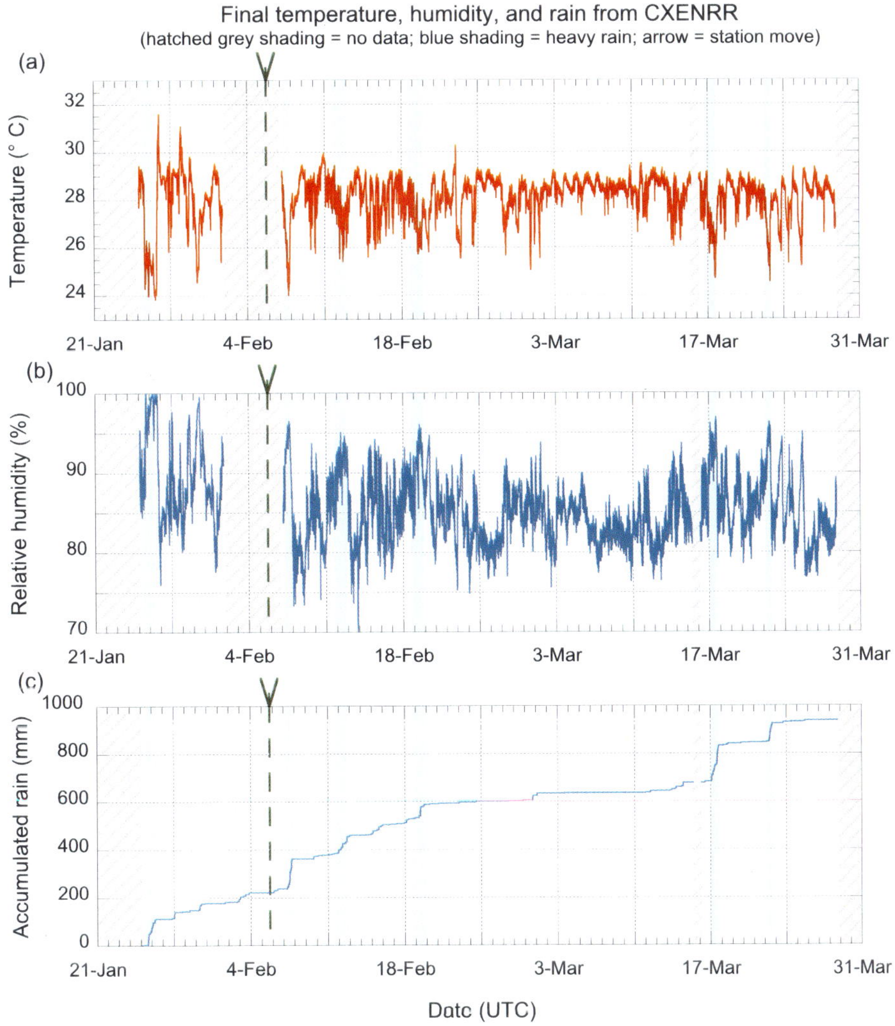

Figure 8. The final time series of temperature **(a)** and relative humidity **(b)** from Kiritimati Island, together with the accumulated rainfall **(c)** during the deployment. Pale blue shading indicates the six largest rainfall events, and is repeated on all graphs for reference, while hatched grey shading indicates times with no data.

3.1.2 Relative humidity and temperature

The radiosonde ground-check procedure serves to calibrate each radiosonde's humidity sensor, so we considered that to be our most trustworthy source of humidity information. In light of that, it became obvious very early in the project that there were problems with the station's original HMP45C humidity probe. These were not entirely solved by the replacement HMP45C. Figure 7a shows the HMP45C humidity readings with the co-located radiosonde data before the 14 February 2016, 12:00 UTC launch. This was a very typical situation, with the HMP45C consistently reporting lower humidity than the radiosonde. Analysis of all co-located data collected after the HMP45C was replaced shows that the mean difference between the HMP45C and the co-located radiosonde humidity was 5.1 % (Fig. 7b). We have therefore corrected the surface relative humidity for this bias by adding 5.1 % to all reported relative humidity values that were considered good. (Put another way, we calibrated our replacement HMP45C against a sequence of calibrated radiosonde humidity sensors.) The final values are shown in Fig. 8b.

During post-deployment evaluation of our data, we also compared the 1 min HMP45C temperature with 1 min averages of radiosonde temperatures collected while the radiosonde was co-located with the instrument tripod. All comparisons are with the replacement sensor. The results (Fig. 9) show that the mean temperature difference between the radiosonde temperatures and those from the HMP45C is less than ±0.1 °C. This is better than expected, given the differences of shielding and aspiration[3]. We concluded that the

[3]The HMP45C was inside a radiation shield, while the radiosonde was merely set in the shade of the solar panel. Neither the HMP45C nor the radiosonde was mechanically aspirated. The

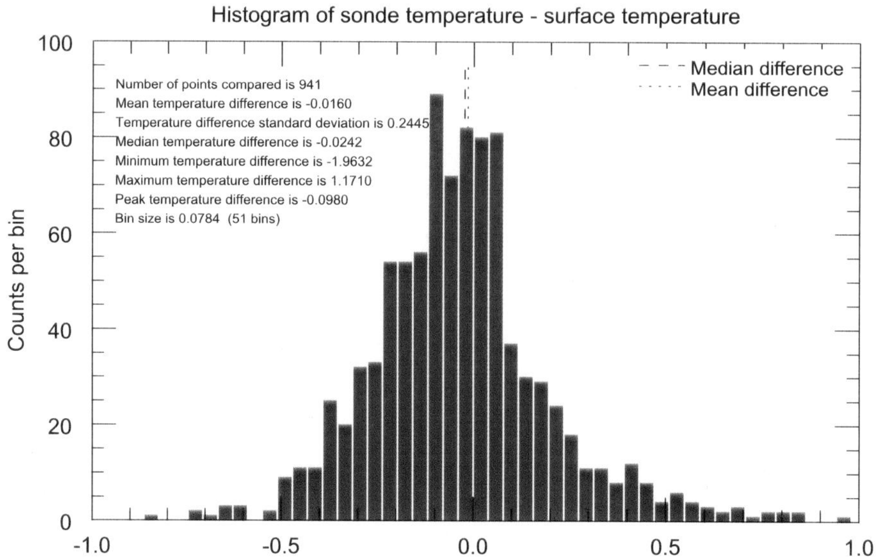

Figure 9. Differences between 1 min average radiosonde temperatures and 1 min values of temperatures from the HMP45C when the radiosonde was co-located with the Kiritimati surface instruments prior to launch. All available good surface data and the corresponding radiosonde data (101 radiosondes) are included.

temperature probe data, except for the initial deployment period when there was a bad probe, could be used as recorded in the field. Figure 8a shows the final time series.

3.1.3 Winds

Several hours after initial setup the base of the anemometer was aligned to face south using a Brunton hand-held compass. The user was standing within a few meters of the tripod and the anemometer was 3.88 m above ground level. Wind observations prior to this are of speed only. The alignment was repeated when the station was relocated. On 28 March 2016 the on-site observers used a hand-held compass and string to determine whether the anemometer was aligned at 168° relative to magnetic north, within a ±3° range. Since the declination for our site is 9.03° east of north (National Centers for Environmental Information, 2016), the values recorded on site will show a 180° wind direction when the wind is $168° + 9° = 177°$. We have therefore subtracted 3° from the recorded wind directions to correct for this bias. This correction was applied to the entire record, since the alignment method was the same during initial deployment and after the move. The measured wind speeds are the best estimate we have of the actual wind speeds, and we have not changed them. The final time series is shown in Fig. 10.

As mentioned above, neither the initial nor final location of the surface instruments was ideal from a wind perspective. The bungalows were approximately 6 m high, and no place near our launch site qualified as "open terrain", i.e., with "the distance between the anemometer and any obstruction ... at least 10 times ... the height of the obstruction" (World Meteorological Organization, 2013, Part III Sect. 3.3.5). Therefore, even after the move there were still blockage issues affecting the measured winds. However, moving the station clearly increased the range of wind directions and speeds we observed. We have analyzed the corrected winds over several time periods, calculating various statistics from the high-resolution time series and also evaluating the directions in both 5 and 10° bins[4]. Before the move, the mean wind speed was 2.23 m s^{-1} with a mean direction of 22.5°; 71 % of the winds came from 0 to 50° (Fig. 11a), with very small secondary clusters from the south (160–180°, 5 %) and southwest (230–275°, 7 %). Only 3.4 % of the measured winds were from the 50–160° sector, which is surprising considering Kiritimati's location in the trade wind region. After the move, the mean wind speed was 4.52 m s^{-1} and the mean direction 61.9°; 85 % of the winds came from 15 to 100° (Fig. 11b), with large clusters at 50–70° (28 %) and 80–95° (19 %). Winds from the remainder of the 50–160° sector, i.e., from 100 to 160°, were uncommon (4 %), while extremely small but distinct secondary clusters were still present from the south (175–195°, 2 %) and southwest (240–265°, 1 %).

Statistical analysis of the station move's impact, e.g., whether or not the winds are different before and after the

radiation shield is designed to allow aspiration by the ambient airflow, but we do not assume that the effects of air flow through the shield and air flow over the exposed radiosonde sensors were the same.

[4]The 10° bins provide a representative visual summary of the slightly more nuanced view the 5° bins show, so we present figures using 10° bins accompanied by quantitative discussion from the 5° binned data.

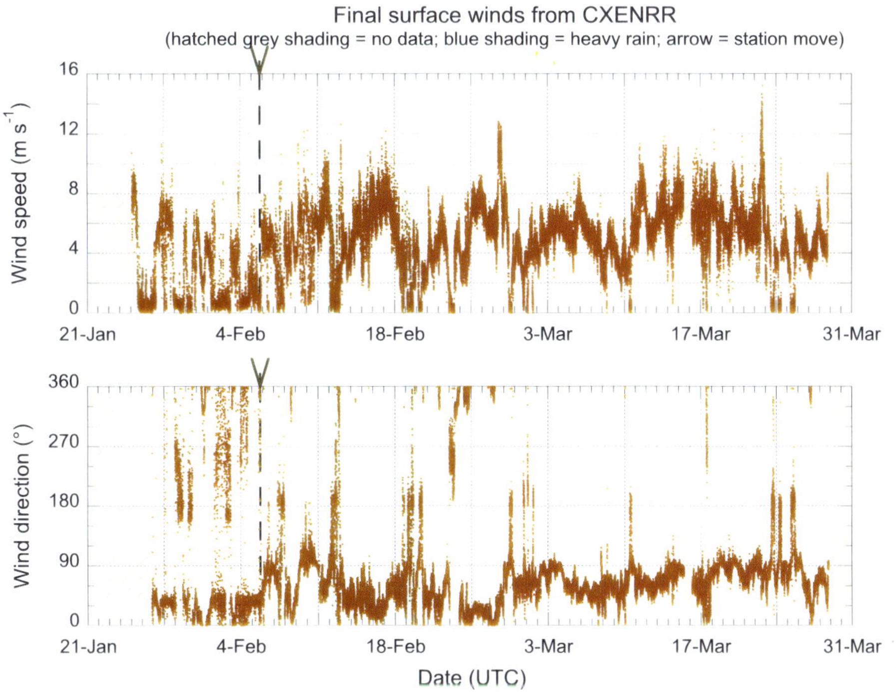

Figure 10. The corrected surface winds from the Kiritimati Island site. Pale blue shading indicates the six largest rainfall events (cf. Fig. 8), while hatched grey shading indicates times with no data.

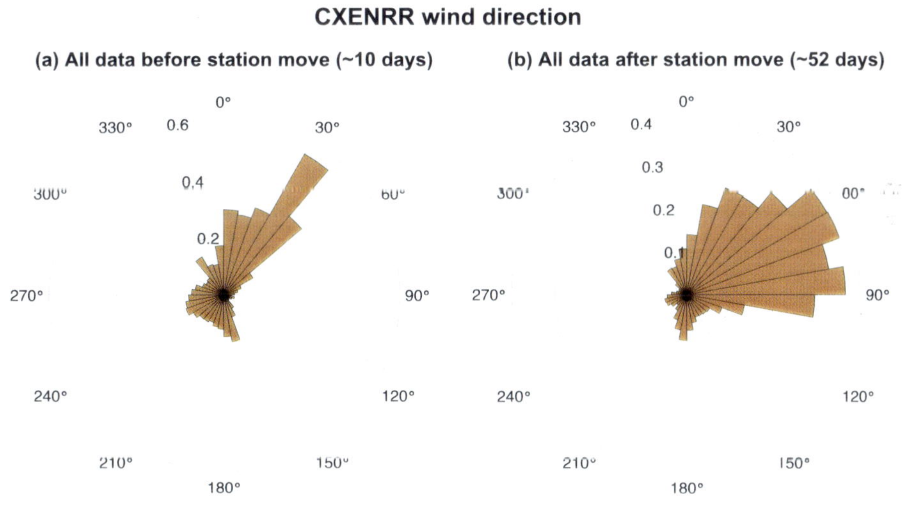

Figure 11. Rose diagrams of all data collected before **(a)** and after **(b)** the CXENRR station was moved on 5 February 2016. Sector lengths are proportional to the square root of the relative frequency of that 10° bin, so the area of each sector is proportional to that bin's frequency. All data before the move are 2 min averages; there are about 5.25 days of such data after the move, and thereafter the data are 1 min averages.

move, is complicated both by the very unequal lengths of the "before" and "after" periods (and the attendant possible large-scale meteorological variations) and by the change in averaging period on 11 February 2016. We have therefore formally compared the 5.25 days' worth of 2 min winds available after the move with 5.25 days' worth of winds immediately before the move. Our analysis employs circular statistics; it follows Fisher (1995) and relies heavily on MATLAB routines described by Berens (2009). Rose diagrams from these equal before and after periods are shown in Fig. 12, key statistical quantities in Table 4, and calculation details in Appendix A. The before and after wind roses show many of the same features noted in Fig. 11, and Fig. 10 shows that the 5.25 days immediately after the move incor-

Table 4. Circular statistics computed from 5.25 days' worth of 2 min average CXENRR winds collected immediately before (left) and after (right) the CXENRR station was moved on 5 February 2016. The 3° directional correction discussed in Sect. 3.1.3 was applied before statistics were calculated. Statistical definitions are given in Appendix A1.

Parameter	Before (31 January 2016, 15:19:43 UTC– 5 February 2016, 21:19:43 UTC)	After (5 February 2016, 22:03:43 UTC– 11 February 2016, 4:03:43 UTC)
Number of observations n	3780	3765
Mean direction $\overline{\theta}$	18.8°	81.1°
Mean resultant length $\overline{R}$	0.63	0.77
Median direction $\widetilde{\theta}$	302.0°	86.6°
Circular dispersion $\hat{\delta}$	0.54	0.45
Concentration parameter $\hat{\kappa}$	1.65	2.56
	Summary	
$\hat{\delta}_{max}/\hat{\delta}_{min}$	1.2	
Test statistic Y_2	4262.4	

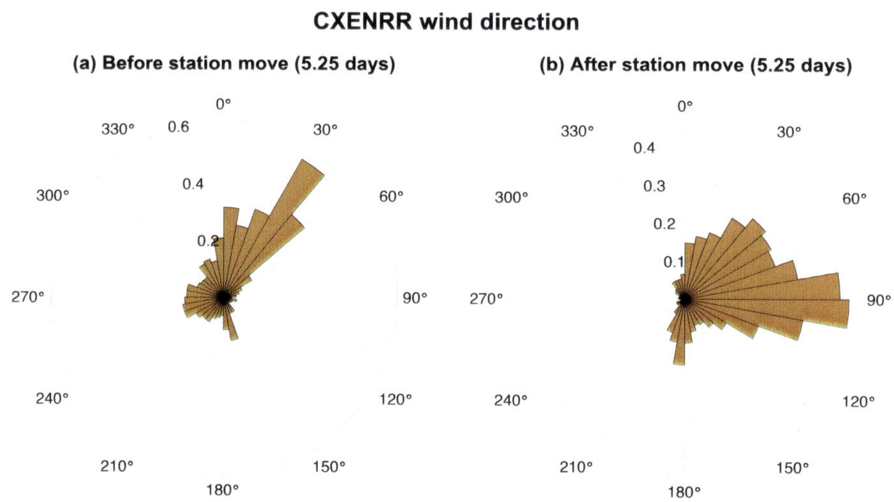

Figure 12. Rose diagrams of 5.25 days' worth of 2 min averages collected immediately before **(a)** and after **(b)** the CXENRR station was moved on 5 February 2016. Sector lengths are proportional to the square root of the relative frequency of that 10° bin, so the area of each sector is proportional to that bin's frequency.

porate variability similar to the full "after" time series. The null hypothesis that the mean directions are the same is to be rejected if the test statistic Y_r is too large when compared to the upper $100(1-\alpha)\,\%$ point on a χ^2_ν distribution with $\nu = r-1$ degrees of freedom. In this case $r=2$, and Y_2 far exceeds the threshold needed to reject the null hypothesis at the 0.1% significance level.

3.2 NOAA Ship *Ronald H. Brown*

Post-processing and quality control of the surface data recorded on NOAA Ship *Ronald H. Brown* made use of the ship's occasional proximity to TAO buoys that were actively collecting data (generally the buoys were offline for maintenance when the ship was nearby), and also took advantage of equipment associated with our radiosonde launches. Before launch the Vaisala RS92-SGP radiosondes were put through a ground-check process inside an air-conditioned space. Because the final surface meteorology data would be needed for the post-processed radiosonde data, some of these ship observations have been adjusted to the height of the Vaisala GC25 ground-check set (3.8 m a.s.l.).

3.2.1 Station pressure

Atmospheric pressure was measured with a Vaisala PTB330 which had been calibrated on 29 December 2015. While the barometer was located on the ship's bridge (15.24 m a.s.l.) the pressure reported in the data set has been adjusted to

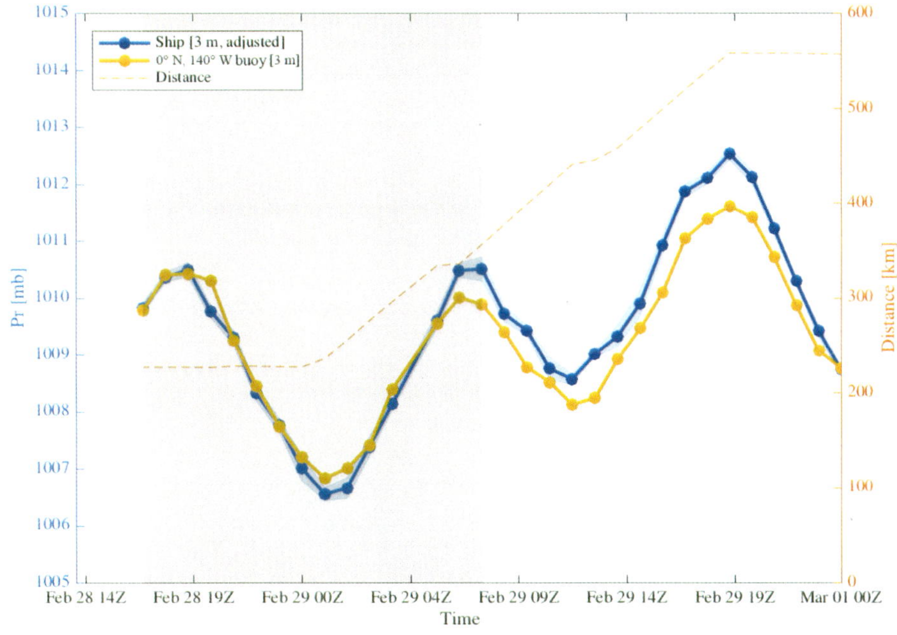

Figure 13. Barometric pressure at 3 m a.s.l. from the ship (blue) and from buoy dm166b (yellow), 28 February through 1 March 2016. Each observation is a 2 min average centered on the top of the hour. Blue shading shows the 1σ variability of the 2 min average from the ship; this information is not available from the buoy. The red-dashed line shows the distance between the ship and the buoy in kilometers. For the time period when the ship was < 350 km from the buoy (grey shading), the difference (ship minus buoy) is -0.03 ± 0.3 hPa.

3.8 m, the height of the Vaisala GC25 ground-check set, via the hypsometric equation.

The ship's pressure sensor is mounted on the port side of the bridge. Airflow distortion caused by the superstructure of the ship can result in localized pressure anomalies. We compared the ship pressure with the pressure observed by the radiosondes while the radiosondes equilibrated outside on the fantail of the ship (see Hartten et al., 2018). This analysis revealed a bias in the ship pressure as a function of relative wind direction that is approximately 0.007 hPa per degree of relative wind direction ($r = -0.46$, $p < 0.001$): positive when the winds are incident upon the port side, negative when they are incident upon the starboard side, and near zero when the relative wind direction is aligned along the ship's length. However, while a correction improves the overall statistical bias of the data set, it introduces discontinuities into the time series when applied to these high temporal resolution data. Therefore, the uncertainty is reported here and a correction was applied to the pressures that were input as surface observations for the soundings, but no such correction was applied to the 1 min average pressure in the surface data set.

The TAO buoy at 0° N, 140° W (dmb166) reports atmospheric pressure at the top of every hour using a Paroscientific MET1 pressure sensor mounted at 3 m (National Data Buoy Center, 2010a, b). Unfortunately, when the ship arrived on location we discovered that dmb166 was damaged, so no data are available during our approach. The closest overlapping pressure observations are from after the buoy was repaired, on 28 and 29 February 2016 while the ship was traveling south but still within several hundred kilometers of the buoy. Comparing the ship pressure, adjusted to the buoy pressure sensor height of 3 m a.s.l., and the buoy pressure during this time (Fig. 13) shows that the two were in close agreement from when the buoy pressure becomes available until around 06:00 UTC on 29 February when the two measurements began to diverge systematically, likely because of the large distance between the sensors (more than 350 km after 07:00 UTC on 29 February). From 14:00 UTC on 28 February through 07:00 UTC on 29 February the mean and standard deviation of the difference (ship minus buoy) is $0.08\,\text{hPa} \pm 0.21\,\text{hPa}$.

3.2.2 Temperature

Atmospheric thermodynamic temperature was measured at 1 Hz by an IMET Rotronic MP-101A mounted on the bow mast at approximately 15 m a.s.l. The sensor, which has been modified by Woods Hole Oceanographic Institution (WHOI, 2010a, b), was last calibrated on 26 July 2014. Most TAO buoys record temperature, so the ship's air temperature can be validated against many measurements made away from the superstructure of the ship. The buoys record temperature at 3 m a.s.l. using Rotronic MP-101A temperature–humidity probes (National Data Buoy Center, 2010a, b). Buoy measurements are recorded every 10 min based on 2 min averages of 2 Hz measurements around the sample

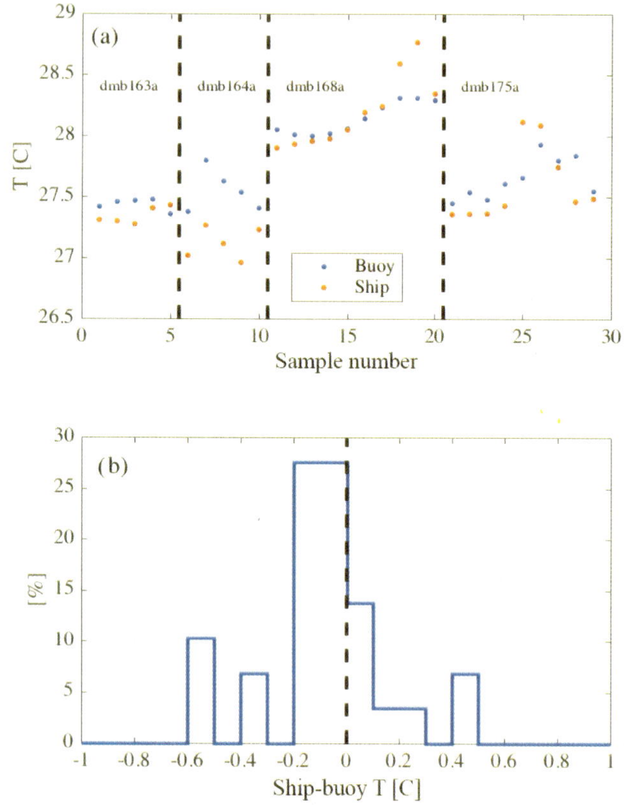

Figure 14. Comparisons between the air temperatures measured by the ship and by a buoy when the ship was within 0.1° latitude/longitude of the buoy. All data are 10 min averages; ship measurements are from 15 m and buoy measurements from 3 m a.s.l. **(a)** All available observation pairs, with dashed vertical lines separating buoys: dmb163a (9° N, 140° W), dmb164a (5° N, 140° W), dmb168a (5° S, 140° W), and dmb175a (8° N, 125° W). **(b)** The distribution of all differences between those pairs.

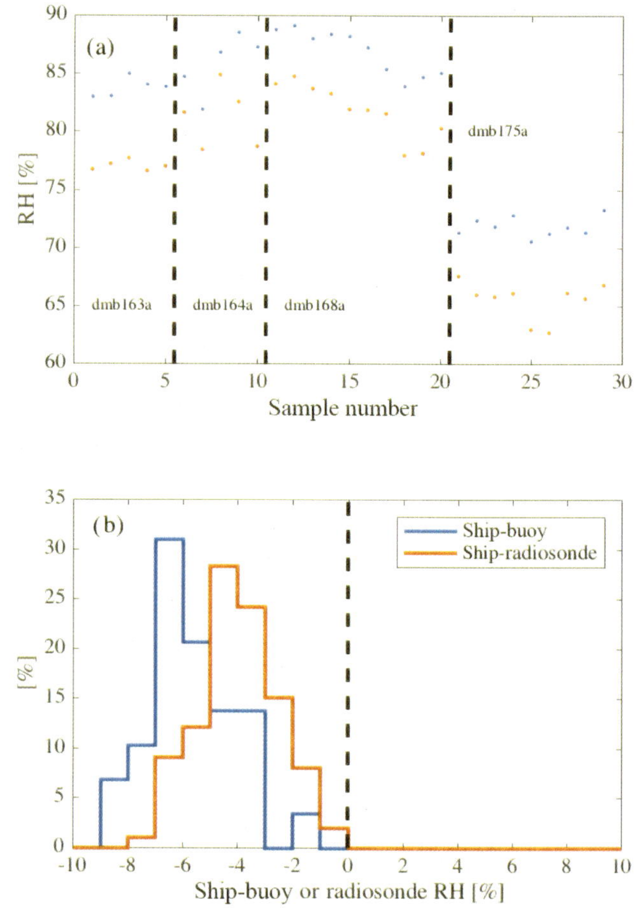

Figure 15. Comparisons between the RH measured by the ship and by a buoy when the ship was within 0.1° latitude/longitude of the buoy. All data are 10 min averages; ship measurements are from 15 m and buoy measurements from 3 m a.s.l. **(a)** All available observation pairs, with dashed vertical lines separating buoys: dmb163a (9° N, 140° W), dmb164a (5° N, 140° W), dmb168a (5° S, 140° W), and dmb175a (8° N, 125° W). **(b)** The distribution of all differences between paired ship and buoy RHs, together with the differences between 99 pairs of ship RH and radiosonde RH, the latter from 4 m a.s.l.

time. Figure 14 shows a comparison between the temperature from the bow mast, averaged to match to buoy sampling, and the temperatures reported from select buoys. Comparisons are only made when the ship was within 0.1° latitude and 0.1° longitude (∼ 10–15 km) of the buoy, i.e., on 29 occasions involving four buoys at (9° N, 140° W), (5° N, 140° W), (5° S, 140° W), and (8° N, 125° W). The difference between the ship and buoy temperatures (ship minus buoy) is $-0.09\,°C \pm 0.25\,°C$, which is within the uncertainty in the sensor calibrations (∼ 0.4 °C). Note that this comparison is between temperatures observed at slightly different heights; for context, the average difference between the temperature observed at 15 m and 3 m from the 93 radiosondes was similar ($-0.1\,°C \pm 0.14\,°C$).

3.2.3 Relative humidity

Atmospheric relative humidity (RH) was also measured at 1 Hz by the IMET Rotronic MP-101A on the bow mast at approximately 15 m a.s.l., and by Rotronic MP-101A probes mounted on the TAO buoys at 3 m a.s.l. We compared the ship RH and buoy RH for the same cases and using the same methodology as for temperature, with a different outcome (Fig. 15). There is a systematic bias of $-5.7\,\% \pm 1.6\,\%$ in the ship's RH observation compared to those from the buoys. Also plotted in Fig. 15b is a comparison between the ship RH and the humidity reported by radiosondes while they were outside on the fantail (∼ 4 m a.s.l.) prior to launch. The comparison was only made for radiosondes launched during night (solar zenith angle > 90°) because of solar heating of the fantail deck during the day (Hartten et al., 2018) and uses radiosonde data averaged over the 1 min just prior to launch. This comparison was again between two different heights, ~15 and ~4 m a.s.l. Edson et al. (2004) documented strong near-surface vertical humidity gradients at approximately (3° S, 122 to 150° W) during February 2001, using measurements made from NOAA Ship *Ronald H. Brown* during GasEx-01, so we expect to see a difference of about 1.5 %. The actual difference between our nighttime radiosonde RH profiles at these heights (15 m minus 4 m) is −1.6 % ± 1.5 %, consistent with this expectation. The ship-to-radiosonde comparison indicates that the ship RH is biased low (−4.0 % ± 1.4 %). Note that sensor response times and mixing of the ambient air by the superstructure of the ship give reason to exercise caution with respect to these results. The results nevertheless are consistent with the interpretation of the observed bias being instrumental error that is not explained by differences in height. We have therefore corrected the ship's recorded RH by adding a constant 4.0 % throughout the cruise, resulting in an RH for an effective height of approximately 3 m. We chose to use the correction based on the comparison to the radiosondes instead of the buoys because the sample size is larger ($n = 99$) and covers a longer range of time, and because the distance between the sensors was always less than 100 m. As noted in Sect. 3.1.2, the radiosonde humidity sensors were calibrated as part of the pre-launch procedure, so this correction can also be described as calibrating the bow mast's IMET Rotronic MP-101A against a sequence of calibrated radiosonde humidity sensors.

3.2.4 Winds

Wind speed and direction during the cruise were measured by sonic anemometers mounted on poles on both the port and starboard sides of the bridge roof, roughly 19 m a.s.l. Three issues affected the measurements, one particular to this cruise and the others more general: one of the bridge anemometers was misaligned; flow is distorted when an anemometer is in the lee of local obstacles; and the ship itself causes distortion of the wind flow at all points. Our corrections were complicated by the fact that the bow mast's anemometer was not operating and the ship was never co-located with a buoy that was collecting winds, both of which deprived us of independent measurements. The procedure we used to obtain what we consider to be a credible time series of wind speed and direction is described below.

We began by addressing the misaligned anemometer. After the field campaign was over we investigated this bias by examining the distribution of the differences in the wind direction when the wind speeds are in agreement and greater than 10 m s^{-1} while the *ship-relative* wind direction recorded by the port sonic was between 320 and 360°. The latter condition ensures that winds are from the port-side bow, since initial analysis indicated possible obstruction when winds were from 0 to 45°. The results indicate that the starboard sonic was misaligned −17.6° *relative to the port sonic measurement*, which is assumed to be properly aligned parallel to the ship's beam with "north" pointing in the direction of the bow. Therefore, all starboard sonic wind directions have been rotated clockwise by 17.6°, and all further discussion of starboard winds is based on the rotated values.

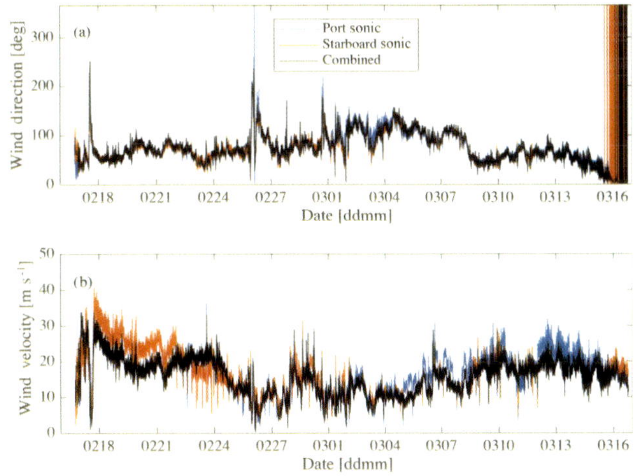

Figure 16. **(a)** Time series of wind directions during the entire cruise, as measured by the port side and starboard side ultrasonic anemometers and the time series constructed by combining the two. These winds have not been corrected for flow distortion. **(b)** Same as in **(a)**, but for wind velocity.

We next combined data from the two bridge anemometers into a single time series by selecting the windward anemometer at all times. Specifically, we used the starboard anemometer when it measured ship-relative winds between 0 and 170° and the port anemometer when it measured ship-relative winds between 190 and 360°. When both were within range, the starboard anemometer's winds were used because our analysis of the data indicates there is less localized distortion near the starboard anemometer. When ship-relative winds were from the stern, which was quite rare, they were treated as missing. The resulting time series for the entire cruise is shown in Fig. 16.

The problems associated with free stream velocity issues were addressed in multiple steps. The typical procedure would be to use the winds measured on the bow mast and correct for flow distortion at that position, which is well characterized for this vessel. Moat et al. (2001) used computational fluid dynamics simulations to calculate the biases in measured wind velocities due to flow distortions caused by NOAA Ship *Ronald H. Brown*'s superstructure. They made calculations for sensor positions on the bow mast, but not for the anemometer positions on the bridge roof. However, even in the absence of calculations for the bridge positions, correction scale factors for an effective height in free stream for the sonic anemometers on the bridge could be estimated from a comparison to the corrected bow-mast wind measurements.

Since we had no bow-mast winds during ENRR, we instead estimated correction factors from data acquired by the

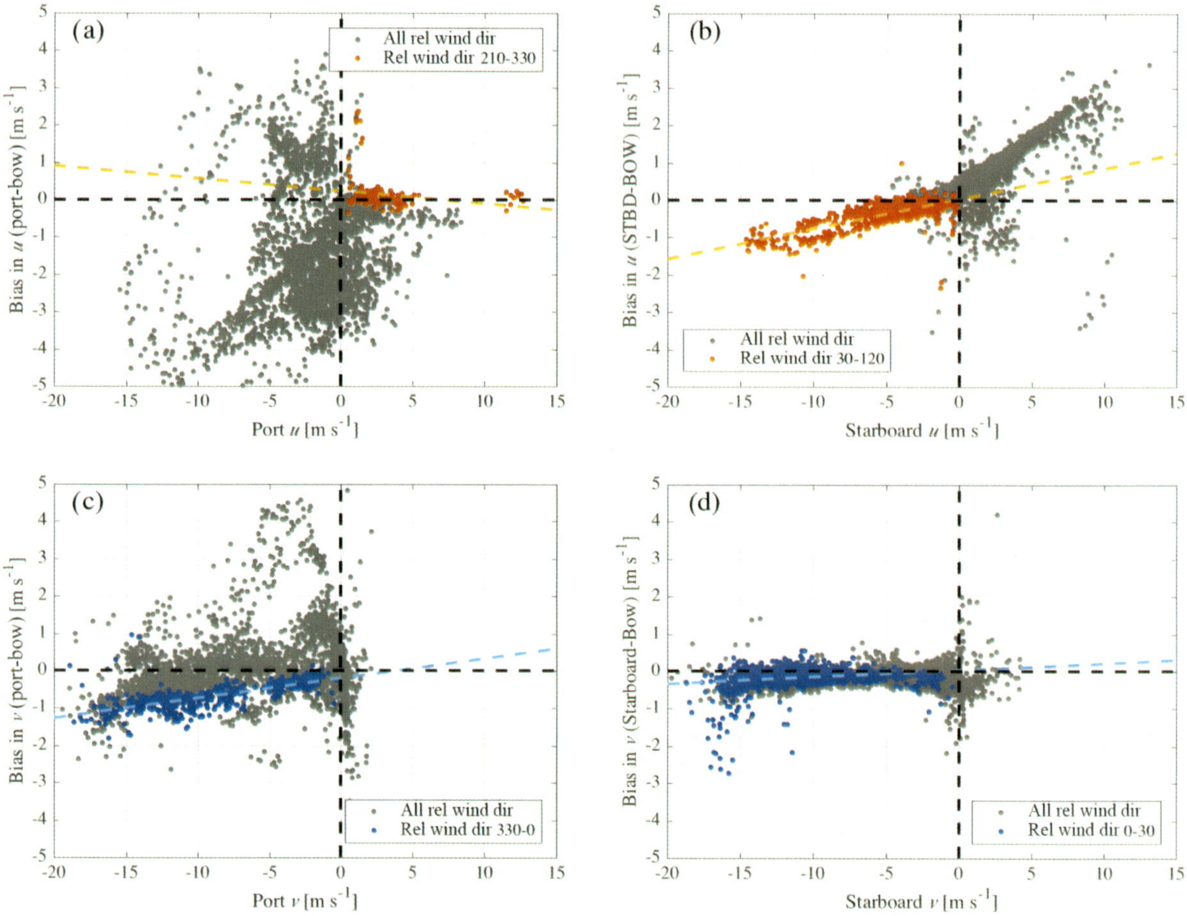

Figure 17. Comparisons between winds measured by the bridge sonic anemometers ("Port" and "Starboard") and the bow-mast anemometer ("Bow") on NOAA Ship *Ronald H. Brown* during CalWater-2, 2015. **(a)** The bias in u (Port – Bow) as a function of observed wind speed at Port, **(b)** similar to **(a)**, but for the Starboard sonic; **(c)** and **(d)** similar to **(a)** and **(b)** for the v wind component. u is defined perpendicular to the ship's orientation and v is defined parallel.

Ronald H. Brown in January and February 2015 during the CalWater-2 field campaign. That cruise occurred in the Pacific, northeast of Hawaii (Ralph et al., 2016), and data from both bridge-mounted sonic anemometers and from a sonic anemometer mounted on the bow mast were recorded. We assume that the bias in the bow mast's sonic is $-5\,\%$ in v (parallel to the ship's orientation) and $+15\,\%$ in u (transverse to the ship's orientation) (Christopher Fairall and Byron Blomquist, personal communication, 2017) and incorporate correction factors accordingly for the CalWater-2 data. The biases in u and v for both of the bridge sonic anemometers are shown in Fig. 17 as functions of measured u and v at the bridge. These relationships are linear and so a single scale factor in u and v, analogous to the corrections applied to the bow-mast data, can be used to correct the bridge data. The correction factors were determined using the mean proportional error (vertical lines in Fig. 18): $-6.1\,\%$ (Port v), $+1.1\,\%$ (Port u), $-2.7\,\%$ (Starboard v), $-7.1\,\%$ (Starboard u). These corrections effectively convert data from the bridge sonic anemometers to match the corrected winds measured by the sonic mounted on the bow, which is at a different height and is in a different position on the ship. Since the bow sonic anemometer is positioned at 17.251 m, this becomes the effective height for the bridge sonic anemometer's data. We applied these correction factors to the wind data collected during ENRR at the bridge, and then scaled them assuming neutral stability conditions using the wind profile power law with an exponent of 0.11 (Hsu et al., 1994) in order to represent winds at the 3.8 m a.s.l. height of the GC25 ground-check set.

3.2.5 Sea surface temperature, downwelling shortwave radiation, and ship position and movement

SSTs were logged at 1 Hz from a Sea Bird SBE45 thermosalinograph operating at a depth of 5 m. Downwelling shortwave radiation was logged at 1 Hz from an Eppley PSP. Latitude and longitude, as well as the ship's course and speed over ground, were recorded by a Furuno GP90, while ship

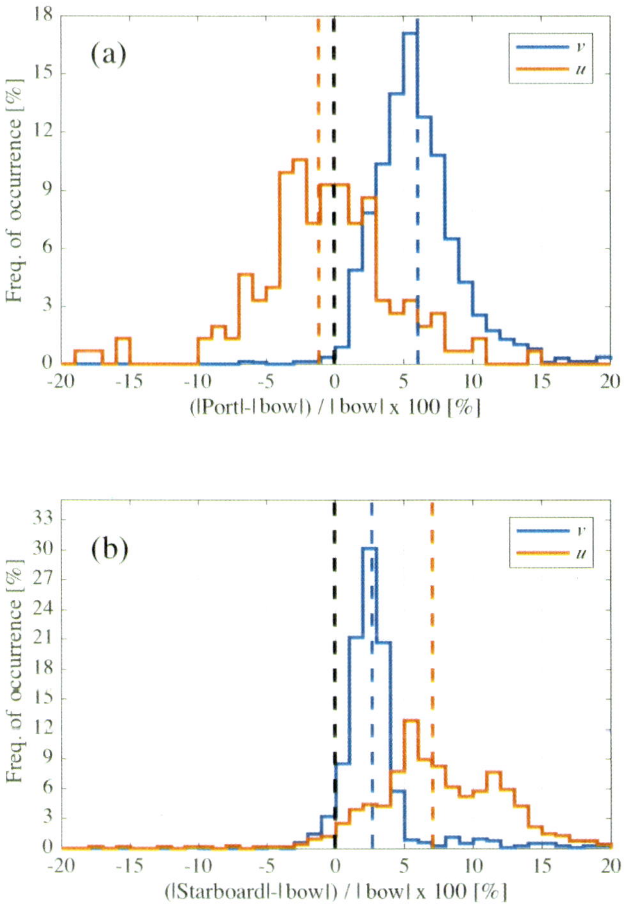

Figure 18. Distributions of the percent error in the **(a)** port and **(b)** starboard bridge sonic anemometers relative to the bow-mast sonic during CalWater-2. The data are the same as in Fig. 14, with u defined as relative wind directions 210–330 and 30–120° and v defined as relative wind directions 330–30°, but here they are shown normalized by wind speed to illustrate the source of the correction factors (denoted by the colored vertical-dashed lines) that are reported in the text.

heading was measured by a Sperry MK37, all at 1 Hz. These collected data were averaged to 1 min; no other changes were made.

4 Surface fluxes at NOAA Ship *Ronald H. Brown*

Based on the ENRR surface observations described in the previous section, we calculated bulk air–sea fluxes (1 h averages) using a recent version of the Tropical Ocean and Global Atmosphere (TOGA) Coupled Ocean–Atmosphere Response Experiment (COARE) flux algorithm (Fairall et al., 1996, 2003; most recently updated in Edson et al., 2013). These calculations include an estimate of the full surface heat budget (see Sect. 4.3). A complete list of the input data and the variables calculated from them by the COARE flux algorithm is in Table 5. The full COARE algorithm requires some atmosphere–ocean quantities that were not measured during the cruises. We have attempted to estimate radiative fluxes that were not measured directly using other variables that were, as described below, but some other variables such as rain rate and direct measurements of turbulent fluxes from eddy covariance methodology are not provided because the ancillary measurements are not available. However, bulk calculations of both latent and sensible heat fluxes are made using COARE and are provided (Table 5). The bulk calculations from COARE have been previously validated in the tropical ocean by comparison to direct measurements from eddy covariance (Fairall et al., 1996). Longwave downwelling radiation (LWD) is needed for the bulk flux calculations within the COARE algorithm if a cool-skin correction is applied to SST; it is required if a full surface heat budget is to be calculated from the COARE results. Unfortunately, LWD was not measured during this cruise and so we estimated it using other means, as described in Sect. 4.2.

4.1 Treatment of measured properties

We constructed 1 h average time series of several quantities needed as input by the COARE algorithm. These are all computed from the 1 min surface data set described in Sect. 3.2: atmospheric pressure (3.8 m a.s.l.); temperature (15 m); relative humidity (effective height 3 m); wind speed (valid at ∼4 m a.s.l.); ocean temperature (also known as SST, valid at 5 m depth); downwelling shortwave radiation; and latitude and longitude.

4.2 Estimation of downwelling longwave radiation

The LWD includes contributions from atmospheric gases and from clouds, and following convention we define those contributions as LWD_{clr}, the "clear-sky" flux from atmospheric gases only, and longwave cloud radiative effect (LWCRE), the enhancement to the LWD caused by clouds (e.g., McFarlane et al., 2013):

$$LWD = LWD_{clr} + LWCRE. \quad (1)$$

To determine LWD, we estimated LWD_{clr} and LWCRE from observations we did collect plus a number of simple assumptions, as described below. The values reported are referenced to the height of the Vaisala GC25 ground-check set, 3.8 m a.s.l.

We first used the longwave Rapid Radiative Transfer Model (RRTM, version 3.3; Mlawer et al., 1997) to calculate LWD_{clr} from each atmospheric sounding collected during the cruise. At each sounding time we provided the model with the "Level 2" radiosonde temperature profile (Cox et al., 2017a) as well as with profiles of the seven most significant radiatively active gases: H_2O, CO_2, O_3, N_2O, CO, CH_4, and O_2. The profile of H_2O came from the same radiosonde profile as the temperature. The CO_2 was set to 398.4 ppm, which was the in situ monthly mean surface value measured

Table 5. Variables, all at hourly resolution, contained in the surface flux data file. The RRTMv3.3 algorithm is described in Mlawer et al. (1997) and the coare35vnWarm algorithm in Edson et al. (2013). The height of the input winds, 3.8 m a.s.l., is denoted by z_u and the Obukhov length by L.

Description	Source	Abbreviation	Units
Select input			
Time	ship observations	Time (UTC)	days
Year, month, day, hour, minute, second	ship observations	Y, M, D, h, m, s	
Latitude	ship observations	Lat	°N
Longitude	ship observations	Lon	°E
Atmospheric pressure corrected to z_u	ship observations	P	hPa
Sky cover	ship observations	SCV	
Clear-sky longwave downward radiation	RRTMv3.3	LWDclr	$W m^{-2}$
Surface energy budget			
Longwave downward radiation at z_u	calculated from LWD_{clr}, LWCRE	LWD	$W m^{-2}$
Longwave upward radiation	coare35vnWarm	LWU	$W m^{-2}$
Shortwave downward radiation	ship observations	SWD	$W m^{-2}$
Shortwave upward radiation	coare35vnWarm	SWU	$W m^{-2}$
Bulk latent heat flux out of the ocean	coare35vnWarm	hlb	$W m^{-2}$
Bulk sensible heat flux out of the ocean	coare35vnWarm	hsb	$W m^{-2}$
Bulk buoyancy flux into the ocean	coare35vnWarm	hbb	$m^2 s^{-3}$
Monin–Obhukov similarity			
Friction velocity that includes gustiness	coare35vnWarm	usr	$m s^{-1}$
Wind stress	coare35vnWarm	tau	$N m^{-2}$
Temperature scaling parameter	coare35vnWarm	tsr	K
Specific humidity scaling parameter	coare35vnWarm	qsr	$g kg^{-2}$
Thermal roughness length	coare35vnWarm	z0t	m
Moisture roughness length	coare35vnWarm	z0q	m
Wind stress transfer (drag) coefficient at z_u	coare35vnWarm	Cd	
Sensible heat transfer coefficient (Stanton number) at z_u	coare35vnWarm	Ch	
Latent heat transfer coefficient (Dalton number) at z_u	coare35vnWarm	Ce	
Obukhov length scale	coare35vnWarm	L	m
Monin–Obukhov stability parameter, z_u/L	coare35vnWarm	zet	
Surface conditions			
Wind speed adjusted to 10 m	coare35vnWarm	U10	$m s^{-1}$
Air temp at 10 m	coare35vnWarm	T10	°C
Specific humidity at 10 m	coare35vnWarm	Q10	$g kg^{-1}$
Relative humidity at 10 m	coare35vnWarm	RH10	%
Neutral value of wind speed at z_u	coare35vnWarm	UN	$m s^{-1}$
Neutral value of wind speed at 10 m	coare35vnWarm	UN10	$m s^{-1}$
Neutral value of drag coefficient at 10 m	coare35vnWarm	Cdn_10	
Neutral value of Stanton number at 10 m	coare35vnWarm	Chn_10	
Neutral value of Dalton number at 10 m	coare35vnWarm	Cen_10	
Sea surface temperature	oare35vnWarm	SST	°C
Cool-skin temperature depression	coare35vnWarm	dter	°C
Surface saturation specific humidity	coare35vnWarm	Qs	$g kg^{-1}$
Latent heat of vaporization	coare35vnWarm	Le	$J kg^{-1}$
Evaporation rate	coare35vnWarm	Evap	$mm h^{-1}$

Table 6. The vertical grid to which all gas profiles (H_2O, CO_2, O_3, N_2O, CO, CH_4, and O_2) were interpolated before being used as input to the Rapid Radiative Transfer Model (RRTMv3.3). Note that the second point (3.8 m) is at the same height as the surface data.

Heights (m)			
0.0	1000.0	5000.0	20 000.0
3.8	1200.0	5500.0	25 000.0
10.0	1400.0	6000.0	30 000.0
25.0	1600.0	6500.0	35 000.0
50.0	1800.0	7000.0	40 000.0
75.0	2000.0	7500.0	50 000.0
100.0	2250.0	8000.0	60 000.0
150.0	2500.0	9000.0	
200.0	2750.0	10 000.0	
250.0	3000.0	11 000.0	
350.0	3250.0	12 000.0	
400.0	3500.0	13 000.0	
450.0	3750.0	14 000.0	
500.0	4000.0	15 000.0	
600.0	4250.0	16 000.0	
700.0	4500.0	17 000.0	
800.0	4750.0		
900.0			

by the NOAA/ESRL's Global Monitoring Division (GMD) at American Samoa in March 2015 (Thoning et al., 2015), and distributed using a constant mixing ratio at all heights. Profiles of the other gases were taken from the US Standard Atmosphere tropical profile (McClatchey et al., 1972). All data were interpolated to a pre-determined height grid consisting of 58 levels from 0 to 60 km (see Table 6). All levels above the termination of the radiosonde were filled using the standard atmosphere. By perturbing profiles with ± 0.5 °C temperature and ± 5 % relative humidity, values taken from the manufacturer uncertainty estimates, we estimate a maximum uncertainty of 11.4 W m^{-2} (< 3 %) in the clear-sky calculation.

For each sounding we also calculated LWD using the measured temperature humidity profile but for a scene containing a hypothetical optically thick cloud positioned at 1 km to represent an estimate of the maximum LWD for that scene. We used RRTM coupled with the Discrete Ordinate Radiative Transfer (DISORT; Stamnes et al., 1988), and considered only absorption, no scattering. The result is a rough estimate of the maximum LWD for that profile, i.e., the LWD in the presence of a low, optically thick stratiform cloud. We were thus able to calculate a maximum estimate of LWCRE, $LWCRE_{max}$, using Eq. (1).

Cloud optical depth was not measured during the cruise, but sky cover (in oktas) is available from hourly observations made from the bridge. We converted oktas to fractional sky cover by dividing by 8, and linearly interpolated these hourly values to the 1 min resolution of the surface observations. Then we multiplied the cloud fraction at each sounding time by the corresponding $LWCRE_{max}$ to yield an estimate of the actual LWCRE. This LWCRE accounts for cloud fraction, including scattered cloud cover, but assumes all clouds are optically thick; this assumption is reasonable since most of the clouds encountered during the cruise were low- to mid-level cumulus and cumulonimbus. Our final estimate of LWD at each sounding time was computed from Eq. (1) using the estimates of LWD_{clr} and LWCRE. This was linearly interpolated to the 1 min resolution of the surface observations and averaged over each hour, and the resulting time series (Fig. 19) was then used as input to the COARE flux algorithm.

Since LWD_{clr} is calculated directly from the observations, it is expected to be much more robust than the estimate of LWCRE, which requires more assumptions. In Fig. 19 it is apparent that the $LWCRE_{max}$ was somewhat larger at the beginning and end of the cruise, when the total column water vapor was lower (not shown; see also Cox et al., 2015), and fairly consistent during the middle part of the cruise when the ship was between 9° N and 8° S (the mean $LWCRE_{max}$ was 38.1 W m^{-2} during this time). Since it was generally cloudy during the cruise (fractional cloud cover = 69 %), the mean LWCRE is estimated to be 31.1 W m^{-2} overall, representing just 7.3 % $\pm$ 3.8 % (1σ) of the LWD. Sensitivity calculations using an assumed cloud base ranging from 0.25 to 5.0 km in 0.25 km increments show that the average LWCRE would decrease by about 2.7 W m^{-2} per extra 1 km of cloud base height (i.e., the mean LWCRE would be 4.3 % smaller (larger) if the assumed cloud base were at 1.5 km (0.5 km)). We performed a second sensitivity study in which we recalculated LWCRE after assigning the cloud bases to reasonable heights based on visual analysis of the radiosonde profiles. Estimating cloud bases in that manner was quite challenging, but while we do not have enough confidence in the estimates to use them in our flux calculation, we think they can put bounds on the possible errors in our 1 km assumption. Comparing those LWCRE values with the ones based on a nominal 1 km cloud height revealed maximum differences of +6 W m^{-2}, nearly 20 % of LWCRE on average, which we take to be a reasonable estimate of the error in LWCRE.

4.3 Bulk flux calculations and final data set

We used the pressure, air temperature, wind, bulk ocean temperature, solar radiation, and longwave radiation time series described above as input to the COARE flux algorithm (version coare35vnWarm; Edson et al., 2013). In keeping with Fairall et al. (1996), we assumed a surface emissivity of 0.97 and a surface albedo of 0.055. The resulting time series, together with a subset of the surface time series used as input, constitute the released bulk flux data set (Table 5). Because we had a bulk ocean temperature rather than a true ocean skin temperature, we had the algorithm do a cool-skin calculation. Figure 20a shows the net surface heat flux Q_0 calculated

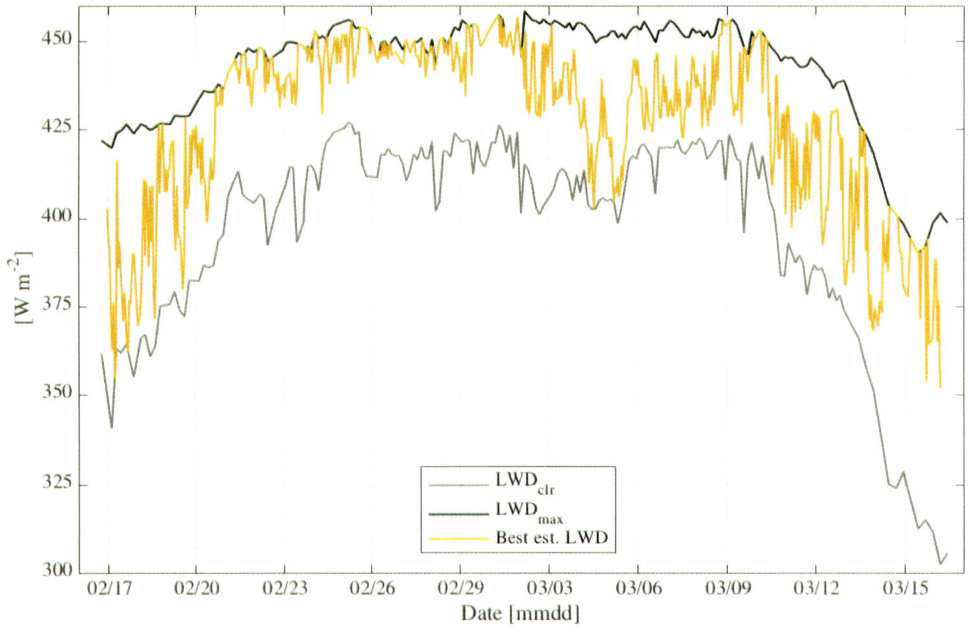

Figure 19. Time series of our best estimate of hourly longwave downwelling radiation (LWD) during the ENRR cruise, compared with hourly estimates during clear-sky conditions, LWD_{clr}, and in the presence of a hypothetical optically thick cloud scaled by the observed cloud area fraction with a base height of 1 km a.s.l., LWD_{max}.

Table 7. Estimates of the uncertainties in the longwave radiation variables plotted in Fig. 19 and in other variables used to calculate the surface heat budget in Fig. 20. Refer to the text for details of the estimates. The combined uncertainty is the estimated uncertainty in the surface heat budget, Q_0, as found by combining the individual estimates in quadrature.

Variable	Source of uncertainty	Uncertainty ($W\,m^{-2}$)	Notes
Q_{sen}	Various	3.0	Fairall et al. (1996)
Q_{lat}	Various	5.0	Fairall et al. (1996)
SWD, (SWU)	Measured	4.4	Relative calibration accuracy of Eppley PSP
SWU	Albedo (clouds)	13.7	Sensitivity study based on Jin et al. (2011)
SWU	Albedo (wind)	2.3	Sensitivity study based on Jin et al. (2011)
LWU	Emissivity	4.5	Based on range reported by Konda et al. (1994)
LWU	Surface skin temperature	1.2	Fairall et al. (1996)
LWD	LW cloud radiative effect	6.0	Sensitivity study (see text)
LWD	Temperature and humidity profiles	11.4	Sensitivity study (see text)
Combined uncertainty		20.8	

from some of the measurements and calculated quantities as

$$Q_0 = Q_{sw} + Q_{lw} - Q_{lat} - Q_{sen}, \quad (2)$$

where Q_{sw} is the net (absorbed) solar irradiance, Q_{lw} is the net longwave flux, Q_{lat} is the bulk latent heat flux out of the ocean, and Q_{sen} is the bulk sensible heat flux out of the ocean. For reference, both the observed air temperature and ocean temperature are plotted in Fig. 20b. The large amplitude of the diurnal cycle (500–1000 $W\,m^{-2}$) indicates that uncertainty associated with the estimate of LWD is quite small relative to the magnitude of the other terms, except for the times of the day when the net surface heat budget is switching from net warming to net cooling, and vice versa (near sunrise and sunset).

Since the true values of surface emissivity and surface albedo are not precisely known, nor would they be constant, our assumed values result in uncertainty in the calculations of upwelling longwave and shortwave radiation (LWU and SWU), respectively. The uncertainty in the calculation of LWU associated with the estimation of skin temperature, as well as a range of emissivity values from the literature, are also shown in Table 7. Sensitivity studies using the ocean albedo parameterization of Jin et al. (2011) suggest 0.055 is a poor approximation for clear skies when albedo is dependent on solar zenith angle but is a close approximation for

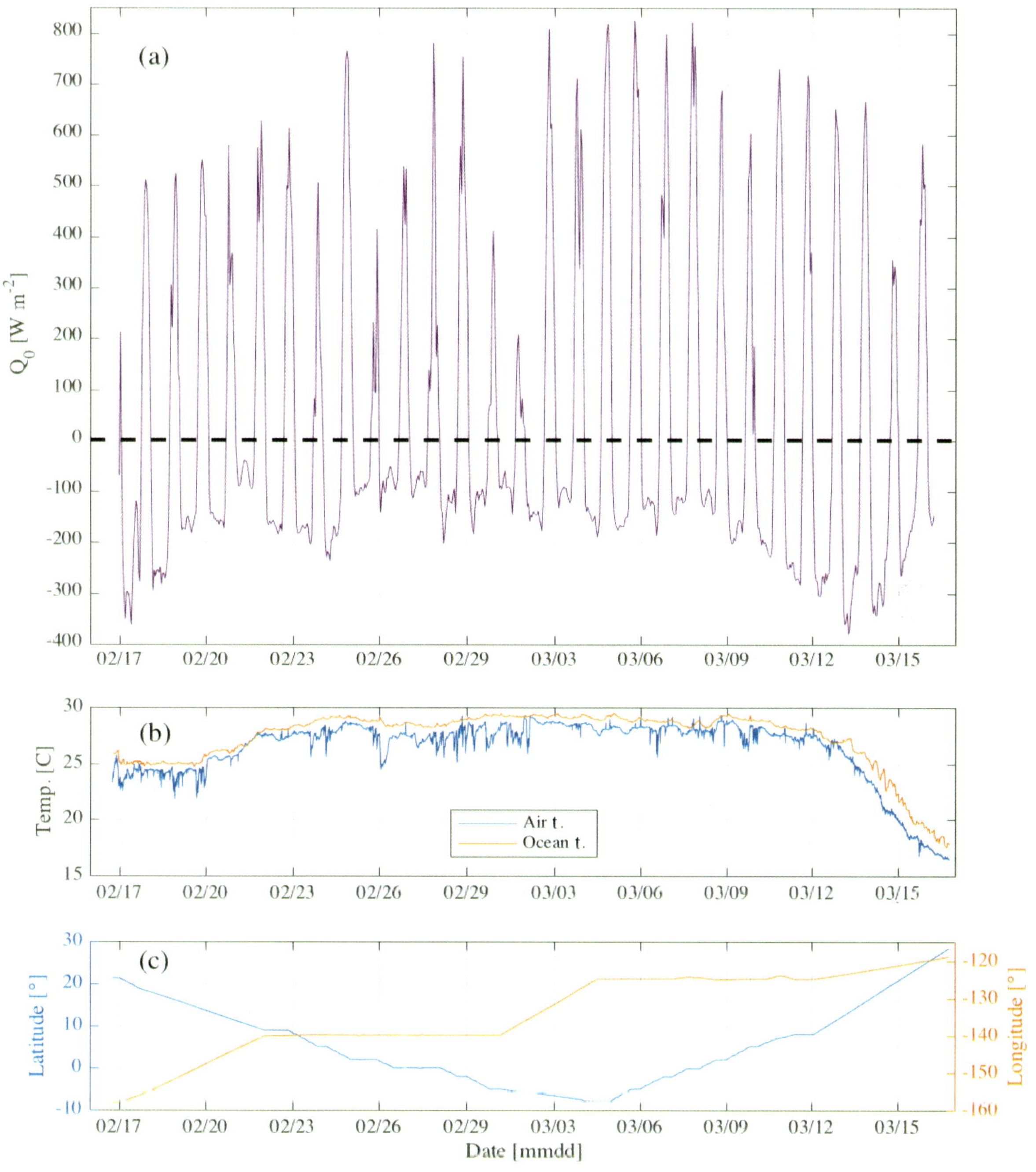

Figure 20. **(a)** 1 h average ocean net surface heat flux (Q_0) along NOAA Ship *Ronald H. Brown*'s track from 16 February to 17 March 2016. The flux is based on a combination of measurements and calculations with the COARE bulk flux algorithm. **(b)** For the same time period, 1 min average ocean temperature at 5 m depth, and air temperature. **(c)** Ship position.

the diffuse regime (overcast skies), so our usage is consistent with the high cloud fraction during the cruise. Unfortunately, we do not have a good estimate of the diffuse fraction to use for a calculation of albedo, but by comparing an overcast and clear-sky albedo parameterization using the Jin et al. (2011) routines, we have added the contribution of cloudiness to the uncertainty in SWU (Table 7). Since albedo is also dependent on wind speed (due to its effect on surface roughness), we also estimated an uncertainty in diffuse-regime albedo by comparing estimates including wind speeds with errors of $\pm 1\,\mathrm{m\,s^{-1}}$ to albedo = 0.055, reporting the largest error (RMSE) of the three-way comparison in calculated SWU (Table 7). Estimates of uncertainty in the other variables used to calculate Q_0 are also shown in Table 7. We estimate the combined uncertainty in Q_0 to be $20.7\,\mathrm{W\,m^{-2}}$.

5 Data availability and use

The data sets described here are archived at the NOAA's National Center for Environmental Information (NCEI), with DOIs as follows: surface meteorology from Kiritimati (https://doi.org/10.7289/V51Z42H4), surface meteorology and SST from NOAA Ship *Ronald H. Brown* (https://doi.org/10.7289/V5SF2T80), and surface fluxes from NOAA Ship *Ronald H. Brown* (https://doi.org/10.7289/V58050VP). They are also available, together with sample code to read the data, from NOAA/ESRL/PSD at https://www.esrl.noaa.gov/psd/enso/rapid_response/data_pub/. Users of these data must cite the appropriate DOI and reference the data as indicated below.

Cox, C., and Hartten, L.: El Niño Rapid Response (ENRR) field campaign: Surface Fluxes from NOAA Ship Ronald H. Brown, 2016-02 to 2016-03 (NCEI Accession 0167875), Version 1.1, NOAA/National Centers for Environmental Information, https://doi.org/10.7289/V58050VP, 2017.

Cox, C., Wolfe, D., Hartten, L., and Johnston, P.: El Niño Rapid Response (ENRR) Field Campaign: Surface Meteorological and Ship Data from the NOAA Ship Ronald H. Brown, February–March 2016 (NCEI Accession 0167875), Version 1.1, NOAA/National Centers for Environmental Information, https://doi.org/10.7289/V5SF2T80, 2017.

Hartten, L., Johnston, P., Cox, C., and Wolfe, D.: El Niño Rapid Response (ENRR) Field Campaign: Surface Meteorological Data from Kiritimati Island, January–March 2016 (NCEI Accession 0161526), Version 1.1, NOAA/National Centers for Environmental Information, https://doi.org/10.7289/V51Z42H4, 2017.

The data from Kiritimati start on 25 January 2016, 03:07:43 UTC and end on 28 March 2016, 19:14:43 UTC, with four gaps in one or more parameters lasting 1 h or more (Table 3). The ship data start on 16 February 2016, 18:32 UTC and end on 16 March 2016 from 18:50 to 18:57 UTC, depending on the parameter.

Each data set is in the form of a single file written in the NASA Ames Format for Data Exchange (hereafter "NASA-Ames Format"); see Gaines and Hipskind (1998) as well as the Centre for Environmental Data Analysis (CEDA) explanatory material (Centre for Environmental Data Analysis, 2002). NASA-Ames Format's plain-text (ascii) nature makes it portable to any machine and easily accessible to users with limited computer resources; the rich metadata in the mandatory and optional header sections make it self-describing. NASA-Ames Format requires that the total number of header lines be the first number in the first line of the file, and that the data following the header lines be space-delimited. Our files use the last header line to provide rudimentary column headers for the data that follow, which makes importing the data into spreadsheets and starting to work with them fairly straightforward. Time is the independent variable in all the files, and is presented as days since 1900-01-01 00:00:00 + 00:00, i.e., since 1 January 1900, 00:00:00 UTC. Users working with the data from Kiritimati Island, especially those interacting with residents and local records, should keep in mind that Kiritimati is in the Line Islands Time (LINT) time zone, UTC+14. Users who prefer to work with data in netCDF format may build their own routines to do so, or may use software such as NASA Ames Processing in PYthon (NAPPy, available at https://github.com/cedadev/nappy). We would caution users to carefully cross-check the output of any conversion software, to make sure that it correctly puts into the netCDF file(s) all the original data as well as the metadata contained in the "special comment" and "normal comment" header sections.

6 Conclusions

As the 2015/2016 El Niño gathered strength, the NOAA's ESRL/Physical Sciences Division conceived the idea of a rapidly deployed multi-platform field campaign to meet a variety of operational and research goals (Dole et al., 2018). Observations during the ENRR field campaign, which involved many partners in and outside of the NOAA, included almost 10 weeks of surface meteorology from Kiritimati Island and 4 weeks of surface meteorology from NOAA Ship *Ronald H. Brown* in the central equatorial Pacific. These data, which were collected at both sites primarily to support radiosonde measurements, have issues caused primarily by instrument failure, out-of-calibration instruments, and less than optimal instrument placement. Some of the issues are a side-effect of the rapid nature of the campaign planning and deployment; others are endemic on crowded research vessels or in remote and geographically challenging locations.

We have carefully vetted the data against expectations and alternate observations when available, identifying issues and minimizing their impacts by making corrections when possible. We have also estimated air–sea surface fluxes from the ship's data. The final data products that are described in this paper have been placed in a long-term repository and are freely accessible for users. Each file contains considerable internal documentation; this article serves as a detailed description of the instrumentation, siting, issues, and corrections. Highlights of the data sets include the high-resolution measurements of nearly a meter of rainfall at Kiritimati and continuous measurements of ocean and air temperatures in the heart of the El Niño warming as the ship visited the TAO buoys along 140 and 125° W. These data represent a rich resource for local studies, for inclusion in models of local/regional processes, or for comparison with satellite observations or model simulations.

Appendix A: Calculation details for circular statistics

The circular statistics presented in Table 4 and discussed in Sect. 3.1.3 are predominantly based on Fisher (1995) and MATLAB code described in Berens (2009), itself strongly reliant on Fisher's book. Since most of the statistics are not yet often used in meteorological work, and since some alterations have been made to Berens' code, the calculations are described here. Discussion of the rationale and details behind the equations is to be found in the cited references.

The sample mean direction $\overline{\theta}$ and mean resultant length $\overline{R}$ are obtained by vector averaging the n winds in the sample (Fisher, 1995, Eq. 2.7–2.9). The sample median $\widetilde{\theta}$ is found by determining "the diameter that divides the data into two equal sized groups", and is either the wind direction nearest "the endpoint of the diameter closer to the center of mass of the data" (for an odd number of wind values) or "half-way between the two closest" wind directions (for an even number of wind values) (Berens, 2009). The sample circular dispersion $\hat{\delta}$, one measure of the spread of the data, is calculated as per an online Erratum (Fisher, 2017) using

$$\hat{\delta} = (1 - \hat{\alpha}_2) / \left(2\overline{R}^2\right) \tag{A1}$$

where

$$\hat{\alpha}_2 = (1/n) \sum_{i=1}^{n} \cos 2\left(\theta_i - \overline{\theta}\right). \tag{A2}$$

This calculation departs from Eq. (2.28) in Fisher (1995),

$$\hat{\delta} = (1 - \hat{\rho}_2) / \left(2\overline{R}^2\right), \tag{A3}$$

where

$$\hat{\rho}_p = \sqrt{\left[(1/n) \sum_{i=1}^{n} \cos p\left(\theta_i - \overline{\theta}\right)\right]^2 + \left[(1/n) \sum_{i=1}^{n} \sin p\left(\theta_i - \overline{\theta}\right)\right]^2} \tag{A4}$$

(Fisher, 1995, Eq. 2.25; also Berens, 2009). However, the Erratum's amendment to this form given above makes sense given Fisher's definition of $\hat{\rho}_2$ as

$$\hat{\rho}_2 = (1/n) \sum_{i=1}^{n} \cos 2\left(\theta_i - \overline{\theta}\right) \tag{A5}$$

in his Eq. (2.27).

In a normal distribution, the dispersion about the mean is discussed in terms of the variance; in circular statistics the dispersion of the wind about the mean direction is discussed in terms of the sample concentration parameter $\hat{\kappa}$. This can be simulated with some difficulty (Fisher, 1995, Sect. 3.3.6) or else estimated as a function of the mean resultant length $\overline{R}$ using a table (e.g., Fisher, 1995, Appendix 3) or a conditional formula (Fisher, 1995, Eq. 4.40):

$$\hat{\kappa} = \begin{cases} 2\overline{R} + \overline{R}^3 + \dfrac{5\overline{R}^5}{6}, & \overline{R} < 0.53 \\ -0.4 + 1.39\overline{R} + \dfrac{0.43}{(1-\overline{R})}, & 0.53 \le \overline{R} < 0.85 \\ \dfrac{1}{\left(\overline{R}^3 - 4\overline{R}^2 + 3\overline{R}\right)}, & \overline{R} \ge 0.85 \end{cases}. \tag{A6}$$

The CircStat MATLAB code (Berens, 2009), which employs the latter, was used to obtain the values in Table 4.

Finally, Fisher (1995, Sect. 5.3.4) outlines a method of testing whether two or more samples have a common mean direction. Since both our samples consist of more than 25 observations and their dispersions are considered comparable because $\frac{\hat{\delta}_{\max}}{\hat{\delta}_{\min}} \le 4$, we used Fisher (1995, Eq. 5.10–5.13) to calculate the test statistic

$$Y_2 = 2(N - R_P)/\hat{\delta}_0 \tag{A7}$$

where N is the sum of the two sample sizes,

$$R_P = \sqrt{\left[\sum_{i=1}^{2} n_i \cos\overline{\theta}_i\right]^2 + \left[\sum_{i=1}^{2} n_i \sin\overline{\theta}_i\right]^2}, \tag{A8}$$

and

$$\hat{\delta}_0 = \sum_{i=1}^{2} n_i \hat{\delta}_i / N. \tag{A9}$$

Author contributions. All the authors spent at least 2 weeks at Kiritimati or 4 weeks aboard NOAA Ship *Ronald H. Brown* collecting data. SA set up and later relocated the Kiritimati surface instruments, PEJ modified the software for the Kiritimati datalogger to give 1 min data and higher-resolution pressure values, and HAM replaced the power supply after the solar panel failed. PEJ corrected the surface data from Kiritimati with input from LMH; CJC corrected the surface data from NOAA Ship *Ronald H. Brown* with input from DEW. CJC generated the surface flux data from NOAA Ship *Ronald H. Brown*. LMH assembled metadata and created the file headers with input from CJC, PEJ, and DEW, and then created the final files from PEJ's and CJC's corrected surface meteorological data and CJC's surface flux data. LMH prepared the manuscript with input from all co-authors.

Competing interests. The authors declare that they have no conflict of interest.

Acknowledgements. The ENRR field campaign was funded by the NOAA through the Office of Oceanic and Atmospheric Research (OAR) as well as the National Weather Service (NWS), the Office of Marine and Aviation Operations (OMAO), and the National Environmental Satellite, Data and Information Service (NESDIS). We thank the Republic of Kiribati for allowing the NOAA to conduct research on Kiritimati Island, and the Kiribati Meteorological Service for their interactions with our observers during and after the field campaign. We appreciate the support we received from the crew of NOAA Ship *Ronald H. Brown*, in particular NOAA/OMAO Survey Technicians Joshua Gunter and Mark Bradley and NOAA/OMAO Lieutenant Adrienne Hopper. The authors were joined in collecting these data by Xiao-Wei Quan (CIRES and ESRL/PSD) on Kiritimati Island and by Matthew Winterkorn (NVision Solutions and NOAA/National Data Buoy Center) on NOAA Ship *Ronald H. Brown*. The Kiritimati-based team is grateful to the management and staff of the Captain Cook Hotel, who welcomed ENRR observers and equipment into their lives and went out of their way to make our stay very pleasant, and to John Bryden (JMB Enterprises), whose logistical and supply support was critical to our success. Thanks to Christopher Fairall (ESRL/PSD), Byron Blomquist (CIRES and ESRL/PSD), and Penny Rowe (Northwest Research Associates) for consultations regarding producing surface flux data, Cathy Smith (CIRES and ESRL/PSD) for helping to navigate archival waters and testing conversion software, Don Hooper (CIRES and ESRL/PSD) for scripting magic, Chris Kreutzer (Cherokee Services Group and ESRL/PSD) for testing conversion software, and Christopher Fairall and Robert Zamora (ESRL/PSD) for carefully reading the manuscript. The colors used in many figures are from sets at http://ColorBrewer.org.

Edited by: David Carlson

References

Berens, P.: CircStat: A MATLAB Toolbox for Circular Statistics, J. Stat. Softw., 31, 1–21, https://doi.org/10.18637/jss.v031.i10, 2009.

Centre for Environmental Data Analysis (CEDA): NASA Ames Format for Data Exchange, available at: https://cedadocs.ceda.ac.uk/73/4/index.html (last access: 18 January 2018), 2002.

Climate Prediction Center (CPC) and International Research Institute for Climate and Society (IRI): El Ni no/Southern Oscillation (ENSO) Diagnostic Discussion, available at: http://www.cpc.ncep.noaa.gov/products/analysis_monitoring/enso_disc_jun2015/ensodisc.html (last access: 2 March 2017), 2015.

Conley, S. A., Faloona, I., Miller, G. H., Lenschow, D. H., Blomquist, B., and Bandy, A.: Closing the dimethyl sulfide budget in the tropical marine boundary layer during the Pacific Atmospheric Sulfur Experiment, Atmos. Chem. Phys., 9, 8745–8756, https://doi.org/10.5194/acp-9-8745-2009, 2009.

Cox, C. and Hartten, L.: El Niño Rapid Response (ENRR) Field Campaign: Surface Fluxes from NOAA Ship Ronald H. Brown, 2016-02 to 2016-03 (NCEI Accession 0167875), Version 1.1, NOAA /National Centers for Environmental Information, https://doi.org/10.7289/V58050VP, 2017.

Cox, C. J., Walden, V. P., Rowe, P. M., and Shupe, M. D.: Humidity trends imply increased sensitivity to clouds in a warming Arctic, Nat. Commun., 6, 10117, https://doi.org/10.1038/ncomms10117, 2015.

Cox, C., Wolfe, D., Hartten, L., and Johnston, P.: El Niño Rapid Response (ENRR) Field Campaign: Radiosonde Data (Level 2) from the NOAA Ship Ronald H. Brown, February–March 2016 (NCEI Accession 0161527), Version 1.1, NOAA/National Centers for Environmental Information, https://doi.org/10.7289/V5X63K15, 2017a.

Cox, C., Wolfe, D., Hartten, L., and Johnston, P.: El Niño Rapid Response (ENRR) Field Campaign: Surface Meteorological and Ship Data from the NOAA Ship Ronald H. Brown, February–March 2016 (NCEI Accession 0161528), Version 1.1, NOAA /National Centers for Environmental Information, https://doi.org/10.7289/V5SF2T80, 2017b.

Dee, D. P., Uppala, S. M., Simmons, A. J., Berrisford, P., Poli, P., Kobayashi, S., Andrae, U., Balmaseda, M. A., Balsamo, G., Bauer, P., Bechtold, P., Beljaars, A. C. M., van de Berg, L., Bidlot, J., Bormann, N., Delsol, C., Dragani, R., Fuentes, M., Geer, A. J., Haimberger, L., Healy, S. B., Hersbach, H., Hólm, E. V., Isaksen, L., Kållberg, P., Köhler, M., Matricardi, M., McNally, A. P., Monge-Sanz, B. M., Morcrette, J.-J., Park, B.-K., Peubey, C., de Rosnay, P., Tavolato, C., Thépaut, J.-N., and Vitart, F.: The ERA-Interim reanalysis: configuration and performance of the data assimilation system, Q. J. Roy. Meteor. Soc., 137, 553–597, https://doi.org/10.1002/qj.828, 2011.

Dole, R. M., Spackman, J. R., Newman, M., Compo, G. P., Smith, C. A., Hartten, L. M., Barsugli, J. J., Webb, R. S., Hoerling, M. P., Cifelli, R., Wolter, K., Barnet, C. D., Gehne, M., Gelaro, R., Kiladis, G. N., Abbott, S., Akish, E., Albers, J., Brown, J. M., Cox, C. J., Darby, L., Boer, G. D., DeLuisi, B., Dias, J., Dunion, J., Eischeid, J., Fairall, C., Gambacorta, A., Gorton, B. K., Hoell, A., Intrieri, J., Jackson, D., Johnston, P. E., Lataitis, R., Mahoney, K. M., McCaffrey, K., McColl, H. A., Mueller, M. J., Murray, D., Neiman, P. J., Otto, W., Persson, O., Quan, X.-W., Rangwala, I., Ray, A. J., Reynolds, D. Dellaripa, E. R., Rosenlof, K., Sakaeda, N., Sardeshmukh, P. D., Slivinski, L. C., Smith, L., Solomon,

A., Swales, D., Tulich, S., White, A., Wick, G., Winterkorn, M. G., Wolfe, D. E., and Zamora, R.: Advancing Science and Services during the 2015/16 El Niño: The NOAA El Niño Rapid Response Field Campaign, B. Am. Meteorol. Soc., 99, 975–1001, https://doi.org/10.1175/bams-d-16-0219.1, 2018.

Edson, J. B., Zappa, C. J., Ware, J. A., McGillis, W. R., and Hare, J. E.: Scalar flux profile relationships over the open ocean, J. Geophys. Res.-Oceans, 109, C08S09, https://doi.org/10.1029/2003JC001960, 2004.

Edson, J. B., Jampana, V., Weller, R. A., Bigorre, S. P., Plueddemann, A. J., Fairall, C. W., Miller, S. D., Mahrt, L., Vickers, D., and Hersbach, H.: On the Exchange of Momentum over the Open Ocean, J. Phys. Oceanogr., 43, 1589–1610, https://doi.org/10.1175/jpo-d-12-0173.1, 2013.

Evans, M. N., Fairbanks, R. G., and Rubenstone, J. L.: The thermal oceanographic signal of El Niño reconstructed from a Kiritimati Island coral, J. Geophys. Res.-Oceans, 104, 13409–13421, https://doi.org/10.1029/1999JC900001, 1999.

Fairall, C. W., Bradley, E. F., Rogers, D. P., Edson, J. B., and Young, G. S.: Bulk parameterization of air-sea fluxes for Tropical Ocean-Global Atmosphere Coupled-Ocean Atmosphere Response Experiment, J. Geophys. Res.-Oceans, 101, 3747–3764, https://doi.org/10.1029/95JC03205, 1996.

Fairall, C. W., Bradley, E. F., Hare, J. E., Grachev, A. A., and Edson, J. B.: Bulk parameterization of air-sea fluxes: Updates and verification for the COARE algorithm, J. Climate, 16, 571–591, https://doi.org/10.1175/1520-0442(2003)016<0571:BPOASF>2.0.CO;2, 2003.

Fisher, N. I.: Statistical Analysis of Circular Data, Cambridge University Press, Cambridge, 277 pp., 1995.

Fisher, N. I.: Further erratum for Statistical Analysis of Circular Data, available at: http://www.valuemetrics.com.au/Downloads/Directional/directional.zip, last access: 28 August 2017.

Gage, K. S., Balsley, B. B., Ecklund, W. L., Carter, D. A., and McAfee, J. R.: Wind profiler-related research in the tropical Pacific, J. Geophys. Res.-Oceans, 96, 3209–3220, https://doi.org/10.1029/90JD01829, 1991.

Gaines, S. E. and Hipskind, R. S.: Format Specification for Data Exchange, Version 1.3, NASA Ames Research Center, 30, available at: http://cloud1.arc.nasa.gov/solve/archiv/archive.tutorial.html (last access: 12 July 2016), 1998.

Hartten, L., Johnston, P., Cox, C., and Wolfe, D.: El Niño Rapid Response (ENRR) Field Campaign: Surface Meteorological Data from Kiritimati Island, January–March 2016 (NCEI Accession 0161526), Version 1.1, NOAA /National Centers for Environmental Information, https://doi.org/10.7289/V51Z42H4, 2017.

Hartten, L. M., Cox, C. J., Johnston, P. E., Wolfe, D. E., Abbott, S., McColl, H. A., Quan, X.-W., and Winterkorn, M. G.: Ship- and island-based soundings from the 2016 El Niño Rapid Response (ENRR) field campaign, Earth Syst. Sci. Data, 10, 1165–1183, https://doi.org/10.5194/essd-10-1165-2018, 2018.

Hosom, D. S., Weller, R. A., Payne, R. E., and Prada, K. E.: The IMET (Improved Meteorology) Ship and Buoy Systems, J. Atmos. Ocean. Tech., 12, 527–540, https://doi.org/10.1175/1520-0426(1995)012<0527:timsab>2.0.co;2, 1995.

Hsu, S. A., Meindl, E. A., and Gilhousen, D. B.: Determining the Power-Law Wind-Profile Exponent under Near-Neutral Stability Conditions at Sea, J. Appl. Meteorol., 33, 757–765, https://doi.org/10.1175/1520-0450(1994)033<0757:DTPLWP>2.0.CO;2, 1994.

Jin, Z., Qiao, Y., Wang, Y., Fang, Y., and Yi, W.: A new parameterization of spectral and broadband ocean albedo, Opt. Express, 19, 26429–26443, https://doi.org/10.1364/OE.19.026429, 2011.

Joint Subcommittee on Ocean Science and Technology (JSOST): Federal Fleet Status Report: Current Capacity and Near-Term Priorities, Washington, DC, 18 pp., available at: http://www.nopp.org/wp-content/uploads/2016/06/federal_fleet_status_report_final_03.2016.pdf, 2016.

Konda, M., Imasato, N., Nishi, K., and Toda, T.: Measurement of the sea surface emissivity, J. Oceanogr., 50, 17–30, https://doi.org/10.1007/BF02233853, 1994.

McClatchey, R. A., Fenn, R. W., Selby, J. E. A., Volz, F. E., and Garing, J. S.: Optical Properties of the Atmosphere, 3rd edn., Air Force Cambridge Research Laboratories, Bedford, Mass, Environmental Research Papers, No. 411, 113 pp., available at: www.dtic.mil/dtic/tr/fulltext/u2/753075.pdf, 1972.

McFarlane, S. A., Long, C. N., and Flaherty, J. A.: A climatology of surface cloud radiative effects at the ARM Tropical Western Pacific sites, J. Appl. Meteorol. Clim., 52, 996–1012, https://doi.org/10.1175/JAMC-D-12-0189.1, 2013.

Mlawer, E. J., Taubman, S. J., Brown, P. D., Iacono, M. J., and Clough, S. A.: Radiative transfer for inhomogeneous atmospheres: RRTM, a validated correlated-k model for the longwave, J. Geophys. Res.-Atmos., 102, 16663–16682, https://doi.org/10.1029/97JD00237, 1997.

Moat, B. I., Berry, D. I., and Yelland, M. J.: Airflow distortion at instrument sites on the RV *Ronald H. Brown*, Southampton Oceanography Centre, University of Southampton, Southampton, Hampshire, UK, 27 pp., 2001.

Morate, O.: 2015 Population and Housing Census Volume 1: Management Report and Basic Tables, National Statistics Office, Ministry of Finance, Bairiki, Tarawa, Republic of Kiribati, 197 pp., available at: http://www.mfed.gov.ki/publications/census-report-2015-volume-i-final-report (last access: 11 January 2017), 2016.

National Centers for Environmental Information (NCEI): Magnetic Field Calculators, available at: https://www.ngdc.noaa.gov/geomag-web/#declination, last access: 18 March 2016.

National Data Buoy Center (NDBC): TAO Mooring Information, available at: http://tao.ndbc.noaa.gov/proj_overview/mooring_ndbc.shtml (last access: 27 March 2017), 2010a.

National Data Buoy Center (NDBC): TAO Sensor Specifications, available at: http://tao.ndbc.noaa.gov/proj_overview/sensors_ndbc.shtml (last access: 27 March 2017), 2010b.

Ralph, F. M., Prather, K. A., Cayan, D., Spackman, J. R., DeMott, P., Dettinger, M., Fairall, C., Leung, R., Rosenfeld, D., Rutledge, S., Waliser, D., White, A. B., Cordeira, J., Martin, A., Helly, J., and Intrieri, J.: CalWater field studies designed to quantify the roles of atmospheric rivers and aerosols in modulating U.S. west coast precipitation in a changing climate, B. Am. Meteorol. Soc., 97, 1209–1228, https://doi.org/10.1175/bams-d-14-00043.1, 2016.

Scott, D. A. (Ed.): A Directory of Wetlands in Oceania, Slimbridge, UK and Kuala Lumpur, Malaysia, 370 pp., 1993.

Stamnes, K., Tsay, S. C., Wiscombe, W., and Jayaweera, K.: Numerically stable algorithm for discrete-ordinate-method radiative transfer in multiple scattering and emitting layered media, Appl. Optics, 27, 2502–2509, https://doi.org/10.1364/AO.27.002502, 1988.

Thoning, K. W., Kitzis, D. R., and Crotwell, A.: Atmospheric Carbon Dioxide Dry Air Mole Fractions from quasi-continuous measurements at American Samoa, Version 2016-8, updated annually, NOAA Earth System Research Laboratory (ESRL) Global Monitoring Division (GMD), https://doi.org/10.7289/V51834DB, 2015.

Vaisala Oyj: User's Guide: Ground Check Set GC25, Vaisala Oyj, M210329EN-E, Helsinki, Finland, 43 pp., available at: www.vaisala.com (last access: 20 July 2017), 2006.

Vaisala Oyj: DigiCORA® User's Guide, M210488EN-F, Vaisala Oyj, Helsinki, Finland, 113 pp., available at: www.vaisala.com (last access: 20 July 2017), 2010.

Vaisala Oyj: Vaisala Radiosonde RS92-SGP User's Guide, Vaisala Oyj, M210295EN-J, Helsinki, Finland, 54 pp., available at: www.vaisala.com (last access: 28 January 2016), 2015.

Wang, J., Cole, H. L., Carlson, D. J., Miller, E. R., Beierle, K., Paukkunen, A., and Laine, T. K.: Corrections of Humidity Measurement Errors from the Vaisala RS80 Radiosonde – Application to TOGA COARE Data, J. Atmos. Ocean. Tech., 19, 981–1002, https://doi.org/10.1175/1520-0426(2002)019<0981:Cohmef>2.0.Co;2, 2002.

Woods Hole Oceanographic Institution (WHOI): ASIMET Module Specifications, available at: http://frodo.whoi.edu/asimet/asimet_module_specs.html#bpr_mod (last access: 31 March 2017), 2010a.

Woods Hole Oceanographic Institution (WHOI): ASIMET Sensors, available at: http://www.whoi.edu/instruments/viewInstrument.do?id=14686 (last access: 31 March 2017), 2010b.

World Meteorological Organization: Manual on the Global Observing System, 2010 edn., WMO (Series) no. 544, Secretariat of the World Meteorological Organization, Geneva, Switzerland, 2013.

Zipser, E. J.: The Line Islands Experiment, its place in tropical meteorology and the rise of the fourth school of thought, B. Am. Meteorol. Soc., 51, 1136–1146, 1970.

3

SPREAD: a high-resolution daily gridded precipitation dataset for Spain – an extreme events frequency and intensity overview

Roberto Serrano-Notivoli[1,2,3], Santiago Beguería[3], Miguel Ángel Saz[1,2], Luis Alberto Longares[1,2], and Martín de Luis[1,2]

[1]Department of Geography and Regional Planning, University of Zaragoza, Zaragoza, 50009, Spain
[2]Environmental Sciences Institute (IUCA), University of Zaragoza, Zaragoza, 50009, Spain
[3]Estación Experimental de Aula Dei, Consejo Superior de Investigaciones Científicas (EEAD-CSIC), Zaragoza, 50059, Spain

Correspondence to: Roberto Serrano-Notivoli (rs@unizar.es)

Abstract. A high-resolution daily gridded precipitation dataset was built from raw data of 12 858 observatories covering a period from 1950 to 2012 in peninsular Spain and 1971 to 2012 in Balearic and Canary islands. The original data were quality-controlled and gaps were filled on each day and location independently. Using the serially complete dataset, a grid with a 5×5 km spatial resolution was constructed by estimating daily precipitation amounts and their corresponding uncertainty at each grid node. Daily precipitation estimations were compared to original observations to assess the quality of the gridded dataset. Four daily precipitation indices were computed to characterise the spatial distribution of daily precipitation and nine extreme precipitation indices were used to describe the frequency and intensity of extreme precipitation events. The Mediterranean coast and the Central Range showed the highest frequency and intensity of extreme events, while the number of wet days and dry and wet spells followed a north-west to south-east gradient in peninsular Spain, from high to low values in the number of wet days and wet spells and reverse in dry spells. The use of the total available data in Spain, the independent estimation of precipitation for each day and the high spatial resolution of the grid allowed for a precise spatial and temporal assessment of daily precipitation that is difficult to achieve when using other methods, pre-selected long term stations or global gridded datasets

1 Introduction

Daily precipitation is a key variable to understanding the behaviour of extreme weather events and the severe impacts they cause on hydrological systems, in natural systems and on human societies. These impacts can be considered in regional and local plans which can help to mitigate major disasters if the correct environmental data are available. Unfortunately, the raw climatic information is usually highly fragmented in time and discontinuous in space. Quality control and reconstruction processes are therefore in high demand, as well as final products such as serially continuous observational series or gridded datasets. High-resolution spatiotemporal precipitation datasets are useful tools for land management, and their availability over a complete region can be useful to many other fields due to the relevance of precipitation in many disciplines as hydrology (for instance in water resources management at catchment scale) (e.g. Werner and Cannon, 2016; Kay et al., 2015; Lorenz et al., 2014; Maurer et al., 2002), environmental risks such as droughts (e.g. Touchan et al., 2011; Vicente-Serrano et al., 2010a; Bordi et al., 2009; Andreadis et al., 2005), floods (e.g. Hartmann and Andresky, 2013; Rojas et al., 2011) or wildfires

(Westerling et al., 2003), groundwater recharge (e.g. Döll and Fiedler, 2008) or agricultural applications (e.g. Mishra and Cherkauer, 2010; Lo et al., 2007). Upon the development of these kinds of datasets, the researchers cannot only improve their studies in analytic climatology but also in environmental studies, habitat modelling, climate change model validations, relations of climate and vector diseases, crop forecasting, agroclimatic selection of plant landraces, palaeoclimatic analysis supported on the observational period, etc. Gridded datasets provide valuable products that can be used for both scientific and decision-making policy purposes.

The number and density of climate observatories are finite. Therefore, there have been many contributions over the last decade to solve this problem by creating high-resolution and world-based gridded datasets based on the (geo)statistical relations of the observatories (New et al., 2002; Hijmans et al., 2005; Di Luzio et al., 2008; Harris et al., 2014; Schamm et al., 2014). However, these global datasets are often not optimum for regional analysis, where a higher resolution both in space and time is required (Herrera et al., 2012). Increasing temporal resolution from monthly or annual to daily scale allows analysis of other relevant components of the climate, such as the extreme precipitation events or the length and intensity of dry spells that precede most of the annual hydrological processes. In this respect, daily precipitation grids of different spatial resolutions have been developed at the global (Piper and Stewart, 1996; Menne et al., 2012) or regional scale (Frei and Schär, 1998; Eischeid et al., 2000; Kyriakidis et al., 2001; Rubel and Hantel, 2001; Klein-Tank et al., 2002; Hewitson and Crane, 2005; Liebmann and Allured, 2005; Haylock et al., 2008; Klok and Klein-Tank, 2009; Mekis and Vincent, 2011; Hwang et al., 2012; Yatagai et al., 2012; Jones et al., 2013; Chaney et al., 2014; Isotta et al., 2014; Hernández et al., 2016).

A few daily datasets have also been made for Spain or some of its regions (Vicente-Serrano et al., 2010b; Herrera et al., 2012). These datasets are useful to analyse global precipitation, which in Spain have a wide range of different spatial and temporal distributions (Esteban-Parra et al., 1998; Rodríguez-Puebla et al., 1998; Belo-Pereira et al., 2011; De Luis et al., 2011; González-Hidalgo et al., 2011; Cortesi et al., 2014). But they are also suitable to analyse spatial patterns in daily characteristics of precipitation through extreme indices that allow comparisons between regions. In Spain, these indices have been applied to mainland Spain (Martín-Vide, 2004; Herrera et al., 2012; Merino et al., 2015) and to specific regions (Casas et al., 2007; Martínez et al., 2007; Rodrigo and Trigo, 2007; Lopez-Moreno et al., 2010). All of these datasets used the precipitation estimations to compute climatic indices. However, unlike temperature, uncertainty in precipitation estimation is considerably higher. Firstly, rain-gauge measurements can be notoriously uncertain, especially in certain circumstances such as windy conditions (Rodda and Dixon, 2012) that can lead to significant undercatches in highly exposed areas (Rodda and Smith, 1986). Secondly, spatial variability of daily precipitation can be extremely high under certain atmospheric conditions such as convective processes that can occur at a very local scale. Whereas the sum of these uncertainties is very difficult to estimate in practice, the uncertainties of these estimations are expected to decrease as the density of observations used to compute the models increases (Tveito et al., 2008; Hofstra et al., 2010). This issue has important implications in subsequent climatic analyses, and this is one consideration that needs to be taken into account (Beguería et al., 2015).

In this paper, we present SPREAD (Spanish PREcipitation At Daily scale), a new high-resolution daily gridded precipitation dataset for Spain. Thirteen daily and extreme precipitation indices were calculated as an example of applicability, characterising daily precipitation distribution (4 indices to characterise the spatial distribution of daily precipitation and 9 for extreme precipitation). The grid was computed for 1950–2012 period covering peninsular Spain and 1971–2012 for Balearic and Canary islands. It was built from 12 858 original stations by creating reference values (RVs) using generalised linear models (GLMs) based on the 10 nearest observations using altitude, longitude and latitude as covariates. All the calculations were developed with *reddPrec* (https://cran.r-project.org/web/packages/reddPrec), an R package containing the required functions to reconstruct original daily precipitation series and create grids (Serrano-Notivoli et al., 2017a).

This paper is organised as follows. Section 2 describes the original dataset. Section 3 shows the methods applied for the reconstruction and the gridding process considering uncertainties of estimates. Also, the statistical procedures are explained here. Section 4 shows the results of the reconstruction, gridding procedure and spatial distribution of daily and extreme precipitation characteristics in Spain. Section 5 specifies the availability of the dataset. The results are discussed in Sect. 6, with some final remarks in Sect. 7.

2 Input data

A total of 12 858 observatories covering Spain were used. The grid was divided in three areas: (1) peninsular Spain (492 175 km^2), with 11 513 stations covering the period 1950–2012; (2) the Balearic Islands (4992 km^2), with 425 stations covering 1971–2012; and (3) the Canary Islands (7493 km^2) covering 1971–2012, using 920 stations (Fig. 1 down).

Most of the data were provided by the Spanish Meteorological Agency (AEMET), but we also used data from regional hydrological and meteorological services, and from the national agronomic network (Table 1). The greatest part of the information comes from manual stations; the automated ones entered service in mid-1990s, making up 23 % of the total in AEMET network in 2012. The mean length of the original series was 18.8 years, and only 17 of the 12 858 orig-

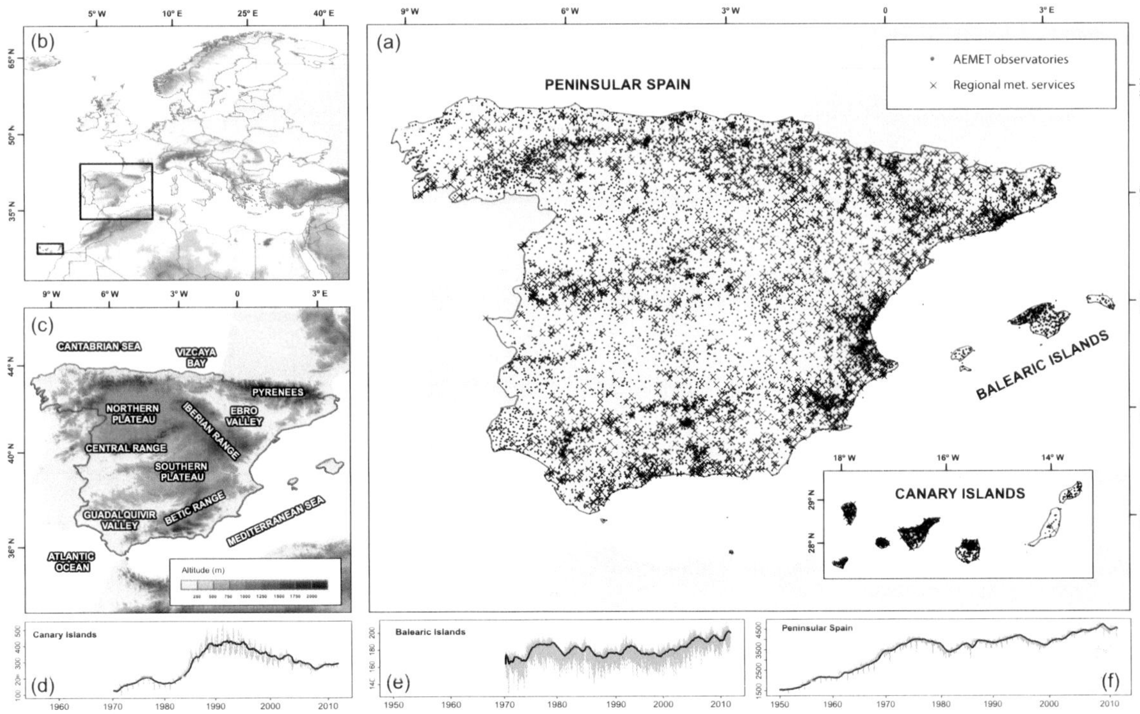

Figure 1. Location of the precipitation stations used **(a)**, location of Spain in Europe context **(b)** and geographical references used in the text **(c)**. Number of daily available observatories (grey lines), and its moving average of 365 days (black lines) in the Canary Islands **(d)**, Balearic Islands **(e)** and peninsular Spain **(f)**.

inal observatories had covered the period 1950–2012. However, the spatial distribution of the observatories showed a remarkable density (Fig. 1 main map), which is useful to make proper reconstructions. Although the recovered information from raw databases of the meteorological offices was precipitation, in some cases this can include both rainfall and the water equivalent of snow if the source did not make the distinction.

3 Methods

3.1 The reference values (RVs) as a general tool for quality control and reconstruction

The key process for climate reconstruction is based on the calculation of individual RVs for each day and location, according to the information available from the closest neighbouring stations. GLMs are used to compute the RVs using the precipitation data (occurrence and magnitude) of the 10 nearest neighbours as the dependent variable and the geographic information of each station (latitude, longitude and altitude) as the independent variables. The average distances to the 10 nearest stations were 24.67, 8.99 and 26.78 km for the Iberian Peninsula, Balearic and Canary islands, respectively, with average standard deviations of 8.46, 3.69 and 17.32 km, and average minimum distances of 10.04, 3.01 and 5.51 respectively.

Table 1. Number of observatories, by source (AEMET: Spanish meteorological agency; MAGRAMA: Ministry of Agriculture and Environment; AHIS: Automatic Hydrological Information System).

Source	Number of observatories
AEMET	10 683 (83.1 %)
MAGRAMA	541 (4.2 %)
Meteorological Service of Catalonia	173 (1.3 %)
Navarra government	38 (0.3 %)
AHIS of Cantábrico Basin	36 (0.3 %)
AHIS of Duero Basin	216 (1.7 %)
AHIS of Ebro Basin	333 (2.6 %)
AHIS of Guadalquivir Basin	212 (1.7 %)
AHIS of Hidrosur Basin	102 (0.8 %)
AHIS of Júcar Basin	182 (1.4 %)
AHIS of Miño-Sil Basin	89 (0.7 %)
AHIS of Segura Basin	66 (0.5 %)
AHIS of Tajo Basin	187 (1.4 %)
TOTAL	12 858 (100 %)

The geographic information of each station (latitude, longitude and altitude) is used as the independent variable for modelling. As the data availability varies from day to day, selected neighbour stations also vary. Since independent models are constructed for each location and day, the estimated parameters of the models (reflecting the influence of the in-

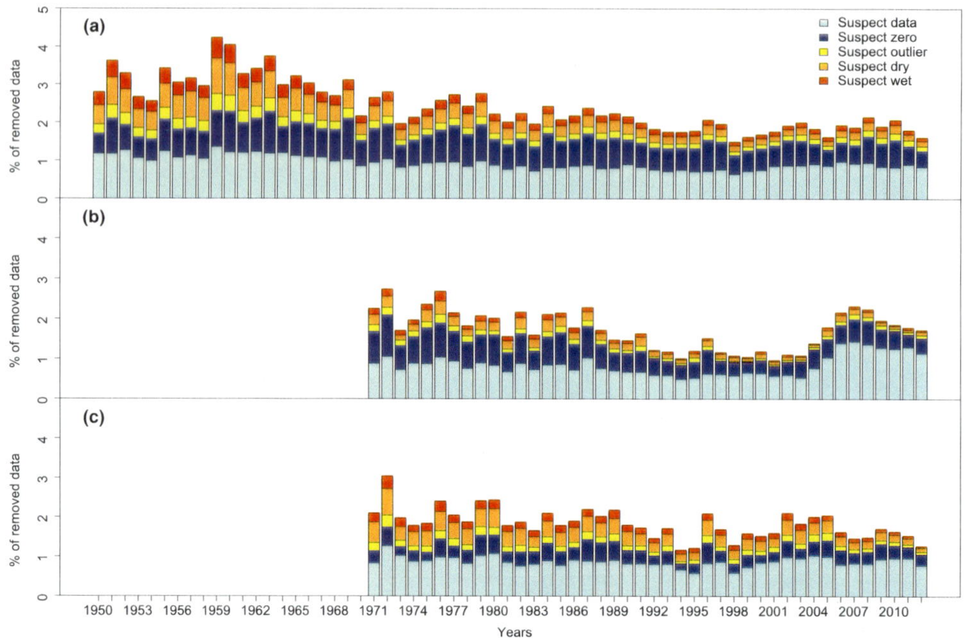

Figure 2. Removed data by criteria (suspect data, suspect zero, suspect outlier, suspect dry day and suspect wet day): **(a)** peninsular Spain; **(b)** Balearic Islands and **(c)** Canary Islands.

dependent variables and their probability to influence the occurrence and magnitude of precipitation) may also vary with measurement day and location.

Including three factors in the model allows for increased sensitivity of the model to be able to reflect the local changes in precipitation patterns. Because this method is based on local and independent data points across time, there are no restrictions imposed due to the length or structural characteristics of the series, allowing efficient use of all of the available information.

The computation of each individual RV is based on two predicted values: (i) a binomial prediction (BP) of the probability of occurrence of a wet day and (ii) a magnitude prediction of precipitation (MP), in the case where a wet day is predicted. The combination of these two values ($RV = MP$ if $BP > 0.5$, else $RV = 0$) produce the estimated (RV) and its corresponding standard error for each day and location. Further details on the statistical procedures are described and discussed in Serrano-Notivoli et al. (2017a, b).

The obtained RVs were first used to develop a quality control (QC) test in order to detect anomalous data in the original dataset, and then to estimate precipitation at all missing locations and days for the whole dataset. Finally, a 5×5 km grid was built for all the whole of Spain and the entire period of reconstructed stations. The reconstruction and gridding processes were applied using the R package *reddPrec* following the methodology described in Serrano-Notivoli et al. (2017a).

The QC process detected and removed suspect data by comparing daily values registered at each station with predicted values calculated from its surrounding observations. Five criteria were used to flag and remove suspect data from the original dataset – (1) suspect data: observed value was over zero and all their 10 nearest observations were zero; (2) suspect zero: observed value was zero and all its 10 nearest observations were over zero; (3) suspect outlier: the magnitude of the observed value was 10 times higher or lower than that predicted by its 10 nearest observations; (4) suspect dry day: observed value was zero, wet probability was over 99 %, and predicted magnitude was over 5 mm; and (5) suspect wet day: observed value was over 5 mm, dry probability was over 99 %, and predicted magnitude was under 0.1 mm.

Once the QC was completed, a new set of RVs were calculated using the curated dataset. Since the RVs were calculated for all days and locations, including those for which an observation exists but without using that observation, the comparison between RVs and the corresponding observed values constitutes a leave-one-out cross-validation (LOO-CV) process. A number of goodness of fit statistics were therefore used for assessing the quality of the estimated values (RVs): the mean absolute error (MAE), as a measure of the error magnitude; the mean error (ME) and the ratio of means (RM = mean of estimations / mean of observations), as a measure of bias; and the ratio of the standard deviations (RSD = SD of estimations / SD of observations), as a measure of bias in the variance. These statistics were computed for monthly aggregates and for 13 daily and extreme precipitation indices (described on Sect. 3.3).

Additionally, scatterplots were made between observations and estimations of daily precipitation at the station loca-

Table 2. Computed indices over daily gridded dataset.

Identifier	Description	Units
Daily precipitation indices		
PMED	Daily mean precipitation intensity (median of daily precipitation in wet days)	mm
NWD	Number of wet days in a year	days
CDDm	Mean length of dry spell (mean number of consecutive dry days)	days
CWDm	Mean length of wet spell (mean number of consecutive wet days)	days
Daily extreme precipitation indices		
SDII	Daily precipitation intensity (annual precipitation/number of wet days)	mm
RX1	Maximum 1-day precipitation	mm
RX5	Maximum 5-day precipitation	mm
R10mm	Number of days with precipitation over 10 mm in a year	days
R20mm	Number of days with precipitation over 20 mm in a year	days
CDD	Maximum length of dry spell (maximum number of consecutive dry days)	days
CWD	Maximum length of wet spell (maximum number of consecutive wet days)	days
R95	95th percentile of daily precipitation in whole series	mm
R95rel	Annual contribution of precipitation over 95th percentile	%

tions using the number of zero precipitation days (dry days), daily means (considering all days), medians of the wet days and the 95th percentile of wet days, and the Pearson's correlation statistic was computed in each case.

Finally, the missing values in the original data series were filled with the reference values (RV). From 12 858 original observatories, we reconstructed those that had more than 10 years of original data (7604 stations). This guaranteed a reliable reconstructed series with enough observations to compute the grid.

3.2 Gridding and uncertainty

The same procedure based on the calculation of RV was used to build a 5×5 km spatial resolution grid. For each point of the grid (x, y and z) and each day of the total period, RV was computed based on the data of the 10 closest reconstructed stations.

As a measure of uncertainty, we computed the standard error (in mm) for each RV. Using the ratio between this error and the RV we obtained the relative error (expressed in percentage) in each index and aggregation (monthly, annual and seasonal).

Table 3. Accuracy of the wet/dry day estimates: percent observed dry ($P = 0$) and wet ($P > 0$) days, and percent predicted dry (RV = 0) and wet (RV > 0) days on observed dry and wet days.

	Peninsular Spain		Balearic Islands		Canary Islands	
	$P = 0$	$P > 0$	$P = 0$	$P > 0$	$P = 0$	$P > 0$
Observed	79.26	20.74	81.59	18.41	89.96	10.04
RV = 0	94.84	20.47	96.35	18.11	96.76	29.46
RV > 0	5.16	79.53	3.65	81.89	3.24	70.54

3.3 Applications: daily mean and extreme precipitation indices

As examples of possible applications of the gridded dataset, four indices using daily precipitation were calculated to characterise the spatial distribution of daily precipitation and its extremes (nine more indices) (Table 2). Most of these indices are included in the suite of extreme precipitation and temperature indices (Zhang et al., 2011) developed by WMO Expert Team on Climate Change Detection and Indices (ETCCDI). They have been applied in previous works to assess the distribution of extreme events in many areas (e.g. Donat et al., 2013, 2014; Keggenhof et al., 2014; Asadieh and Krakauer, 2015; Sanogo et al., 2015; Yin et al., 2015; Sigdel and Ma, 2016). All the indices were computed at the annual scale and the average annual values were calculated.

4 Results

4.1 Reconstruction of the observational dataset and grids

The quality control process flagged and removed an annual average of 2.4 % of data in peninsular Spain, 1.7 % in the Balearic Islands and 1.8 % in the Canary Islands. There were no major differences in the number of removed data by years. A brief increase was observed in the first 20 years in peninsular Spain data (Fig. 2a), while from 1971 to the end of the period the number of removed data barely changed.

In the Balearic Islands the number of removed data was more variable (Fig. 2b). Suspect data and zeros were usually detected because they represented low precipitation values with estimates from 0 to 1 or 2 mm, so they were included in one of these two criteria. The low detected values of outliers, suspect dry and wet data were probably due to the configuration of the islands, small with a high density of observatories even at high elevations. In the Canary Islands, although the dataset contained more data series than in the Balearic Islands, and the number of available daily stations was very variable, the quality control process removed a similar number of data over the collection period (Fig. 2c). This was also

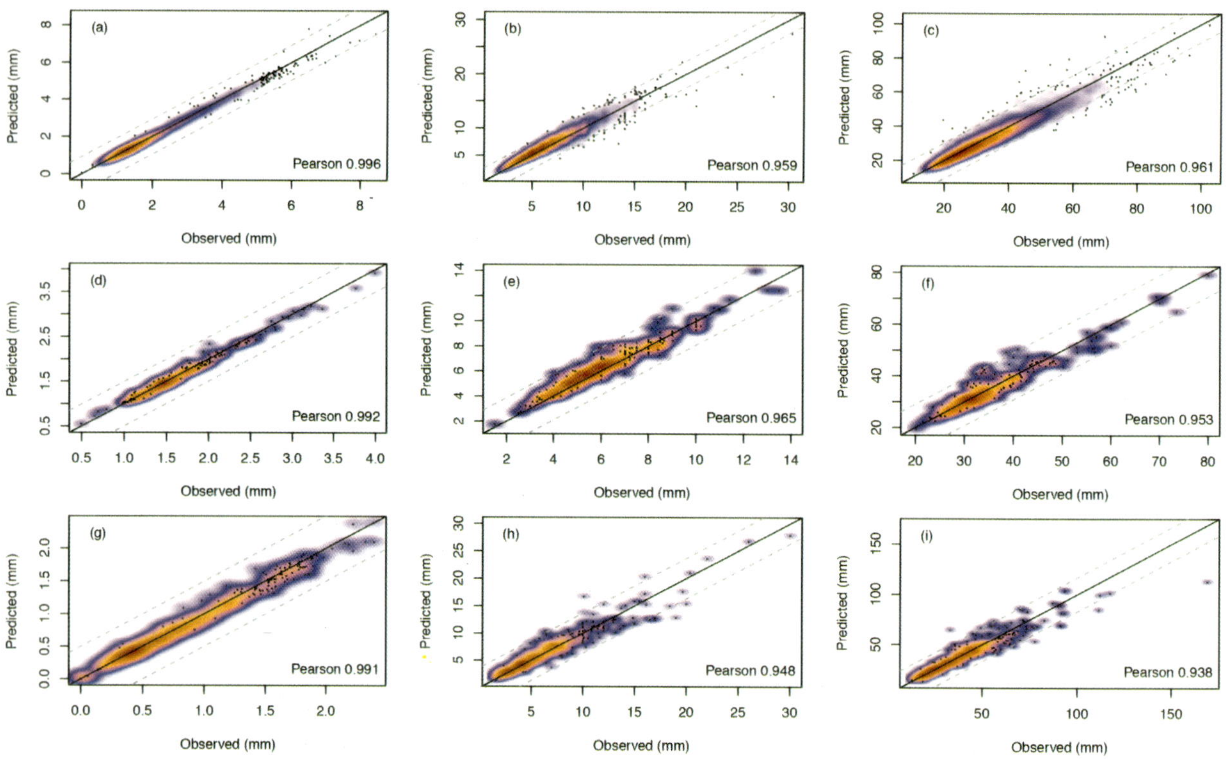

Figure 3. Scatterplots and Pearson correlation coefficients between observations and estimations of daily precipitation in peninsular Spain (upper line; **a, b c**), Balearic Islands (midline: **d, e, f**) and Canary Islands (bottom line: **g, h, i**). Dots represent the stations and colours indicate the density. Daily precipitation mean (left column: **a, d, g**), daily precipitation medians in wet days (central column: **b, e, h**) and daily precipitation over 95th percentile (right column: **c, f, i**) are shown.

the area with the least number of suspect zeros of the three localities.

A complete 5×5 km spatial resolution grid was calculated based on the reconstructed station series. Daily precipitation was estimated from 1950 to 2012 in peninsular Spain and from 1971 to 2012 in Balearic and Canary Islands. The standard error of the model used to compute the estimations was calculated as a measure of uncertainty for each day and grid point.

4.2 Observations – estimation comparison

Daily precipitation values were estimated at the same locations and days as the observed dataset for comparison purposes. This section shows the results of this comparison.

4.2.1 Wet/dry estimation

The number of observed zero precipitation days (dry days) in the entire study area was 57 761 815 and the number of estimated ones (only for corresponding days with observations) was 57 773 250 (a ratio of 0.9998), so it can be concluded that this method is not biased in the prediction of wet/dry days. The comparison between the original dataset and the corresponding estimates showed a high correlation in peninsular Spain, the Balearic Islands and the Canary Islands (Pearson correlation coefficients of 0.83, 0.85 and 0.73, respectively), with similar frequency distribution by station (see Fig. S1 in the Supplement). Terming the wet days as positive (observed $P > 1$) and the dry days as negative (observed $P = 0$), the true negative rate p ($\mathrm{RV} = 0 \mid P = 0$) was over 94 % in all cases (Table 3), and the true positive rate or precision p ($\mathrm{RV} > 0 \mid P > 0$) was over 79 % except in the Canary Islands, where it decreased to 70 %. To a large extent, the false negatives p ($\mathrm{RV} = 0 \mid P > 0$) and false positives p ($\mathrm{RV} > 0 \mid P = 0$) were due to the prediction of precipitation in days with low amounts. In events with very low precipitation amounts, the estimate of the probability of occurrence was likely to be dry, despite the fact that the station could register a minimum quantity of rain (usually under 1 or 2 mm). This causes a brief difference in amounts, becoming more distinct in the dry/wet accuracy assessment.

4.2.2 Magnitude estimation

The comparison between amounts of observed and estimated precipitation showed a high correlation both in daily means (daily mean precipitation by stations, considering the whole series), daily medians in wet days (only considering days with $P > 0$ in observations and estimations) and in the 95th

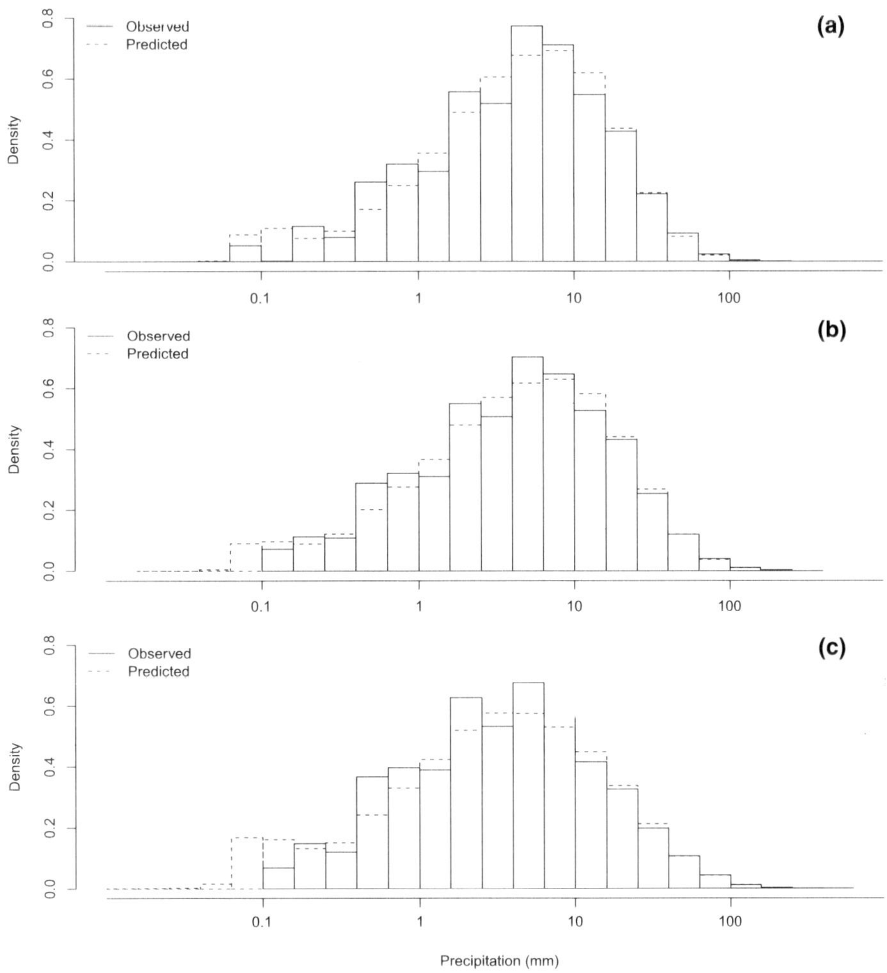

Figure 4. Histograms of observed and predicted daily precipitation frequency in **(a)** peninsular Spain, **(b)** the Balearic Islands and **(c)** the Canary Islands.

percentile of wet days (Fig. 3). Daily precipitation means (considering dry and wet days) reached the maximum correlation between observations and estimations (Fig. 3a, d, g), decreasing in daily precipitation medians on wet days ($P > 0$) (Fig. 3b, e, h) and considering only the daily precipitation over the 95th percentile on wet days (Fig. 3c, f and i). However, in all cases the values of the Pearson correlation coefficient were over 0.93. In the Canary Islands the goodness of fit was lower than in the other areas. The Canary Islands experience high orographic rainfall and climatic variability, which is thought to have contributed to the lower availability of data. Most of these islands have their own climate, with high differences between both sides (north–south in eastern islands and east–west in western islands). As the number of observatories on the Canary Islands was limited and the estimates were from a low density of stations, a greater radius was used to get the minimum number of stations required to run the model (maximum 100 km). This can lead to a more inaccurate estimation, although the aggregation by days instead of stations showed a better agreement, with correlations over 0.96 in most of the cases (see Fig. S2).

The histograms of estimated and observed precipitation (Fig. 4) showed a good general agreement. There was a slight overestimation of the values below 1 mm, and a slightly flatter distribution around the mean. The agreement between the histograms was high above 10 to 20 mm. The differences found in the lowest values (under 0.1 mm) were due in large part to the fact that estimates below 0.1 mm were allowed. This value was the minimum measurement in the observed dataset. These results were similar in peninsular Spain, the Balearic Islands and the Canary Islands datasets.

One common problem to most observational datasets is the uneven distribution of the stations with respect to the altitude. Overall, the high-elevation areas tend to be under represented in the datasets due to a lower spatial density of stations. This could result in biases in any derived datasets. For instance, in Spain only around 2 % of the stations are above 1500 m a.s.l., which represents 4 % of the Spanish ter-

Table 4. The leave-one-out cross-validation (LOO-CV) statistics showing the goodness of fit between observations and estimations of daily precipitation separated by altitudes (m a.s.l.). IP: Iberian Peninsula; BI: Balearic Islands; CI: Canary Islands; N: number of stations; MAE: mean absolute error; ME: mean error; %OBS: percentage of observed precipitation; %PRE: percentage of predicted precipitation; RM: ratio of means; RSD: ratio of standard deviations. Results were constrained to 2 decimal places.

		0–100	> 100–300	> 300–500	> 500–700	> 700–900	> 900–1100	> 1100–1300	> 1300–1500	> 1500–2000	> 2000
IP	N	918	1054	1077	1210	1271	766	415	128	65	11
	%OBS	12.50	14.40	14.20	17.20	17.30	12.70	7.90	2.30	1.30	0.20
	%PRE	12.60	14.40	14.20	17.20	17.30	12.70	7.80	2.30	1.40	0.20
	RM	1.00	0.99	0.99	0.99	0.99	0.99	0.98	1.01	1.06	1.07
	RSD	0.95	0.96	0.96	0.96	0.97	0.96	0.96	1.03	1.10	1.19
	ME	0.04	−0.02	−0.02	−0.01	0.01	0.00	−0.03	0.26	0.64	1.47
	MAE	4.83	4.55	4.55	4.51	4.3	4.86	5.48	6.32	6.92	9.61
BI	N	128	93	20	4	2	1	0	0	0	0
	%OBS	43.80	39.20	12.80	2.70	1.10	0.40				
	%PRE	44.00	39.00	12.90	2.60	1.10	0.40				
	RM	1.02	0.99	1.03	0.84	1.09	1.43				
	RSD	1.00	0.96	0.97	0.96	0.99	1.34				
	ME	0.17	−0.10	0.11	−0.48	0.52	2.65				
	MAE	3.78	3.82	5.71	7.76	10.11	8.12				
CI	N	78	96	79	70	40	37	13	9	9	10
	%OBS	8.40	15.00	18.70	20.50	13.40	11.00	5.20	4.30	2.20	1.30
	%PRE	8.80	15.20	18.40	20.40	13.20	10.90	5.20	4.20	2.30	1.40
	RM	1.02	1.02	0.97	1.00	1.02	0.99	0.98	1.02	1.01	0.92
	RSD	1.02	1.02	0.98	0.99	0.98	0.99	1.05	0.96	1.03	0.93
	ME	0.39	0.17	−0.15	0.00	−0.13	−0.15	−0.02	−0.16	0.48	1.85
	MAE	4.49	4.18	5.28	5.91	5.94	5.95	6.26	5.09	5.48	14.28

ritory. For this reason, it is relevant to evaluate the goodness of fit of the estimated values by altitudinal ranges (Table 4). The RM showed that there were no substantial biases (values close to 1) until 1500 m a.s.l., and only a slight overestimation of 6–7 % above this altitude in peninsular Spain. In the Balearic Islands there were under- and overestimations above 500 m a.s.l., but the number of stations is too low to consider these values representative. In the Canary Islands precipitation was slightly underestimated (−8 %) at higher altitudes (> 2000 m a.s.l.).

Assessing the monthly aggregates of daily precipitation (Table 5), we found the results were very similar to the ratio of means (RM), indicating the absence of systematic biases, with the exception of November and December which showed a slight underestimation. The ratio of the standard deviations (RSD) was also very close to one in peninsular Spain and the Balearic Islands, indicating no biases in the variance estimation, although there was a small underestimation in November and December. The RSD was more variable in the Canary Islands, with an overestimation between 10 and 20 % in the summer (June to August), and underestimation of around 10 % in November and December. Very low values of MAE were also found (average of 9.89, 8.98 and 5.99 in peninsular Spain, the Balearic and Canary islands, respectively), as well as ME (average of −0.41, −0.29 and −0.73). Summer months usually receive low precipitation (most of them zero) and the events are typically of small spatial extent, leading to higher uncertainty of the estimations, thus producing low values of fit between observed and estimated precipitation. However, the RM was near to 1 in all cases, indicating the absence of bias in the estimations despite the variable uncertainty.

The development of a spatially and temporally complete gridded dataset allowed the assessment of the characteristics of daily precipitation over Spain. For that, 13 daily and extreme precipitation indices were computed. The comparison between the observed and the estimated values of the indices showed values of RM and RSD near to 1 in all the considered spatial units and indices (Table 6), indicating no substantial biases in the mean and variance of the estimated indices. However, RX1 showed a slight underestimation in peninsular Spain and the Balearic Islands, as did R20mm in peninsular Spain and SDII in the Canary Islands, where the number of wet days (NWD) was overestimated. Overall, all the indices were similar at observed stations and their corresponding estimates in the Canary Islands showed the largest differences, as shown at the monthly scale (Table 5).

Table 5. The leave-one-out, cross-validation (LOO-CV) statistics, showing the goodness of fit between observations and estimates of monthly aggregates. IP: Iberian Peninsula; BI: Balearic Islands; CI: Canary Islands; MAE: mean absolute error; ME: mean error; RM: ratio of means; RSD: ratio of standard deviations. Results were constrained to 2 decimal places.

		Jan	Feb	Mar	Apr	May	Jun	Jul	Aug	Sep	Oct	Nov	Dec
IP	MAE	10.18	9.66	9.28	10.88	11.05	8.39	4.49	6.00	9.68	12.10	13.47	13.59
	ME	0.45	0.40	0.44	0.60	0.53	0.13	−0.07	−0.01	0.32	0.66	−4.04	−4.30
	RM	1.01	1.01	1.01	1.01	1.01	1.00	0.99	1.00	1.01	1.01	0.94	0.93
	RSD	1.02	1.02	1.03	1.03	1.03	1.03	1.02	1.03	1.03	1.02	0.94	0.94
BI	MAE	8.82	7.40	7.35	8.10	7.32	4.27	2.72	6.39	13.25	14.32	14.73	13.18
	ME	0.42	0.31	0.49	0.38	0.24	0.15	0.04	0.01	0.40	1.07	−3.72	−3.30
	RM	1.01	1.01	1.01	1.01	1.01	1.01	1.01	1.00	1.01	1.01	0.96	0.95
	RSD	1.02	1.02	1.01	1.01	1.01	1.02	1.02	1.02	1.01	1.03	0.96	0.96
CI	MAE	8.91	8.61	8.33	5.44	2.52	1.07	0.52	0.79	2.89	6.85	12.10	13.95
	ME	0.00	0.10	0.01	0.00	−0.08	0.00	0.00	−0.03	−0.10	−0.01	−4.5	−4.18
	RM	1.00	1.00	1.00	1.00	0.99	1.00	1.01	0.98	0.99	1.00	0.90	0.93
	RSD	1.01	1.02	1.06	1.06	1.08	1.11	1.19	1.10	1.07	1.05	0.93	0.91

Table 6. The leave-one-out, cross-validation (LOO-CV) statistics, showing the goodness of fit between observations and estimates of daily and extreme precipitation indices. IP: Iberian Peninsula; BI: Balearic Islands; CI: Canary Islands; MAE: mean absolute error; ME: mean error; RM: ratio of means; RSD: ratio of standard deviations. Results were constrained to 2 decimal places.

		PMED	NWD	CDDm	CWDm	SDII	RX1	RX5	R10mm	R20mm	CDD	CWD	P95	R95rel
IP	MAE	−0.09	−0.02	0.00	0.00	−0.31	−2.42	0.15	−0.01	−0.02	0.02	−0.01	0.02	0.00
	ME	1.29	0.38	0.08	0.01	1.27	4.64	2.49	0.05	0.03	0.37	0.07	0.57	0.25
	RM	1.09	1.02	1.01	1.00	1.00	0.96	1.01	0.99	0.95	1.02	1.01	1.00	1.00
	RSD	1.14	1.01	1.03	1.02	1.04	0.98	1.01	1.00	0.97	1.03	1.01	0.97	1.08
BI	MAE	−0.12	−0.07	0.01	0.00	−0.22	−2.18	−0.35	−0.01	−0.01	−0.06	−0.01	0.14	0.02
	ME	1.33	0.52	0.09	0.01	1.35	5.25	3.89	0.05	0.03	0.47	0.06	0.45	0.2
	RM	1.06	1.01	1.02	1.00	1.00	0.97	1.00	1.00	0.99	1.00	1.00	1.01	1.00
	RSD	1.09	1.02	1.24	1.01	1.00	0.99	1.01	1.02	1.01	1.02	1.01	1.02	1.03
CI	MAE	−1.17	−0.05	−0.07	0.00	−1.23	−1.22	0.47	−0.01	0.00	−0.22	−0.01	−0.1	0.10
	ME	2.20	0.37	0.27	0.01	2.40	4.80	3.79	0.04	0.02	0.86	0.06	0.48	0.30
	RM	0.98	1.13	0.99	0.99	0.95	1.00	1.04	1.01	1.00	0.99	1.01	0.97	1.01
	RSD	1.19	1.07	1.16	0.97	1.01	1.01	1.05	1.01	1.02	1.01	0.98	0.99	1.40

4.3 Spatial distribution and uncertainty in daily precipitation

The daily mean precipitation intensity (PMED) (computed as the median precipitation in wet days) map (Fig. 5a) showed three areas with maximum values (Central Range, Pyrenees and Baetic Range). Overall, PMED was higher (> 8 mm) in the south-western sector of peninsular Spain, in the north-west and in the Pyrenees. The rest of the territory reached values between 4 and 6 mm.

The NWD showed a strong gradient from the northwest (Fig. 5c), with more than 100 days of precipitation, to the south-east, where the lowest was fewer than 30 days of precipitation per year. This value was the most frequent in the Canary Islands except in the island of La Palma (extreme west) and in the highest areas of the other islands (over 1500 m a.s.l.), where rainfall occurrence reached almost 90 days per year. The rest of peninsular Spain and the Balearic Islands showed average values between 50 and 70 days.

The mean length of the dry spells (CDDm) (Fig. 5e) showed a similar gradient. The northern sector reached values under 5 days, increasing to the southeast, where the average length of consecutive dry days was more than 30 days, like in the largest part of the Canary Islands.

NWD, CDDm and mean consecutive wet days (CWDm) characterised daily precipitation frequency, showing that in the southwest of peninsular Spain and the Canary Islands the average number of rainy days per year is fewer than 30 and, when rain occurs, the mean length of the events was fewer than 1.5 days. Conversely, the mean number of wet days in almost the entire Cantabric fringe was over 120 days per year, and the mean consecutive number of dry days was under 5 days.

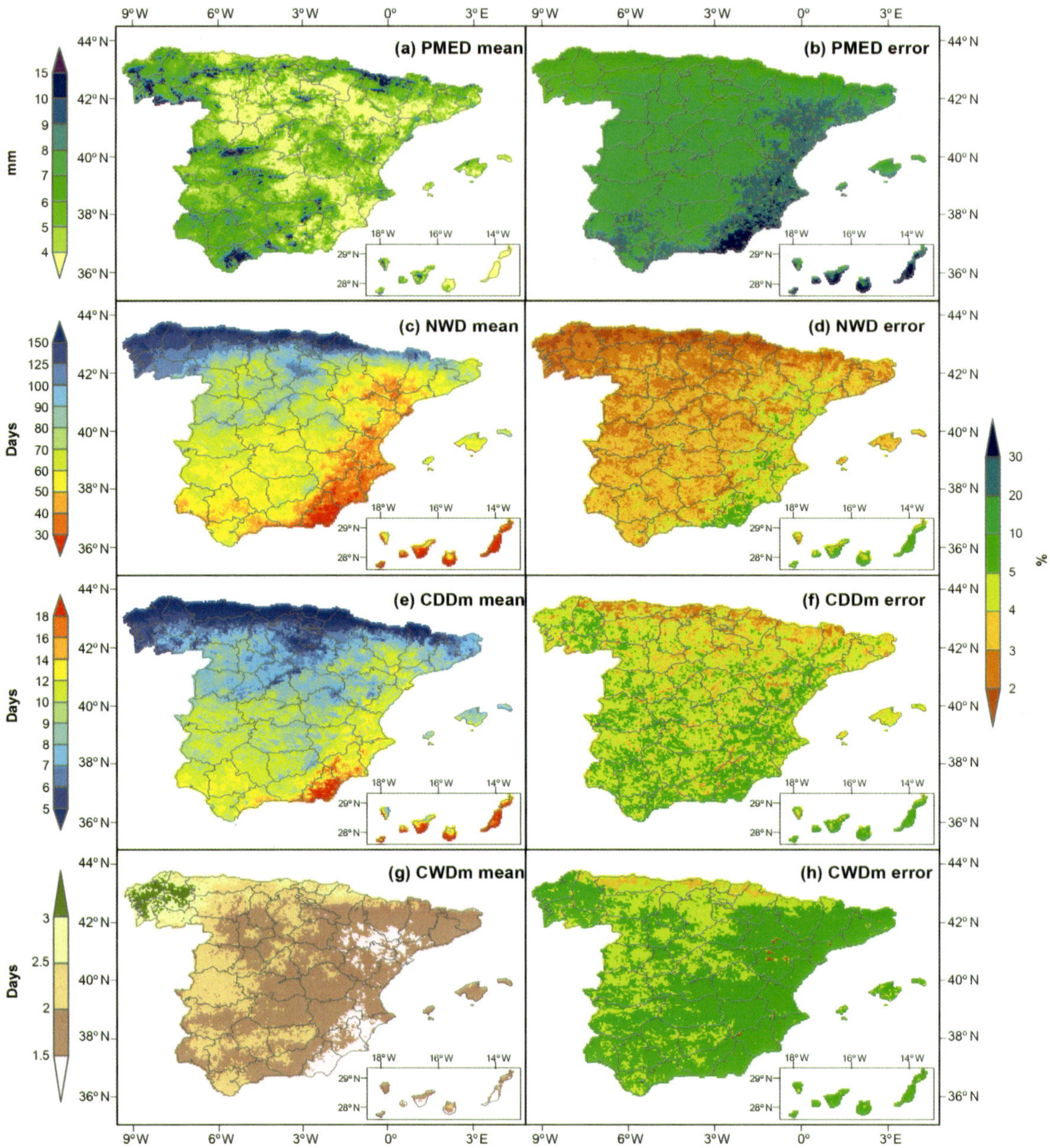

Figure 5. Daily precipitation indices **(a, d, e, g)** and their uncertainty **(b, d, f, h)**. PMED: daily mean precipitation intensity; NWD: number of wet days; CDDm: mean consecutive dry days; CWDm: mean consecutive wet days.

The uncertainty of estimates in daily precipitation indices was spatially variable, but in all cases it had an increasing gradient from north-west to south-east in peninsular Spain, especially in PMED and NWD. This uncertainty, which was not necessarily similar to the distribution of its corresponding variable, informed about the reliability of the results of the indices. The higher values in most of the indices occurred in the south-west.

4.4 Spatial distribution and uncertainty in daily extreme precipitation

Mean precipitation in wet days (SDII) (computed as the mean precipitation in wet days) in Spain ranged between 5 and more than 25 mm (Fig. 6). The lowest values were distributed in the northern and southern plateaus and at the bottom of the Ebro Valley, unlike the south-east of peninsular Spain, which had higher values similar to the eastern coast, where the total rainfall was low but daily precipitation intensity was very high. The normal values in the Pyrenees were

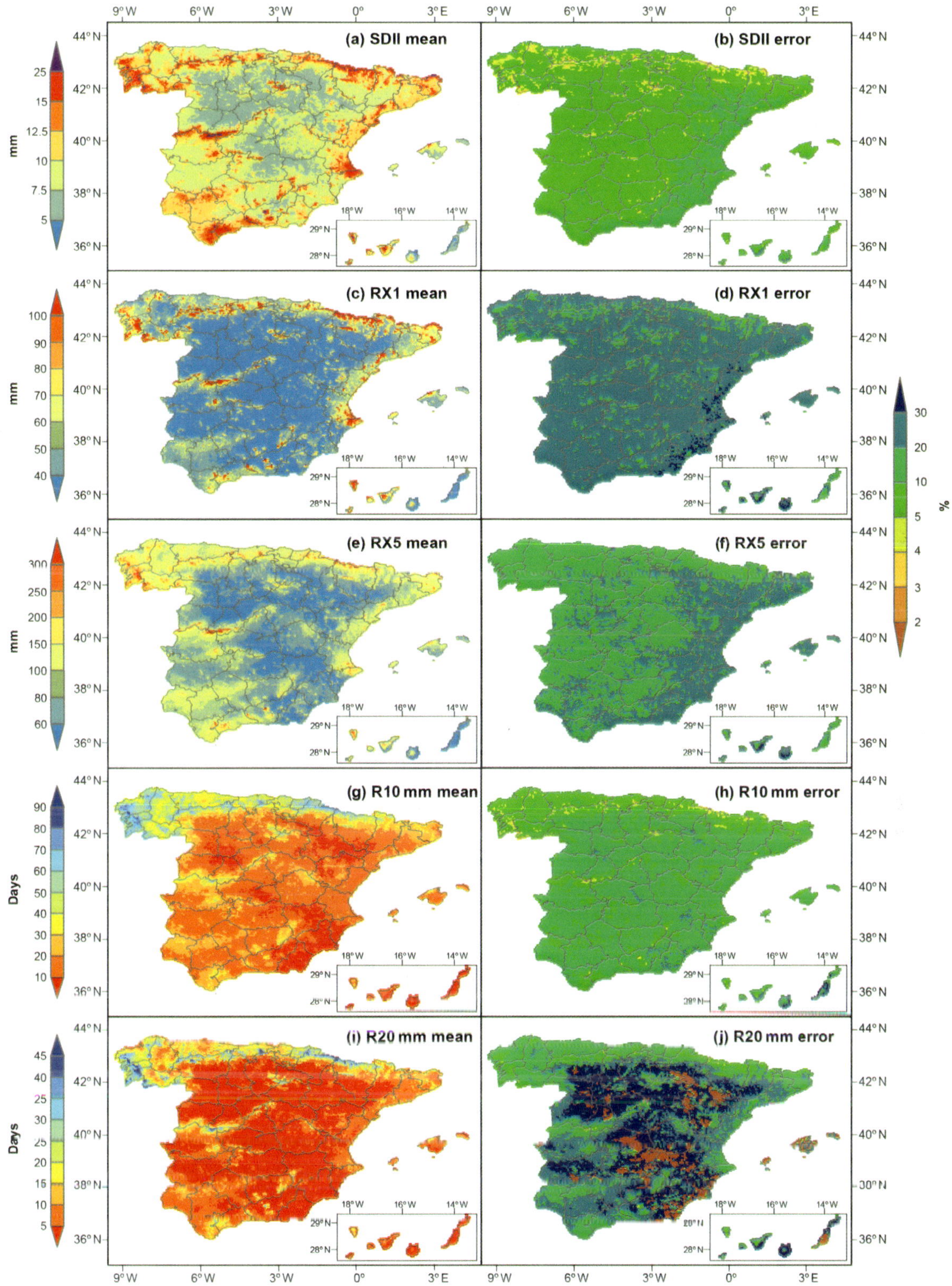

Figure 6. Daily extreme precipitation indices **(a, c, e, g, i)** and their uncertainty **(b, d, f, h, j)**. SDII: daily precipitation intensity; RX1: maximum 1-day precipitation; RX5: maximum 5-day precipitation; R10mm: number of days with precipitation over 10 mm; R20mm: number of days with precipitation over 20 mm.

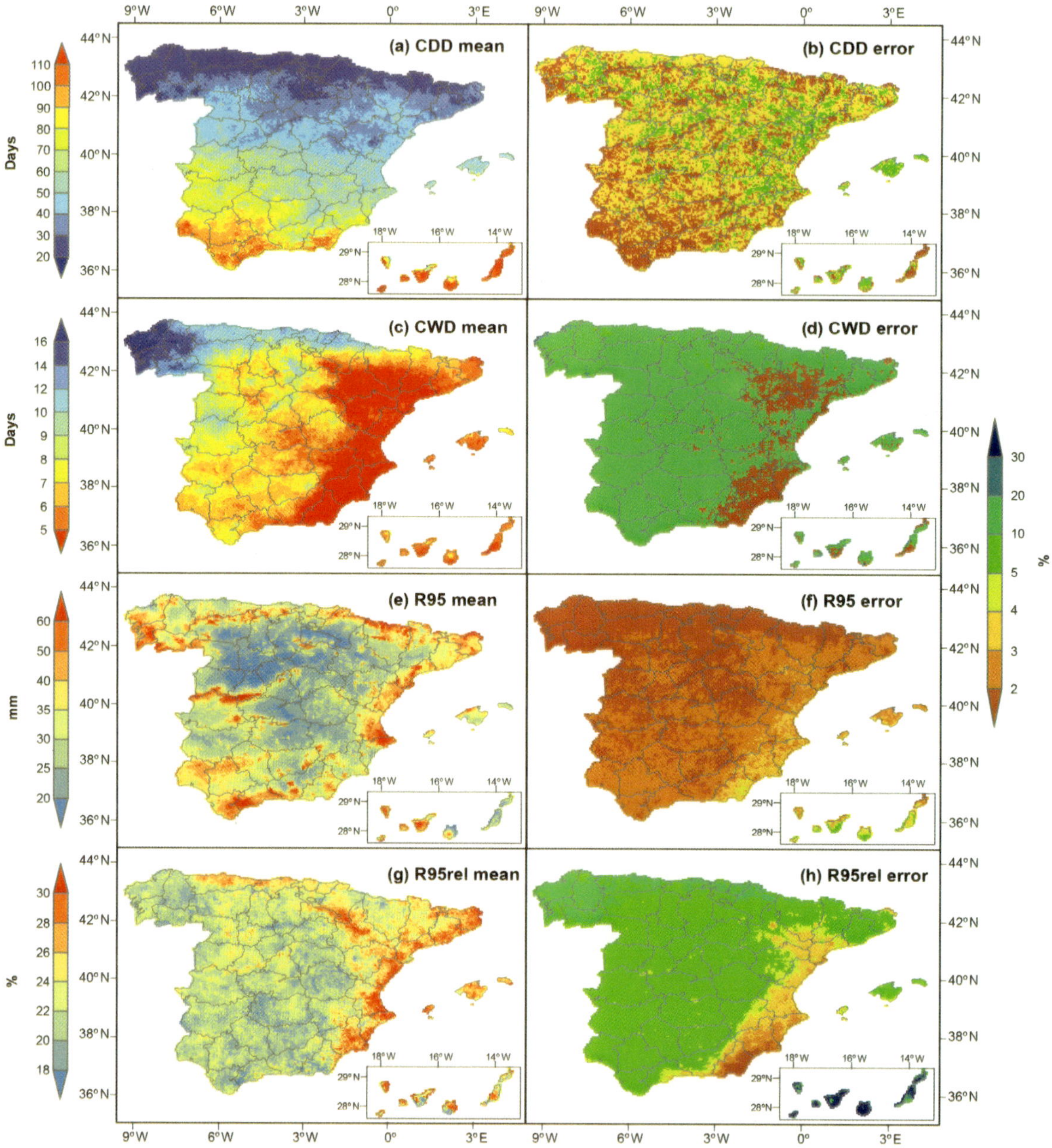

Figure 7. Daily extreme precipitation indices **(a, c, e, g)** and their uncertainty **(b, d, f, h)**. CDD: maximum consecutive dry days; CWD: maximum consecutive wet days; R95: 95th percentile of precipitation; R95rel: contribution of precipitation over 95th percentile.

higher than 15 mm for each wet day, especially in the western half, where the Atlantic influence is strongest. Despite this, the highest values were in the Central Range, with more than 25 mm per wet day, being the area with most precipitation on wet days, in the whole country.

The mean maximum precipitation in one day (RX1) (Fig. 6) was concentrated mainly in the highest areas, which create orographic barriers. The Central Range was the only zone in Spain that reached values higher than 200 mm, decreasing with elevation until 40 mm. This pattern was replicated in most of the high-elevation areas of peninsular Spain and islands, but also in the Mediterranean coast, which is characterised by a high frequency of extreme events. The distribution of the maximum precipitation in five days (RX5) (Fig. 6) was very similar, but with a smoother gradient. As the variability in the extreme nature of the RX5 was less intense, the spatial distribution was more homogeneous with milder differences between the regions.

A high number of days per year with more than 10 mm of precipitation (R10mm, Fig. 6) is relatively frequent in Spain, and especially so in a Mediterranean climate as it corresponds to a large part of peninsular Spain and the Balearic Islands. Overall, the spatial distribution of this index was very similar to the mean annual precipitation in Spain, with the highest values in the north-west and in high elevations. Considering events with precipitation over 20 mm (R20mm), the spatial pattern mimicked that of the R10mm.

The uncertainty distribution was very similar for SDII, RX1, and RX5, with higher values along the Mediterranean coast, where intense precipitation is more frequent and, consequently, the differences between neighbouring observatories were also higher. R10mm had a low and homogeneous uncertainty all over Spain and R20mm had very low values over central and southern peninsular Spain, in the Balearic Islands and in the eastern Canary Islands.

A clear latitudinal gradient over peninsular Spain was evident for the mean annual maximum consecutive dry days (CDD) index, with values exceeding 100 days in the south in contrast with fewer than 20 days in the north (Fig. 7 up). The maximum consecutive wet days (CWD) extreme (Fig. 7) had a strong longitudinal gradient, with fewer than 5 days in the east to more than 16 days in the western. The Balearic Islands showed a latitudinal gradient in both indices with more CDD (> 60 days) and less CWD (< 5 days) in the south, coinciding with lower elevations. The Canary Islands had a similar behaviour in all individual islands with the maximum values of CDD (> 110 days) and minimum of CWD (< 5 days).

The 95th percentile of precipitation (R95) showed the maximum values at high-elevation areas and in the eastern and southern sides of the Mediterranean coast. This region is considered the central region of peninsular Spain and was more homogeneous with lower values, coinciding with a more continental precipitation regime. The uncertainty values were very low here, which showed the reliability of the estimations of this index. The percentage of precipitation over the 95th percentile contribution to the annual precipitation total (R95rel) showed different patterns, more extreme at eastern peninsular Spain and western Canary Islands. More than the 30 % of the precipitation along the Mediterranean coast corresponds to events with amounts of precipitation over the 95th percentile. These values are also common in the Balearic Islands, where all areas had R95rel values over 20 %. This spatial distribution represents the extreme character of daily precipitation, especially in Mediterranean areas and in the Ebro Valley.

5 Discussion

High-resolution gridded datasets are useful for regional analysis of daily precipitation, but the accuracy of the estimates depends mainly on the number of available observatories and on the estimation method. Although the method used to build this grid makes independent calculations of each grid point and day, the results showed coherent patterns in spatial distributions of all indices at regional scales.

Some basic parameters of the reconstruction methodology have a key influence on the dissimilarity between different datasets. The selection of one specific method from the many available gridding interpolation methods that can be applied to precipitation may change the final result (e.g. Creutin and Obled, 1982; Hartkamp et al., 1999; Vicente-Serrano et al., 2003; Dobesch et al., 2007; Hofstra et al., 2008; Hwang et al., 2012; Brunetti et al., 2014; Militino et al., 2015; Contractor et al., 2015; Herrera et al., 2016). For example, Robeson and Ensor (2006) and Ensor and Robeson (2008) argued that the use of geostatistical interpolators for daily precipitation leads to a higher frequency of low-precipitation values while greatly reducing the extreme events. In addition, the high flexibility in the independent variables across the sites allows for a reasonable estimation of the uncertainty, which is very important for producing datasets that will feed further analyses. Local regressions have been used widely to model daily precipitation with different approaches (Buishand and Tank, 1996; Rajagopalan and Lall, 1998; Marquínez et al., 2003; Simolo et al., 2010; Tardivo and Verti, 2014; Partal et al., 2015), and in this case we used them to compute, from all reconstructed stations, a high-resolution grid estimating separately the probability of a wet/dry day occurrence and the precipitation amount. This two-step procedure avoids an excessive smoothness of the estimated precipitation fields (Robeson and Ensor, 2006). Furthermore, this individualised calculation allows for an easy update of the dataset, since single days can be reconstructed individually and added to the pre-existing dataset. We also added a measure of uncertainty, which is a big improvement over previous gridded datasets. Uncertainty (which we express by means of the standard error) informs in a quantitative way about the reliability of the estimated data, in a way that can be translated to further calculations such as the daily precipitation indices explored in this article. Our uncertainty estimation arises from a local interpolation, so it varies spatially and from one day to the next, reflecting the changes in conditions that affect the estimates.

Although some previous datasets exist that obtain daily and/or extreme precipitation indices for Spain, the different methodologies to compute them are a key influence in these differences. The use of the longest precipitation series and the spatial resolution of the final grid produces very different results. For instance, considering the global datasets, Sillmann et al. (2013) showed for Spain (eight grid points covering the whole of peninsular Spain) RX5 values from 50 and 200 mm using the HadEX2 dataset and from 40 and 75 mm using CMIP5 dataset. In the present work, these values ranged between 50 and more than 300 in more than 20 000 grid points. These values were the maximums in HadEX2 and CMIP5 for monsoon areas in southern Asia. Similarly, May (2007), using the HIRHAM model, showed SDII values for Spain between 4.5 and 12.5, while in the present work we found that this index can reach values over 25 mm, especially in the Central Range. Schamm et al. (2014) used the GPCC (Global Precipitation Climatology Centre) dataset to show values of daily mean precipitation intensity (PMED) between 0.5 and less than 10 mm, while we found a wider range between 3 and more than 15 mm.

Some previous works in Spain that used a lower spatial resolution and a lower number of observatories (Herrera et al., 2012; Merino et al., 2015) showed overestimated values in extreme precipitation indices in some areas (especially in the north-west area of the Iberian Peninsula) but smoothing in others (e.g. Central Range). López-Moreno et al. (2010) showed similar values and spatial patterns in north-east Spain, compared to the ones in the present work for NWD, SDII, CDD and CWD. Despite the relatively high-density station dataset (217 stations) used in their study, they rejected most of the original stations in favour of only the longest series, resulting in a smoothed spatial distribution of the indices, probably also due to the interpolation method (not indicated). Martínez et al. (2007), using 75 stations, showed for Catalonia (northeast peninsular Spain) the 95th percentile of precipitation values ranging from 20 to 70 mm, which are very similar to those from the SPREAD dataset. All of these works made sub-optimal use of the available data. The values obtained in all of them were correct considering a global conception of precipitation distribution, but daily precipitation requires the highest possible density of observations in order to obtain a proper characterisation of its spatial distribution, especially for extreme precipitation. This work provides a representation of the local variability of extremes by using all the available information and applying a local reconstruction method. If the spatial resolution is amplified for a regional study, based on the use of more precipitation data, the results are less smoothed as shown in Pereira et al. (2016), which used 36 stations in Sierra Nevada (southern peninsular Spain) to compute NWD, R10mm and R20mm, with similar values to this work specifying a precise spatial distribution. The use of the complete information of the precipitation network in Spain provided a more detailed precipitation distribution over time and space. Although only a few stations covered the complete period, the use of short data series helped to estimate the missing precipitation values in longer ones, which were used to build the whole grid. A high number of grid points (higher spatial resolution) in combination with a low-density stations network could lead to higher uncertainties. This work aimed to set a compromise between both factors by using a high number of stations and a medium-high spatial resolution. In addition, the magnitude of the uncertainty informed about the reliability of each estimate. A higher uncertainty means more differences between the data used to estimate precipitation and these differences can be increased with a lower number of stations.

6 Data availability

The SPREAD dataset is freely available in the web repository of the Spanish National Research Council (CSIC). It can be accessed through https://doi.org/10.20350/digitalCSIC/7393, and cited as Serrano-Notivoli et al. (2016). The data are arranged in six files (daily precipitation estimations and their uncertainties for peninsular Spain, Balearic Islands and Canary Islands) in NetCDF format that allows for easy processing in scientific analysis software (e.g. R, Python) and GIS (list of compatible software at http://www.unidata.ucar.edu/software).

7 Conclusions

A high-resolution daily precipitation dataset for Spain (SPREAD) is presented. Based on all the available daily precipitation information, a 5×5 km spatial resolution grid was built using the *reddPrec* R package (Serrano-Notivoli et al., 2017a). The original dataset of observations was quality-controlled and the missing values were fitted using the 10 surrounding stations for each day and location to obtain a serially complete dataset from 1950 to 2012 in peninsular Spain and from 1971 to 2012 in the Balearic and Canary islands. From this dataset, individual daily precipitation estimations were computed for each grid point, resulting in a gridded dataset which was consequently used to compute four daily precipitation indices and nine extreme precipitation indices.

PMED showed the highest values in the Central Range and other elevated areas, while NWD, CDDm and CWDm followed a north-west to south-east gradient in peninsular Spain, from high to low values in NWD and CWDm and reverse in CDDm. The south-east of the Iberian Peninsula and the Canary Islands were the driest areas in Spain, with fewer than 30 wet days per year and more than 18 days of the average maximum annual dry spell length. These regions registered fewer than 2 days of the mean wet event duration.

Extreme precipitation indices showed that the Mediterranean coast is more active in these kinds of events, but also that the highest values of SDII, RX1, RX5, R10mm, R20mm and R95 are concentrated in a north–south band of north-west peninsular Spain and, especially, in the Central Range. These results have revealed areas with maximum values not detected in former studies, emphasising the importance of the use of all available observatories and a sensible methodology that do not produce excessive smoothing while being able to capture local and day-to-day variability.

Author contributions. MdL and SB designed the methodological approach in collaboration with RSN, who applied it to the reconstruction of the climate data and the development of the gridded dataset. MAS and LAL contributed to the validation process and performed the climatic analysis of the results. RSN prepared the manuscript with contributions from all co-authors.

Competing interests. The authors declare that they have no conflict of interest.

Acknowledgements. This study was supported by research projects CGL2015-69985-R and CGL2014-52135-C3-1-R, financed by the Spanish Ministerio de Economía y Competitividad (MINECO) and FEDER-ERDF funds. The researchers were supported by the government of Aragón through the "Programme of research groups" (groups H38, "Clima, Cambio Global y Sistemas Naturales" and "E68, Geomorfología y Cambio Global").

Edited by: David Carlson

References

Andreadis, K. M., Clark, E. A., Wood, A. W., Hamlet, A. F., and Lettenmaier, D. P.: Twentieth-century drought in the conterminous United States, J. Hydrometeorol., 6, 985–1001, 2005.

Asadieh, B. and Krakauer, N. Y.: Global trends in extreme precipitation: climate models versus observations, Hydrol. Earth Syst. Sci., 19, 877–891, https://doi.org/10.5194/hess-19-877-2015, 2015.

Beguería, S., Vicente-Serrano, S. M., Tomás-Burguera, M., and Maneta, M.: Bias in the variance of gridded data sets leads to misleading conclusions about changes in climate variability, Int. J. Climatol., 36, 3413–3422, https://doi.org/10.1002/joc.4561, 2015.

Belo-Pereira, M., Dutra, E., and Viterbo, P.: Evaluation of global precipitation data sets over the Iberian Peninsula, J. Geophys. Res.-Atmos., 116, D20101, https://doi.org/10.1029/2010JD015481, 2011.

Bordi, I., Fraedrich, K., and Sutera, A.: Observed drought and wetness trends in Europe: an update, Hydrol. Earth Syst. Sci., 13, 1519–1530, https://doi.org/10.5194/hess-13-1519-2009, 2009.

Brunetti, M., Maugeri, M., Nanni, T., Simolo, C., and Spinoni, J.: High-resolution temperature climatology for Italy: interpolation method intercomparison, Int. J. Climatol., 34, 1278–1296, https://doi.org/10.1002/joc.3764, 2014.

Buishand, T. and Tank, K. Regression model for generating time series of daily precipitation amounts for climate change impact studies, Stoch. Hydrol. Hydraul., 10, 87–106, https://doi.org/10.1007/bf01581761, 1996.

Casas, M. C., Herrero, M., Ninyerola, M., Pons, X., Rodríguez, R., Rius, A., and Redano, A.: Analysis and objective mapping of extreme daily rainfall in Catalonia, Int. J. Climatol., 27, 399–409, https://doi.org/10.1002/joc.1402, 2007.

Contractor, S., Alexander, L. V., Donat, M. G., and Herold, N.: How Well Do Gridded Datasets of Observed Daily Precipitation Compare over Australia?, Adv. Met., 2015, 325718, https://doi.org/10.1155/2015/325718, 2015.

Chaney, N. W., Sheffield, J., and Villarini, G.: Developement of a High-Resolution Gridded Daily Meteorological Dataset over Sub-Saharan Africa: Spatial Analysis of Trends in Climate Extremes, J. Climate, 27, 5815–5835, https://doi.org/10.1175/JCLI-D-13-00423.1, 2014.

Cortesi, N., Gonzalez-Hidalgo, J. C., and Brunetti, M.: Spatial variability of precipitation in Spain, Reg. Environ. Chang., 14, 1743–1749, https://doi.org/10.1007/s10113-012-0402-6, 2014.

Creutin, J. and Obled, C.: Objective analyses and mapping techniques for rainfall fields: an objective comparison, Water Resour. Res. 18, 413–431, https://doi.org/10.1029/WR018i002p00413, 1982.

de Luis, M., González-Hidalgo, J. C., Brunetti, M., and Longares, L. A.: Precipitation concentration changes in Spain 1946–2005, Nat. Hazards Earth Syst. Sci., 11, 1259–1265, https://doi.org/10.5194/nhess-11-1259-2011, 2011.

Di Luzio, M., Johnson, G. L., Daly, C., Eischeid, J. K., and Arnold, J. G.: Constructing retrospective gridded daily precipitation and temperature datasets for the conterminous United States, J. Appl. Meteorol. Climatol., 47, 475–497, https://doi.org/10.1175/2007JAMC1356.1, 2008.

Dobesch, H., Dumolard, P., and Dyras, I. (Eds): Spatial interpolation for climate data: the use of GIS in climatology and meteorology, ISTE, London, 2007.

Döll, P. and Fiedler, K.: Global-scale modeling of groundwater recharge, Hydrol. Earth Syst. Sci., 12, 863–885, https://doi.org/10.5194/hess-12-863-2008, 2008.

Donat, M. G., Alexander, L. V., Yang, H., Durre, I., Vose, R., Dunn, R. J. H., Willet, K. M., Aguilar, E., Brunet, M., Caesar, J., Hewitson, B., Jack, C., Klein-Tank, A. M. G., Kruger, A. C., Marengo, J., Peterson, T. C., Renom, M., Oria-Rojas, C., Rusticucci, M., Salinger, J., Elrayah, A. S., Sekele, S. S., Srivastava, A. K., Trewin, B., Villaroel, C., Vincent, L. A., Zhai, P., Zhang, X., and Kitching, S.: Updated analyses of temperature and precipitation extreme indices since the beginning of the twentieth century: The HadEX2 dataset, J. Geophys. Res.-Atmos., 118, 2098–2118, https://doi.org/10.1002/jgrd.50150, 2013.

Donat, M. G., Sillmann, J., Wild, S., Alexander, L. V., Lippmann, T., and Zwiers, F. W.: Consistency of temperature and precipitation extremes across various global gridded in situ and reanalysis datasets, J. Climate, 27, 5019–5035, https://doi.org/10.1175/JCLI-D-13-00405.1, 2014.

Eischeid, J. K., Pasteris, P. A., Diaz, H. F., Plantico, M. S., and Lott, N. J.: Creating a serially complete, national daily time series of temperature and precipitation for the western United States, J. Appl. Meteorol., 39, 1580–1591, https://doi.org/10.1175/1520-0450(2000)039<1580:CASCND>2.0.CO;2, 2000.

Ensor, L. and Robeson, S.: Statistical Characteristics of Daily Precipitation: Comparisons of Gridded and Point Datasets, J. Appl. Meteorol. Clim., 47, 2468–2476, https://doi.org/10.1175/2008JAMC1757.1, 2008.

Esteban-Parra, M. J., Rodrigo, F. S., and Castro-Diez, Y.: Spatial and temporal patterns of precipitation in Spain for the period 1880–1992, Int. J. Climatol., 18, 1557–1574, https://doi.org/10.1002/(SICI)1097-0088(19981130)18:14<1557::AID-JOC328>3.0.CO;2-J, 1998.

Frei, C. and Schär, C.: A precipitation climatology of the alps from high-resolution rain-gauge observations, Int. J. Climatol., 18, 873–900, https://doi.org/10.1002/(SICI)1097-0088(19980630)18:8<873::AID-JOC255>3.0.CO;2-9, 1998.

González-Hidalgo, J. C., Brunetti, M., and De Luis, M.: A new tool for monthly precipitation analysis in Spain: MOPREDAS database (Monthly precipitation trends December 1945–November 2005), Int. J. Climatol., 31, 715–731, https://doi.org/10.1002/joc.2115, 2011.

Harris, I., Jones, P. D., Osborn, T. J., and Lister, D. H.: Updated high-resolution grids of monthly climatic observations – the CRU TS3.10 Dataset, Int. J. Climatol., 34, 623–642, https://doi.org/10.1002/joc.3711, 2014.

Hartmann, H. and Andresky, L.: Flooding in the Indus River basin – A spatiotemporal analysis of precipitation records, Global Planet. Change, 107, 25–35, 2013.

Hartkamp, A. D., De Beurs, K., Stein, A., and White, J. W. (Eds.): Interpolation Techniques for Climate Variables, NRG-GIS Series 99-01, Mexico D.F., CIMMYT, 1999.

Haylock, M., Hofstra, N., Tank, A., Klok, E., Jones, P., and New, M.: A European daily high-resolution gridded data set of surface temperature and precipitation for 1950–2006, J. Geophys. Res., 113, D20119, https://doi.org/10.1029/2008JD010201, 2008.

Hernández, A., Kutiel, H., Trigo, R. M., Valente, M. A., Sigró, J., Cropper, T., and Esprito-Santo, F.: New Azores archipelago daily precipitation dataset and its links with large-scale modes of climate variability, Int. J. Climatol., 36, 4439–4454, https://doi.org/10.1002/joc.4642, 2016.

Herrera, S., Gutiérrez, J. M., Ancell, R., Pons, M. R., Frías, M. D., and Fernández, J.: Development and analysis of a 50 year high-resolution daily gridded precipitation dataset over Spain (Spain02), Int. J. Climatol., 32, 74–85, https://doi.org/10.1002/joc.2256, 2012.

Herrera, S., Fernández, J., and Gutiérrez, J.: Update of the Spain02 gridded observational dataset for EURO-CORDEX evaluation: assessing the effect of the interpolation methodology, Int. J. Climatol., 36, 900–908, https://doi.org/10.1002/joc.4391, 2016.

Hewitson, B. C. and Crane, R. G.: Gridded area-averaged daily precipitation via conditional interpolation, J. Climate, 18, 41–57, https://doi.org/10.1175/JCLI3246.1, 2005.

Hijmans, R. J., Cameron, S. E., Parra, J. L., Jones, P. G., and Jarvis, A.: Very high resolution interpolated climate surfaces for global land areas, Int. J. Climatol., 25, 1965–1978, https://doi.org/10.1002/joc.1276, 2005.

Hofstra, N., Haylock, M., New, M., Jones, P., and Frei, C.: Comparison of six methods for the interpolation of daily, European climate data, J. Geophys. Res.-Atmos., 113, D21110, https://doi.org/10.1029/2008jd010100, 2008.

Hofstra, N., New, M., and McSweeney, C.: The influence of interpolation and station network density on the distributions and trends of climate variables in gridded daily data, Clim. Dynam., 35, 841–858, https://doi.org/10.1007/s00382-009-0698-1, 2010.

Hwang, Y., Clark, M., Rajagopalan, B., and Leavesley, G.: Spatial interpolation schemes of daily precipitation for hydrologic modelling, Stoch. Env. Res. Risk A., 26, 295–320, https://doi.org/10.1007/s00477-011-0509-1, 2012.

Isotta, F. A., Frei, C., Weilguni, V., Tadic, M. P., Lassègues, P., Rudolf, B., Pavan, V., Cacciamani, C., Antolini, G., Ratto, S. M., Munari, M., Micheletti, S., Bonati, V., Lussana, C., Ronchi, C., Panettieri, E., Marigo, G., and Vertacnik, G.: The climate of daily precipitation in the Alps: development and analysis of a high-resolution grid dataset from pan-Alpine rain-gauge data, Int. J. Climatol., 34, 1657–1675, https://doi.org/10.1002/joc.3794, 2014.

Jones, P. D., Lister, D. H., Harpham, C., Rusticucci, M., and Penalba, O.: Construction of a daily precipitation grid for southeastern South America for the period 1961–2000, Int. J. Climatol., 33, 2508–2519, https://doi.org/10.1002/joc.3605, 2013.

Kay, A. L., Rudd, A. C., Davies, H. N., Kendon, E. J., and Jones, R. G.: Use of very high resolution climate model data for hydrological modelling: baseline performance and future flood changes, Climatic Change, 133, 193-208, 2015.

Keggenhoff, I., Elizbarashvili, M., Amiri-Farahani, A., and King, L.: Trends in daily temperature and precipitation extremes over Georgia, 1971–2010, Weather and Climate Extremes, 4, 75–85, https://doi.org/10.1016/j.wace.2014.05.001, 2014.

Klein-Tank, A. M. G., Wijngaard, J. B., Können, G. P., Böhm, R., Demarée, G., Gocheva, A., Mileta, M., Pashiardis, S., Hejkrlik, L., Kern-Hansen, C., Heino, R., Bessemoulin, P., Müller-Westermeier, G., Tzanakou, M., Szalai, S., Pálsdóttir, T., Fitzgerald, D., Rubin, S., Capaldo, M., Maugeri, M., Leitass, A., Bukantis, A., Aberfeld, R., van Engelen, A. F. V., Forland, E., Mietus, M., Coelho, F., Mares, C., Razuvaev, V., Nieplova, E., Cegnar, T., López, J. A., Dahlström, B., Moberg, A., Kirchhofer, W., Ceylan, A., Pachaliuk, O., Alexander, L. V., and Petrovic, P.: Daily dataset of 20th-century surface air temperature and precipitation series for the European Climate Assessment, Int. J. Climatol., 22, 1441–1453, https://doi.org/10.1002/joc.773, 2002.

Klok, E. and Klein-Tank, A.: Updated and extended European dataset of daily climate observations, Int. J. Climatol., 29, 1182–1191, https://doi.org/10.1002/joc.1779, 2009.

Kyriakidis, P. C., Kim, J., and Miller, N.: Geostatistical mapping of precipitation from rain gauge data using atmospheric and terrain characteristics, J. Appl. Meteorol., 40, 1855–1877, https://doi.org/10.1175/1520-0450(2001)040<1855:GMOPFR>2.0.CO;2, 2001.

Liebmann, B. and Allured, D.: Daily precipitation grids for South America, B. Am. Meteorol. Soc., 86, 1567–1570, https://doi.org/10.1175/BAMS-86-11-1567, 2005.

Lo, F., Wheeler, M. C., Meinke, H., and Donald, A.: Probabilistic forecasts of the onset of the north Australian wet season, Mon. Weather Rev., 135, 3506–3520, 2007.

López-Moreno, J. I., Vicente-Serrano, S. M., Angulo-Martínez, M., Beguería, S., and Kenawy, A.: Trends in daily precipitation on the northeastern Iberian Peninsula, 1955–2006, Int. J. Climatol., 30, 1026–1041, https://doi.org/10.1002/joc.1945, 2010.

Lorenz, C., Kunstmann, H., Devaraju, B., Tourian, M. J., Sneeuw, N., and Riegger, J.: Large-scale runoff from landmasses: A global assessment of the closure of the hydrological and atmospheric water balances, J. Hydrometeorol., 15, 2111–2139, 2014.

Marquinez, J., Lastra, J., and Garcia, P.: Estimation models for precipitation in mountainous regions: the use of gis and multivariate analysis, J. Hydrol., 270, 1–11, https://doi.org/10.1016/S0022-1694(02)00110-5, 2003.

Martín-Vide, J.: Spatial distribution of a daily precipitation concentration index in peninsular Spain, Int. J. Climatol., 24, 959–971, https://doi.org/10.1002/joc.1030, 2004.

Martínez, M. D., Lana, X., and Burgueño, A.: Spatial and temporal daily rainfall regime in Catalonia (NE Spain) derived from four precipitation indices, years 1950–2000, Int. J. Climatol., 27, 123–138, https://doi.org/10.1002/joc.1369, 2007.

May, W.: The simulation of the variability and extremes of daily precipitation over Europe by the HIRHAM regional climate model, Global Planet. Change, 57, 59–82, https://doi.org/10.1016/j.gloplacha.2006.11.026, 2007.

Maurer, E. P., Wood, A. W., Adam, J. C., Lettenmaier, D. P., and Nijssen, B.: A long-term hydrologically based dataset of land surface fluxes and states for the conterminous United States, J. Climate, 15, 3237–3251, 2002.

Mekis, É. and Vincent, L.: An Overview of the Second Generation Adjusted Daily Precipitation Dataset for Trend Analysis in Canada, Atmos. Ocean, 49, 163–177, https://doi.org/10.1080/07055900.2011.583910, 2011.

Menne, M. J., Durre, I., Vose, R. S., Gleason, B. E., and Houston, T. G.: An overview of the global historical climatology network-daily database, J. Atmos. Ocean. Tech., 29, 897–910, https://doi.org/10.1175/JTECH-D-11-00103.1, 2012.

Merino, A., Fernández-Vaquero, M., López, L., Fernández-González, S., Hermida, L., Sánchez, J., García-Ortega, E., and Gascón, E.: Large-scale patterns of daily precipitation extremes on the Iberian Peninsula, Int. J. Climatol., 36, 3873–3891, https://doi.org/10.1002/joc.4601, 2015.

Militino, A., Ugarte, M., Goicoa, T., and Genton, M.: Interpolation of daily rainfall using spatiotemporal models and clustering, Int. J. Climatol., 35, 1453–1464, https://doi.org/10.1002/joc.4068, 2015

Mishra, V. and Cherkauer, K. A.: Retrospective droughts in the crop growing season: Implications to corn and soybean yield in the Midwestern United States, Agr. Forest. Meteorol., 150, 1030–1045, 2010.

New, M., Lister, D., Hulme, M., and Makin, I.: A high-resolution data set of surface climate over global land areas, Clim. Res., 21, 1–25, https://doi.org/10.3354/cr021001, 2002.

Partal, T., Cigizoglu, K., and Kahya, E.: Daily precipitation predictions using three different wavelet neural network algorithms by meteorological data, Stoch. Env. Res. Risk A., 29, 1317, https://doi.org/10.1007/s00477-015-1061-1, 2015.

Pereira, P., Oliva, M., and Misiune, I.: Spatial interpolation of precipitation indexes in Sierra Nevada (Spain): comparing the performance of some interpolation methods, Theor. Appl. Climatol., 126, 683–698, https://doi.org/10.1007/s00704-015-1606-8, 2016.

Piper, S. C. and Stewart E. F.: A gridded global data set of daily temperature and precipitation for terrestrial biospheric modeling, Global Biogeochem. Cy., 10, 757–782, https://doi.org/10.1029/96GB01894, 1996.

Rajagopalan, B. and Lall, U.: Locally weighted polynomial estimation of spatial precipitation, GIDA, 2, 44–51, 1998.

Robeson, S. M. and Ensor, L. A.: Daily precipitation grids for South America, B. Am. Meteorol. Soc., 86, 1567–1570, https://doi.org/10.1175/BAMS-86-11-1567, 2006.

Rodda, J. C. and Dixon, H.: Rainfall measurement revisited, Weather, 67, 131–136, 2012.

Rodda, J. C. and Smith, S. W.: The significance of the systematic error in rainfall measurement for assessing wet deposition, Atmos. Environ., 20, 1059–1064, 1986.

Rodrigo, F. S. and Trigo, R. M.: Trends in daily rainfall in the Iberian Peninsula from 1951 to 2002, Int. J. Climatol., 27, 513–529, https://doi.org/10.1002/joc.1409, 2007.

Rodríguez-Puebla, C., Encinas, A., and Nieto, S.: Spatial and temporal patterns of annual precipitation variability over the Iberian Peninsula, Int. J. Climatol., 18, 299–316, https://doi.org/10.1002/(SICI)1097-0088(19980315)18:3<299::AID-JOC247>3.0.CO;2-L, 1998.

Rojas, R., Feyen, L., Dosio, A., and Bavera, D.: Improving pan-European hydrological simulation of extreme events through statistical bias correction of RCM-driven climate simulations, Hydrol. Earth Syst. Sci., 15, 2599–2620, https://doi.org/10.5194/hess-15-2599-2011, 2011.

Rubel, F. and Hantel, M.: BALTEX 1/6-degree daily precipitation climatology 1996– 1998, Meteorol. Atmos. Phys., 77, 155–166, https://doi.org/10.1007/s007030170024, 2001.

Sanogo, S., Fink, A. H., Omotosho, J. A., Ba, A., Redl, R., and Ermert, V.: Spatio-temporal characteristics of the recent rainfall recovery in West Africa, Int. J. Climatol., 35, 4589–4605, https://doi.org/10.1002/joc.4309, 2015.

Schamm, K., Ziese, M., Becker, A., Finger, P., Meyer-Christoffer, A., Schneider, U., Schröder, M., and Stender, P.: Global gridded precipitation over land: a description of the new GPCC First Guess Daily product, Earth Syst. Sci. Data, 6, 49–60, https://doi.org/10.5194/essd-6-49-2014, 2014.

Serrano-Notivoli, R., De Luis, M., Beguería, S., and Saz, M. A.: SPREAD (Spanish PREcipitation At Daily scale) [Dataset], https://doi.org/10.20350/digitalCSIC/7393, 2016.

Serrano-Notivoli, R., De Luis, M., and Beguería, S.: An R package for daily precipitation climate series reconstruction, Environ. Model. Softw., 89, 190–195, https://doi.org/10.1016/j.envsoft.2016.11.005, 2017a.

Serrano-Notivoli, R., de Luis, M., Saz, M. A., Beguería, S.: Spatially-based reconstruction of daily precipitation instrumental data series, Clim. Res., 73, 167–186, https://doi.org/10.3354/cr01476, 2017b.

Sigdel, M. and Ma, Y.: Variability and trends in daily precipitation extremes on the northern and southern slopes of the central Himalaya, Theor. Appl. Climatol., 1–11, https://doi.org/10.1007/s00704-016-1916-5, 2016.

Sillmann, J., Kharin, V. V., Zhang, X., Zwiers, F. W., and Bronaugh, D.: Climate extremes indices in the CMIP5 multimodel ensemble: Part 1. Model evaluation in the present climate, J. Geophys. Res.-Atmos., 118, 1716–1733, https://doi.org/10.1002/jgrd.50203, 2013.

Simolo, C., Brunetti, M., Maugeri, M., and Nanni, T.: Improving estimation of missing values in daily precipitation series by a probability density function-preserving approach, Int. J. Climatol., 30, 1564–1576, https://doi.org/10.1002/joc.1992, 2010.

Tardivo, G. and Berti, A.: The selection of predictors in a regression-based method for gap filling in daily temperature datasets, Int. J. Climatol., 34, 1311–1317, https://doi.org/10.1002/joc.3766, 2014.

Touchan, R., Anchukaitis, K. J., Meko, D. M., Sabir, M., Attalah, S., and Aloui, A.: Spatiotemporal drought variability in northwestern Africa over the last nine centuries, Clim. Dynam., 37, 237–252, 2011.

Tveito, O. E., Wegehenkel, M., van der Wel, F., and Dobesch, H.: COST Action 719: the use of geographic information systems in climatology and meteorology, EUR-OP, Luxembourg, 2008.

Vicente-Serrano, S. M., Saz-Sánchez, M. A., and Cuadrat, J. M.: Comparative analysis of interpolation methods in the middle Ebro Valley (Spain): Application to annual precipitation and temperature, Clim. Res., 24, 161–180, https://doi.org/10.3354/cr024161, 2003.

Vicente-Serrano, S. M., Beguería, S., López-Moreno, J. I., Angulo, M., and El Kenawy, A.: A new global 0.5° gridded dataset (1901–2006) of a multiscalar drought index: Comparison with current drought index datasets based on the palmer drought severity index, J. Hydrometeorol., 11, 1033–1043, 2010a.

Vicente-Serrano, S. M., Beguería, S., López-Moreno, J., García-Vera, M., and Stepanek, P.: A complete daily precipitation database for northeast Spain: reconstruction, quality control, and homogeneity, Int. J. Climatol., 30, 1146–1163, https://doi.org/10.1002/joc.1850, 2010b.

Werner, A. T. and Cannon, A. J.: Hydrologic extremes – an intercomparison of multiple gridded statistical downscaling methods, Hydrol. Earth Syst. Sci., 20, 1483–1508, https://doi.org/10.5194/hess-20-1483-2016, 2016.

Westerling, A. L., Gershunov, A., Brown, T. J., Cayan, D. R., and Dettinger, M. D.: Climate and wildfire in the western United States, B. Am. Meteorol. Soc., 84, 595–604, 2003.

Yatagai, A., Kamiguchi, K., Arakawa, O., Hamada, A., Yasutomi, N., and Kitoh, A.: APHRODITE: Constructing a Long-Term Daily Gridded Precipitation Dataset for Asia Based on a Dense Network of Rain Gauges, B. Am. Meteorol. Soc., 93, 1401–1415, https://doi.org/10.1175/bams-d-11-00122.1, 2012.

Yin, H., Donat, M. G., Alexander, L. V., and Sun, Y.: Multi-dataset comparison of gridded observed temperature and precipitation extremes over china, Int. J. Climatol., 35, 2809–2827, https://doi.org/10.1002/joc.4174, 2015.

Zhang, X., Alexander, L., Hegerl, G., Jones, P., Tank, A., Peterson, T., Trewin, B., and Zwiers, F.: Indices for monitoring changes in extremes based on daily temperature and precipitation data, Wiley Interdiscip, Rev. Clim. Change, 2, 851–870, https://doi.org/10.1002/wcc.147, 2011.

4

Daily temperature records from a mesonet in the foothills of the Canadian Rocky Mountains

Wendy H. Wood[1], **Shawn J. Marshall**[1], **Terri L. Whitehead**[1], **and Shannon E. Fargey**[2]

[1]Department of Geography, University of Calgary, Calgary AB, T2N 1N4, Canada
[2]Department of Geography, University of Victoria, Victoria BC, Canada

Correspondence: Wendy H. Wood (whwood@ucalgary.ca)

Abstract. Near-surface air temperatures were monitored from 2005 to 2010 in a mesoscale network of 230 sites in the foothills of the Rocky Mountains in southwestern Alberta, Canada. The monitoring network covers a range of elevations from 890 to 2880 m above sea level and an area of about 18 000 km^2, sampling a variety of topographic settings and surface environments with an average spatial density of one station per 78 km^2. This paper presents the multiyear temperature dataset from this study, with minimum, maximum, and mean daily temperature data available at https://doi.org/10.1594/PANGAEA.880611. In this paper, we describe the quality control and processing methods used to clean and filter the data and assess its accuracy. Overall data coverage for the study period is 91 %. We introduce a weather-system-dependent gap-filling technique to estimate the missing 9 % of data. Monthly and seasonal distributions of minimum, maximum, and mean daily temperature lapse rates are shown for the region.

1 Introduction

Air temperature is a critical environmental variable across a wide range of disciplines and processes, affecting physical, ecological, and human systems. While temperature fields can be relatively homogeneous in simple topography and surface environments, the same generalizations cannot be made over complex terrain, such as mountain environments and where there are strong horizontal gradients in topography or surface cover. However, it is commonly necessary to estimate or model temperature in such environments, where direct data are unavailable. As examples, catchment-scale hydrological models require temperature estimates to calculate snow melt (e.g., Förster et al., 2014) and glacier mass balance (Hock, 2005), landscape models of vegetation phenology or agricultural yield need distributed temperature fields (e.g., Jochner et al., 2016; Schönhart et al., 2016), and temperature is one of the regulators of species range (e.g., Logan and Powell, 2001; Deutsch et al., 2008; Comte et al., 2014). Applications such as these foster tremendous interest in landscape-scale temperature patterns and their structure under different weather systems.

Temperature variability across the landscape also needs to be understood to support the growing demand to model climate change impacts on Earth systems (e.g., Thomas et al., 2006; Deutsch et al., 2008; Clarke et al., 2015; Zuliani et al., 2015; Yospin et al., 2015; Franklin et al., 2016). General circulation models that are used for climate change scenarios or global climate reanalyses typically operate on scales of tens to hundreds of kilometres (e.g., Dee et al., 2011; Taylor et al., 2012; Murakami et al., 2016). Grid-cell temperatures from these models need to be distributed over the subgrid terrain for local- or regional-scale ecological, hydrological, or climate change impact analyses. Typically, few ground control stations are available on this subgrid scale to inform the interpolation schemes or the accuracy of such downscaling efforts.

These considerations motivated the establishment of the Foothills Climate Array (FCA) in the foothills of the Rocky Mountains in southwestern Alberta, Canada. An area of roughly 18 000 km^2 (120 km × 150 km) was instrumented with a network of automatic weather stations recording near-surface temperature, relative humidity, and rainfall. The FCA

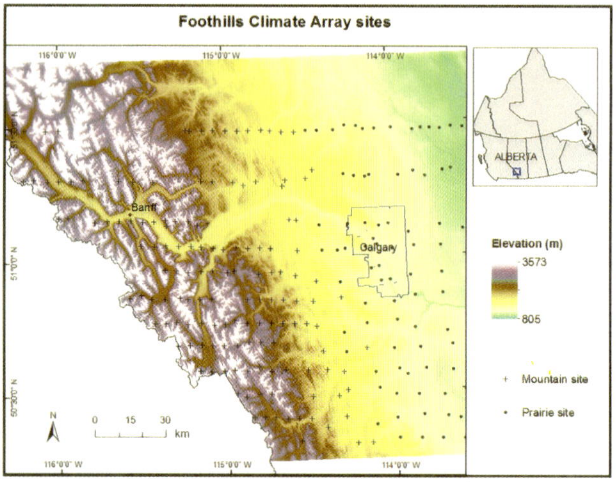

Figure 1. Foothills Climate Array study area. Crosses indicate mountain sites and dots are prairie sites. The City of Calgary municipal boundary is shown as a black outline. Sites within the boundary are classified as urban sites.

was set up in a series of east–west transects, spaced roughly 10 km apart and running from the continental divide on the western end of the study region to the flat, prairie grasslands on the eastern edge (Fig. 1). Station spacing along the east–west transects was about 5 km in the mountains and 10 km for the prairie sites. The grid was designed to be roughly the size of a global climate model (GCM) grid cell.

The main objective with the FCA was to understand and quantify spatial patterns of temperature as a function of elevation, topographic characteristics, surface type, and weather systems in this region. This includes conventional air temperature lapse rates (the change in minimum, maximum, and mean daily temperature with elevation) and their daily and geographic variability, but also other coherent patterns of temperature variance, to inform interpolation and downscaling models of temperature as a function of terrain and weather conditions. The FCA covers a large range of remote environments, with a high spatial density and 5 years of relatively complete data, spanning an elevation range of 890 to 2880 m. We are not aware of other mesonet arrays with this kind of spatial and temporal coverage in a mountain environment.

Here we present this unique multiyear temperature dataset, to support the public availability of this data and illustrate its relevance for environmental modelling and temperature downscaling. We describe the quality control and processing methods used to clean and filter the data, assess its accuracy, and present the distribution of daily temperature lapse rates in the region as a function of month and season. These lapse rate results can improve hydrological, ecological, or glaciological modelling applications in the region, where there is a need for distributed temperature estimates based on climate models or low-elevation reference stations.

2 Study region, design, and instrumentation

The FCA study area extends from the continental divide of the Rocky Mountains in the west to the relatively flat prairie farmlands about 50 km east of the city of Calgary, Alberta, centred on approximately 51° N and 114.5° W (Fig. 1). The Rocky Mountains, with altitudes up to 3500 m, straddle the border between British Columbia and Alberta in this region with a northwest-to-southeast alignment. The Bow River is a major river basin in the study area, cutting a southeast path through the mountains before turning east and exiting the mountains east of Canmore, Alberta. In the mountains, the floor of the Bow Valley has an altitude between 1300 and 1600 m, with peaks rising more than 1000 m on either side. There are also numerous alpine lakes and low-order creeks and tributaries of the Bow River that drain the steep mountain valleys.

The upper slopes of the mountains are above the treeline and consist of rock and rubble, with sparse vegetation (see Table 1). Coniferous forest and alpine meadows occur below the treeline. The foothills are comprised of lower-elevation rolling hills, with coniferous and aspen forests interspersed with grassy meadows. The prairies are mostly shrub, grassland, and cultivated cropland, with farming the dominant land use. Sites are classified as mountain or prairie based on their elevation and terrain variability surrounding the site. Prairie sites are generally situated below 1250 m with little terrain variability. A total of nine sites are situated within the Calgary municipal boundary. Elevations drop to below 900 m at the eastern edge of the survey. Thus the area and site locations comprise a wide range of elevation, topography, and surface types, including grassland, cultivated farmland, urban, forest, shrub, meadow, and bare rock.

Each FCA installation consisted of rainfall, temperature, and relative humidity data-logging sensors mounted on a pole that was either pounded into the ground or supported with a cairn. Rainfall was recorded using HOBO RG2 tipping bucket rain gauges manufactured by Onset Computer Corporation. Temperature and relative humidity (RH) were recorded at 1 h intervals, with instantaneous measurements taken at the top of the hour, using SP-2000 temperature-relative humidity data loggers manufactured by Veriteq Instruments Inc[1]. Daily mean temperatures are calculated as the arithmetic mean of hourly values. The data loggers were mounted inside radiation shields manufactured by Onset Scientific Ltd to protect the loggers from direct sunlight and allow air circulation. The manufacturer-reported accuracy for the sensors is ±0.25 °C between −25 and +70 °C, with a resolution of 0.02 °C at +25 °C. Where possible, temperature and RH were recorded at a height of 1.5 m above the ground. At sites in areas of significant snow accumulation (commonly at elevations above 2000 m), pole extensions were

[1] Veriteq has subsequently been bought out by Vaisala, who continue to make an adapted version of this data logger (DL2000).

Table 1. The number of sites having different elevation, slope, aspect, and land surface types.

Elevation (m)		Slope (°)		Aspect		Land surface	
800–999	21	flat	66	E	55	forest	66
1000–1199	56	1–5	71	N	33	grass–shrub–crops	127
1200–1399	45	5–10	38	NE	44	rock–rubble	22
1400–1599	28	10–15	22	NW	9	urban	9
1600–1799	26	15–20	21	S	19	wetland	8
1800–1999	22	20–25	6	SE	31		
2000–2199	18	25–30	6	SW	24		
2200–2399	10	> 30	2	W	17		
2400–2599	4						
> 2800	2						

added to help keep the sensors above the winter snowpack and instrument heights were between 2 and 3 m.

Between 200 and 230 stations were in operation during the main recording period from July 2005 to June 2010. Site locations sampled the varying topography and land surfaces of the area, with an attempt to establish sites on a grid that would lead to a representative sampling of the landscape. Logistical realities meant that the FCA realization was not a perfectly regular grid. The local farming and ranching communities were co-operative, but permission to establish FCA sites on private land in the eastern part of the domain was not always granted, leading to gaps and irregularities. There are also numerous irregularities and gaps in coverage for the mountain sites, as some locations on the grid were inaccessible. Nonetheless, the sites effectively capture a range of elevations, slopes, aspects, and surface types (Table 1) and are statistically representative of the terrain in the region (Cullen and Marshall, 2011).

Given the remote environment, the FCA design criteria necessitated portable data-logging sensors that could be left in place for many years. We included sites that could be reached within a day's travel on foot or on bike (or commonly a combination of the two), and most of the mountain sites involved an off-trail approach from the nearest trail or road. It required about 160 person-days per year (80 days for a team of two) to complete the annual rounds. The infrequent site visits and sometimes hostile conditions (e.g., high winds, tree-fall, avalanches, flooding, vandalism, snow burial, wildlife interference) led to some challenges with data quality and missing data. The next section discusses data-processing and quality control procedures in detail.

The FCA installation began in spring 2004 and was completed in summer 2005. Data recording continued until autumn 2010, when all of the stations were taken down except for line 4, an east–west transect through the city of Calgary. Prairie sites were more accessible and were visited twice annually for site maintenance and data downloads, during the spring and autumn field seasons. Mountain sites were only visited once per year. Site maintenance included data collection, battery replacement, exchange of data-loggers as needed, instrument cleaning, and basic maintenance as required (e.g., reinstalled fallen or leaning instruments). The rain gauges, radiation shields, and sensors become dirty over time through exposure to dust, pollen, insects, and forest detritus (e.g., pine needles). Field notes and photographs were taken to document the physical location and condition of the sites during each visit.

Sites do not conform to World Meteorological Organization (WMO) standards, which specify that climate recording sites should be level, away from vegetation and buildings, and not in areas of variable topography (WMO, 2008). This prescription is not consistent with the purpose of the study, which is to examine topographic and surface environmental influences on weather. In fact, it was part of the project design to sample different slopes, aspects, and degrees of forest closure, to quantify deviances from flat, open control settings. This contributes to the complex patterns of spatial temperature variability that we recorded, but in a manner that is realistic with respect to landscape-scale temperature modelling. Examples of site locations are shown in Fig. 2.

3 Sensor accuracy and quality control

3.1 Instrument calibration

Instruments were tested at the University of Calgary weather research station (WRS) before being set up in the field, and on an ongoing basis during the study to assess systematic bias or drift of the sensors. These tests were to determine whether it was necessary to apply bias or drift corrections to any sensors. Calibration tests consisted of sensors set up at the WRS and recording instantaneous temperature at 1, 2, or 5 min intervals for 1- to 2-week periods. Data were aggregated to hourly intervals and compared with aggregated hourly temperature measurements at the WRS reference station, which uses a Campbell Scientific HMP35CF sensor mounted in a ventilated, Gill Model 41004-5 12 plate radiation shield and records data at 1 min intervals. We used more frequent sampling which was then aggregated, in order to reduce the im-

Figure 2. Site location examples: prairie, forest, urban, and mountain.

Table 2. Test result statistics by year showing the total number of tests performed, the average difference between the hourly logger and WRS temperatures for all hours 00:00 to 24:00 MST, between 10:00 and 16:00 MST, and 00:00 and 06:00 MST.

		Average temperature difference: logger–WRS (°C)		
Year	No. of tests	00:00–24:00	10:00–16:00	00:00–06:00
2004	216	−0.02	0.13	−0.09
2005	67	−0.01	0.10	−0.08
2006	6	−0.13	−0.04	−0.20
2007	38	−0.05	0.16	−0.16
2008	48	−0.20	−0.25	−0.16
2009	73	−0.22	−0.26	−0.20
2010	140	−0.21	−0.23	−0.19
2011	83	−0.23	−0.36	−0.20

pact of once-off erratic readings which may have an undue influence during the short testing period.

Average hourly differences were calculated for all sensors for all hours in a day (24 h), for hours between 00:00 and 06:00 MST (night), and between 10:00 and 16:00 MST (day). During the test period, if no absolute value of the hourly temperature difference between logger and WRS exceeded 3 °C, a test was considered "good" and no further investigation of sensor performance was required. This was the result for 94 % of calibration runs. Average differences by year are shown in Table 2. The average difference between the Veriteq loggers and WRS is −0.1 °C.

To more closely represent the actual data recording methodology, a similar analysis was performed whereby instantaneous measurements were extracted for each sensor and compared with corresponding WRS values. From the instantaneous measurements, daily minimum, maximum, and mean temperatures were extracted for each sensor and compared with WRS temperatures. The average differences between the Veriteq loggers and WRS are −0.26, −0.17, and −0.16 °C for daily maximum, mean, and minimum temperatures, respectively. The average daily mean difference using instantaneous measurements was −0.2 °C, compared with −0.1 °C using aggregated hourly measurements.

Calibration tests after 2007 show more negative offsets relative to the earlier WRS values, possibly due to more tests being conducted during winter, where there may be less daytime heating effect or an overnight cooling effect in the naturally ventilated sensors. However, there was no apparent significant drift for individual sensors, nor did any sensors show consistent bias. Therefore, no corrections were applied.

Daytime heating can be an issue in naturally ventilated radiation shields, with a potential warm bias under calm, sunny conditions (Nakamura and Mahrt, 2005). Temperature sensors and shields have different associated errors, depending on the design of the shield, the type of sensor, and environmental conditions. The design of the shield should ensure that the air within the shield is at the same temperature as the surrounding air (WMO, 2008). Shields may rely on natural ventilation from prevailing winds or may be artificially ventilated using a fan.

Studies have investigated the magnitude of temperature differences for naturally vs. artificially ventilated sensors under different wind and solar radiation conditions (e.g., Georges and Kaser, 2002; Huwald et al., 2009). Daytime temperature differences are the greatest, due to solar heating, and can be problematic at wind speeds less than 1 m s^{-1}. This heating effect may be from direct radiative heating of the sensor or indirect heating of the shield, which then heats the air inside the shelter. We performed an experiment at the University of Calgary WRS from October 2012 to September 2013 to quantify the effect of using naturally ventilated sensors, as used in the FCA study, compared to a mechanically ventilated sensor used at the WRS. The Veriteq loggers were set to record instantaneous temperature at 5 min intervals, and hourly averages were calculated.

Table 3. The average hourly temperature difference (°C) between Veriteq loggers and the WRS temperature sensor at different wind speeds and incoming shortwave radiation values.

	Incoming shortwave radiation ($W\,m^{-2}$)		
Wind speed ($m\,s^{-1}$)	< 300	300–600	> 600
< 1	0.3	0.4	0.7
1–2	0.3	0.4	0.5
2–3	0.2	0.3	0.3
3–5	0.1	0.0	0.1
> 5	−0.2	0.0	0.0

For this set of loggers, the average difference between hourly average logger and WRS temperatures was 0.1 °C for the test period. Differences averaged for the whole year for the combined wind speed and incoming shortwave radiation categories are shown in Table 3. The maximum difference of 0.7 °C occurs at high solar radiation (greater than $600\,W\,m^{-2}$) and winds speeds less than $1\,m\,s^{-1}$. For high wind speeds and low solar radiation, the average difference is −0.2 °C. Results in Table 3 are systematic and point to a relatively simple correction if hourly shortwave radiation and wind data are available. These are not available at the FCA sites, so this is a source of error that we must tolerate. However, the mean error in hourly temperature associated with the worst-case conditions is +0.7 °C, and the daily average errors associated with naturally ventilated sensors are much less (solar radiation is less than $600\,W\,m^{-2}$ for most of the day, and throughout the winter). This source of error is likely insignificant for daily mean and minimum (overnight) temperatures, but it might affect maximum temperature measurements in sheltered locations. Sheltered sites may experience additional warming relative to exposed locations due to reduced ventilation, particularly in the summer months, but differences are expected to seldom exceed 1 °C.

In the 6 % of calibration tests with suspect logger performance, the logger was flagged for further examination and field records for these loggers were manually inspected. Calibration experiments exhibited a common form of sensor failure associated with periods of heavy rainfall and sustained high humidity. Under these circumstances, some sensors experienced errant, short-term behaviour, typically recording temperatures that were too high for a period of hours to days, then returning to normal. Careful examination confirms that values before and after the observed aberrations are reliable, so we retain these sensors, but quality control measures described below are designed in part to identify this behaviour. The sensors are adequately sheltered, such that rainfall itself (liquid water) is not likely the problem; rather, we suspect that water vapour diffuses into the data loggers and internal condensation can compromise the circuit board, until such time as humidity drops and the sensor dries out.

3.2 Quality control

Assessment of data quality prior to data analysis is an essential step to ensure that only valid data are used. While sensors generally functioned well, periodically they malfunctioned or were damaged in the field, resulting in unreliable data. One challenge in identifying questionable data is to differentiate between extreme events and actual compromised data. Wade (1987) identified four general sources of measurement error, namely: instrument failure, drift, bias, and random error. Calibration tests identified that loggers do not show significant bias or drift, so these were not corrected for. Instrument failure is readily identifiable as missing data or unrealistic values. Random errors are more difficult to identify. We introduced a multi-step quality control/quality assurance (QC/QA) procedure for objective detection of errors.

Durre et al. (2010) detail comprehensive automated quality assurance procedures for daily meteorological measurements, as are being applied operationally in the Global Historical Climatology Network (GHCN). Quality control procedures are designed to identify as many errors as possible with few false positives (that is, valid data flagged as unreliable data). Tests used in the procedure include the following: (i) physically reasonable bounds; (ii) internal consistency – the daily value should be within statistical bounds for that day in the year; (iii) external consistency – the value should lie within reasonable limits of surrounding stations; (iv) multiple duplicate or repetitive values; (v) unusually large changes in daily minimum and maximum values. We use similar tests in this study.

Quality control procedures for the FCA data include a sequence of automated (A) and manual (M) data checks applied to an entire file, hourly, daily, or monthly data, and run in the order shown below.

- Field checks (M – entire file)
- Time shifts (M – hourly)
- Spikes (A – hourly)
- Snow burial (A/M – monthly)
- Neighbourhood consistency (A/M – daily)
- Review questionable loggers based on field notes and calibration tests (M – daily/hourly)
- Final review (M – daily/hourly)

3.2.1 Field checks

During the download process, each logger's data was compared with one or more near neighbours. Downloads were characterized as "good", "some bad", or "bad" using this comparison. Where a download was characterized as bad, the entire file was excluded. Files characterized as "some bad" were included and sections readily identified as erroneous based on a visual review were deleted.

3.2.2 Time shifts

The loggers adopt the start and end time from the computer doing the setup and download. The Spectrum v3.7c software uses the computer download time to assign a time to each measurement. Multiple machines were used for field downloads and at times the clocks were set to the wrong time zone, or alternated between daylight savings time (DST) and mountain standard time (MST). Times were commonly out by several minutes. In addition, on occasion some loggers malfunctioned and missed recording data for hours or days at a time, as seen in comparisons with neighbours. This was noted in the field notes.

The data loggers are able to store approximately 18 months of data recorded at 1 h intervals. If the time between site visits was too long, or a logger was inadvertently set to record at a shorter time interval, the memory filled up and no more data were recorded. By comparing multiple neighbours, and reviewing field notes which included the actual download time and whether time was DST or MST, files with time shifts and missing data were identified. Once the data were loaded into the database, the necessary time shifts were applied. Where possible, periods of missing data were identified and measurement times adjusted to align data with neighbours.

3.2.3 Spikes

A directional step test used by Hall et al. (2008) identifies consecutive measurements exceeding a user-defined limit. Limits vary by location (climate region), measurement interval, and direction (rise or fall). Temperature changes of up to 9 °C in 5 min have been observed in the Oklahoma mesonet. Due to such events, Graybeal et al. (2004) found step tests were capturing actual frontal events, whereas data problems were predominantly 1 h spikes rather than step changes in the data. In southern Alberta, chinook winds and cold fronts cause both large rises and falls in hourly temperature measures, but these conditions usually persist after the step change. Temperature swings of greater than 30 °C in a day have been observed at the onset of chinook winds (Nkemdirim, 1988).

A review of FCA data identified by step tests indicates spikes, either up or down, lasting 1 h or more can be real errors. A brief spike may not influence daily mean values enough to be identified as unusual relative to neighbours (cf. neighbourhood consistency check described later), but spikes lasting 4 h or more will. Therefore, a spike test to identify spikes lasting from 1–3 h was applied to FCA data. A spike was defined as a value exceeding subsequent or previous measures up to 3 h apart by 5 °C, either up or down.

3.2.4 Snow burial

Some of the high mountain sites were prone to burial by snow during late winter. Snow burial is apparent through a small diurnal temperature range. For the snow burial test, we flag entire months where at least 25 days have a diurnal range of less than 3 °C. Days on either side of the flagged months are examined for snow burial signal and manually flagged. Commonly, snow burial was identified during field checks, and blocks of data were deleted based on a visual comparison with neighbouring sites.

3.2.5 Neighbourhood consistency

Neighbourhood consistency checks are used to identify unusual values at a site relative to neighbouring stations. The method, threshold values, and number of stations used depend on station density, topography of the area, and the weather variable being checked. In all cases, estimated values calculated from neighbouring sites are compared to observed values and large deviations are flagged as potential errors.

Shafer et al. (2000) use a weighted average of neighbouring stations to calculate a value for the station being checked, with differences exceeding 3 standard deviations flagged as suspicious. In mountain regions, Lanzante (1996) suggests using vertical neighbours (closest station with a similar elevation) as an alternative to horizontal proximity.

As a means of identifying the most appropriate neighbours, we specified two spatial neighbourhoods for the FCA data. A horizontal-proximity neighbourhood was defined using all sites within a 25 km radius of a site (50 km for sites at the edge of the survey), and a vertical neighbourhood was defined using sites in 200 m elevation bands, e.g., 1400 to 1600 m, 1600 to 1800 m. All sites above 2200 m are grouped together; therefore, this group includes the two high-elevation sites above 2800 m, with the remainder of the sites being less than 2500 m. In all cases, groups consist of at least 10 sites.

The spatial proximity test accounts for local variability and the elevation band accounts for elevation consistency. The test examines daily minimum, maximum, and mean values for each site compared to the average and standard deviations calculated from all sites within elevation and horizontal-proximity neighbourhoods. In calculating the neighbourhood average and standard deviation for each day, the site(s) with the lowest and highest values within the group are excluded, as are all site-days flagged as errors in previous quality control tests.

As an initial screening procedure, site-days are flagged as suspect if their daily minimum, maximum, or mean value differs from both horizontal-proximity and elevation-band neighbourhood means by more than 5 standard deviations. All suspect site-days are then manually reviewed and flagged as either natural variability or unreliable data. For sites identified as unreliable data, manual review generally indicated erratic sensor performance, and the threshold for questionable data was reduced to 3 standard deviations. All days for these sites are flagged as unreliable where any of the daily minimum, maximum, or mean exceeds the group mean values by 3 or more standard deviations.

3.2.6 Manual review of field notes

Field notes indicated any obvious issues with the site itself or unusual data. For instance, stations were sometimes found to have fallen or been knocked over. In this situation, station data were examined in relation to elevation and proximity neighbours as in the neighbour consistency check, to identify the time at which the station fell over. Sensors at ground level record too-high maxima and too-low minima during summer, and would typically become snow-buried during the winter. All data from the time of station compromise were removed.

3.2.7 Final review

Apart from entire files or obvious bad data blocks excluded from the compiled file, data are flagged as bad rather than being deleted. A different flag is assigned for each quality control test failure. The final review looks at groups of data with flags turned on or off to verify that tests have correctly identified errant data, and no suspect data remains. Any further bad data identified in this final review, most commonly seen in days on either side of failed test days or very few days remaining in a month, are flagged.

As the topography of the FCA and the climate of southwestern Alberta show high variability, quality control checks were applied leniently in order to retain interesting data; therefore, some questionable data may be retained as well. On average, 91 % of data were good and 9 % of data were flagged as unreliable and excluded from the analysis. Missing data are distributed randomly in the study area, with missing or suspect data affecting 70 % of the sites.

4 Gap-filling of missing data and monthly lapse rates

While overall data coverage is excellent, given the remote nature of the FCA and the infrequent site visits, missing data do compromise the utility of the dataset, e.g., in determination of monthly or annual means. Stooksbury et al. (1999) show that 3-day data gaps can result in errors of $\pm 1\,^{\circ}$C in the calculation of monthly means, with larger errors during winter and in continental interior locations. Missing or erroneous data can also cause poor performance such as spatial or temporal discontinuities in interpolation or modelling of temperature surfaces. We therefore introduce gap-filling measures for the 9 % of the data that is missing, to make for straightforward and reliable application of this dataset.

Approaches to gap-filling depend on the environment and the neighbourhood of stations that are available for modelling of missing data. Nkemdirim (1996) created monthly regression equations using closely correlated stations to recreate daily minimum and maximum temperatures in southern Alberta. Eischeid and Pasteris (2000) used between one and four most closely correlated neighbouring stations to estimate daily minimum and maximum temperatures for the western United States, using a version of the general linear least squares regression estimation with least absolute deviations criteria. Both studies calculated correlations on a monthly basis. However, Courault and Monestiez (1999) note that station correlations vary with wind direction and topographic location, indicating that the most correlated station may vary during any month.

We tested gap-filling methods using the most closely correlated stations, calculated by month and by weather type for each of daily mean, minimum, and maximum temperatures. The latter requires a classification system for regional weather systems, which we describe briefly below. Days are grouped by month and by weather type. For each group, correlation coefficients are calculated between all possible station pairs for each of daily mean, minimum, and maximum temperatures. For each station, temperature measure, and group (month or weather type), neighbour stations are ordered by correlation with the target station from high to low and the most highly correlated neighbour station for each target station and group is identified. Station temperatures are highly correlated, but the coefficients and most highly correlated station do vary by month and weather type. A regression–prediction equation is generated for each station–neighbour pair for each group. Missing daily temperature measures are calculated using the prediction equation for the group to which the day belongs (month and weather type), and, for each station, the most highly correlated neighbour station with available data is used to calculate the missing temperature. The next two sections briefly describe the weather classification system and the accuracy of the gap-filling methods. Readers are referred to Wood (2017) for further details.

4.1 Weather system classification

We classify daily weather type based on a discriminant function analysis (DFA), seeded with the surface characteristics of the primary weather patterns that occur in our study region (Table 4). Daily weather conditions are characterized from the long-term data available at the Meteorological Service of Canada station at Calgary International Airport (Environment Canada, 2015). We explored a wide array of daily weather variables, including temperature, pressure, humidity, wind speed, and 500 hPa geopotential height, along with measures of the 24 h trends and variance in these variables (Wood, 2017). Mean daily conditions for the period 1970–2010 are calculated using an 11-day moving window. Daily weather anomalies are then calculated from these mean values in order to remove the influence of seasonal cycles and provide a year-round classification system (Wood, 2017).

The control groups for the DFA weather types are based on manually classified daily weather conditions from October 2013 to September 2014. DFA provides a multivariate method to characterize a given weather type in our region, based on daily weather conditions. There is an im-

Table 4. Selected qualitative surface meteorological characteristics associated with different weather types in southern Alberta.

Weather type	500 hPa pattern	Surface temperature anomaly	Diurnal temperature range	Pressure anomaly	Wind direction	Relative humidity	Specific humidity
Cold dry (CD)	strong trough	strong negative	low–average	positive	N/NE		low
Chinook (Ch)	ridge	strong positive			W/SW strong	low	
Cold-front transition (Tr)	variable		high		variable		
Cool wet (CW)	large cyclone in southern Alberta	negative	low	negative	E/SE	high	high
Warm, high-pressure (Ht)	strong ridge	positive	high	positive	S/SW light		high

plicit assumption that a given weather type, e.g., a chinook or an anticyclonic ridge over the region, will have similar weather anomalies throughout the year, and from one year to the next. The DFA classification can then be applied across time. The period used for creating the discriminant functions does not overlap with the FCA data collection period, but these periods are only a few years apart, so mean climate conditions are not expected to have changed much and the discriminant functions should still be appropriate. The DFA method also requires that the dominant weather-system types are all represented in the calibration period (2013–2014), as per our seed groups. Exotic weather types during the FCA period will not be captured through this approach, but our interest here is to characterize the most common regional weather systems. The identified weather classes include the following: (i) cold, dry (i.e., continental polar) air masses (CD); (ii) chinook conditions (Ch); (iii) cool, wet (cyclonic) weather systems (CW); (iv) warm, high-pressure conditions (Ht); (v) cold-front transition days (Tr); and (vi) normal conditions (Nl), which are defined as days with surface weather characteristics within 1 standard deviation of their long-term mean value. Selected characteristics of these weather types are listed in Table 4. Wood (2017) provides additional details.

The accuracy for the final DFA model is 81 % using jackknife cross-validation, where each day is left out in turn; prediction functions are calculated with the remaining data, and these functions are used to predict the omitted day. The CD and Nl types are consistently well-predicted, with accuracies between 80 and 90 %, while the remaining types vary from 50 to 90 % accuracy, with transition days proving to be the most difficult to capture.

The spatial characteristics of weather systems were found to vary seasonally, so weather types were further divided into seasonal subgroups. Three seasons were defined as follows: winter (November to February), summer (May to August), and autumn–spring (September, October, March, April). As the normal weather type (Nl) comprises more than 50 % of days, and correlations and topographic influences show an annual cycle, correlations were also calculated for Nl days grouped by month.

4.2 Accuracy of the gap-filling models

Errors associated with the gap-filling models are shown in Table 5. Mean absolute errors by month and weather type are shown for minimum (T_{min}), maximum (T_{max}), and mean daily (T_{mean}) temperatures. Values are further divided into mountain and prairie subsets of the data. Weather-type estimates are better than those based on monthly correlations, with T_{mean} having the lowest errors, followed by T_{max} and T_{min}. Errors for all temperature measures are larger in the cold months, November to February. Although there is still a lot of inherent variability within weather types, seasonal-weather-type correlations improve estimates by $\sim 7\,\%$ over monthly correlations. Error reductions based on weather type are most significant for cold-front transition, cool–wet, and chinook days.

Missing temperature data in the FCA are gap-filled using regression equations generated using the most closely correlated station for each site, where correlations are calculated by seasonal weather type. The accuracy of the gap-filling equations was assessed using jackknife cross-validation, where daily temperature measures for each site and day with data are estimated from temperatures from the most closely correlated station using regression equations. Mean absolute errors based on seasonal-weather-type correlations range from 0.40 °C (T_{mean} in the prairies) to 0.84 °C (T_{min} in the prairies). Of site-day errors for all methods, 90, 95, and 98 % are less than 2 °C for T_{min}, T_{max}, and T_{mean}, respectively. Figure 3 plots mean site error estimates vs. elevation. Errors in T_{max} and T_{mean} increase with elevation and are lowest for the prairie sites. Errors for T_{min} show a weaker negative relationship with elevation.

4.3 Monthly lapse rates in the foothills region

Temperature lapse rates are commonly required in ecological and hydrological models and for downscaling of scenarios from climate models. The FCA data provide a 5-year record of daily temperature lapse rates, calculated from regressions of T_{min}, T_{max}, and T_{mean} vs. elevation for the FCA mountain data. Mean values and standard deviations are summarized in Table 6. Monthly values are based on the mean of individ-

Table 5. Estimated gap-filling errors shown as mean absolute temperature errors (°C) calculated as the difference between actual temperature and temperature estimated using the most highly correlated station by month (mnth) and weather type (wt), shown for each **(a)** month and **(b)** weather type for prairie and mountain sites.

Mean absolute error (°C) for gap-filling using correlated stations calculated by **(a)** month and **(b)** weather type.

(a) Month		Mountain		Prairie		**(b)** Weather type		Mountain		Prairie	
	T_{var}	mnth	wt	mnth	wt		T_{var}	mnth	wt	mnth	wt
Jan	T_{max}	0.87	0.80	0.79	0.73	CD	T_{max}	0.88	0.86	0.62	0.61
	T_{mean}	0.69	0.63	0.59	0.55		T_{mean}	0.56	0.53	0.41	0.39
	T_{min}	1.06	0.98	1.11	1.04		T_{min}	0.86	0.79	0.85	0.79
Feb	T_{max}	0.83	0.80	0.76	0.72	Ch	T_{max}	0.75	0.65	0.61	0.54
	T_{mean}	0.63	0.60	0.53	0.50		T_{mean}	0.59	0.52	0.51	0.46
	T_{min}	1.05	0.97	1.05	0.99		T_{min}	1.01	0.94	1.06	0.99
Mar	T_{max}	0.76	0.71	0.65	0.61	CW	T_{max}	0.78	0.70	0.58	0.53
	T_{mean}	0.50	0.47	0.46	0.43		T_{mean}	0.43	0.36	0.36	0.30
	T_{min}	0.93	0.85	0.93	0.85		T_{min}	0.69	0.55	0.65	0.52
Apr	T_{max}	0.75	0.71	0.60	0.53	Ht	T_{max}	0.72	0.70	0.55	0.52
	T_{mean}	0.45	0.43	0.40	0.36		T_{mean}	0.55	0.53	0.43	0.41
	T_{min}	0.79	0.75	0.85	0.80		T_{min}	0.91	0.87	0.94	0.92
May	T_{max}	0.74	0.67	0.50	0.46	Nl	T_{max}	0.76	0.71	0.58	0.55
	T_{mean}	0.42	0.38	0.34	0.29		T_{mean}	0.52	0.49	0.42	0.40
	T_{min}	0.72	0.64	0.80	0.73		T_{min}	0.86	0.81	0.89	0.85
Jun	T_{max}	0.73	0.69	0.50	0.47	Tr	T_{max}	0.86	0.74	0.71	0.63
	T_{mean}	0.42	0.40	0.31	0.29		T_{mean}	0.53	.47	0.42	0.36
	T_{min}	0.72	0.67	0.72	0.67		T_{min}	0.87	0.73	0.85	0.71
Jul	T_{max}	0.68	0.66	0.49	0.47						
	T_{mean}	0.41	0.40	0.29	0.29						
	T_{min}	0.75	0.71	0.70	0.68						
Aug	T_{max}	0.70	0.65	0.51	0.48						
	T_{mean}	0.43	0.40	0.34	0.32						
	T_{min}	0.75	0.68	0.76	0.71						
Sep	T_{max}	0.71	0.67	0.46	0.45						
	T_{mean}	0.50	0.47	0.36	0.35						
	T_{min}	0.77	0.74	0.81	0.78						
Oct	T_{max}	0.78	0.73	0.51	0.49						
	T_{mean}	0.54	0.50	0.42	0.41						
	T_{min}	0.83	0.78	0.86	0.83						
Nov	T_{max}	0.79	0.75	0.59	0.58						
	T_{mean}	0.61	0.57	0.50	0.46						
	T_{min}	0.99	0.93	1.01	0.94						
Dec	T_{max}	0.94	0.88	0.72	0.69						
	T_{mean}	0.74	0.68	0.55	0.52						
	T_{min}	1.07	1.00	1.08	1.03						

Table 6. Mean and standard deviation (SD) of temperature lapse rates (°C km^{-1}) and inversion frequency (%) for daily minimum, maximum, and mean temperature, by month and season.

	T_{max}			T_{mean}			T_{min}		
Period	lapse rate	SD	inv (%)	lapse rate	SD	inv (%)	lapse rate	SD	inv (%)
January	−6.6	4.0	7.7	−3.7	4.1	19.4	−0.9	4.5	34.8
February	−6.2	4.7	9.2	−3.0	4.0	19.1	−0.1	4.2	41.1
March	−8.4	3.2	1.9	−6.0	2.6	3.2	−3.3	2.8	11.0
April	−8.2	3.4	2.7	−6.2	2.2	1.3	−3.8	2.4	7.3
May	−8.9	2.5	0.0	−6.5	1.4	0.0	−3.0	2.1	9.0
June	−8.3	2.1	0.0	−6.1	1.5	0.0	−2.7	2.2	14.0
July	−7.3	2.4	0.0	−4.9	1.6	1.0	−1.7	2.3	25.2
August	−6.8	2.6	1.9	−4.2	1.7	0.6	−1.1	2.6	34.8
September	−7.9	2.3	0.0	−4.2	1.8	1.3	−0.9	2.6	35.3
October	−8.2	3.0	1.9	−5.1	2.0	2.6	−2.3	2.5	18.1
November	−7.7	3.4	4.7	−5.1	3.1	8.0	−2.6	3.0	20.0
December	−5.5	4.2	11.0	−3.1	4.3	25.2	−0.9	4.1	38.1
DJF	−6.1	4.3	9.3	−3.3	4.1	21.3	−0.6	4.3	37.9
MAM	−8.5	3.1	1.5	−6.2	2.1	1.5	−3.4	2.4	9.1
JJA	−7.4	2.4	0.7	−5.0	1.8	0.4	−1.8	2.5	24.8
SON	−7.9	2.9	2.2	−4.8	2.4	4.0	−2.0	2.8	24.2
Annual	−7.5	3.4	3.4	−4.8	2.9	6.7	−2.0	3.2	24.0

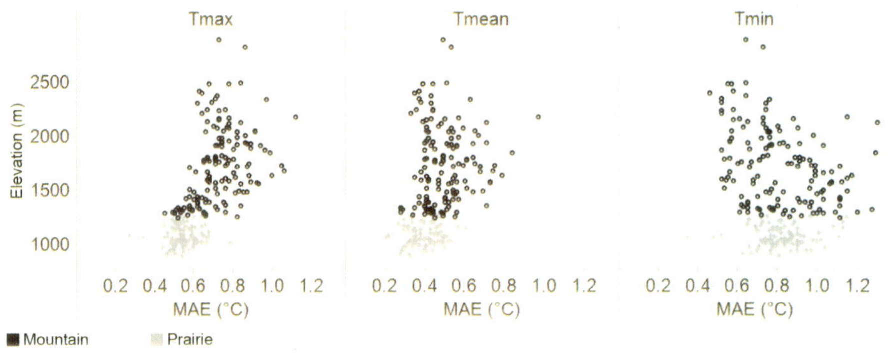

Figure 3. Estimated gap-filling errors shown as mean absolute error (MAE) for each site as a function of elevation. Errors are calculated as the differences between the actual temperature and the temperature estimated from the most closely correlated station.

ual days; over 5 years, this gives $N \approx 150$ for each month in Table 6. Inversion frequency is also reported for each period and temperature measure, calculated from the percentage of days with positive lapse rates. The strength of the linear regression (R^2 value) for monthly lapse rates gives an indication of how well a linear lapse rate represents the data, and this varies systematically by season and by temperature measure. Mean values vary from 0.5 to 0.8 for T_{max}, 0.4 to 0.8 for T_{mean}, and 0.15 to 0.4 for T_{min} and are weakest during the winter months. R^2 is also weak during summer months for T_{min}.

There are significant differences in lapse rate between minimum, mean, and maximum temperatures and the monthly inversion frequencies for each. Maximum temperature has the steepest lapse rates in all months and is the least prone to inversions with altitude. Monthly values range from −5.5 to −8.9 °C km^{-1} and are steepest in the spring and autumn. On average, maximum temperature inversions occur only 1.5 % of the time outside of the winter months (i.e., from March through November), and 9.3 % of the time from December through February (DJF). Winter inversions are common in this region in association with cold, continental polar air masses (Cullen and Marshall, 2011), and these air mass inversions are sometimes strong enough to persist through the day, affecting the maximum temperatures. Under these conditions, comparatively warm, westerly air commonly overrides the shallow, cold air mass that is in place, supporting the inversion structure.

Mean and minimum temperatures show similar seasonal structure to maximum temperatures, with the steepest and shallowest values in spring and winter, respectively. Inversion frequency and lapse rate variability are much greater in the winter season, and are lowest in the summer. Monthly lapse rates for mean temperature vary from −3.0 to −6.5 °C km^{-1}, with an annual mean of −4.8 °C km^{-1}. Regional-scale inversions occur on 21 % of days in the winter season, but are relatively rare in the spring, summer, and autumn (2 % of days). Minimum temperature lapse rates are shallower, varying from −0.1 to −3.8 °C km^{-1}, with an annual mean of −2.0 °C km^{-1}. This is a large departure from free-air lapse rates, and is driven by the overnight development of cold-air drainage and pooling at lower elevations, as well as the prevalence of cold, shallow air masses in the study region, particularly in the winter season. Minimum temperature inversions are observed on 38 % of days in the winter months and 24 % of days annually and are common in all seasons.

5 Data availability

Data described are available on the PANGAEA repository https://doi.org/10.1594/PANGAEA.880611 (Wood et al., 2017). Minimum, maximum, and mean daily temperature data are available as well as site attributes – latitude, longitude, elevation, land surface type, site type, sensor height above ground, and whether the temperature values are measured or estimated.

6 Summary

Data gathered within the Foothills Climate Array offer a unique collection of high-density, multiyear observations in complicated mountain terrain. These data provide a detailed three-dimensional view of the near-surface temperature structure and its temporal variability as a function of season and weather type.

The Veriteq temperature instruments used in this study performed exceptionally well, with high accuracy and no drift over time. Multistage quality control procedures were successful in identifying and removing questionable data. In total, 9 % of data collected at the ~230 sites between 2005 and 2010 are missing or unreliable. Missing and unreliable data are distributed randomly in the study area. While the dense station network provides some redundancy and the percentage of missing data is not high, gap-filling to create a complete dataset has benefits for applications requiring monthly means or for creating interpolated temperature surfaces. We therefore gap-filled the data that is presented here, with a flag to denote that these data have been estimated.

Daily temperature lapse rates in the southwestern Alberta foothills show a strong seasonal cycle, with shallower values and greater variability in the winter months. Lapse rates of maximum temperature are steeper and are similar to free-air lapse rate values (e.g., −7.5 °C km^{-1}) that are commonly used in temperature downscaling or extrapolation in mountainous terrain, but lapse rates of mean and minimum temperature are much shallower, with mean annual values of −4.8 and −2.0 °C km^{-1}, respectively. Minimum temperature inversions are frequent year-round, and are present on 38 % of days in the winter months.

The temperature and lapse rate structure reported here are specific to our region, but the general behaviour of temperature variations over the terrain and the observation of systematic daily and seasonal lapse rate variability are expected to be relevant to all mountain regions. In mid-latitude regions like our study area, there is additional complexity due to air mass and frontal interactions, and their interaction with the terrain. These effects will be examined in more detail elsewhere, through consideration of daily weather types and their influence on regional temperature patterns and lapse rates.

Competing interests. The authors declare that they have no conflict of interest.

Acknowledgements. The Foothills Climate Array was funded by the Canada Foundation for Innovation, the Natural Sciences and Engineering Research Council (NSERC) of Canada, and the Canada Research Chairs program. We thank Rick Smith at the University of Calgary weather research station for his invaluable help over the lifetime of the FCA study. Graduate and undergraduate students, summer research assistants, friends, and volunteers that contributed to the FCA study are too numerous to list, but prominent within this group are Kelly Racz, Kara Przeczek, and Bridget Linder. We could not have managed this study without their dedication and enthusiasm. We also acknowledge the input from three anonymous reviewers who provided thorough and constructive suggestions.

Edited by: Chris DeBeer

References

Clarke, G. K. C., Jarosch, A. H., Anslow, F. S., Radic, V., and Menounos, B.: Projected deglaciation of western Canada in the twenty-first century, Nat. Geosci., 8, 372–377, https://doi.org/10.1038/ngeo2407, 2015.

Comte, L., Murienne, J., and Grenouillet, G.: Species traits and phylogenetic conservatism of climate-induced range shifts in stream fishes, Nat, Commun., 5, 5023, https://doi.org/10.1038/ncomms6053, 2014.

Courault, D. and Monestiez, P.: Spatial interpolation of air temperature according to atmospheric circulation patterns in southeast France, Int. J. Climatol., 19, 365–378, https://doi.org/10.1002/(SICI)1097-0088(19990330)19:4<365::AID-JOC369>3.0.CO;2-E, 1999.

Cullen, R. M. and Marshall, S. J.: Mesoscale temperature patterns in the Rocky Mountains and foothills region of southern Alberta, Atmos.-Ocean, 49, 189–205, https://doi.org/10.1080/07055900.2011.592130, 2011.

Dee, D. P., Uppala, S. M., Simmons, A. J., Berrisford, P., Poli, P., Kobayashi, S., Andrae, U., Balmaseda, M. A., Balsamo, G., Bauer, P., and Bechtold, P.: The ERA-Interim reanalysis: configuration and performance of the data assimilation system, Q. J. Roy. Meteorol. Soci., 137, 553–597, https://doi.org/10.1002/qj.828, 2011.

Deutsch, C. A., Tewksbury, J. J., Huey, R. B., Sheldon, K. S., Ghalambor, C. K., Haak, D. C., and Martin, P. R.: Impacts of climate warming on terrestrial ectotherms across latitude, P. Natl. Acad. Sci. USA, 105, 6668–6672, https://doi.org/10.1073/pnas.0709472105, 2008.

Durre, I., Menne, M. J., Gleason, B. E., Houston, T. G., and Vose, R. S.: Comprehensive automated quality assurance of daily surface observations, J. Appl. Meteorol. Climatol., 49, 1615–1633, https://doi.org/10.1175/2010JAMC2375.1, 2010.

Eischeid, J. K. and Pasteris, P. A.: Creating a Serially Complete, National Daily Time Series of Temperature and Precipitation for the western United States, J. Appl. Meteorol., 39, 1580–1591, doi.org/10.1175/1520-0450(2000)039<1580:CASCND>2.0.CO;2, 2000.

Environment Canada: Historical climate data for Calgary International Airport, available at: http://climate.weather.gc.ca/climate_data/daily_data_e.html?StationID=50430, last access: 2 November 2015.

Förster, K., Meon, G., Marke, T., and Strasser, U.: Effect of meteorological forcing and snow model complexity on hydrological simulations in the Sieber catchment (Harz Mountains, Germany), Hydrol. Earth Syst. Sci., 18, 4703–4720, https://doi.org/10.5194/hess-18-4703-2014, 2014.

Franklin, J., Serra-Diaz, J. M., Syphard, A. D., and Regan, H. M.: Global change and terrestrial plant community dynamics, P. Natl. Acad. Sci. USA, 113, 3725–3734, https://doi.org/10.1073/pnas.1519911113, 2016.

Georges, C. and Kaser, G.: Ventilated and unventilated air temperature measurements for glacier-climate studies on a tropical high mountain site, J. Geophys. Res., 107, 4775, https://doi.org/10.1029/2002JD002503, 2002.

Graybeal, D. Y., DeGaetano, A. T., and Eggleston, K. L.: Improved quality assurance for historical hourly temperature and humidity: Development and application to environmental analysis, J. Appl. Meteorol., 43, 1722–1735, https://doi.org/10.1175/JAM2162.1, 2004.

Hall Jr., P. K., Morgan, C. R., Gartside, A. D., Bain, N. E., Jabrzemski, R., and Fiebrich, C. A.: Use of climate data to further enhance quality assurance of Oklahoma Mesonet observations, 20th Conf. on Climate Variability and Change, 20–24 January, 2008, New Orleans, USA, 2008.

Hock, R.: Glacier melt: a review of processes and their modelling, Prog. Phys. Geogr., 29, 362–391, https://doi.org/10.1191/0309133305pp453ra, 2005.

Huwald, H., Higgins, C. W., Boldi, M., Bou-Zeid, E., Lehning, M., and Parlange, M. B.: Albedo effect on radiative errors in air temperature measurements, Water Resour. Res., 45, W08431, https://doi.org/10.1029/2008WR007600, 2009.

Jochner, S., Sparks, T. H., Laube, J., and Menzel, A.: Can we detect a nonlinear response to temperature in European plant phenology?, Int. J. Biometeorol., 60, 1551–1561, https://doi.org/10.1007/s00484-016-1146-7, 2016.

Lanzante, J.: Resistant, robust and non-parametric techniques for the analysis of climate data: Theory and examples, including applications to historical radiosonde station data, Int. J. Climatol., 16, 1197–1226, https://doi.org/10.1002/(SICI)1097-0088(199611)16:11<1197::AID-JOC89>3.0.CO;2-L, 1996.

Logan, J. A. and Powell, J. A.: Ghost forests, global warming, and the mountain pine beetle (Coleoptera Scolyridae), Am. Entemol., 47, 160–172, https://doi.org/10.1093/ae/47.3.160, 2001.

Murakami, H., Vecchi, G. A., Villarini, G., Delworth, T. L., Gudgel, R., Underwood, S., Yang, X., Zhang, W., and Lin, S.: Seasonal forecasts of major hurricanes and landfalling tropical cyclones using a high-resolution GFDL coupled climate model, J. Climate, 29, 7977–7989, 2016.

Nakamura, R. and Mahrt, L.: Air temperature measurement errors in naturally ventilated radiation shields, J. Atmos. Ocean. Technol., 22, 1046–1058, https://doi.org/10.1175/JTECH1762.1, 2005.

Nkemdirim, L.: On the Frequency of Precipitation-Days in Calgary, Canada, Professional Geographer., 40, 65–76, 1988.

Nkemdirim, L.: Canada's chinook belt, Int. J. Climatol., 16, 441–462, 1996.

Schönhart, M., Schauppenlehner, T., Kuttner, M., Mirchner, M., and Schmid, E.: Climate change impacts on farm production, landscape appearance, and the environment: Policy scenario results from an integrated field-farm-landscape model in Austria, Agr. Syst., 145, 39–50, https://doi.org/10.1016/j.agsy.2016.02.008, 2016.

Shafer, M., Fiebrich, C., Arndt, D., Fredrickson, S., and Hughes, T.: Quality assurance procedures in the Oklahoma Mesonetwork, J. Atmos. Ocean. Technol., 17, 474–494, https://doi.org/10.1175/1520-0426(2000)017<0474:QAPITO>2.0.CO;2, 2000.

Stooksbury, D., Idso, C., and Hubbard, K.: The effects of data gaps on the calculated monthly mean maximum and minimum temperatures in the continental United States: A spatial and temporal study, J. Climate, 12, 1524–1533, https://doi.org/10.1175/1520-0442(1999)012<1524:TEODGO>2.0.CO;2, 1999.

Taylor, K. E., Stouffer, R. J., and Meehl, G. A.: An Overview of CMIP5 and the experiment design, B. Am. Meteorol. Soc., 93, 485–498, https://doi.org/10.1175/BAMS-D-11-00094.1, 2012.

Thomas, C. D., Franco, A. M. A., and Hill, J. K.: Range retractions and extinction in the face of climate warming, Trends Ecol. Evol., 21, 415–416, https://doi.org/10.1016/j.tree.2006.05.012, 2006.

Wade, C. G.: A quality control program for surface mesometeorological data, J. Atmos. Ocean. Technol., 4, 435–453, https://doi.org/10.1175/1520-0426(1987)004<0435:AQCPFS>2.0.CO;2, 1987.

Wood, W. H.: Topographic and geographic influences on near-surface temperatures under different seasonal weather types in southwestern Alberta, unpublished PhD Thesis, University of Calgary, 2017.

Wood, W. H., Marshall, S. J., Fargey, S. E., and Whitehead, T. L.: Daily temperature data from the Foothills Climate Array mesonet, Canadian Rocky Mountains, 2005–2010, PANGAEA, https://doi.org/10.1594/PANGAEA.880611, 2017.

World Meteorological Organization: Guide to meteorological instruments and methods of observations, 7th edition, World Meteorological organization, Geneva, Switzerland, 2008.

Yospin, G. I., Bridgham, S. D., Neilson, R. P., Bolte, J. P., Bachelet, D. M., Gould, P. J., Harrington, C. A., Kertis, J. A., Evers, C., and Johnson, B. R.: A new model to simulate climate-change impacts on forest succession for local land management, Ecol Appl., 25, 226–242, https://doi.org/10.1890/13-0906.1, 2015.

Zuliani, A., Massolo, A., Lysyk, T., Johnson, G., Marshall, S., Berger, K., and Cork, S. C.: Modelling the northward expansion of Culicoides Sonorensis (Diptera Ceratopogonidea) under future climate scenarios, PLoS ONE, 10, e0130294, https://doi.org/10.1371/journal.pone.0130294, 2015.

5

History of chemically and radiatively important atmospheric gases from the Advanced Global Atmospheric Gases Experiment (AGAGE)

Ronald G. Prinn[1], Ray F. Weiss[2], Jgor Arduini[3], Tim Arnold[4], H. Langley DeWitt[1], Paul J. Fraser[5], Anita L. Ganesan[6], Jimmy Gasore[7], Christina M. Harth[2], Ove Hermansen[8], Jooil Kim[2], Paul B. Krummel[5], Shanlan Li[9], Zoë M. Loh[5], Chris R. Lunder[8], Michela Maione[3], Alistair J. Manning[10,11], Ben R. Miller[12], Blagoj Mitrevski[5], Jens Mühle[2], Simon O'Doherty[11], Sunyoung Park[9], Stefan Reimann[13], Matt Rigby[11], Takuya Saito[14], Peter K. Salameh[2], Roland Schmidt[2], Peter G. Simmonds[6], L. Paul Steele[5], Martin K. Vollmer[13], Ray H. Wang[15], Bo Yao[16], Yoko Yokouchi[14], Dickon Young[11], and Lingxi Zhou[16]

[1]Center for Global Change Science, Massachusetts Institute of Technology, Cambridge, MA, USA
[2]Scripps Institution of Oceanography, University of California San Diego, La Jolla, CA, USA
[3]Department of Pure and Applied Sciences, University of Urbino, Urbino, Italy
[4]National Physical Laboratory, Teddington, Middlesex, UK and School of GeoSciences, University of Edinburgh, Edinburgh, UK
[5]Climate Science Centre, Oceans and Atmosphere, Commonwealth Scientific and Industrial Research Organization (CSIRO), Aspendale, Victoria, Australia
[6]School of Geographical Sciences, University of Bristol, Bristol, UK
[7]Rwanda Climate Observatory Secretariat, Ministry of Education of Rwanda, Kigali, Rwanda
[8]Norwegian Institute for Air Research (NILU), Kjeller, Norway
[9]Department of Oceanography, Kyungpook National University, Daegu, Republic of Korea
[10]Hadley Centre, The Met Office, Exeter, UK
[11]School of Chemistry, University of Bristol, Bristol, UK
[12]National Oceanic and Atmospheric Administration (NOAA), Earth System Research Laboratory, Boulder, CO, USA
[13]Laboratory for Air Pollution and Environmental Technology (Empa), Swiss Federal Laboratories for Materials Science and Technology, Dübendorf, Switzerland
[14]National Institute for Environmental Studies (NIES), Tsukuba, Japan
[15]Georgia Institute of Technology, Atlanta, GA, USA
[16]China Meteorological Administration (CMA), Beijing, China

Correspondence: Ronald G. Prinn (rprinn@mit.edu)

Abstract. We present the organization, instrumentation, datasets, data interpretation, modeling, and accomplishments of the multinational global atmospheric measurement program AGAGE (Advanced Global Atmospheric Gases Experiment). AGAGE is distinguished by its capability to measure globally, at high frequency, and at multiple sites all the important species in the Montreal Protocol and all the important non-carbon-dioxide (non-CO_2) gases assessed by the Intergovernmental Panel on Climate Change (CO_2 is also measured at several sites). The scientific objectives of AGAGE are important in furthering our understanding of global chemical and climatic phenomena. They are the following: (1) to accurately measure the temporal and spatial distributions of anthropogenic gases that contribute the majority of reactive halogen to the stratosphere and/or are strong infrared absorbers (chlorocarbons, chlorofluorocarbons – CFCs, bromocarbons, hydrochlorofluorocarbons – HCFCs, hydrofluorocarbons – HFCs and polyfluorinated compounds (perfluorocarbons – PFCs), nitrogen trifluoride – NF_3, sulfuryl fluoride – SO_2F_2, and sulfur hexafluoride – SF_6) and use these measurements to determine the global rates of their emission and/or destruction (i.e., lifetimes); (2) to accurately measure the global distributions and temporal behaviors and determine the sources and sinks of non-CO_2 biogenic–anthropogenic gases important to climate change and/or ozone depletion (methane – CH_4, nitrous oxide – N_2O, carbon monoxide – CO, molecular hydrogen – H_2, methyl chloride – CH_3Cl, and methyl bromide – CH_3Br); (3) to identify new long-lived greenhouse and ozone-depleting gases (e.g., SO_2F_2, NF_3, heavy PFCs (C_4F_{10}, C_5F_{12}, C_6F_{14}, C_7F_{16}, and C_8F_{18}) and hydrofluoroolefins (HFOs; e.g., $CH_2=CFCF_3$) have been identified in AGAGE), initiate the real-time monitoring of these new gases, and reconstruct their past histories from AGAGE, air archive, and firn air measurements; (4) to determine the average concentrations and trends of tropospheric hydroxyl radicals (OH) from the rates of destruction of atmospheric trichloroethane (CH_3CCl_3), HFCs, and HCFCs and estimates of their emissions; (5) to determine from atmospheric observations and estimates of their destruction rates the magnitudes and distributions by region of surface sources and sinks of all measured gases; (6) to provide accurate data on the global accumulation of many of these trace gases that are used to test the synoptic-, regional-, and global-scale circulations predicted by three-dimensional models; and (7) to provide global and regional measurements of methane, carbon monoxide, and molecular hydrogen and estimates of hydroxyl levels to test primary atmospheric oxidation pathways at midlatitudes and the tropics. Network Information and Data Repository: http://agage.mit.edu/data or http://cdiac.ess-dive.lbl.gov/ndps/alegage.html (https://doi.org/10.3334/CDIAC/atg.db1001).

1 Introduction

The Advanced Global Atmospheric Gases Experiment (AGAGE: 1993–present) and its predecessors (Atmospheric Lifetime Experiment, ALE: 1978–1981; Global Atmospheric Gases Experiment, GAGE: 1982–1992) have measured the greenhouse gas and ozone-depleting gas composition of the global atmosphere continuously since 1978. The ALE program was instigated to measure the then five major ozone-depleting gases (CFC-11 ($CFCl_3$), CFC-12 (CCl_2F_2), CCl_4, CH_3CCl_3, N_2O) in the atmosphere four times per day using automated gas chromatographs with electron-capture detectors (GC-ECDs) at four stations around the globe and to determine the atmospheric lifetimes of the purely anthropogenic of these gases from their measurements and industry data on their emissions (Prinn et al., 1983a). The GAGE project broadened the global coverage to five stations, the number of gases being measured to eight (adding CFC-113 (CCl_2FCClF_2), $CHCl_3$, and CH_4 to the ALE list), and the frequency to 12 per day by improving the GC-ECDs and adding gas chromatographs with flame-ionization detectors (GC-FIDs; Prinn et al., 2000). The AGAGE program then significantly improved upon the GAGE instruments by increasing their measurement precision and frequency (to 36 per day) and adding gas chromatographs with mercuric oxide reduction detectors, to measure 10 biogenic and/or anthropogenic gases overall (adding H_2 and CO to the GAGE list). AGAGE also introduced powerful new gas chromatographs with mass spectrometric detection and cryogenic pre-concentration measuring over 50 trace gases 20 times per day. In this overview paper, while we address the entire 1978–present database and its public availability, we focus more on the evolution of the network after 2000, details of the period before that are addressed in the previous comprehensive overviews provided by Prinn et al. (2000) and Prinn et al. (1983a). The case for high-frequency measurement networks like AGAGE with data available to operators in real time is strong, and the observations and their interpretation are important inputs to the scientific understanding of ozone depletion and climate change. AGAGE is characterized by its capability to measure globally the trends at high frequency and estimate emissions from these trends for all of the important species in the Montreal Protocol on Substances that Deplete the Ozone Layer, and all of the important non-carbon-dioxide (non-CO_2) trace gases assessed by the Intergovernmental Panel on Climate Change. More recently, AGAGE has also been measuring CO_2 using high-frequency optical spectroscopy (focusing on sites where such measurements are not made by other groups; Sect. 2.3 and 2.4). The scientific objectives of AGAGE (summarized in the Abstract) are of considerable significance in furthering our understanding of important global chemical and climatic phenomena. The remainder of this Introduction is devoted to describing the network of stations (Sect. 1.1), the measurements (Sect. 1.2), and the place of AGAGE in the global observing system (Sect. 1.3). Then Sect. 2 addresses the instrumentation, calibration, and station infrastructure, Sect. 3 the data analysis and modeling, Sect. 4 the scientific accomplishments, and Sect. 5 the AGAGE data availability.

1.1 A Global network of stations

The ALE/GAGE/AGAGE stations are coastal or mountain sites around the world, chosen primarily to provide accurate

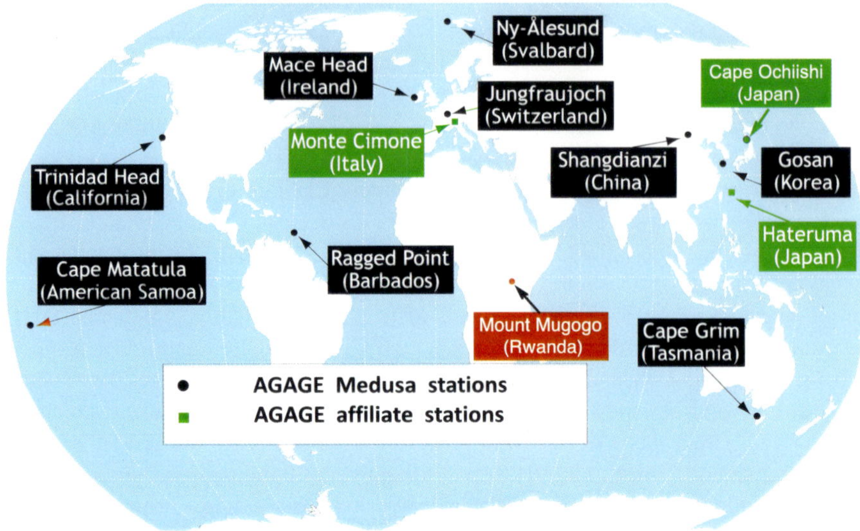

Figure 1. Locations of the 10 current AGAGE primary stations (red highlighted stations) that have Medusa gas chromatograph–mass spectrometer (GC-MS) instruments and the 3 current AGAGE affiliate stations (green highlighted stations) that have alternative pre-concentration GC-MS instruments. AGAGE and the other major global air sampling network, NOAA-ESRL-GMD, are independent but closely cooperating, including frequent data intercomparisons, especially at the American Samoa shared site.

measurements of trace gases whose lifetimes are long compared to global atmospheric circulation times (Fig. 1). The 10 "primary" AGAGE stations that all share common calibrations and gas chromatographic–mass spectrometric instrumentation (see Sect. 1.2) are the following: (a) on Ireland's west coast, first at Adrigole (52° N, 10° W; 50 m (inlet height a.s.l. here and for all other stations), 1978–1983), then at Mace Head (53° N, 10° W; 25 m 1987 to present); (b) on the US west coast, first at Cape Meares, Oregon (45° N, 124° W; 30 m, 1979–1989), then at Trinidad Head, California (41° N, 124° W; 140 m, 1995 to present); (c) at Ragged Point, Barbados (13° N, 59° W; 42 m, 1978 to present); (d) at Cape Matatula, American Samoa (14° S, 171° W; 77 m, 1978 to present); (e) at Cape Grim, Tasmania, Australia (41° S, 145° E; 164 m, 169 m, 1978 to present); (f) on the Jungfraujoch, Switzerland (47° N, 8° E; 3580 m, 2000 to present); (g) on Zeppelin Mountain, Ny-Ålesund, Svalbard, Norway (79° N, 12° E; 489 m, 2001 to present); (h) at Gosan, Jeju Island, Korea (33° N, 126° E; 89 m, 2007 to present); (i) at Shangdianzi, China (41° N, 117° E; 383 m, 2010 to present with gap) and (j) Mt. Mugogo, Rwanda (1.6° S, 29.6° E; 2640 m, 2015 to present). The AGAGE network also includes three AGAGE-compatible (but not identical) instruments in the following locations: (k) Hateruma Island, Japan (24° N, 123.8° E; 47 m, 2004 to present); (l) Cape Ochiishi, Japan (43° N, 145.5° E; 100 m, 2006 to present), and (m) Monte Cimone, Italy (44° N, 10° E; 2165 m, 2004 to present). These are called AGAGE "affiliate" stations in Fig. 1. There are also "secondary", usually continental and some urban, stations that are linked to and complement the primary and affiliate stations (discussed below).

1.2 Measurements

At its primary stations, AGAGE uses in situ gas chromatography with mass spectrometry (GC-MS) in the "Medusa" system (Miller et al., 2008; Arnold et al., 2012) to measure over 50 largely synthetic gases including hydrochlorofluorocarbons (e.g., HCFC-22; $CHClF_2$) and hydrofluorocarbons (e.g., HFC-134a; CH_2FCF_3), which are interim or long-term alternatives to chlorofluorocarbons (CFCs) now restricted by the Montreal Protocol, other hydrohalocarbons (e.g., methyl chloride; CH_3Cl), halons (e.g., Halon-1211; $CBrClF_2$), perfluorocarbons (e.g., PFC-14; CF_4), and trace chlorofluorocarbons, all of which, except CH_3Cl, are involved in the Montreal or Kyoto Protocol. Affiliate stations use similar but not identical cryogenic pre-concentration GC-MS systems (Maione et al., 2013; Yokouchi et al., 2006).

At its Mace Head, Trinidad Head, Ragged Point, Cape Matatula, and Cape Grim primary stations, AGAGE also uses in situ gas chromatographs (GC) with electron-capture detection (ECD), flame-ionization detection (FID), mercuric oxide reduction detection (MRD, at Mace Head and Cape Grim only), and pulsed discharge detection (PDD, at Cape Grim only) to measure five biogenic–anthropogenic gases (methane – CH_4, nitrous oxide – N_2O, and chloroform – $CHCl_3$ at all sites; carbon monoxide – CO and hydrogen – H_2 at Mace Head and Cape Grim only) and five anthropogenic gases at all five sites: CFC-11 (CCl_3F), CFC-12 (CCl_2F_2), and CFC-113 (CCl_2FCClF_2), methyl chloroform

(CH_3CCl_3), and carbon tetrachloride (CCl_4) 36 times per day (Prinn et al., 2000). The list of gases measured with these gas chromatography "multidetector" (GC-MD) systems includes the three major chlorofluorocarbons (CFCs) restricted by the Montreal Protocol and the four major long-lived non-CO_2 greenhouse gases (GHGs). Table 1 lists all the major gases being measured in AGAGE using the Medusa GC-MS and GC-MD instruments, their 2016 global average mole fractions, and their typical measurement precisions.

The precisions for each species are determined from the interspersed measurements of the on-site station calibration tanks and are reported along with the mole fractions of the interspersed atmospheric measurements in the AGAGE data archives. In general the precisions in Table 1 are highest (< 0.1 %) for the species with the highest absolute mole fractions and lowest (~ 10 %) for those with the lowest mole fractions; there are also more subtle differences depending on the species behavior in the trapping (Medusa), separation (GC), and detection (MS, MD; ECD, FID, MRD) stages. The accuracy of the measurements is determined by calibration scale and tertiary tank accuracies that are discussed in Sect. 2.6.

Recent developments have enabled precise analyses of CH_4, CO_2, CO, and N_2O by laser spectroscopic detection to begin in AGAGE. These optical instruments are now expanding the measurement capabilities within AGAGE, and there are advantages in switching from the GC-MD approach for measuring CH_4, N_2O, and CO to these less operationally demanding optical spectroscopy methods resulting in near-continuous measurements of comparable or better precision. As discussed in Sect. 2.3 and 2.4, this transition is happening already at several AGAGE stations. The GC-MD and optical spectroscopy instruments will follow the AGAGE protocol used for all cases in which a new improved instrument replaces an earlier one; namely, the two instruments are run together for at least several months (and years for gases currently measured on both the GC-MD and Medusa GC-MS) to ensure data comparability and verify improvements.

Each instrument system is automated and under computer control. All chromatograms, instrumental data, and operator logs are transmitted via the internet to the data processing sites. AGAGE includes timely public archiving and publication of all data, regular intercomparisons of AGAGE measurements, absolute calibrations with other networks (e.g., NOAA's Global Monitoring Division, GMD), and contributions to national and international assessments of ozone depletion and climate change. The data are calibrated against on-site air standards, which are calibrated relative to off-site parent standards before and after use at each station. AGAGE depends upon well-defined absolute gravimetric calibration procedures that are repeated periodically to ensure the accuracy of the long-term measured trends (Prinn et al., 2000).

To emphasize the need for very frequent real-time measurements we show data for several trace gases (Fig. 2a–d) for the years 2004 and 2016. These GC-MD and GC-MS data demonstrate the existence of regional pollution-induced or local sink-induced (e.g., for H_2; shown in red) and large-scale transport-induced (shown in black) variability, which are not captured with weekly flask measurements typically designed to avoid local pollution. Our approach for identifying these pollution events is discussed in Sect. 3.1. Note also the evolution of the sizes of these pollution events between 2004 and 2016 associated with the decreases in the emissions of regulated gases and the growth of emissions of unregulated ones. This high-frequency sampling enables the pollution events in particular to be used to estimate emissions from nearby source regions (e.g., Cape Grim station for SE Australian emissions; e.g., Dunse et al., 2005; Stohl et al., 2009; O'Doherty et al., 2009; Fraser et al., 2014; Lunt et al., 2015), Trinidad Head for the west coast US emissions (e.g., Li et al., 2005; O'Doherty et al., 2009; Lunt et al., 2015; Fortems-Cheiney et al., 2015), Mace Head and the other European stations for European and in some cases eastern USA emissions (e.g., O'Doherty et al., 2009; Stohl et al., 2009; Keller et al., 2012; Simmonds et al., 2015; Lunt et al., 2015; Fortems-Cheiney et al., 2015; Graziosi et al., 2017), and Hateruma, Shangdianzi, and Gosan for East Asian emissions (e.g., Stohl et al., 2009, 2010; Kim et al., 2010; Li et al., 2011; Yao et al., 2012a, b; Saito et al., 2015; Fang et al., 2015; Lunt et al., 2015; Fortems-Cheiney et al., 2015). The sources of many anthropogenic and natural trace gases measured in AGAGE are often colocated so that measurement of a wide range of gases enhances the ability to accurately estimate their sources and sinks. The AGAGE data in graphical and digital forms are available for most stations at the AGAGE website: http://agage.mit.edu (last access: 21 May 2018) (Sect. 3.2).

1.3 Integral element of the global observing system

AGAGE is part of a powerful complementary observing system that is measuring various aspects of the evolving composition of Earth's atmosphere and providing the fundamental understanding needed to preserve this vital sphere of life on our planet. Sharing the AGAGE surface-based perspective are, for example, the remote-sensing Network for Detection of Atmospheric Composition Change (NDACC; see De Mazière et al., 2018) supported by NASA and other agencies and nations (AGAGE is an NDACC Cooperating Network) and the NOAA ESRL Global Monitoring Division in situ and flask networks. AGAGE contributes to the World Meteorological Organization's Global Atmosphere Watch (WMO-GAW) and regularly provides its data to the WMO-GAW's World Data Center for Greenhouse Gases (WDCGG) website (see Sect. 5). AGAGE European stations provide data to the Integrated Carbon Observation System (ICOS)that coordinates pan-European observations of GHGs, and Monte Cimone, Jungfraujoch, and Ny Ålesund are now formally joining. Also measuring atmospheric composition (as column profiles or abundances) are instruments onboard the NASA

Table 1. Primary AGAGE measured species using Medusa GC-MS and GC-MD systems. Gases measured with Medusa GC-MS and GC-MD only are in black regular font; those measured with both systems are in italic font. Calibrations are on AGAGE SIO gravimetric scales (Sect. 2.6) unless otherwise noted.

Compound	Global mean 2016 conc. (ppt[c])	Typical precision (%)	Compound	Global mean 2016 conc. (ppt[c])	Typical precision (%)
PFC-14	82.7	0.15	CFC-114[h]	16.3	0.3
PFC-116	4.56	1	CFC-115	8.48	0.7
PFC-218	0.63	3	Halon-1211	3.59	0.4
PFC-c318	1.56	1.5	Halon-1301	3.37	1.7
PFC-5-1-14	0.31	3	Halon-2402	0.41	2
SF_6	8.88	0.6	CH_3Cl	552	0.2
SF_5CF_3	0.17	7	CH_3Br	6.96	0.6
SO_2F_2	2.26	2	CH_3I[e]	0.58	2
NF_3	1.44	1	CH_2Cl_2	31.1	0.5
HFC-23	28.9	0.7	CH_2Br_2[e]	1.08	1.5
HFC-32	12.6	3	*$CHCl_3$*	*8.78*	*0.4*
HFC-134a	89.3	0.5	$CHBr_3$[e]	1.84	0.6
HFC-152a	6.71	1.4	*CCl_4*	*79.9*	*1*
HFC-125	20.8	0.7	*CH_3CCl_3*	*2.61*	*0.7*
HFC-143a	19.3	1	$CHCl=CCl_2$	~0.11	3
HFC-227ea	1.24	2.2	$CCl_2=CCl_2$[e]	1.07	0.5
HFC-236fa	0.15	10	COS[e]	543	0.5
HFC-245fa	2.42	3	C_2H_6[d]	586	0.3
HFC-365mfc	1.00	5	C_3H_8[f]	9.04	0.6
HFC-43-10mee	0.27	3	C_6H_6[d]	17.9	0.3
HCFC-22	237	0.3	C_7H_8[d]	4.19	0.6
HCFC-141b	24.5	0.5			
HCFC-142b	22.6	0.4			
HCFC-124[d]	1.11	2	GC-MD only	(ppb[c])	
CFC-11	*230*	*0.2*	CH_4	1842	0.2
CFC-12	*516*	*0.1*	N_2O	329.3	0.05
CFC-13[g]	3.28	2	CO[a]	54 to 115	0.2
CFC-113	*71.4*	*0.2*	H_2[a]	515 to 550	0.6 (0.08)[b]

[a] CO and H_2 measured at Mace Head and Cape Grim only (range for annual means of these two stations given). [b] GC-PDD system at Cape Grim. [c] ppt: parts per trillion and ppb: parts per billion. [d] Preliminary (AGAGE) scale (Sect. 2.6), [e] preliminary (transfer of NOAA) scale (Sect. 2.6), [f] preliminary (Empa) scale (Sect. 2.6), [g] METAS-2017 (Empa) scale (Sect. 2.6), [h] quasi-linear sum of CFC-114 and CFC-114a.

TERRA and AURA satellites and the ESA ENVISAT satellite. Aircraft- and balloon-borne instruments provide vital in situ measurements in the middle troposphere and lower stratosphere. The combination of all of these complementary data with state-of-the-art global chemistry and circulation models is providing major advances in our understanding of the global sources, chemistry, transport, and sinks of atmospheric trace substances and allows for the determination of atmospheric composition and air quality, the radiative forcing of climate change, and impacts on stratospheric ozone.

2 Instruments, calibration, and infrastructure

The AGAGE program has placed a strong emphasis on instrumental innovation and the gravimetric preparation of primary standards to obtain high-frequency and high-precision automated trace gas measurements at all the AGAGE measurement sites. In this section, the first four subsections discuss the AGAGE GC-MD (Sect. 2.1), Medusa GC-MS (Sect. 2.2), optical spectroscopy (Sect. 2.3), and isotopic (Sect. 2.4) instruments. Then we address data acquisition and processing (Sect. 2.5), instrumental calibration (Sect. 2.6), primary and affiliate station facilities and infrastructure (Sect. 2.7), secondary stations (Sect. 2.8), and stored air archives (Sect. 2.9).

In the early 1990s the GC-MD instruments were developed and deployed at the Mace Head, Trinidad Head, Ragged Point, Cape Matatula, and Cape Grim stations and at the Scripps Institution of Oceanography (SIO) calibration laboratory (Prinn et al., 2000). In the late 1990s, AGAGE pioneered the deployment of automated GC-MS instruments at our stations in Mace Head and Cape Grim and at the University of Bristol. These instruments featured an adsorption–desorption system (ADS) with cryogenic (−50 °C) pre-

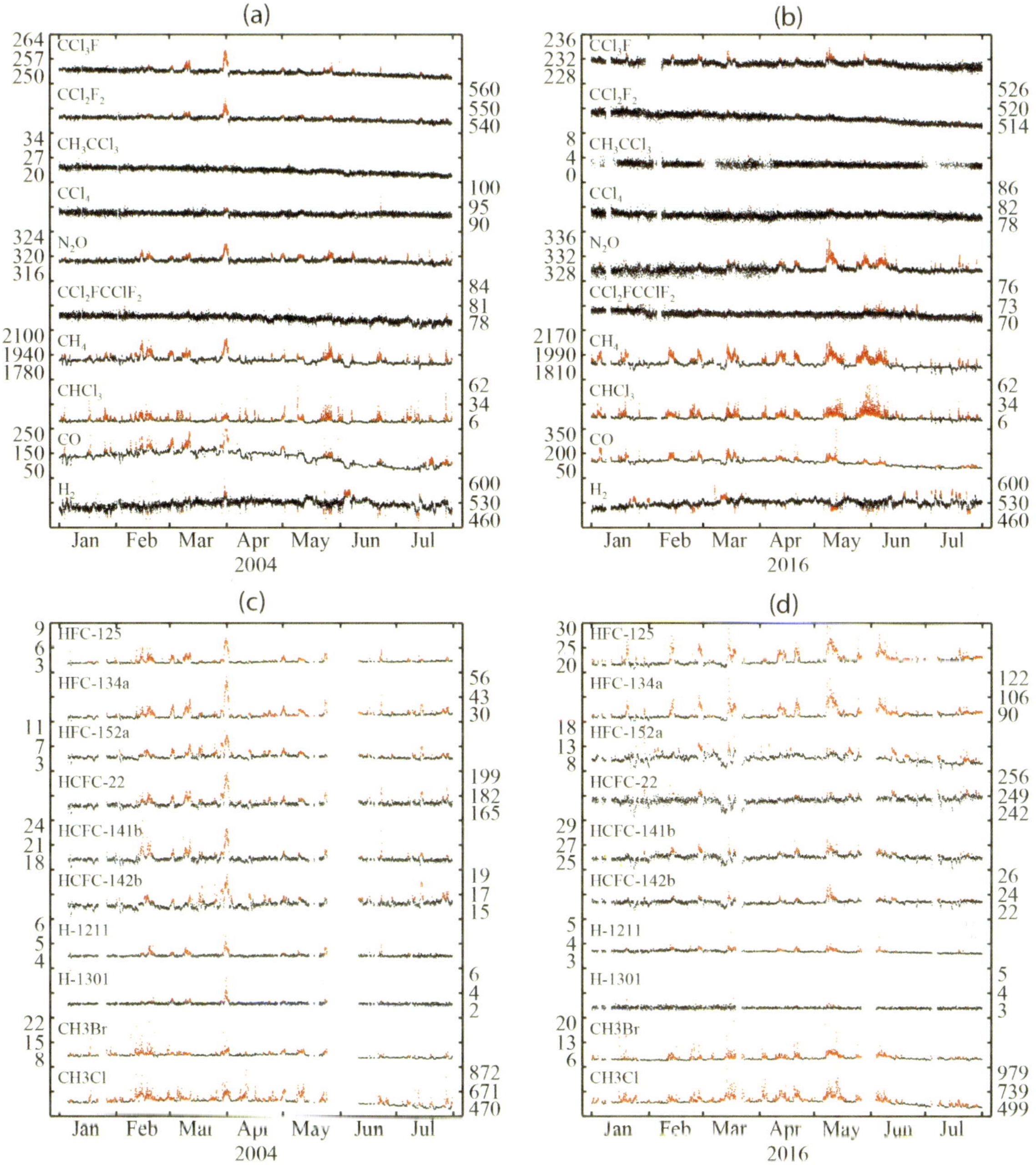

Figure 2. A total of 7 months of data for gases measured at Mace Head, Ireland: (1) with the GC-MD in **(a)** 2004 and **(b)** 2016 (units: mole fractions; ppb for N_2O, CH_4, H_2, and CO; ppt for all others) and (2) with the Medusa GC-MS for selected gases in **(c)** 2004 and **(d)** 2016 (units: mole fractions in ppt for all gases). In all four panels, measurements in polluted air originating from Europe (also in air affected by local sinks; see text) are shown in red, while those in clean air off the Atlantic Ocean are shown in black. Note that pollution events are defined separately for each gas due to their often differing sources.

concentration of analytes from 2 L air samples (Simmonds et al., 1995). The technological developments incorporated into these instruments, the methods of data collection, transmission, and processing, the primary and secondary calibration standards produced at the SIO calibration laboratory, and the on-site tertiary (from SIO) and quaternary (calibrated on-site from the tertiary) standards necessary to sustain the AGAGE network are partly described in the first AGAGE overview (Prinn et al., 2000), but updated here in Sect. 2.6 and 2.7.

Beginning in the early 2000s, the AGAGE team recognized that modern refrigeration technology made it possible to make major improvements to the ADS concept and

Table 2. GC–multidetector instruments at current AGAGE primary and secondary stations. Detectors: ECD for CFC-11, CFC-12, CFC-113, CH_3CCl_3, CCl_4, N_2O, and $CHCl_3$; FID for CH_4; MRD for CO and H_2; and PDD for H_2.

GC-ECD-FID	GC-ECD-FID-MRD	GC-ECD-FID-MRD-PDD
Trinidad Head, CA, USA	Mace Head, Ireland	Cape Grim, Tasmania
Ragged Point, Barbados	Tacolneston[a], UK	
Cape Matatula, Samoa	Aspendale[b], Australia	
La Jolla, CA, USA		
Ridge Hill[a], UK		
Bilsdale[a], UK		
Heathfield[a], UK		

[a] Modified version of the GC-MD without FID channel. [b] Uses three individual GC systems with ECD, FID, and MRD detectors.

to greatly extend the range of compounds that could be measured by enhanced cryogenic pre-concentration at −165 °C. As a result, the AGAGE GC-MS effort was redirected to the development of the new Medusa instrument (Miller et al., 2008; Arnold et al., 2012).

2.1 GC–multidetector instruments

The current AGAGE GC-MD instruments replaced the earlier GAGE GC-MD instruments in 1993–1996 (Table 2). These Agilent© GC instruments employ two electron-capture detector (ECD) channels and one flame-ionization detection (FID) channel to measure the principal chlorine-bearing anthropogenic ozone-depleting compounds now banned by the Montreal Protocol (CFC-11, CFC-12, CFC-113, CCl_4, and CH_3CCl_3), as well as the both natural and anthropogenic compounds N_2O, CH_4, and $CHCl_3$ (see Table 1). The GC-MDs at Mace Head and Cape Grim include an extra channel for the measurement of CO and H_2 by a mercuric oxide reduction detector (MRD; Prinn et al., 2000). In early 2015, the GC-MD system at Cape Grim also added a further extra channel for the measurement of H_2 by pulsed discharge detector (PDD), bringing a more than 10-fold improvement in precision. The GC-MD measurements are made on dried whole-air samples automatically injected by a computer-controlled sampling module. Each analysis cycle takes 20 min.

Compared to its ALE and GAGE predecessors, the AGAGE GC-MD provides greatly enhanced precision and measurement frequency, custom software (GCWerks©, http://www.gcwerks.com, last access: 21 May 2018) for instrument control and digital acquisition of all chromatograms and measurement parameters, and use of the internet for data transmission and remote diagnosis and control (Prinn et al., 2000, Sect. 2.5). These instruments can also carry out pressure-programmed injections to assess their own non-linearities and use flexible custom algorithms for the post-analysis quantitative interpretation of chromatograms. The performance and reliability of these instruments have been and continue to be exceptional, leading to important advances in scientific interpretation, as discussed below. For some of the species that the GC-MDs measure, AGAGE is now also beginning to deploy new technologies including GC-MS, cavity ring-down spectroscopy (CRDS), and quantum cascade laser (QCL; optical) methods that offer improved sensitivity as discussed in the following sections. The GC-MD instruments will continue to be operated until such time as they can be phased out after careful overlap in the field using these newer technologies.

2.2 Medusa GC-MS instruments

The AGAGE Medusa GC-MS instruments have become the major instruments of the AGAGE network and collaborating measurement laboratories. Instrument development work beyond that described by Miller et al. (2008) continues, with enhanced operational parameters, upgrades, and new species being added over time. For example, subsequent important changes were made in the Medusa flow scheme and column configuration that add the powerful greenhouse gas NF_3 emitted by the electronics industry to its measurement capability without sacrificing any of its other capabilities (Arnold et al., 2012). The reader is directed to these two papers for a full description of the current Medusa configuration – only a brief overview is given here.

A complement of 19 AGAGE Medusas has now been deployed (Table 3), with one at each of the 10 primary stations (red labels in Fig. 1), two at the SIO calibration and instrument development laboratory, and seven more at other secondary stations or laboratories in the UK (Tacolneston & Bristol), Switzerland (Dübendorf), Australia (two at Aspendale), Norway (Kjeller), and China (Beijing).

At the heart of the Medusa is a Polycold© "Cryotiger" cold end that maintains a temperature of about −175 °C within the Medusa's vacuum chamber, even with a substantial heat load, using a simple single-stage compressor with a proprietary mixed-gas refrigerant. This cold end conductively cools dual micro-traps to about −165 °C. By using standoffs of limited thermal conductivity to connect the traps to the cold head, each trap can independently be heated re-

Table 3. GC-MS instruments at AGAGE primary, affiliate, and secondary monitoring stations and at laboratories.

Primary or affiliate station (by latitude)	Instrument	Secondary station or laboratory (by country)	Instrument
Ny-Ålesund	Medusa	La Jolla, USA (laboratory[a] and secondary)	Two Medusas
Mace Head	Medusa	Tacolneston, UK	Medusa
Jungfraujoch	Medusa	Bristol, UK (laboratory)	Medusa
Monte Cimone	Affiliate	Dübendorf, Switzerland (laboratory)	Medusa
Cape Ochiishi	Affiliate	Aspendale, Australia (laboratory and secondary)	Two Medusas
Shangdianzi	Medusa	Kjeller, Norway (laboratory)	Medusa
Trinidad Head	Medusa	Beijing, China (laboratory)	Medusa
Gosan	Medusa		
Hateruma	Affiliate		
Ragged Point	Medusa		
Mount Mugogo	Medusa[b]		
Cape Matatula	Medusa		
Cape Grim	Medusa		

[a] Central AGAGE Calibration Laboratory. [b] Installed in spring 2018.

sistively to any temperature from -165 to $+100\,^{\circ}C$ or more, while the cold end remains cold. The use of two traps with extraordinarily wide programmable temperature ranges, coupled with the development of appropriate trap adsorbents and the use of separating columns between traps, permits the desired analytes from 2 L air samples to be effectively separated from more abundant gases that would otherwise interfere with chromatographic separation or mass spectrometric detection, such as nitrogen (N_2), oxygen (O_2), argon (Ar), water vapor (H_2O), CO_2, CH_4, krypton (Kr), and xenon (Xe). Importantly, the dual micro-trap and revised column configuration also permit the analytes to be purified of interfering compounds from the larger first-stage trap (T1) by fractional distillation, chromatographic separation, and refocusing onto a smaller trap (T2) at very low temperatures so that the resulting injections to the main chromatographic column in the Agilent© 5975C quadrapole GC-MS are sharp and reproducible. By trapping and eluting analytes at very low temperatures, the range of compounds that can be measured is greatly extended to include a number of important volatile compounds, and problems with the reaction of analytes on the traps at higher temperatures are avoided. The Medusa system uses high-precision integrating mass-flow controllers for the measurement of sample volumes. In addition, significant advances have been made in the software (GCWerks) to control and acquire data from the Medusa and the GC-MS itself so that the entire system has programmability, versatility, and ease of operation comparable to that of the AGAGE GC-MD instruments. The original Agilent 5973 mass-selective detectors (MSDs) used in the six early Medusas have been replaced with newer and more sensitive Agilent 5975C MSDs. As a result, sensitivities on the Medusas with the new MSDs increased 1.5- to 2-fold over those with the old MSDs, which has especially benefitted measurements of the lowest-abundance species.

As noted above, instrument development work on the Medusas continues. The species routinely measured at Medusa field stations are listed in Table 1. Compounds added only recently to routine Medusa measurements (and therefore not yet in Table 1) are HCFC-133a and CF_3CFOCF_2, while the light hydrocarbons C_2H_2 and C_2H_4, although still measured, are also not included in Table 1 because co-elution compromises their measurement as the GC column ages. The AGAGE Medusas were the first instruments monitoring in situ the global distributions and trends of the high-GWP industrial gases CF_4, NF_3, and SO_2F_2 (Mühle et al., 2009, 2010; Weiss et al., 2008; Arnold et al., 2013). In addition to the compounds listed in Table 1, additional species (e.g., CFC-112) are in various stages of being added to the station measurements. Recently, the "fourth-generation" halocarbons HFC-1234yf, HFC-1234ze(E), and HCFC-1233zd(E), as well as HCFC-31 and four inhalation anesthetics have been measured in the atmosphere using the Medusa system (Vollmer et al., 2015a, b; Schoenenberger et al., 2015). The development work on the Medusa utilizes the two instruments in this central laboratory. These instruments allow a wide range of development work to be undertaken while maintaining the important functions of primary and secondary calibration of the global AGAGE network and also continuing "urban" AGAGE ambient measurements of air pumped from the SIO pier at La Jolla. At CSIRO Aspendale, one Medusa instrument is deployed in an urban air monitoring mode and the other is generally deployed for flask sample measurements, in particular analyses of the Cape Grim air archive. The Medusas at the other five secondary stations listed in Table 3 are deployed either for monitoring or laboratory functions.

The Medusa technology continues to evolve in response to the needs of AGAGE researchers to measure new compounds, improvements in software, including data processing, diagnostics and alarms, and improvements in available technology. Most notably, the Polycold Cryotiger cold-end technology that was so revolutionary at the outset of the Medusa program is nearing the end of its useful life, but very fortunately Stirling cooling technology has advanced considerably with improved performance and reliability and reduced cost during the same time period. One Medusa at the SIO laboratory has been retrofitted to Stirling cooling (Sunpower CryoTel-GT) and is performing extremely well, as well as offering increased flexibility in trapping parameters. At the Empa and SIO laboratories, efforts are also underway to upgrade current Medusa technology to time-of-flight mass spectrometry (TOF-MS) in place of quadrupole mass spectrometric detection. This offers the advantage of very high mass resolution ($\sim$4000) that is capable of separating gases with the same integer masses but different actual masses that interfere with each other in the chromatograms using quadrupole technology (e.g., Obersteiner et al., 2016).

There are also three AGAGE-affiliated stations that use similar but not identical automated GC-MS measurements with cryogenic pre-concentration (stations denoted "affiliate" in Table 3), but are tied to AGAGE standards, at Hateruma Island and Cape Ochiishi, Japan (NIES) and at Monte Cimone, Italy (University of Urbino). Monte Cimone uses a GC-MS (Agilent 6850 and 5975, respectively) with an autosampling and pre-concentration device (Markes International©, UNITY2-Air Server2©) to enrich the halocarbons on a focusing adsorbent trap (Maione et al., 2013) and AGAGE-derived calibrations. Hateruma and Ochiishi both use a GC-MS (Agilent 6890 and 5973, respectively) with a unique cryogenic pre-concentration module (Yokouchi et al., 2006, 2012) and independently produced gravimetric standards that are intercompared with AGAGE standards to provide intercalibration factors.

2.3 Optical spectroscopic instruments

Recent advances in wavelength-scanned cavity ring-down spectroscopy (CRDS) have enabled precise analyses of CH_4, CO_2, CO, N_2O, and H_2O without chromatographic separation to begin in AGAGE. The analyzed air sample needs to be dried or, if not dried, corrections applied using the ancillary H_2O measurement. The Nafion sample drying and gas sampling approach used in AGAGE has been adapted to a sampling module with an MKS Instruments© inlet pressure controller for CRDS instruments that has been designed by SIO and built by Earth Networks© (Welp et al., 2013). These optical instruments are now expanding the measurement capabilities within AGAGE. There are several advantages in switching from the GC-FID approach for measuring CH_4, the GC-ECD approach for N_2O, and the GC-MRD approach for CO in AGAGE to these optical spectroscopy methods: no chromatography (so no carrier gases needed), essentially continuous, reduced costs including ongoing instrument maintenance, and improved linearity of response (for N_2O, CO). This transition is happening already at several AGAGE stations (see Table 4).

The CSIRO Picarro© G2301 for CO_2, CH_4, and H_2O at Cape Grim (which is being operated at present without drying the sample gas) has been compared with the AGAGE GC-MD CH_4 data at Cape Grim and the agreement is very good, with a mean offset of only $\sim$0.26 ppb ($\sim$0.02 %) when reported on the same calibration scale. The AGAGE group at SIO, in collaboration with the laboratory of R. F. Keeling, the company Earth Networks©, and the California Air Resources Board (CARB), has been evaluating the performance of various CRDS instruments, including calibration optimization, using Allan variance analyses (Allan, 1966; Werle et al., 1993). This has included the Picarro G2301, the Picarro G2401 for CO_2, CO, CH_4, and H_2O, the Picarro G5205 (prototype) and G5310 mid-IR for N_2O and H_2O, and the Los Gatos Research (LGR©) high-precision mid-IR instrument for N_2O, CO, and H_2O. For CO, the LGR mid-IR instrument is an order of magnitude more precise than the Picarro G2401, but to take full advantage of the LGR's precision requires frequent calibration (hourly or less) that is impractical for long-term atmospheric monitoring. With only daily calibration this difference is reduced to about a factor of 2. The precisions of the G5310 (and G5205) and to a lesser extent of the G2401 are improved by drying the air sample to minimize the H_2O correction using the aforementioned sampling modules built by Earth Networks, and these modules have been adopted at the Ragged Point, Mt. Mugogo, and Cape Matatula stations. Finally, CSIRO is operating high-precision Aerodyne Research© quantum cascade laser (QCL) spectroscopy systems for CO and N_2O at Aspendale, Australia.

2.4 Isotopomer–isotopologue instruments

For GHGs that have natural, anthropogenic, industrial, and biogenic sources, such as CO_2, CH_4, and N_2O, measurements of atmospheric abundances alone are often inadequate to precisely differentiate among these different sources. High-frequency in situ measurements of not just the total mole fractions of these gases, but also their stable isotopic compositions (^{12}C, ^{13}C, ^{14}N, ^{15}N, ^{16}O, ^{18}O, H, D) are a new frontier in global monitoring and hold the promise of revolutionizing our understanding of the global cycles of these gases (e.g., Rigby et al., 2012). High-frequency in situ isotopic measurements are now feasible using optical (laser) detection.

MIT and Aerodyne Research have codeveloped and deployed (2015–2017) at the Mace Head station an automated high-frequency instrument for the analysis of the isotopic composition of N_2O using tunable infrared laser differential absorption spectroscopy (TILDAS) with mid-infrared quantum cascade lasers (Harris et al., 2013). This instrument is fully automated and can be accessed and controlled via the internet. The new instrument monitors the

Table 4. CRDS spectroscopic instruments at AGAGE primary stations and secondary stations (including the UK Deriving Emissions related to Climate Change (DECC) network and UK National Physical Laboratory (NPL) stations). Instruments with Earth Networks (EN) driers lower the sample water vapor mole fractions to decrease H_2O interferences.

Instrument	Gases	Stations				
Picarro G1301	CH_4, CO_2, H_2O	Jungfraujoch (G2401 after 2011)	Mace Head			
Picarro G2301	CH_4, CO_2, H_2O	La Jolla (+EN drier), Trinidad Head (+EN drier)	Cape Grim	Mace Head	Bristol, Tacolneston (+EN drier), Ridge Hill (UK DECC)	Aspendale
Picarro G2401	CH_4, CO_2, CO, H_2O	Ragged Point (+EN drier)	Cape Matatula (+EN drier)	Mt. Mugogo (+EN drier)	Heathfield (UK NPL), Bilsdale (UK DECC)	Ny-Ålesund
Picarro G5205 or G5310	N_2O, H_2O	Mt. Mugogo (+EN drier)	Ny-Ålesund (G5310)			
LGR high performance	N_2O, CO, H_2O	La Jolla (+EN drier)			Tacolneston	
High-precision Aerodyne QCL	CO, N_2O	Aspendale, Australia				

four major isotopologues and isotopomers of nitrous oxide ($^{15}N^{14}N^{16}O$, $^{14}N^{15}N^{16}O$, $^{14}N^{14}N^{18}O$, and $^{14}N^{14}N^{16}O$) with a precision of at least 0.3 per mil (‰) for individual measurements spanning 28 min. For at least 0.1 per mil (‰) precision, we need to average 3–11 such measurements depending on the isotope (Harris et al., 2013). The needed pre-concentration was achieved through the development of a new high-efficiency cryo-focusing trap and sample transfer module (called Stheno) using concepts from the AGAGE Medusa module (Potter et al., 2013).

Similar automated N_2O isotope instrumentation has been developed at Empa (Wächter et al., 2008; Heil et al., 2014) and has been used for analyzing flask samples from Jungfraujoch. Also, a similar pre-concentration system has been developed by Mohn et al. (2010) and their pre-concentration TILDAS system has shown excellent compatibility with isotope ratio MS in an interlaboratory comparison campaign (Mohn et al., 2014). The pre-concentration technique has been further developed at Empa by implementing a more powerful Stirling cooler and a moveable trap design for quantitative CH_4 adsorption (Eyer et al., 2016). Also, CSIRO operates an Aerodyne Research quantum cascade laser system for the three stable isotopologues of CO_2 ($^{12}CO_2$, $^{13}CO_2$, and $^{18}O^{12}C^{16}O$) at Cape Grim.

Further developments in these instruments will facilitate their future deployment at AGAGE stations for continuous high-frequency in situ isotopic composition measurements of CO_2, CH_4, and N_2O.

2.5 Data acquisition and processing

The custom data acquisition and processing software (GCWerks) used in AGAGE for both the GC-MD and Medusa GC-MS instruments and run under the Linux operating system is described in moderate detail by Miller et al. (2008) and Prinn et al. (2000). There are many benefits to using this custom software approach, including complete source-code control over all instrument operation software, integration and data processing algorithms, and the ability to improve the software interactively. All AGAGE stations (except Hateruma and Ochiishi) and laboratories are linked via the internet so that functions such as instrument control and software updating can be done remotely. The strength of this approach is illustrated by the fact that, in addition to being used for all Medusa instruments in the AGAGE network, portions of the GCWerks software have been adopted by other leading laboratories engaged in non-AGAGE atmospheric and oceanic trace gas measurements, including NOAA/ESRL, CSIRO, the University of Bristol, and Empa.

Chromatograms are acquired and displayed in real time and are stored in a highly compressed format. Electronic strip charts record critical instrument parameters and a multitude of log files are generated as well, which contain parameters critical for data quality control. The GCWerks software allows operators and data processors to quickly review and batch-integrate chromatograms and produce time series and diagnostic plots of integration results to assess instrumental performance. The AGAGE data processing system relies on having identical software and databases at the field stations and at the data processing sites. This allows the station operators and investigators to review identical chromatograms and instrumental data in a timely manner and fosters constructive exchanges among the AGAGE investigators. The SIO server maintains a complete database for all stations and produces final results for all sites once the periodic data reviews have been completed. Data are routinely reviewed at regular intervals, and a final review is done approximately every 6 months prior to and at each AGAGE team meeting, with all the data processing sites involved concurrently.

New software (GCCompare, http://www.gcwerks.com, last access: 21 May 2018) continues to be developed for data processing, quality control, and visualization. This software has greatly streamlined the review and editing of AGAGE data that takes place over the internet and at AGAGE meetings twice a year. This software is highly interactive and has features such as being able to click on individual measurements and display back trajectories from the UK Met Office's NAME model (Jones et al., 2007) to help diagnose observed departures from background values. Recent station software developments continue, including enhancements of automated alarms to improve the oversight of day-to-day field operations and, importantly, to protect the instrumentation from damage when key components fail. Software for the correction of occasional drifts in more reactive gases in the on-site tertiary and quaternary calibration standards continues to be improved and implemented. Working in collaboration with NOAA/GMD, the software has also been modified to remove the need to divide the acquisition of peak data into time "windows". This had caused problems in optimizing dwell times on certain masses and in following small drifts in retention times of peaks located near transitions between windows. This change also allows for a reduction, to some degree, in the numbers of ions acquired at a given time, thereby improving precisions and detection limits, especially for the less abundant emerging compounds. GCWerks also keeps all of the raw data, including the chromatograms, thus enabling the routine reprocessing of the entire record for each species at each station whenever needed (e.g., when calibration scales are updated (see Sect. 2.6) or when peak integration methods are improved).

Finally, this GCWerks software is becoming an increasingly important "spin-off" from the AGAGE project. In particular, considerable progress has been made in adapting AGAGE data acquisition, visualization, and quality-control software for discrete sample GC and GC-MS instruments to applications involving continuous optical instruments such as the cavity ring-down spectrometer (CRDS) instruments of Picarro and Los Gatos Research (LGR) and the quantum cascade laser (QCL) instruments of Aerodyne Research.

2.6 Calibration

One of the strengths of AGAGE is its dependence upon well-defined internal absolute gravimetric calibration procedures that can be repeated periodically to ensure the accuracy of the long-term measured trends. During the period of AGAGE there have been seven absolute primary calibration efforts, SIO-93, SIO-98, SIO-05, SIO-07, SIO-12, SIO-14, and SIO-16, named after the SIO laboratory and the year in which the scale was completed. The "bootstrap" methods used to prepare primary gravimetric standards at ppt levels and the way in which these standards are integrated to define a calibration scale are described in the AGAGE "history paper" (Prinn et al., 2000). The methods used to propagate these scales to the species measured by the Medusa GC-MS are discussed by Miller et al. (2008). At present, ambient-level SIO primary calibration scales have been prepared for 42 AGAGE species: N_2O, PFC-14 (CF_4), PFC-116 (C_2F_6), PFC-218 (C_3F_8), PFC-318 (c-C_4F_8), PFC-3-1-10 (C_4F_{10}), PFC-4-1-12 (C_5F_{12}), PFC-5-1-14 (C_6F_{14}), PFC-6-1-16 (C_7F_{16}), PFC-7-1-18 (C_8F_{18}), SF_6, SF_5CF_3, SO_2F_2, NF_3, HFC-23, HFC-32, HFC-125, HFC-134a, HFC-143a, HFC-152a, HFC-227ea, HFC-236fa, HFC-245fa, HFC-356mfc, HFC-43-10mee, HCFC-22, HCFC-141b, HCFC-142b, CFC-11, CFC-12, CFC-113, CFC-114, CFC-115, Halon-1211, Halon-1301, Halon-2402, CH_3Br, CH_3Cl, CH_2Cl_2, $CHCl_3$, CH_3CCl_3, and CCl_4. Among them, NF_3, C_4F_{10}, C_5F_{12}, C_6F_{14}, C_7F_{16}, and C_8F_{18} were calibrated by the method of internal additions, which is by spiking real air with gravimetrically determined amounts of the analyte (Arnold et al., 2012), while the remaining gases were calibrated by the conventional AGAGE method of adding gravimetrically determined amounts of the analytes to analyte-free artificial "zero air". For CF_4, the primary calibrations have been made both ways with excellent agreement. For the volatile gases like CF_4 and NF_3, the use of the internal additions method is particularly valuable to avoid biases in their separation or detection due to interferences from the presence of krypton and other inert gases in real air but not in artificial zero air. The precisions of these calibration scales, based on the internal consistency among the individual primary standards, range from about 2 % for the least abundant compounds to < 0.1 % for the more abundant compounds. The absolute accuracies of these scales, based on estimates of maximum systematic uncertainties, including the purities of the reagents used in their preparation and possible systematic analytical interferences, are between 0.3 and 2 % greater than the statistical uncertainties depending on the compound and its atmospheric abundance.

The evolution of GC-MS techniques in AGAGE has greatly increased the number of species that are measured in the program and has thus exceeded, at least temporarily, our capacity to prepare and maintain gravimetric primary calibration scales. To bridge this gap and, very importantly, to decouple the long-term measurement program for the evolving and independent primary calibration process, AGAGE has adopted a relative calibration scale for all Medusa and GC-MD measurements. This scale, designated R1, is defined by regular intercomparisons of trace gas concentrations in a suite of whole-air secondary ("gold") tanks maintained at the SIO laboratory. These tanks are compared against each other to assess possible drift and against primary standards for those species for which we have primary standard calibrations. Every year, this suite of secondary tanks is extended with at least one new tank filled under clean air conditions in winter or spring and the intercomparison is repeated. Other tanks filled at the same time are calibrated against this suite of tanks and sent to each station as calibration "tertiary" standards, where they are either directly measured (GC-MD) or used to calibrate working "quaternary" standards (Medusa) at each measurement site. As primary calibration scales evolve at SIO, NOAA/ESRL, Bristol, Empa, NCAR, NIES, or any other laboratory, the relationships of their scales to the R1 scale can be measured to obtain a set of factors by which our R1 values can be multiplied to report Medusa data on any of these calibration scales. The R1 scale is flexible to designate tanks other than R1 as a reference tank for individual compounds, which were not present at sufficient concentrations or were not measured in the original R1 tank. Looking to the future, this enables us to keep pace with the changing atmospheric concentrations of many species and to incorporate corrections for possible nonlinearities in the calibration process and for possible drifts in standard mixtures. This technique has been used to provide calibrations for species not on an SIO scale such as CFC-13 (METAS-2017), $CHBr_3$ (NOAA-2009P), PCE (NOAA-2003B), and HCFC-133a (Empa-2013; Vollmer et al., 2015c).

AGAGE gravimetric calibration activities are independent from those in other laboratories (except for the CO_2 calibrations used in the bootstrap method that come from the Keeling laboratory at SIO), but there are also strong synergies, especially with NOAA/ESRL. For example, the SIO-14 calibrations showed excellent agreement with NOAA for Halon-2402 (Vollmer et al., 2016), while AGAGE atmospheric CH_2Cl_2 mole fractions based on the SIO-14 scale are significantly higher than those reported by NOAA (Carpenter et al., 2014). This subject of intercalibration is discussed further in Sect. 3.2.

Whole-air and synthetic mixture calibration standards used in AGAGE are stored in 34 L high-pressure (60 bar) electropolished stainless steel canisters designed at SIO and manufactured by Essex Industries© that are legal for international shipment. Although the adoption of a single primary calibration scale from a central calibration facility for each measured species has been advocated by some researchers, AGAGE does not favor this approach. The existence of more than one independent high-precision traceable calibration scale for each measured species, with frequent intercomparisons among independently calibrated field measurements (see Table 5, Sect. 3.2) and with direct intercomparison of the calibration standards themselves (Hall et al., 2014), reduces vulnerability to systematic errors and long-term calibration drifts for all participating primary calibration and measurement programs.

2.7 Primary and affiliate station facilities and infrastructure

While the individual station size and infrastructure varies depending on their location and the presence of other complementary gas and aerosol measurement programs, all stations consist of permanent buildings (wood, concrete, steel, fiberglass) with air samples drawn using non-contaminating pumps through lines with inlets located on adjacent high towers. The details about the general air sampling setup for each instrument are provided in Miller et al. (2008) and Prinn et al. (2000). The sampling lines are either stainless steel or layered polyethylene–aluminum–Mylar (Dekabon© or Synflex©). For more information on individual stations, we refer the reader to the AGAGE website (http://agage.mit.edu (last access: 21 May 2018). All stations (except Hateruma and Cape Ochiishi) periodically exchange stainless steel on-site Essex calibration tanks (tertiary standards) calibrated at SIO linking the measurements to the AGAGE SIO primary and secondary standards. Some stations also use modified RIX© oil-free air compressors and the tertiary standards to prepare quaternary standards either on-site, in their home laboratories, or supplied by SIO to extend the lifetime of the tertiary standards. At Cape Grim and Ny-Ålesund, the quaternary standards are prepared by a cryogenic collection of whole air with subsequent ejection of condensed water.

2.8 Secondary stations

In addition to the primary and affiliate stations in AGAGE, there are complementary secondary stations, usually at either more polluted urban locations or at more remote sites that share some or all of the AGAGE technology and calibrations.

SIO carries out continuous measurements of all AGAGE gases in La Jolla in conjunction with its extensive calibration (Sect. 2.6) and instrument development operations.

The University of Bristol runs the UK DECC (Deriving Emissions related to Climate Change) network of tall towers at Ridge Hill, Angus (now decommissioned), Tacolneston (in collaboration with the University of East Anglia), Heathfield (UK National Physical Laboratory), and Bilsdale in the UK measuring CO_2, CO, CH_4, N_2O, and SF_6 and linked to the AGAGE Mace Head station and to AGAGE calibrations and some technologies. Tacolneston also includes measurements of H_2 and CO via MRD and a Medusa GCMS.

CSIRO is operating two Medusa GCMSs at Aspendale, and Picarro CRDS CH_4 and CO_2 (and CO at one station) instruments at Burncluith (26° S, G2401), Ironbark (27° S, G2301), Aspendale (38° S, G2301), Macquarie Is-

Table 5. Scale conversion factors between NOAA and AGAGE (SIO) expressed as a NOAA / AGAGE ratio based on a comparison of NOAA/ESRL/GMD flask data to AGAGE in situ data at common sites. For CH_4, N_2O, and SF_6, NOAA flask data from the carbon cycle and greenhouse gases (CCGG) group have been used; for all other species NOAA flask data from the halocarbons and other atmospheric trace species (HATS) group are used. The respective scales used in each network are indicated in the table along with the instrumental method used for the analysis. The sites used in the comparisons are listed in column five, followed by the length of the comparison period. Lastly, comments on the consistency of the comparisons for each species are given.

Species	Ratio (NOAA / AGAGE)	NOAA scale method	AGAGE (SIO) scale method	Sites	Time period	Comment
CH_4	1.0001 ± 0.0007	NOAA-2004A GC-FID	Tohoku University GC-FID (GC-MD)	Five sites (CGO, SMO, RPB, THD, MHD)	1993–2017	0.1 % consistency over time
N_2O	0.9985 ± 0.0005	NOAA-2006A GC-ECD	SIO-16 GC-ECD (GC-MD)	Five sites (CGO, SMO, RPB, THD, MHD)	1997–2017	0.1–0.2 % consistency over time, slight increasing trend of 0.08 % per decade
SF_6	1.0049 ± 0.0029	NOAA-2014 GC-ECD	SIO-05 GC-MS Medusa	Six sites (CGO, SMO, RPB, THD, MHD, ZEP)	2004–2017	Small step in 2010, 0.5 % consistency over time
CFC-11	0.9993 ± 0.0009	NOAA-2016 GC-ECD	SIO-05 GC-ECD (GC-MD)	Four sites (CGO, SMO, THD, MHD)	1993–2017	~ 1 % consistency over time
CFC-12	0.9962 ± 0.0010	NOAA-2008 GC-ECD	SIO-05 GC-ECD (GC-MD)	Four sites (CGO, SMO, THD, MHD)	1993–2017	0.5 % consistency over time
CFC-113	1.0003 ± 0.0023	NOAA-2003MS GC-MS	SIO-05 GC-ECD–GC-MS Med	Four sites (CGO, SMO, THD, MHD)	1993–2017	~ 1 % consistency over time
CCl_4	1.015–1.038 (not constant, see comments)	NOAA-2008 GC-ECD	SIO-05 GC-ECD (GC-MD)	Four sites (CGO, SMO,THD, MHD)	1995–2017	Trend: 3.5–4.0 % difference in 1995–2000, to approximately 1.5 % difference in 2013–2017
CH_3CCl_3	1.0055 ± 0.0109	NOAA-2003 GC-MS	SIO-05 GC-ECD–GC-MS Med	Four sites (CGO, SMO, THD, MHD)	1993–2017	Initial trend during 1993–2000, from 3 % down to 0.5 % difference, then good agreement within 1 %
HCFC-22	0.9971 ± 0.0027	NOAA-2006 GC-MS	SIO-05 GC-MS-ADS Med	Four sites (CGO, SMO, THD, MHD)	1998–2017	1–2 % consistency over time
HCFC-141b	0.9941 ± 0.0049	NOAA-1994 GC-MS	SIO-05 GC-MS-ADS Med	Four sites (CGO, SMO, THD, MHD)	1998–2017	~ 2 % consistency over time
HCFC-142b	0.9743 ± 0.0052	NOAA-1994 GC-MS	SIO-05 GC-MS-ADS Med	Four sites (CGO, SMO, THD, MHD)	1998–2017	~ 2 % consistency over time
HFC-134a	1.0015 ± 0.0048	NOAA-1995 GC-MS	SIO-05 GC-MS-ADS Med	Four sites (CGO, SMO, THD, MHD)	1998–2017	~ 2 % consistency, better recently
HFC-152a	0.9976 ± 0.0227	NOAA-2004 GC-MS	SIO-05 GC-MS-ADS Med	Four sites (CGO, SMO, THD, MHD)	1998–2017	2–3 % consistency over time
H-1211	0.9799 ± 0.0050	NOAA-2006 GC-MS	SIO-05 GC-MS-ADS Med	Four sites (CGO, SMO, THD, MHD)	1998–2017	~ 2 % consistency over time
H-1301	0.9766 ± 0.0098	NOAA-2006 GC-MS	SIO-05 GC-MS Medusa	Three sites (CGO, SMO, THD)	2004–2015	~ 2 % consistency over time
H-2402	1.0208 ± 0.0100	NOAA-1992 GC-MS	SIO-14 GC-MS Medusa	Four sites (CGO, SMO, THD, MHD)	2004–2017	Small step change 2008–2009, 3–4 % consistency over time
CH_3Cl	1.0074 ± 0.0073	NOAA-2003 GC-MS	SIO-05 GC-MS-ADS Med	Four sites (CGO, SMO, THD, MHD)	1998–2017	2 % consistency over time

Table notes: comparisons between NOAA HATS data and AGAGE in situ were performed based on the NOAA data posted on the ftp site: ftp://ftp.cmdl.noaa.gov/hats/ (last access: 21 May 2018).
GC-MS-ADS Med indicates data from the ADS instruments at Cape Grim and Mace Head used from 1998–2003, with Medusa data used from 2004 onwards at the sites indicated.
GC-ECD–GC-MS Med indicates a combined data record from the GC-ECD (GC-MD) instruments with the GC-MS Medusa data used for the latter part of the record.
Sites: CGO – Cape Grim, Australia; SMO – Cape Matatula, Samoa; RPB – Ragged Point, Barbados; THD – Trinidad Head, USA; MHD – Mace Head, Ireland; ZEP – Zeppelin Mountain, Ny-Ålesund, Norway.
Some species are measured by multiple instruments and/or flask samples; selected results shown here.

land (55° S, G2301), Casey Station, Antarctica (66° S, originally a G1301 now replaced by a G2301), and onboard the new CSIRO research vessel the RV *Investigator* (G2301). Picarro CRDS CH_4 and CO_2 instruments were also previously operated at Gunn Point, northern tropical Australia (11° S, G1301, 2010–2017, currently suspended), Arcturus (22° S, G1301 replaced by G2301, 2010–2014), and Otway (38° S, ESP1000, 2009–2012). CSIRO is also operating high-precision Aerodyne Research QCL systems for CO and N_2O and another for the stable isotopes of CO_2 at Aspendale. All of these instruments are configured to run with AGAGE–GCWerks software (see Sect. 3.3).

2.9 Air archives

CSIRO has been collecting and archiving pressurized 34 L electropolished canisters of cryo-trapped air collected during clean air conditions at Cape Grim since the mid-1970s, and plans to continue into the future (Fraser et al., 2017). This "Southern Hemisphere air archive" has proven to be an invaluable resource to the international atmospheric chemistry community, including AGAGE, because a wide range of species that could not be measured at the time of collection can be measured retrospectively in the archive as long as those species are conserved in these canisters. Until 2013 a target of four Cape Grim air archive samples were collected each year, while from 2014 onwards six air archive tanks are collected each year. Measurements from this Southern Hemisphere archive have made significant contributions to several recent AGAGE papers by addressing the following: HFC-23 (Miller et al., 2010); PFCs (Mühle et al., 2010; Trudinger et al., 2016); SF_6 (Rigby et al., 2010); CFC-13, CFC-114, and CFC-115 (Vollmer et al., 2018); Halon-1211, Halon-1301, and Halon-2402 (Vollmer et al., 2016); and HFC-365mfc, HFC-245fa, HFC-227ea, and HFC-236fa (Vollmer et al., 2011). There was a parallel "Northern Hemisphere archive" collected by Rei Rasmussen at Cape Meares, Oregon during the ALE and GAGE programs, but these samples are no longer accessible to this program and are mostly used up. The SIO AGAGE group has been storing a Northern Hemisphere archive of air compressed at Trinidad Head and La Jolla since the mid-1990s and has collected a series of Northern Hemisphere air samples from various sources (e.g., SIO laboratories of Charles D. Keeling and Ray F. Weiss, NOAA-GMD, and NILU) and of varying integrity for trace gas measurements that extends this record back to the early 1970s. Measurements from this Northern Hemisphere archive have made significant contributions to several recent AGAGE papers, especially for more inert species such as the PFCs, NF_3, and SF_6 (e.g., Mühle et al., 2009, 2010; Rigby et al., 2010; Weiss et al., 2008; Arnold et al., 2013).

Additional air archive samples used in AGAGE studies were derived from firn air collections in Greenland and Antarctica obtained by international consortia. The AGAGE analyses of firn air used Medusa GC-MS instruments and substantially extended mole fraction data back in time along with emission estimates derived from the data, specifically for Halons (Vollmer et al., 2016), PFCs (Trudinger et al., 2016), and minor CFCs (Vollmer et al., 2018).

3 Data analysis and modeling

In this section, the seven subsections address the following: meteorological interpretation of data (Sect. 3.1), data intercomparisons (Sect. 3.2), flux estimation using data and models (Sect. 3.3), and flux estimation using 3-D Eulerian models (Sect. 3.4), 3-D Lagrangian models (Sect. 3.5), merged 3-D Eulerian and Lagrangian models (Sect. 3.6), and simplified (2-D) models (Sect. 3.7).

3.1 Meteorological interpretation

As part of processing the AGAGE data, we place an identification flag on each measured value in an attempt to separate regional and/or local pollution events from background measurements. The current, objective (statistically based) algorithm has been successfully implemented and uniformly applied to the entire ALE/GAGE/AGAGE time series including data from all AGAGE primary and affiliate stations (except Hateruma and Cape Ochiishi) and all instruments (GC-MS, GC-MD, Picarro). Moreover, the algorithm has been designed to be easily reapplied to the entire dataset in the event of (minor) modifications to the algorithm. The concept of the algorithm is to examine the statistical distributions of 4-month bins of measurements (approximately 4320 GC-MD or 1440 Medusa GC-MS values) of any species at a specified site and centered on one day at a time after removing the trend over the period (O'Doherty et al., 2001; Cunnold et al., 2002). The algorithm can be applied to the results from 3-D models to separate the background and polluted values (Ryall et al., 2001; Simmonds et al., 2005). We also use a 3-D Lagrangian back-trajectory model driven by reanalyzed meteorology, specifically the UK Met Office's Numerical Atmospheric dispersion Modelling Environment (NAME; Ryall et al., 1998; Jones et al., 2007), to further evaluate the statistical pollution algorithm (O'Doherty et al., 2001; Cunnold et al., 2002) and include this evaluation as part of the pollution and background identification flag associated with each measurement. NAME is Lagrangian (Sect. 3.5). In NAME, large numbers of particles at the station are effectively advected backwards in time by 3-D reanalysis meteorological fields, with turbulent dispersion represented by a random walk technique. Particles first encountering the surface or surface boundary layer in known trace-gas-emitting regions are then flagged as polluted. An observation is also considered potentially polluted if the atmosphere at the station is stable with very low winds and known nearby trace gas sources. NAME back trajectories are automatically computed for every AGAGE measurement and used extensively in the semiannual AGAGE data reviews.

3.2 Data intercomparisons

AGAGE cooperates with other groups carrying out flask sampling and/or in situ real-time tropospheric measurements in order to produce harmonized global datasets for use in modeling. Toward this end, AGAGE routinely collaborates with NOAA/ESRL/GMD to develop best estimates of the differences in absolute calibrations and field site calibrations between them and the AGAGE–SIO scales (see Elkins et al., 2015, and the NOAA/ESRL/GMD website for the NOAA/ESRL/GMD database). This is undertaken in several ways: comparisons involving exchanges of tanks (checking absolute calibration); comparisons of hemispheric and global mean trends estimated by the two networks; examination of differences between the AGAGE and GMD in situ instruments at our common in situ site, Cape Matatula (checking the propagation of standards to remote sites); and ongoing extensive comparisons between AGAGE in situ GC-MD and GC-MS data and GMD flask data at the six AGAGE sites where GMD flasks are filled (Zeppelin, Mace Head, Trinidad Head, Ragged Point, Cape Matatula, and Cape Grim), with the results reported at the semiannual AGAGE meetings. To help ensure progress on this and other cooperative endeavors, leaders and members of the relevant NOAA/GMD group regularly attend the semiannual AGAGE meetings; other joint meetings with GMD personnel are held from time to time. Examples of the scale conversion factors determined from the comparison of AGAGE in situ data to NOAA flask results are given in Table 5. There is generally good consistency with time for these with some exceptions, most notably CCl_4. The CCl_4 comparison shows a trend with time from around 3.5–4.0 % in 1995–2000 to approximately 1.5 % in 2013–2017. Because these factors are updated when additional intercomparisons occur, we advise data users to consult the AGAGE website (http://agage.mit.edu, last access: 21 May 2018) for possible updates.

Also, comparisons between AGAGE in situ GC-MD and GC-MS data at Cape Grim and flask data from other groups (CSIRO, NIES, U. East Anglia, SIO, U. Heidelberg, Max Planck Inst. Mainz) have been and continue to be made. Exchanges of tanks between the collaborating NIES group and AGAGE–SIO are also performed to compare absolute calibrations. Also, there are routine data intercomparisons carried out within AGAGE for those gases measured on both the AGAGE Medusa GC-MS and AGAGE GC-MD instruments. Finally, three AGAGE sites (SIO, Mace Head, and Cape Grim) participated in the WMO-organized IHALACE (International HALocarbon in Air Comparison Experiment), round robin intercomparisons (Hall et al., 2014).

3.3 Flux estimation using measurements and models

A major goal of AGAGE is to estimate surface fluxes and/or atmospheric sinks (lifetimes) of trace gases by merging measurements and models using advanced statistical methods (Prinn et al., 2000; Weiss and Prinn, 2011). Specifically, we use a range of Bayesian methods, in which a priori estimates of atmospheric sinks and surface fluxes (or uncertain parameters in flux models) are adjusted to improve agreement with the trace gas observations within estimated uncertainties, and it is important to ensure that the problems are well posed, that the ill-conditioning inherent in our emission estimations is minimized, and that model and measurement imperfections are accounted for properly (e.g., Prinn, 2000; Tarantola, 2005). A basic requirement for all our inverse schemes is an accurate and realistic atmospheric chemical transport model (CTM). Even small transport errors can lead to significant errors in estimated sources or sinks (Hartley and Prinn, 1993; Mahowald et al., 1997; Mulquiney et al., 1998). We use a range of CTMs to estimate trace gas budgets at different spatial scales: two-dimensional "box" models provide global source and sink estimates using baseline observations, global three-dimensional Eulerian models are used for estimating fluxes at national to continental scales, and high-resolution regional Lagrangian models provide fine-scale source estimations close to AGAGE monitoring sites.

Here, and in Sect. 3.4–3.7, we summarize the methods and models actually used by AGAGE scientists to interpret AGAGE measurements. There are alternative methods and models that may give differences in estimated emissions, especially at regional scales. The AGAGE publications generally address the issue of differences, if any, between the estimated emissions and those reported in prior studies by other non-AGAGE scientists. Some of these alternative methods are addressed in Sect. 4.6 and 4.8, but it is beyond the scope of this paper to review all the alternatives. Instead we refer the reader to two comprehensive books that provide in-depth summaries of most of the major models and methods used to estimate sources and sinks from measurements (Enting, 2002; Kasibhatla et al., 2000).

We relate the vector of measured atmospheric mole fractions ($\boldsymbol{y}$) to emissions or initial conditions in a "parameters vector" ($\boldsymbol{x}$) using the "measurement" equation $\boldsymbol{y}_{\text{obs}} = \mathbf{H}\boldsymbol{x}+\mathbf{e}$. Here $\mathbf{H}$ is a matrix of sensitivities, or partial derivatives, of simulated measurements in $\boldsymbol{y}$ $(=\mathbf{H}\boldsymbol{x})$ to each element in $\boldsymbol{x}$ and is derived using the CTMs; $\mathbf{e}$ describes the random component of the error due to errors in the measurements and in the CTM. These errors form the error covariance matrix $\mathbf{R}$. A prior estimate of $\boldsymbol{x}$ ($\boldsymbol{x}_{\text{prior}}$) is generally needed, with uncertainties contained in the error covariance matrix $\mathbf{P}_{\text{prior}}$. Formally, since the chemical lifetime for a reactive trace gas can depend on emissions of that gas, then $\mathbf{H} = \mathbf{H}(\boldsymbol{x})$ so the equation $\boldsymbol{y} = \mathbf{H}\boldsymbol{x}$ is nonlinear in $\boldsymbol{x}$. For many ozone-depleting and greenhouse gases this nonlinearity is negligibly weak and is ignored. Exceptions exist, as discussed later. There are a number of statistical approaches that have been developed and implemented to make these estimations (e.g., Kasibhatla et al., 2000; Prinn, 2000; Rigby et al., 2011; Ganesan et al., 2014).

A common Bayesian statistical approach is "optimal estimation" (e.g., Kasibhatla et al., 2000) in which one minimizes a "cost" function (J) that is the sum of two quadratic forms: $(\boldsymbol{y}_{\text{obs}}-\boldsymbol{y})^T\mathbf{R}^{-1}(\boldsymbol{y}_{\text{obs}}-\boldsymbol{y})$ that minimizes the weighted difference between measured and modeled mole fractions and $(\boldsymbol{x}-\boldsymbol{x}_{\text{prior}})^T\mathbf{P}_{\text{prior}}^{-1}(\boldsymbol{x}-\boldsymbol{x}_{\text{prior}})$ that minimizes the weighted difference between the estimated parameters and their prior. This minimization yields analytical solutions to $\boldsymbol{x}=\boldsymbol{x}_{\text{prior}}+\mathbf{G}(\boldsymbol{y}_{\text{obs}}-\boldsymbol{y})$, $\mathbf{P}=(\mathbf{I}-\mathbf{GH})\mathbf{P}_{\text{prior}}$, and the "gain" matrix $\mathbf{G}=\mathbf{P}_{\text{prior}}\mathbf{H}^T(\mathbf{HP}_{\text{prior}}\mathbf{H}^T+\mathbf{R})^{-1}$. Examples of this approach using global 3-D Eulerian models are provided by Chen and Prinn (2005, 2006) for CH_4, Xiao et al. (2010a) for CH_3Cl, Xiao et al. (2010b) for CCl_4, Rigby et al. (2010, 2011) for SF_6, Saikawa et al. (2012, 2014b) for HCFC-22, and Huang et al. (2008) and Saikawa et al. (2014a) for N_2O. Weak nonlinearities may occur when lifetimes vary with emissions (e.g., OH depends on CO and CH_4 emissions). This problem can be addressed by recalculating the time-dependent partial derivative (sensitivity) **H** matrix after inversion of all the data and then repeating the inversion with the new **H** matrix to ensure convergence (Prinn, 2000).

Random measurement imperfections are associated with in situ instrument precision, satellite retrieval errors, and inadequate sampling in space and time. If known, random model errors can also be incorporated into the model–measurement error covariance matrix (**R**). It is also important to recognize that correlated model–measurement errors, which comprise **R**, and errors in the prior contained in $\mathbf{P}_{\text{prior}}$ are often poorly known quantities. Ganesan et al. (2014) explicitly allow such uncertainties to be derived in the inversion to minimize the effect of subjective assumptions on derived fluxes. This hierarchical Bayesian method (Ganesan et al., 2014) incorporates "hyper parameters" that describe the model–measurement and/or prior uncertainty covariance matrices (**R** and **P**) in the inversion. This approach leads to solutions that are less sensitive to the (often subjective) assumptions that are required about uncertainties in traditional Bayesian approaches. The hierarchical inversion scheme cannot, in general, be solved analytically, and therefore Markov chain Monte Carlo (MCMC) methods must be applied to that sample from the posterior distribution using a large number ($\sim 10^4$–10^5) of realizations of the parameter space (e.g., Rigby et al., 2011). Recently, this MCMC approach has been extended to include problems in which the dimension of the parameter space is itself considered unknown using a so-called "reversible jump" MCMC algorithm (Lunt et al., 2016). This method has been applied to high-resolution regional inversions using a Lagrangian model to sample from a range of possible basis function decompositions of the flux space, objectively determining the level of decomposition that is appropriate to effectively minimize "aggregation error" (i.e., an inflexibility in the space that could lead to errors in the prior distribution unduly influencing the outcome of the inversion; Kaminski et al., 2001), while maintaining an acceptable level of uncertainty reduction.

We also address model structural errors and random and systematic transport errors (i.e., errors in **H**) through the utilization of multiple model versions (Locatelli et al., 2013) and Monte Carlo methods (Prinn et al., 2001, 2005; Huang et al., 2008). The Monte Carlo methods also include systematic errors in measurement calibration.

For the determination of the regional sources of trace gases, beginning with Chen and Prinn (2006) we now frequently merge measurements from the AGAGE and NOAA/ESRL/GMD stations and also aircraft and satellites whenever appropriate (e.g., Ganesan et al., 2017). Because source and sink estimation is very sensitive to errors in time and space gradients, we ensure intercalibration among instruments of the same type and intercomparison between different instruments measuring the same quantity. We also objectively determine the accuracy and precision of each measurement when combining data, since data are weighted inversely to their variances (contained in **R**). Nonzero values for the off-diagonal elements of **R** and **P** can occur. Because the AGAGE measurement stations are well separated, off-diagonal elements (covariances) of **R** should be much smaller than the diagonal elements (variances) and are usually ignored. Also, **P** element covariances should be much smaller than the variances except when state vector elements are correlated, which can be avoided when choosing the elements. These covariances (off-diagonal elements of **P**) are discussed in the individual papers where they are relevant.

3.4 Flux estimation using 3-D Eulerian models

For our inverse studies we initially used the 3-D MATCH of the National Center for Atmospheric Research, NCAR (Mahowald et al., 1997; Rasch et al., 1997; Lawrence et al., 1999). MATCH was driven by data from the NCEP, ECMWF, and GSFC/NASA/DAO reanalyses (Rasch et al., 1997; Mahowald et al., 1997). Subgrid mixing processes, which include dry convective mixing, moist convective mixing, and large-scale precipitation processes, were computed in the model. MATCH was used at a horizontal resolution as fine as T62 ($1.8° \times 1.8°$), with either 42 or 28 levels in the vertical. Utilizing MATCH with AGAGE, ESRL, and other data, we estimated monthly regional and global emissions for many AGAGE species (e.g., Chen and Prinn, 2005, 2006; Huang et al., 2008; Xiao et al., 2010a, b). The ability of MATCH to accurately simulate the effects of transport on long-lived trace gases is well illustrated by CH_4 simulations (Chen and Prinn, 2005). The need to use reanalysis meteorology in MATCH that captures the actual circulation was evident from the observed CH_3CCl_3 seasonal cycle at the tropical South Pacific station (Samoa) that showed remarkable sensitivity to the El Niño–Southern Oscillation (ENSO). This sensitivity was attributed to the modulation of cross-equatorial transport during the Northern Hemisphere winter by the interannually varying upper tropospheric winds in the equatorial Pacific; this was a previously unappreciated aspect of tropical atmospheric tracer transport (Prinn et al., 1992).

More recently we use the newer NCAR Model for Ozone and Related Tracers (MOZART) that also simulates global three-dimensional mole fractions of atmospheric trace species (Emmons et al., 2010). Like MATCH, MOZART can be run off-line and driven by a variety of state-of-the-art reanalysis meteorological fields, including the National Center for Environmental Prediction/NCAR reanalysis (Kalnay et al., 1996) and the NASA Modern Era Retrospective Analysis for Research and Applications, NASA-MERRA (Bosilovich et al., 2008). We have specifically used MOZART inversions to estimate regional emissions for SF_6 (Rigby et al., 2010), heavy PFCs (Ivy et al., 2012a, b), HCFC-22 (Saikawa et al., 2012, 2014b), and N_2O (Saikawa et al., 2014a).

3.5 Flux estimation using 3-D Lagrangian models

Another modeling approach that we have used utilizes air histories or "footprints" computed from Lagrangian models driven by analyzed observed winds. These air histories, computed over a predefined region, quantify the time and locations that air masses have interacted with the surface (and therefore fluxes from the surface) prior to measurement at a station. Using this information and the measurements, we can solve for fluxes from these predefined regions. The method requires accurate simulation of both advective back trajectories and diffusion. We had examined earlier the use of the HYSPLIT model (Draxler and Hess, 1997) for this purpose (Kleiman and Prinn, 2000), and now we also use additional Lagrangian particle dispersion models (LPDMs). In particular, the LPDM NAME of the UK (Ryall et al., 1998) has been used to determine source strengths for observed species on regional scales (e.g., Cox et al., 2003; O'Doherty et al., 2004, 2009; Reimann et al., 2005; Derwent et al., 2007; Ganesan et al., 2015; Manning et al., 2011; Rigby et al., 2011; Lunt et al., 2015). The LPDM FLEXPART has also been applied to the inversion of AGAGE data for several species (Stohl et al., 2009, 2010; Maione et al., 2014; Graziosi et al., 2015, 2016, 2017; Fang et al., 2014).

3.6 Flux estimation using merged Eulerian and Lagrangian models

Given the high-frequency nature of the AGAGE measurements, we can extract a great deal of information on sources close to the monitoring sites. LPDMs like NAME have the useful property that they directly calculate the sensitivity of the measurements to emissions from every grid cell in the domain. However, one limitation of these models is that boundary conditions must be specified or estimated (e.g., Stohl et al., 2009). In contrast, inversions using global Eulerian CTMs, such as MOZART, do not usually require boundary conditions but can only estimate emissions from a limited number of regions (unless an adjoint model of the CTM is available; Meirink et al., 2008). In addition, these models are sensitive to uncertainties in species lifetimes.

To combine the Eulerian and Lagrangian approaches, we can decompose the sensitivity matrix $\mathbf{H}$ into components that represent the sensitivities of the observations to initial conditions ($\mathbf{H}_{IC}$), emissions from model grid cells close to AGAGE stations ($\mathbf{H}_{LE}$), and emissions from aggregated regions that are farther from the AGAGE sites ($\mathbf{H}_{NLE}$): $\mathbf{H} = (\mathbf{H}_{IC}, \mathbf{H}_{NLE}, \mathbf{H}_{LE}$; Rigby et al., 2011). $\mathbf{H}_{IC}$ and $\mathbf{H}_{NLE}$ can be estimated using the Eulerian model at reasonable computational cost, while the term $\mathbf{H}_{LE}$ can be determined using the Lagrangian model. Consideration must be made of the fate of emissions close to AGAGE sites that leave the LPDM region and gradually become mixed into the global atmosphere. $\mathbf{H}_{LE}$ must therefore be decomposed into a short-timescale term $\mathbf{H}_{LE,LAM}$, for which the Lagrangian model is used, and a long-timescale term $\mathbf{H}_{LE,EUM}$, which can be approximated using the Eulerian model. Once $\mathbf{H}$ is constructed, the inversion can be solved using any Bayesian inverse method incorporating measurement, model, and state error covariance matrices (Sect. 3.4). This approach has the advantage over previous global emissions estimates that only used an LPDM in that constant background mole fractions do not have to be assumed (e.g., Stohl et al., 2009). Further, by solving for regional and global emissions and covariance in a single step, we can avoid many of the problems encountered in two-step "nested" inverse methods (e.g., covariance between emissions in the "Lagrangian region" and those outside, as in the method of Rödenbeck et al., 2009).

Inverse estimates of global sulfur hexafluoride (SF_6) emissions have been carried out using this method (Rigby et al., 2011). The derived global total emission rate agrees well with previous CTM-based estimates by Rigby et al. (2010), and the regional emissions qualitatively agree with their findings.

3.7 Application of simplified models

The 3-D models, being computationally expensive, do not always lend themselves well to doing very long time integrations and multiple runs to address uncertainty (e.g., thousands of runs for Monte Carlo treatments of model, rate constant, and absolute calibration errors). Therefore, 2-D models have been widely used to analyze long-term trends in AGAGE data. The AGAGE 12-box model (Cunnold et al., 1994; Prinn et al., 2001, 2005; Rigby et al., 2013, 2014) uses transport parameters that have been "tuned" using AGAGE observations of trace gas trends and latitudinal gradients (e.g., Cunnold et al., 1994; Rigby et al., 2013) so that the model can simulate monthly mean observations at background AGAGE stations with pollution events removed. From these simulations, multi-decadal AGAGE time series have been used to estimate trace gas global emissions and atmospheric lifetimes. For example, this 2-D model has been

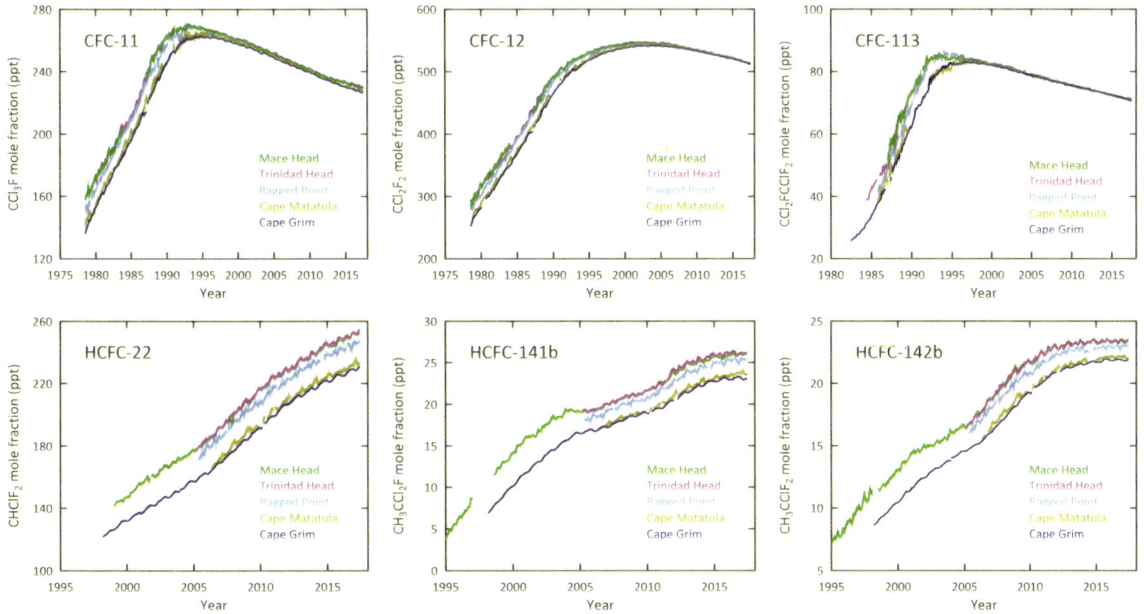

Figure 3. Monthly mean mole fractions (ppt) and their standard deviations (vertical bars) for selected AGAGE Montreal Protocol gases through 2017.

used to estimate emissions of three light PFCs (PFC-14, PFC-116, and PFC-218; Mühle et al., 2010) and NF_3 (Arnold et al., 2013), and combined with the 3-D MATCH has provided estimates of the influence of model errors on overall emission uncertainties for N_2O (Huang et al., 2008). A more simplified three-box 2-D model has also been employed to simultaneously estimate CH_3CCl_3 and CH_4 lifetimes and emissions using AGAGE observations of interhemispheric differences and growth rates (Rigby et al., 2017).

4 Sample scientific accomplishments

In this section, the nine subsections discuss the following: trends in Montreal Protocol gases and their replacements (Sect. 4.1), whether the Montreal Protocol is working (Sect. 4.2), trends in Kyoto Protocol gases (Sect. 4.3), the recent rise of powerful synthetic greenhouse gases (Sect. 4.4), trends in radiative forcing (Sect. 4.5), the determination of OH concentrations using models and multiple gases (Sect. 4.6), AGAGE emission estimates for all gases (Sect. 4.7), emission estimates from multiple networks, measurement platforms and alternative models (Sect. 4.8), and tabulation and access to AGAGE publications (Sect. 4.9). We focus on the greenhouse and ozone-depleting gases in this section, but note that the four non-methane hydrocarbons listed in Table 1 (ethane, propane, benzene, toluene) are optional compounds measured at some of the stations (for examples, see Yates et al., 2010; Grant et al., 2011; Derwent et al., 2012; Lo Vullo et al., 2016a, b). Also, CO and H_2 are measured at two stations (e.g., Xiao et al., 2007). When correlated with the other gases in Table 4, these species can be used as indicators of the sources of these other gases, and they are all also relevant to the fast photochemistry of OH.

4.1 Trends in Montreal Protocol gases and their replacements

The Montreal Protocol on Substances that Deplete the Ozone Layer, enacted to protect the ozone layer, regulates many ozone-depleting gases for the primary purpose of lowering stratospheric chlorine and bromine concentrations. From AGAGE measurements (Fig. 3), two of the major CFCs (CFC-11, CFC-113) have both been decreasing in the atmosphere since the mid-1990s. While their emissions have decreased very substantially in response to the Montreal Protocol, their long lifetimes of around 50 and 90 years, respectively, mean that their sinks can reduce their levels only at about 2 and 1 % per year, respectively. The other major CFC (CFC-12) has a somewhat longer lifetime (about 100 years) and a slower phase-out of emissions, and consequently its atmospheric levels have reached a plateau more recently and are now decreasing.

The three major HCFCs (HCFC-22, HCFC-141b, and HCFC-142b) are replacements for the CFCs and have continued to rise in recent years. The rates of rise decreased somewhat in the late 1990s for HCFC-141b (9-year lifetime) and HCFC-142b (18-year lifetime), which is consistent with decreases in their emissions from developed countries. They then increased again. which is consistent with increases in developing country emissions. In contrast, rates of rise have slowly declined post-2008 for HCFC-22 (12-year lifetime).

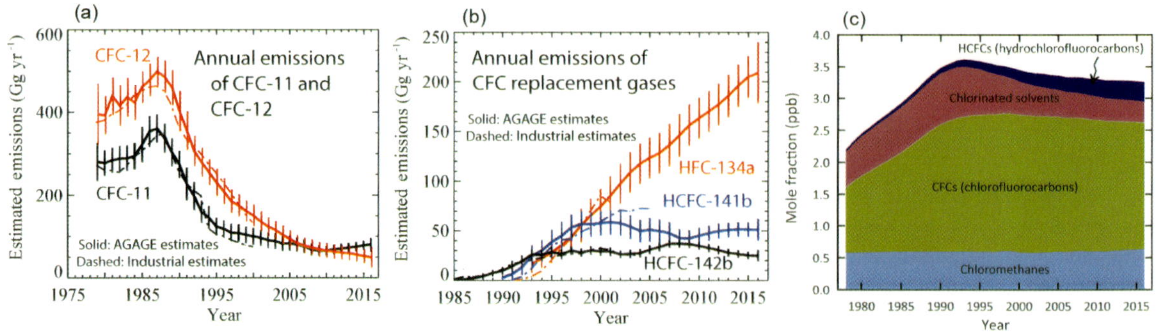

Figure 4. Inversely estimated emissions using our Bayesian statistical approach (Sect. 3.3) and our 12-box model (Sect. 3.7) of the following: selected AGAGE regulated gases **(a)** and selected AGAGE replacement gase **(b)** compared to estimates from industrial, national, and/or UNEP reports. Estimates of total tropospheric chlorine from all AGAGE data (**c**; chlorinated solvents are CCl_4 and CH_3CCl_3; chloromethanes are CH_3Cl, CH_2Cl_2, and $CHCl_3$; see Table 1 for a full list of AGAGE chlorine-containing compounds).

AGAGE mole fraction data and derived emissions from a wide range of ozone-depleting species have been published in multiple recent papers (Fraser et al., 2014; Graziosi et al., 2015, 2016; Keller et al., 2011; Kim et al., 2010, 2012; Li et al., 2011, 2014; Lunt et al., 2015; Maione et al., 2013, 2014; Miller et al., 1998; Rigby et al., 2013; Saikawa et al., 2012, 2014b; Stohl et al., 2010; Vollmer et al., 2016, 2017; Xiang et al., 2014; Xiao et al., 2010b).

4.2 Is the Montreal Protocol working?

The global abundance of tropospheric chlorine and emissions, via inverse methods, of ozone-depleting gases are estimated from AGAGE measurements (Fig. 4). Some specific conclusions are as follows.

1. International compliance with the Montreal Protocol is so far resulting in CFC and chlorocarbon abundances comparable to the target levels – the Protocol is working although estimated global CFC-11 emissions post-2010 are rising (Fig. 4), but the method used does not provide the regional-level emission estimates needed to identify the causes of this rise. Montzka et al. (2018) recently concluded that East Asia was the source.

2. The abundance of total chlorine in long-lived CFCs and other chlorocarbons (CFC-11, CFC-12, CFC-13, CFC-113, CFC-114, CFC-115, HCFC-22, HCFC-141b, HCFC-142b, $CHCl_3$, CH_3CCl_3, CH_3Cl, CCl_4, CH_2Cl_2, CCl_2CCl_2) in the lower troposphere reached a maximum of about 3.6 ppb in 1993 and is beginning slowly to decrease in the global lower atmosphere driven initially by CH_3CCl_3 and later by CFC decreases (note that CH_3Cl, CH_2Cl_2, and CCl_2CCl_2 are not regulated in the Montreal Protocol, yet CH_2Cl_2 is increasing).

3. The CFCs have atmospheric lifetimes consistent with destruction in the stratosphere being their principal removal mechanism.

4. Multi-annual variations in measured CFC, HCFC, HFC, and other chlorocarbon emissions deduced from ALE/GAGE/AGAGE data are approximately consistent with variations estimated independently from industrial production and sales data where available. HCFC-141b shows the greatest discrepancies. The processes producing the deduced CCl_4 emissions are not fully understood. The 2010 and 2014 WMO Scientific Assessments of Ozone Depletion noted that emissions of CCl_4 inferred from AGAGE and NOAA observations were substantially higher ($\sim 50\,Gg\,yr^{-1}$) than estimates based on consumption reported to UNEP (Montzka et al., 2011a; Carpenter et al., 2014). Recent studies have attempted to reevaluate the global CCl_4 budget (Liang et al., 2017). Estimates of the soil and ocean partial lifetimes have been revised upward (Rhew and Happell, 2016; Butler et al., 2016) and several new industrial sources have been identified (Sherry et al., 2017), substantially reducing the gap between top-down and bottom-up estimates (Chipperfield et al., 2016).

5. The mole fractions of the HCFCs, which are interim replacements for CFCs, rose very rapidly in the atmosphere until the early 2000s, but are now only rising relatively slowly; the exception is HCFC-22, which has been in use almost as long as the CFCs. HCFC-22 continues to increase rapidly in the atmosphere and contributes significantly to atmospheric chlorine loading.

6. The mole fractions of HFCs, which are long-term replacements for CFCs and HCFCs, continue to rise rapidly in the atmosphere and are the major Kyoto synthetic greenhouse gases contributing to increased radiative forcing. They were added to the Montreal Protocol in the 2016 Kigali Amendment.

AGAGE scientists, AGAGE data, and AGAGE modeling results played a prominent role in all the WMO-UNEP Ozone Assessments, most recently the WMO-UNEP 2010

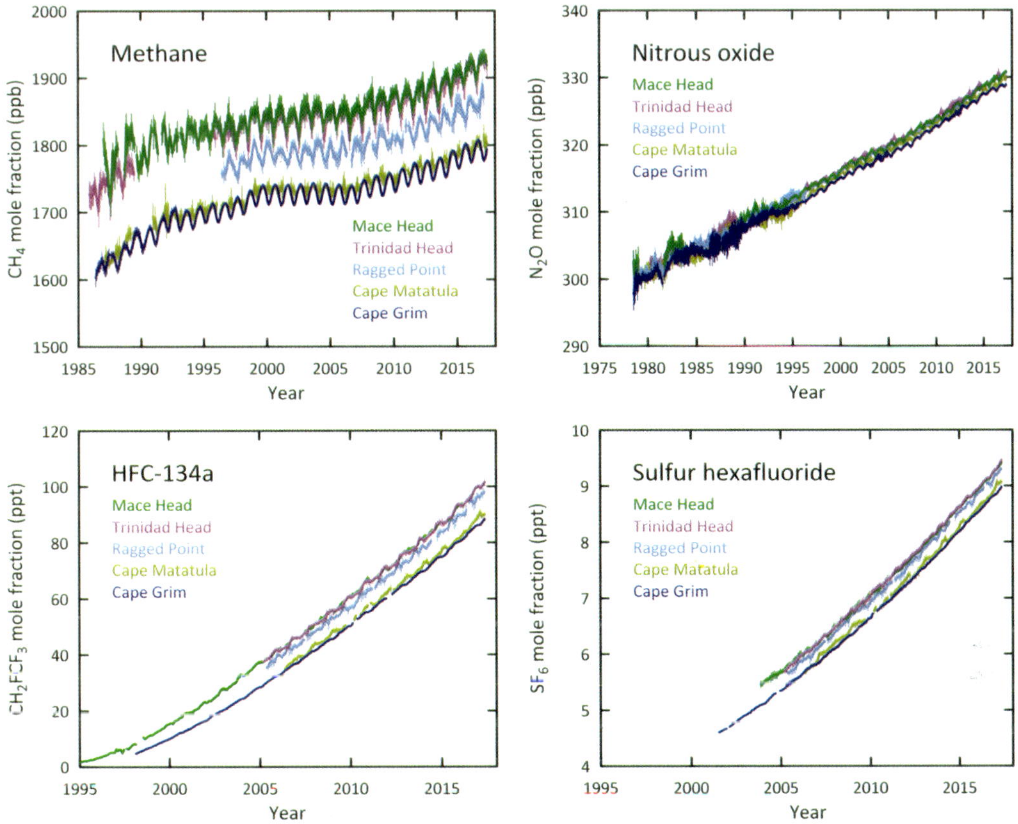

Figure 5. Monthly mean mole fractions and standard deviations for selected Kyoto Protocol gases through 2017.

(Montzka et al., 2011a) and WMO-UNEP 2014 (Carpenter et al., 2014) Ozone Assessments, also providing many coordinating and lead authors, coauthors, contributing authors, and reviewers. The AGAGE-led paper on the reevaluation of the lifetimes of the major CFCs and CH_3CCl_3 using atmospheric trends (Rigby et al., 2013) was an important input into the 2014 Ozone Assessment (Carpenter et al., 2014).

4.3 Trends in Kyoto Protocol gases

The Kyoto Protocol, followed now by the Paris Accord, regulates several powerful GHGs in addition to CO_2. Methane is the second most important long-lived anthropogenic GHG. AGAGE measurements (Fig. 5) show that its concentration has been rising in recent decades with large year to year variations. Its multiyear average rate of increase had been decelerating, with no significant increase over a 9-year period, perhaps as a result of an approach to a state in which its multiple sources are balanced by a roughly constant sink rate (reaction with OH). Methane then began to rise again around 2006. AGAGE data and emission estimates for methane have appeared in multiple recent papers (Rigby et al., 2008; Kirschke et al., 2013; Loh et al., 2015; Manning et al., 2011; Patra et al., 2011; Saito et al., 2013; Thompson et al., 2015; Saunois et al., 2016, 2017). Nitrous oxide is the third most important long-lived greenhouse gas (after CO_2 and CH_4) and the major source of ozone-depleting nitric oxide (NO) and nitrogen dioxide (NO_2) in the stratosphere (Ravishankara et al., 2009). The atmospheric N_2O concentration has been increasing almost linearly over recent decades. Estimated preindustrial N_2O levels are around 270 parts per billion (ppb; see MacFarling-Meure et al., 2006) compared to the 22 % higher levels of 329.3 ppb in 2016. The primary cause of its recent increase and the reasons for its atmospheric cycles are addressed by Huang et al. (2008), Nevison et al. (2011), Thompson et al. (2013, 2014a, b, c), and Saikawa et al. (2014a) using AGAGE and NOAA-ESRL data.

AGAGE measurements and estimated emissions of the purely synthetic Kyoto-Protocol-type gases (HFCs, PFCs, SF_6, NF_3) have been published in many recent AGAGE papers (Arnold et al., 2013, 2014; Graziosi et al., 2017; Ivy et al., 2012a, b; Keller et al., 2011, Kim et al., 2010, 2012, 2014; Li et al., 2011, 2014; Miller et al., 2010; Mühle et al., 2010; O'Doherty et al., 2014; Rigby et al., 2010, 2011, 2014; Saikawa et al., 2014b; Simmonds et al., 2015, 2016; Stohl et al., 2009, 2010; Vollmer et al., 2011; Xiang et al., 2014). Two examples of these are given here: HFC-134a, the most abundant HFC, and sulfur hexafluoride. The atmospheric abundance of the air-conditioning refrigerant HFC-134a is increasing at a rapid rate in response to its growing emissions arising from its role as the major replacement for the refrigerant CFC-12. With a lifetime of about 14 years, its current atmospheric abundance is determined primarily by its emissions and secondarily by its atmospheric destruction. SF_6 is produced largely for use as an insulating fluid in electrical distribution equipment. Its concentrations have been increasing continuously since in situ AGAGE measurements began in the 2000s and archive tanks began to be filled in the 1970s. Its very long lifetime ensures that its emissions accumulate essentially unabated in the atmosphere. AGAGE data have also been used to quantify the recent decline of HCFC emissions and rise in its replacement HFC emissions (Simmonds et al., 2017).

AGAGE scientists and AGAGE data and modeling results played a significant role in multiple IPCC Climate Change Assessments, most recently the IPCC 4th Assessment: Climate Change 2007 (WG1, chap. 2; Forster et al., 2007), and the IPCC 5th Assessment: Climate Change 2013 (WG1, chap. 2; Hartmann et al., 2013), also providing lead authors, contributing authors, and reviewers. AGAGE data also contributed significantly to the recent history of greenhouse gas mole fractions to drive climate model simulations for use in the IPCC 6th Assessment (Meinshausen et al., 2017).

4.4 Recent rise of powerful synthetic greenhouse gases

While the radiative forcing of purely synthetic greenhouse gases (SGHGs) regulated by the Montreal Protocol has decreased substantially since around 1993, newer SGHGs with global warming potentials (GWPs) of many thousands have become more and more important in recent years, and unabated they are expected to become even more so in the future (Rigby et al., 2014). These gases are used in many high-technology applications (e.g., HFCs in refrigeration and air-conditioning, PFCs as solvents and emitted from aluminum, semiconductor, and rare-earth metal production, SF_6 in electric power distribution, and NF_3 in flat-screen displays and semiconductor production). Regulations forcing their recycling or their replacement may be needed.

AGAGE measures all of the significant SGHGs and Fig. 6 show global radiative forcing by each of these gases based on observations (Rigby et al., 2014, extended to 2017). CO_2-equivalent emissions using 100-year GWPs have been derived from AGAGE observations for HFCs and PFCs plus SF_6, NF_3, and SO_2F_2 and compared to reported emissions from Annex-1 countries that are signatories to the United Nations Framework Convention on Climate Change (UNFCCC). Unreported emissions from non-Annex-1 countries (i.e., AGAGE-derived total emissions minus Annex-1 reported emissions) have been rapidly increasing since 1990 for both these classes of SGHGs and are now 35 % more than Annex-1 for the HFCs and 600 % more for the PFCs plus SF_6. The mole fractions and derived emissions of AGAGE-measured heavy HFCs have all been increasing rapidly since

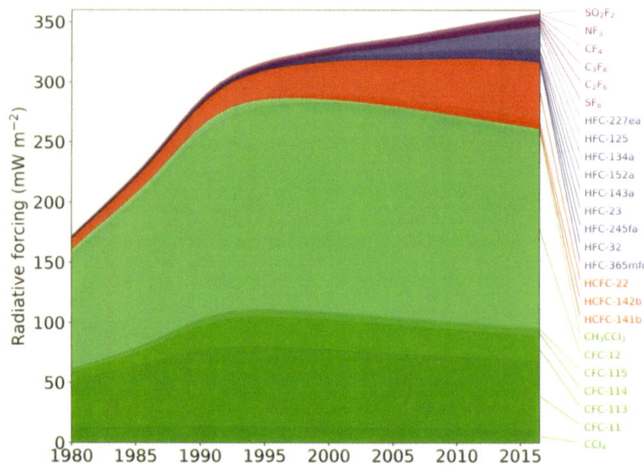

Figure 6. Global radiative forcing due to long-lived SGHGs derived from AGAGE observations from 1980 to 2017 (update of Rigby et al., 2014).

the early 2000s for HFC-365mfc and HFC-245fa and since 1995 for HFC-227ea and HFC-236fa (Vollmer et al., 2011).

4.5 Trends in total radiative forcing

By adding the radiative forcing ($W\,m^{-2}$) of the Montreal Protocol, Kyoto Protocol, and recent unregulated synthetic greenhouse gases, the overall radiative forcing due to all long-lived substances is obtained. Figure 7 shows that radiative forcing by CO_2 still dominates, and the percentage of the total forcing due to the non-CO_2 AGAGE greenhouse gases is slowly decreasing, reaching $\sim 36\,\%$ by the end of 2016. However, the emissions, mole fractions, and absolute radiative forcing of non-CO_2 gases continue to rise.

4.6 Determination of OH concentrations using models and multiple species

The hydroxyl free radical is the major oxidizing chemical in the atmosphere, destroying about 3.7 petagrams of trace gases each year, including many gases involved in ozone depletion, the greenhouse effect, and urban air pollution. The large-scale concentrations and long-term trends in OH can in principle be measured indirectly using global measurements of trace gases whose emissions are well known and whose primary sink is OH. The best trace gas for this purpose is the industrial chemical CH_3CCl_3. First, there are accurate long-term measurements of CH_3CCl_3 beginning in 1978 in the ALE/GAGE/AGAGE network (Prinn et al., 1983b, 2000, 2001, 2005, Rigby et al., 2008, 2013, 2017) and beginning in 1992 in the NOAA/CMDL network (Montzka et al., 2000, 2011). Second, CH_3CCl_3 has fairly simple end uses as a solvent, and voluntary chemical industry reports since 1970, along with the national reporting procedures under the Montreal Protocol in more recent years, have produced reason-

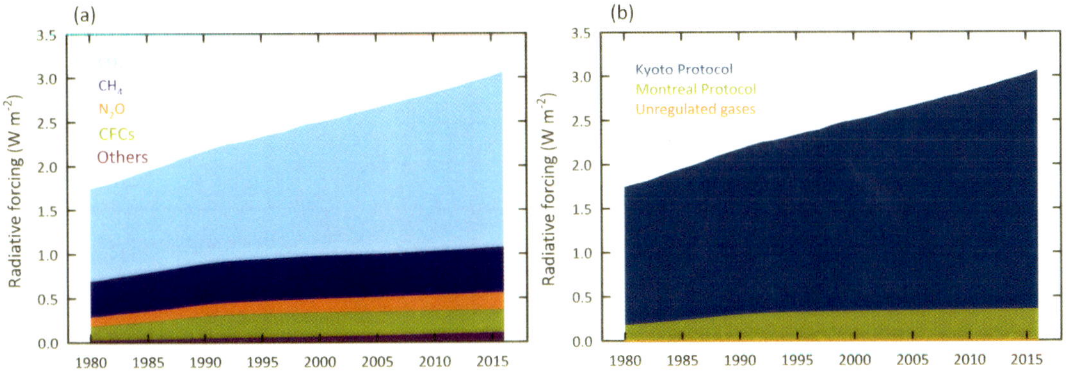

Figure 7. Global total radiative forcing due to long-lived greenhouse gases derived from NOAA-GMD measurements for CO_2 and AGAGE observations for all others **(a)**. Also shown are the contributions from the gases in the Kyoto and Montreal Protocol and those not regulated by either protocol **(b)**.

ably accurate emissions estimates for this chemical (McCulloch and Midgley, 2001). The use of CH_3CCl_3 for OH concentration and trend estimation has been extensive (Prinn et al., 1987, 1995, 2001, 2005; Spivakovsky et al., 2000; Montzka et al., 2000, 2011b; Krol and Lelieveld, 2003; Bousquet et al., 2005). Generally, interannual variability in OH inferred from CH_3CCl_3 inversions is larger than those calculated in atmospheric photochemical models (e.g., Montzka et al., 2011b), and the reasons are currently unresolved. Other gases that are useful OH indicators include ^{14}CO, which is produced primarily by cosmic rays (Manning et al., 2005). Using HCFC-22 measurements for estimating the average OH yields similar results to those derived from CH_3CCl_3 but with less accuracy (Miller et al., 1998; Fortems-Cheiney et al., 2013). The industrial gases HFC-134a, HCFC-141b, and HCFC-142b are potentially useful OH estimators but the accuracy of their emission estimates needs improvement (Huang and Prinn, 2002; Fortems-Cheiney et al., 2015). At the present time, to augment CH_3CCl_3, the potential OH estimation species (the major tropospheric sink is reaction with OH, and industrial emissions estimations are relatively good) are HFC-134a, HCFC-141b, HCFC-142b, and possibly some of the newly introduced HFCs (Liang et al., 2017).

AGAGE data (Fig. 8) show that CH_3CCl_3 levels and latitudinal gradient rose steadily from 1978 to reach a maximum in 1992 and have both since rapidly decreased as the Montreal Protocol drove emissions to near zero. In 2010 the levels were about 3 % of those when AGAGE measurements began in 1978. Analysis of these observations shows that global average OH levels vary from year to year only occasionally significantly, but exhibit no significant long-term trend (Prinn et al., 2001, 2005; Rigby et al., 2008, 2013, 2017, latter updated in Fig. 8). This analysis includes the effects of observationally derived corrections to emissions and model and measurement errors. The 1997–1999 OH minimum coincides with, and is perhaps caused by, major global wildfires and an intense El Niño event at that time. Recent CH_3CCl_3 inversions have proposed a role for a rise and fall in OH in the pause and renewed growth of atmospheric methane (McNorton et al., 2016; Rigby et al., 2017; Turner et al., 2017). However, these trends were not found to be statistically significant when all uncertainties were considered.

4.7 AGAGE emissions estimates for all gases

A major objective of all the AGAGE GC-MD and GC-MS measurements is to produce estimates of global emissions, spatial distributions of emissions, and their trends. These results are given in a large number of AGAGE publications (see Sects. 4.9, 5, and the references) and a selected few will be reviewed here. These AGAGE estimates are then critically compared against estimates provided from manufacturing and sales information for anthropogenic chemicals and from independently derived estimates for natural emissions to improve emission estimates and models. The error bars on the inferred emissions of trace gases in Fig. 4 reflect the uncertainties in the estimates that are generally dominated by uncertainties in point measurement to grid box model extrapolations and in chemical transport models including the species lifetimes.

AGAGE data have helped resolve some important emission controversies. For example, CH_3CCl_3 is an ozone-depleting industrial solvent whose phase-out was introduced under the Montreal Protocol. However, as the phase-out continued the reported emissions appeared too low to explain observations, and unreported European emissions were claimed to be a major cause (Krol et al., 2003). Long-term high-frequency AGAGE data from Mace Head and Jungfraujoch were used to infer European CH_3CCl_3 emissions to better quantify these unreported emissions. European emission estimates declined from about 60 gigagrams per year in the mid-1990s to 0.3–3.4 gigagrams per year in 2000–2003 based on Mace Head and Jungfraujoch data, respectively. These European CH_3CCl_3 emission estimates were higher than calculated from consumption data, but were considerably lower

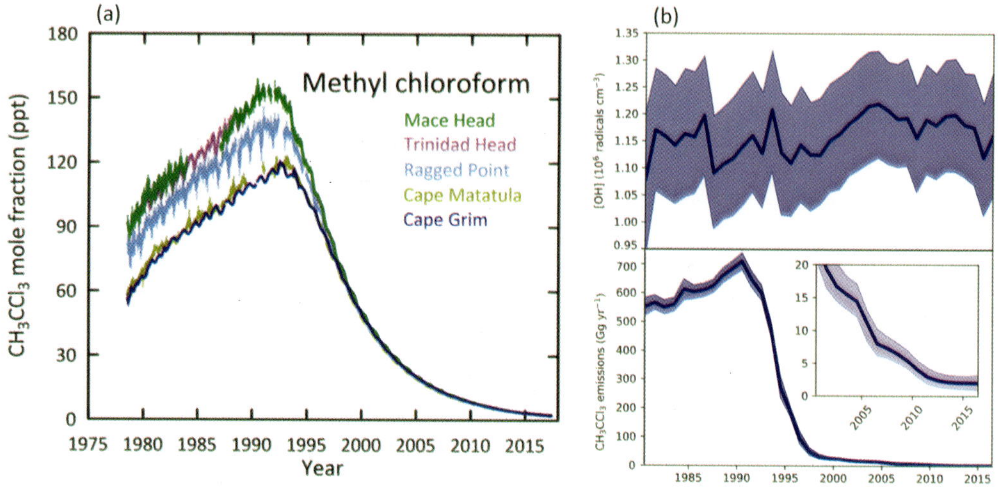

Figure 8. CH_3CCl_3 monthly mean mole fractions and 1σ standard deviations at selected AGAGE stations **(a)**. Global 12-month running mean OH concentrations and CH_3CCl_3 emissions from the AGAGE data and AGAGE 2-D model inversion. Shaded areas give 1σ uncertainty **(b)** (Rigby et al., 2017; updated here).

than those derived for 2000 in the Krol et al. (2003) study (Reimann et al., 2005). AGAGE is unusual amongst global networks in that 30 % of its in situ Medusa GC-MS observational capacity is located in the tropics (Fig. 1). A consistent feature that has emerged from AGAGE research over the period 2011–2015 is the importance of the tropics as the major source region for several important trace gases of biological origin: methane, nitrous oxide, methyl chloride, and hydrogen. Rigby et al. (2008) showed that the 2007 increase in methane growth rate in the atmosphere was likely due to a combination of emissions from unusually warm boreal summers and unusually wet tropical regions. Xiao et al. (2010a) confirmed the major role (> 50 %) that tropical plants play as a source of methyl chloride, the largest natural source of chlorine for the stratosphere. Huang et al. (2008) showed the importance of tropical regions and the Indian subcontinent as major source regions (> 80 %) for nitrous oxide, and Xiao et al. (2007) demonstrated the importance of tropical regions as the major (70 %) source (oxidation of formaldehyde, biomass burning) and major (70 %) sink (surface uptake, oxidation by OH) region for atmospheric hydrogen.

4.8 Emission estimates from multiple networks and measurement platforms and alternative models

In the last decade there has been a distinct move toward trace gas emission estimations using measurements from multiple networks and platforms. A number of the multi-network studies also applied alternative models and inverse methods to those used in AGAGE (Sect. 3.4–3.7). The methane flux estimations by Chen and Prinn (2006) merged for the first time the high-frequency AGAGE data with the low-frequency NOAA/ESRL/GMD, CSIRO, Environment Canada, NIES, and Japan Meteorological Agency flask data. The intercalibration process proved to be very important to this merger and showed that when done correctly the merger increases the precision and accuracy of the fluxes significantly. A formal intercalibration exercise began between the AGAGE, NOAA/ESRL/GMD, and other networks that used intercomparisons between instruments and flask sampling at the same station led by Paul B. Krummel (CSIRO) and intercomparisons of tanks of compressed air circulated among laboratories (IHALACE, Hall et al., 2014). This has enabled a significant number of subsequent studies that involve merging AGAGE data with data from other surface networks and platforms (towers, aircraft, satellites). AGAGE data and GMD (flask, tower, aircraft) data were used to obtain sources and/or sinks of SF_6 (Rigby et al., 2010), CFCs and CH_3CCl_3 (Rigby et al., 2013), HCFC-22 (Saikawa et al., 2012, 2014b), CFCs and N_2O (Simmonds et al., 2013), N_2O (Nevison et al., 2011; Thompson et al., 2013, 2014a, b, c), methane (Thompson et al., 2015), CH_3Cl (Xiao et al., 2010a), and CCl_4 (Xiao et al., 2010b). HIPPO aircraft, AGAGE, and ESRL data were used for seasonal emissions of HCFC-22 and HFC-134a (Xiang et al., 2014) and for OH estimation (Patra et al., 2014). Kirschke et al. (2013) used AGAGE, GMD flask, CSIRO flask, and UCI aircraft data for estimating methane emissions. MIPAS, AGAGE, and GMD data were used by Chirkov et al. (2016) for estimating HCFC-22 emissions, and AGAGE and GMD data were used for estimating CCl_4 emissions (Chipperfield et al., 2016); GOSAT, AGAGE, and GMD data were used for estimating regional methane emissions (Fraser et al., 2013). Saunoir et al. (2016, 2017) used multi-network data and alternative models to elucidate the 2000–2012 methane budget and its multiyear variability. Finally, Rigby et al. (2017) used AGAGE and GMD data for the estimation of OH concentrations and CH_4 emissions, and Ganesan et al. (2017) used

GOSAT satellite, CARIBIC aircraft, and AGAGE-calibrated surface measurements to estimate Indian subcontinent CH_4 emissions.

4.9 AGAGE publications

The central accomplishments of the ALE/GAGE/AGAGE program are documented in several hundred journal publications and theses. A full list of all ALE/GAGE/AGAGE publications in the 1983–2017 time period supported by and/or collaborating with AGAGE is available on the official AGAGE website: http://agage.mit.edu (last access: 21 May 2018); click on RESEARCH, then AGAGE PUBLICATIONS, then AGAGE ACCOMPLISHMENTS (for abstracts). For AGAGE publications with "et al.", the complete author list can be seen by clicking on the paper title in orange text. ALE/GAGE/AGAGE measurements and derived lifetimes, OH concentrations, and emissions are of considerable policy significance and are widely used in international and national ozone layer and climate assessments. AGAGE team members have specifically contributed as authors to almost all of the major international assessments under the IPCC and WMO.

5 Data availability

After calibration, validation, and conversion to a prescribed format, AGAGE data for nine stations (Ny-Ålesund, Mace Head, Trinidad Head, Jungfraujoch, Monte Cimone, Gosan (monthly means), Ragged Point, Cape Matatula, Cape Grim) are made available on the AGAGE public website (http://agage.mit.edu/data, last access: 21 May 2018). The data from the newest station, Mt. Mugogo, will be added to this site once internally validated and the first data are published in peer-reviewed journals. Data from Shangdianzi (Bo Yao; yaob@cma.gov.cn), Hateruma, and Cape Ochiishi (Takuya Saito; saito.takuya@nies.go.jp) can be obtained by contacting the indicated station scientists. Data files for individual measurements and for monthly mean summaries are updated at approximately 6-month intervals, following the semiannual meetings of the international AGAGE team. Data considered a pollution event or a local sink event are flagged. Monthly means and standard deviations of the data with and without these events are included.

The data are currently available through March 2017.

Data files for measurements are updated and archived at approximately 6-month intervals, following quality control reviews before and at the semiannual meetings of the AGAGE team. For scientific credibility, we do not submit data on new gases until at least one peer-reviewed AGAGE paper on that new gas has appeared. Because AGAGE is an international research (not operational) endeavor and because data validation for some gases can sometimes take longer than others, there is not a strict timetable between data acquisition and data submission, but generally we aim to archive data 12–18 months after acquisition.

The data on the AGAGE website are also made available on the US Department of Energy (DOE) Carbon Dioxide Information Analysis Center (CDIAC) website for public access (http://cdiac.ess-dive.lbl.gov/ndps/alegage.html, last access: 21 May 2018; Prinn et al., 2018). Note that data previously stored at the CDIAC archive are being transitioned to the new DOE ESS-DIVE archive. The above website will continue to provide access to the CDIAC data during the transition. Please contact ess-dive-support@lbl.gov for further information on the transition. CDIAC also passes on these data to the World Data Center for Greenhouse Gases (WDCGG) in Japan (http://ds.data.jma.go.jp/gmd/wdcgg/, last access: 21 May 2018). The AGAGE data in the WDCGG data center, however, are further processed by WDCGG staff and are converted to a different format from that used by the CDIAC and AGAGE websites. Thus, we do not recommend this site as a primary source of AGAGE data.

Competing interests. The authors declare that they have no conflict of interest.

Acknowledgements. We express our deep appreciation of the exceptional multi-decadal contributions of the late Prof. Derek M. Cunnold (GaTech) to all aspects of the AGAGE program, especially in the area of inverse modeling for source and sink estimation. We note with sadness the significant contributions to AGAGE of the late Laurie Porter (Bureau of Meteorology, Cape Grim) and the late Dr. Brian Greally (U. Bristol, Mace Head). We specifically acknowledge the cooperation and efforts of the following on-site station operators: Gerry Spain, Mace Head, Ireland; Randy Dickau, Trinidad Head, California; Peter Sealy, Ragged Point, Barbados; NOAA officers in charge, Cape Matatula, American Samoa; Sam Cleland, Jeremy Ward, and Nigel Somerville, Cape Grim, Tasmania; the Käser, Fischer, Otz, Seiler, and Hemund families, Jungfraujoch, Switzerland; Kieran Stanley, Ridge Hill, Bilsdale, and Heathfield, UK; and Stephen Humphery and Andy McDonald, Tacolneston, UK. Nada Derek is gratefully thanked for her assistance with the preparation of figures in this paper and her long-term support of the AGAGE activities. We thank Kat Potter and Arnico Panday, who played significant roles in establishing the Mt. Mugogo, Rwanda station. Finally we thank Jim Butler, Ed Dlugokencky, Jim Elkins, Brad Hall, and Steve Montzka of NOAA-ESRL-GMD for their valuable cooperation over many years in measurement intercomparisons and other scientific activities. The operation of the Mace Head, Trinidad Head, Barbados, American Samoa, and Tasmania AGAGE stations and the MIT theory and inverse modeling and SIO calibration activities are supported by the National Aeronautics and Space Administration (NASA, USA; grants NAG5-12669, NNX07AE89G, NNX11AF17G, and NNX16AC98G to MIT; grants NAG5-4023, NNX07AE87G, NNX07AF09G, NNX11AF15G, and NNX11AF16G to SIO). Additional support comes from the following: the UK Department for Business, Energy and Industrial Strategy (BEIS, UK; formerly the Department of Business, Energy, Industry and Strategy) contract GA01103 to the University of Bristol and the UK Met Office for Mace Head and the UK DECC tall tower Network; the National Oceanic and Atmospheric Administration (NOAA, USA) contract RA133R15CN0008 to the University of Bristol for Barbados; NOAA for the building operations of the American Samoa station; and the Commonwealth Scientific and Industrial Research Organisation (CSIRO, Australia), the Bureau of Meteorology (Australia), the Department of the Environment and Energy (DoEE, Australia), and Refrigerant Reclaim Australia for Cape Grim. Anita Ganesan and Matt Rigby are supported by NERC. The Hateruma and Cape Ochiishi stations are supported fully by the Ministry of Environment of Japan and NIES. Observations at the Jungfraujoch are supported by the Swiss Federal Office for the Environment (FOEN) within the project HALCLIM, by ICOS-CH (Integrated Carbon Observation System Research Infrastructure), and by the International Foundation High Altitude Research Stations for Jungfraujoch and Gornergrat (HFSJG). Operations at Zeppelin, Ny-Ålesund are funded by the Norwegian Environment Agency. Monte Cimone research is supported by the Italian National Research Council (CNR) under the NEXDATA project. Sunyoung Park (and the Gosan AGAGE station) is supported by the Basic Science Research Program through the National Research Foundation of Korea (NRF) funded by the Ministry of Education (no. NRF-2016R1A2B2010663). The Shangdianzi AGAGE station is fully supported by the Chinese Meteorological Administration. Funding for the instruments at the new AGAGE Mt. Mugogo station comes from the MIT Alumni Donors to the Rwanda-MIT Climate Observatory Project and the MIT Center for Global Change Science (CGCS). Funding for the infrastructure and operations comes from the Ministry of Education of Rwanda. Finally, development costs for the MIT high-end 3-D models and inverse techniques used in the theoretical analysis are supported significantly by grants from other federal agencies and by the sponsors of the MIT CGCS and Joint Program on the Science and Policy of Global Change.

Edited by: Vinayak Sinha

References

Allan, D. W.: Statistics of atomic frequency standards, P. IEEE, 54, 221–230, https://doi.org/10.1109/PROC.1966.4634, 1966.

Arnold, T., Mühle, J., Salameh, P. K., Harth, C. M., Ivy, D. J., and Weiss, R. F.: Automated Measurement of Nitrogen Trifluoride in Ambient Air, Anal. Chem., 84, 4798–4804, https://doi.org/10.1021/ac300373e, 2012.

Arnold, T., Harth, C. M., Mühle, J., Manning, A. J., Salameh, P. K., Kim, J., Ivy, D. J., Steele, L. P., Petrenko, V. V., Severinghaus, J. P., Baggenstos, D., and Weiss, R. F.: Nitrogen trifluoride global emissions estimated from updated atmospheric measurements, P Natl. Acad. Sci. USA, 110, 2029 2034, https://doi.org/10.1073/pnas.1212346110, 2013.

Arnold, T., Ivy, D. J., Harth, C. M., Vollmer, M. K., Mühle, J., Salameh, P. K., Steele, L. P., Krummel, P. B., Wang, J., Young, D., Lunder, C. R., Hermansen, O., Rhee, T. S., Kim, J., Reimann, S., O'Doherty, S., Fraser, P. J., Simmonds, P. G., Prinn, R. G., and Weiss, R. F.: HFC-43-10mee atmospheric abundances and global emission estimates, Geophys. Res. Lett., 41, 2228–2235, https://doi.org/10.1002/2013GL059143, 2014.

Bosilovich, M. G., Chen, J., Robertson, F. R., and Adler, R. F.: Evaluation of Global Precipitation in Reanalyses, J. Appl. Meteorol. Clim., 47, 2279–2299, 2008.

Bousquet, P., Hauglustaine, D. A., Peylin, P., Carouge, C., and Ciais, P.: Two decades of OH variability as inferred by an inversion of atmospheric transport and chemistry of methyl chloroform, Atmos. Chem. Phys., 5, 2635–2656, https://doi.org/10.5194/acp-5-2635-2005, 2005.

Butler, J. H., Yvon-Lewis, S. A., Lobert, J. M., King, D. B., Montzka, S. A., Bullister, J. L., Koropalov, V., Elkins, J. W., Hall, B. D., Hu, L., and Liu, Y.: A comprehensive estimate for loss of atmospheric carbon tetrachloride ($CCl + 4$) to the ocean, Atmos. Chem. Phys., 16, 10899–10910, https://doi.org/10.5194/acp-16-10899-2016, 2016.

Carpenter, L. J., Reimann, S., Burkholder, J. B., Clerbaux, C., Hall, B. D., Hossaini, R., Laube, J. C., Yvon-Lewis, S. A., Blake, D. R., Dorf, M., Dutton, G. S., Fraser, P. J., Froidevaux, L., Hendrick, F., Hu, J., Jones, A., Krummel, P. B., Kuijpers, L. J. M., Kurylo, M. J., Liang, Q., Mahieu, E., Mühle, J., O'Doherty, S., Ohnishi, K., Orkin, V. L., Pfeilsticker, K., Rigby, M., Simpson, I. J., Yokouchi, Y., Engel, A., and Montzka, S. A.: Update on Ozone-Depleting Substances (ODSs) and Other Gases of Interest to the Montreal Protocol (Chapter 1), in: Scientific Assessment of Ozone Depletion: 2014, Global Ozone Research and Monitoring Project-Report No. 55, World Meteorological Organization, Geneva, Switzerland, 2014.

Chen, Y.-H. and Prinn, R. G.: Atmospheric modeling of high-frequency methane observations: Importance of interannually varying transport, J. Geophys. Res., 110, D10303, https://doi.org/10.1029/2004JD005542, 2005.

Chen, Y.-H. and Prinn, R. G.: Estimation of atmospheric methane emissions between 1996–2001 using a 3D global chemical transport model, J. Geophys. Res., 111, D10307, https://doi.org/10.1029/2005JD006058, 2006.

Chipperfield, M. P., Liang, Q., Rigby, M., Hossaini, R., Montzka, S. A., Dhomse, S., Feng, W., Prinn, R. G., Weiss, R. F., Harth, C. M., Salameh, P. K., Mühle, J., O'Doherty, S., Young, D., Simmonds, P. G., Krummel, P. B., Fraser, P. J., Steele, L. P., Happell, J. D., Rhew, R. C., Butler, J., Yvon-Lewis, S. A., Hall, B., Nance, D., Moore, F., Miller, B. R., Elkins, J. W., Harrison, J. J., Boone, C. D., Atlas, E. L., and Mahieu, E.: Model sensitivity studies of the decrease in atmospheric carbon tetrachloride, Atmos. Chem. Phys., 16, 15741–15754, https://doi.org/10.5194/acp-16-15741-2016, 2016.

Chirkov, M., Stiller, G. P., Laeng, A., Kellmann, S., von Clarmann, T., Boone, C. D., Elkins, J. W., Engel, A., Glatthor, N., Grabowski, U., Harth, C. M., Kiefer, M., Kolonjari, F., Krummel, P. B., Linden, A., Lunder, C. R., Miller, B. R., Montzka, S. A., Mühle, J., O'Doherty, S., Orphal, J., Prinn, R. G., Toon, G., Vollmer, M. K., Walker, K. A., Weiss, R. F., Wiegele, A., and Young, D.: Global HCFC-22 measurements with MIPAS: retrieval, validation, global distribution and its evolution over 2005–2012, Atmos. Chem. Phys., 16, 3345–3368, https://doi.org/10.5194/acp-16-3345-2016, 2016.

Cox, M. L., Sturrock, G. A., Fraser, P. J., Siems, S. T., Krummel, P. B., and O'Doherty, S.: Regional sources of methyl chloride, chloroform and dichloromethane identified from AGAGE observations at Cape Grim, Tasmania, 1998–2000, J. Atmos. Chem., 45, 79–99, 2003.

Cunnold, D. M., Fraser, P. J., Weiss, R. F., Prinn, R. G., Simmonds, P. G., Miller, B. R., Alyea, F. N., and Crawford, A. J.: Global trends and annual releases of CCl_3F and CCl_2F_2 estimated from ALE/GAGE and other measurements from July 1978 to June 1991, J. Geophys. Res., 99, 1107–1126, 1994.

Cunnold, D. M., Steele, L. P., Fraser, P. J., Simmonds, P. G., Prinn, R. G., Weiss, R. F., Porter, L. W., Langenfelds, R. L., Krummel, P. B., Wang, H. J., Emmons, L., Tie, X. X., and Dlugokencky, E. J.: In-Situ measurements of atmospheric methane at GAGE/AGAGE sites during 1985–1999 and resulting source inferences, J. Geophys. Res., 107, ACH 20-1–ACH 20-18, https://doi.org/10.1029/2001JD001226, 2002.

De Mazière, M., Thompson, A. M., Kurylo, M. J., Wild, J. D., Bernhard, G., Blumenstock, T., Braathen, G. O., Hannigan, J. W., Lambert, J.-C., Leblanc, T., McGee, T. J., Nedoluha, G., Petropavlovskikh, I., Seckmeyer, G., Simon, P. C., Steinbrecht, W., and Strahan, S. E.: The Network for the Detection of Atmospheric Composition Change (NDACC): history, status and perspectives, Atmos. Chem. Phys., 18, 4935–4964, https://doi.org/10.5194/acp-18-4935-2018, 2018.

Derwent, R. G, Simmonds, P. G., Greally, B. R., O'Doherty, S., McCulloch, A., Manning, A., Reimann, S., Folini, D., and Vollmer, M. K.: The phase-in and phase-out of European emissions of HCFC-141b and HCFC-142b under the Montreal Protocol: Evidence from observations at Mace Head, Ireland from 1994–2004, Atmos. Environ., 41, 757–767, 2007.

Derwent, R. G., Simmonds, P. G., O'Doherty, S., Grant, A., Young, D., Cooke, M. C., Manning, A. J., Utembe, S. R., Jenkin, M. E., and Shallcross, D. E.: Seasonal cycles in short-lived hydrocarbons in baseline air masses arriving at Mace Head, Ireland, Atmos. Environ., 62, 89–96, https://doi.org/10.1016/j.atmosenv.2012.08.023, 2012.

Draxler, R. D. and Hess, G. D.: Description of the HYSPLIT-4 modelling system, NOAA Tech. Memo. ERL ARL-224, 24 pp., Air Resour. Lab., Silver Spring, MD, 1997.

Dunse, B., Steele, P., Wilson, S., Fraser, P., and Krummel, P.: Trace gas emissions from Melbourne Australia, based on AGAGE observations at Cape Grim, Tasmania, 1995–2000, Atmos. Environ., 39, 6334–6344, 2005.

Elkins, J. W., Butler, J. H., Hall, B., Montzka, S. A., and Moore, F. L.: Halocarbon and Other Trace Gases Group/Global Monitoring Division/Earth Systems Research Laboratory (NOAA/ESRL) website: available atg: http://www.esrl.noaa.gov/gmd/, Boulder, CO, updated data available on anonymous ftp site at: http://www.esrl.noaa.gov/gmd/dv/ftpdata.html (last access: 21 May 2018), 2015.

Enting, I. G.: Inverse Problems in Atmospheric Constituent Transport, Cambridge University Press, Cambridge CB2 2RU, UK, 2002.

Emmons, L. K., Walters, S., Hess, P. G., Lamarque, J.-F., Pfister, G. G., Fillmore, D., Granier, C., Guenther, A., Kinnison, D., Laepple, T., Orlando, J., Tie, X., Tyndall, G., Wiedinmyer, C., Baughcum, S. L., and Kloster, S.: Description and evaluation of the Model for Ozone and Related chemical Tracers, version 4 (MOZART-4), Geosci. Model Dev., 3, 43–67, https://doi.org/10.5194/gmd-3-43-2010, 2010.

Eyer, S., Tuzson, B., Popa, M. E., van der Veen, C., Röckmann, T., Rothe, M., Brand, W. A., Fisher, R., Lowry, D., Nisbet, E. G., Brennwald, M. S., Harris, E., Zellweger, C., Emmenegger, L., Fischer, H., and Mohn, J.: Real-time analysis of $\delta^{13}C$- and $\delta D-CH_4$ in ambient air with laser spectroscopy: method development and first intercomparison results, Atmos. Meas. Tech., 9, 263–280, https://doi.org/10.5194/amt-9-263-2016, 2016.

Fang, X., Thompson, R. L., Saito, T., Yokouchi, Y., Kim, J., Li, S., Kim, K. R., Park, S., Graziosi, F., and Stohl, A.: Sulfur hexafluoride (SF_6) emissions in East Asia determined by inverse modeling, Atmos. Chem. Phys., 14, 4779–4791, https://doi.org/10.5194/acp-14-4779-2014, 2014.

Fang, X., Stohl, A., Yokouchi, Y., Kim, J., Li, S., Saito, T., Park, S., and Hu, J.: Multiannual Top-Down Estimate of HFC-23 Emissions in East Asia, Environ. Sci. Technol., 49, 4345–4353, https://doi.org/10.1021/es505669j, 2015.

Forster, P., Ramaswamy, V., Artaxo, P., Berntsen, T., Betts, R., Fahey, D., Haywood, J., Lean, J., Lowe, D., Myhre, G., Nganga, J., Prinn, R., Raga, G., Schulz, M., Van Dorland, R.: Changes in Atmospheric Constituents and in Radiative Forcing, in: Climate Change 2007: The Physical Science Basis. Contribution of Working Group I to the Fourth Assessment Report of the Intergovernmental Panel on Climate Change, edited by: Solomon, S., Qin, S., Manning, M., Chen, Z., Marquis, M., Averyt, K., Tignor, M., and Miller, H.: Cambridge University Press, Cambridge, UK and New York, USA, chap. 2, 129–234, 2007.

Fortems-Cheiney, A., Chevallier, F., Saunois, M., Pison, I., Bousquest, P., Cressot, C., Wang, R. H. J., Yokouchi, Y., and Artuso, F.: HCFC-22 emissions at global and regional scales between 1995 and 2010: Trends and variability. J. Geophys. Res.-Atmos., 118, 7379–7388, 2013.

Fortems-Cheiney, A., Saunois, M., Pison, I., Chevallier, F., Bousquet, P., Cressot, C., Montzka, S. A., Fraser, P. J., Vollmer, M. K., Simmonds, P. G., Young, D., O'Doherty, S., Weiss, R. F., Artuso, F., Barletta, B., Blake, D. R., Li, S., Lunder, C., Miller, B. R., Park, S., Prinn, R., Saito, T., Steele, L. P., and Yokouchi, Y.: Increase in HFC-134a emissions in response to the success of the Montreal Protocol, J. Geophys. Res.-Atmos., 120, 11728–11742, https://doi.org/10.1002/2015JD023741, 2015.

Fraser, A., Palmer, P. I., Feng, L., Boesch, H., Cogan, A., Parker, R., Dlugokencky, E. J., Fraser, P. J., Krummel, P. B., Langenfelds, R. L., O'Doherty, S., Prinn, R. G., Steele, L. P., van der Schoot, M., and Weiss, R. F.: Estimating regional methane surface fluxes: the relative importance of surface and GOSAT mole fraction measurements, Atmos. Chem. Phys., 13, 5697–5713, https://doi.org/10.5194/acp-13-5697-2013, 2013.

Fraser, P. J., Dunse, B. L., Manning, A. J., Walsh, S., Wang, R. H. J., Krummel, P. B., Steele, L. P., Porter, L. W., Allison, C., O'Doherty, S., Simmonds, P. G., Mühle, J., Weiss, R. F., and Prinn, R. G.: Australian carbon tetrachloride (CCl_4) emissions in a global context, Environ. Chem., 11, 77–88, https://doi.org/10.1071/EN13171, 2014.

Fraser, P. J., Pearman, G. I., and Derek, N.: CSIRO non-carbon dioxide greenhouse gas research. Part 1: 1975–1990, Hist. Rec. Aust. Sci., 29, 1–13, https://doi.org/10.1071/HR17016, 2017.

Ganesan, A. L., Rigby, M., Zammit-Mangion, A., Manning, A. J., Prinn, R. G., Fraser, P. J., Harth, C. M., Kim, K.-R., Krummel, P. B., Li, S., Mühle, J., O'Doherty, S. J., Park, S., Salameh, P. K., Steele, L. P., and Weiss, R. F.: Characterization of uncertainties in atmospheric trace gas inversions using hierarchical Bayesian methods, Atmos. Chem. Phys., 14, 3855–3864, https://doi.org/10.5194/acp-14-3855-2014, 2014.

Ganesan, A. L., Manning, A. J., Grant, A., Young, D., Oram, D. E., Sturges, W. T., Moncrieff, J. B., and O'Doherty, S.: Quantifying methane and nitrous oxide emissions from the UK and Ireland using a national-scale monitoring network, Atmos. Chem. Phys., 15, 6393–6406, https://doi.org/10.5194/acp-15-6393-2015, 2015.

Ganesan, A. L., Rigby, M., Lunt, M. F., Parker, R. J., Boesch, H., Goulding, N., Umezawa, T., Zahn, A., Chatterjee, A., Prinn, R. G., Tiwari, Y. K., van der Schoot, M., and Krummel, P. B.: Atmospheric observations show accurate reporting and little growth in India's methane emissions, Nat. Commun., 8, 1–7, https://doi.org/10.1038/s41467-017-00994-7, 2017.

Grant, A., Yates, E. L., Simmonds, P. G., Derwent, R. G., Manning, A. J., Young, D., Shallcross, D. E., and O'Doherty, S.: A five year record of high-frequency in situ measurements of non-methane hydrocarbons at Mace Head, Ireland, Atmos. Meas. Tech., 4, 955–964, https://doi.org/10.5194/amt-4-955-2011, 2011.

Graziosi, F., Arduini, J., Furlani, F., Giostra, U., Kuijpers, L. J. M., Montzka, S. A., Miller, B. R., O'Doherty, S. J., Stohl, A., Bonasoni, P., and Maione, M.: European emissions of HCFC-22 based on eleven years of high frequency atmospheric measurements and a Bayesian inversion method, Atmos. Environ., 112, 196–207, https://doi.org/10.1016/j.atmosenv.2015.04.042, 2015.

Graziosi, F., Arduini, J., Bonasoni, P., Furlani, F., Giostra, U., Manning, A. J., McCulloch, A., O'Doherty, S., Simmonds, P. G., Reimann, S., Vollmer, M. K., and Maione, M.: Emissions of carbon tetrachloride from Europe, Atmos. Chem. Phys., 16, 12849–12859, https://doi.org/10.5194/acp-16-12849-2016, 2016.

Graziosi, F., Arduini, J., Furlani, F., Giostra, U., Cristofanelli, P., Fang, X., Hermanssen, O., Lunder, C., Maenhout, G., O'Doherty, S., Reimann, S., Schmidbauer, N., Vollmer, M. K., Young, D., and Maione, M.: European emissions of the powerful greenhouse gases hydrofluorocarbons inferred from atmospheric measurements and their comparison with annual national reports to UNFCCC, Atmos. Environ., 158, 85–97, 2017.

Hall, B. D., Engel, A., Mühle, J., Elkins, J. W., Artuso, F., Atlas, E., Aydin, M., Blake, D., Brunke, E.-G., Chiavarini, S., Fraser, P. J., Happell, J., Krummel, P. B., Levin, I., Loewenstein, M., Maione, M., Montzka, S. A., O'Doherty, S., Reimann, S., Rhoderick, G., Saltzman, E. S., Scheel, H. E., Steele, L. P., Vollmer, M. K., Weiss, R. F., Worthy, D., and Yokouchi, Y.: Results from the International Halocarbons in Air Comparison Experiment (IHALACE), Atmos. Meas. Tech., 7, 469–490, https://doi.org/10.5194/amt-7-469-2014, 2014.

Harris, E., Nelson, D. D., Olszewski, W., Zahniser, M., Potter, K. E., McManus, B. J., Whitehill, A., Prinn, R. G., and Ono, S.: Development of a Spectroscopic Technique for Continuous Online Monitoring of Oxygen and Site-Specific Nitrogen Isotopic Composition of Atmospheric Nitrous Oxide, Anal. Chem., 86, 1726–1734, https://doi.org/10.1021/ac403606u, 2013.

Hartley, D. E. and Prinn, R. G.: On the feasibility of determining surface emissions of trace gases using an inverse method in a three-dimensional chemical transport model, J. Geophys. Res., 98, 5183–5198, 1993.

Hartmann, D. L., Klein Tank, A. M. G., Rusticucci, M., Alexander, L. V., Brönnimann, S., Charabi, Y., Dentener, F. J., Dlugokencky, E. J., Easterling, D. R., Kaplan, A., Soden, B. J., Thorne, P. W., Wild, M., and Zhai, P. M.: Observations: Atmosphere and Surface, in: Climate Change 2013: The Physical Science Basis, Contribution of Working Group I to the Fifth Assessment Report of the Intergovernmental Panel on Climate Change, edited by: Stocker, T. F., Qin, D., Plattner, G.-K., Tignor, M., Allen, S. K., Boschung, J., Nauels, A., Xia, Y., Bex, V., and Midgley, P. M., Cambridge University Press, Cambridge, United Kingdom and New York, NY, USA, 2013.

Heil, J., Wolf, B., Brüggemann, N., Emmenegger, L., Tuzson, B., Vereecken, H., and Mohn, J.: Site-specific ^{15}N isotopic signatures of aiotically produced N_2O, Geochim. Cosmochim. Ac., 139, 72–82, https://doi.org/10.1016/j.gca.2014.04.037, 2014.

Huang, J. and Prinn, R. G.: Critical evaluation of emissions for potential new OH titrating gases, J. Geophys. Res., 107, 4784, https://doi.org/10.1029/2002JD002394, 2002.

Huang, J., Golombek, A., Prinn, R., Weiss, R., Fraser, P., Simmonds, P., Dlugokencky, E. J., Hall, B., Elkins, J., Steele, P., Langenfelds, R., Krummel, P., Dutton, G., and Porter, L.: Estimation of regional emissions of nitrous oxide from 1997 to 2005 using multinetwork measurements, a chemical transport model, and an inverse method, J. Geophys. Res., 113, D17313, https://doi.org/10.1029/2007JD009381, 2008.

Ivy, D. J., Arnold, T., Harth, C. M., Steele, L. P., Mühle, J., Rigby, M., Salameh, P. K., Leist, M., Krummel, P. B., Fraser, P. J., Weiss, R. F., and Prinn, R. G.: Atmospheric histories and growth trends of $C_4F_{10}, C_5F_{12}, C_6F_{14}, C_7F_{16}andC_8F_{18}$, Atmos. Chem. Phys., 12, 4313–4325, https://doi.org/10.5194/acp-12-4313-2012, 2012a.

Ivy, D. J., Rigby, M., Baasandorj, M., Burkholder, J. B., and Prinn, R. G.: Global emission estimates and radiative impact of $C_4F_{10}, C_5F_{12}, C_6F_{14}, C_7F_{16}$ and C_8F_{18}, Atmos. Chem. Phys., 12, 7635–7645, https://doi.org/10.5194/acp-12-7635-2012, 2012b.

Jones, A., Thomson, D., Hort, M., and Devenish, B.: The UK Met Office's next-generation atmospheric dispersion model, NAME III, Air Pollution Modeling and Its Applications Xvii, 17, 580–589, https://doi.org/10.1007/978-0-387-68854-1_62, 2007.

Kalnay, E., Kanamitsu, M., Kistler, R., Collins, W., Deaven, D., Gandin, L., Iredell, M., Saha, S., White, G., Woollen, J., Zhu, Y., Leetmaa, A., Reynolds, R., Chelliah, M., Ebisuzaki, W., Higgins, W., Janowiak, J., Mo, K. C., Ropelewski, C., Wang, J., Jenne, R., and Joseph, D.: The NCEP/NCAR 40-Year Reanalysis Project, B. Am. Meteorol. Soc., 77, 437–471, 1996.

Kaminski, T., Rayner, P. J., Heimann, M., and Enting, I. G.: On aggregation errors in atmospheric transport inversions, J. Geophys. Res., 106, 4703–4715, 2001.

Kasibhatla, P., Heimann, M., Rayner, P., Mahowald, N., Prinn, R. G., and Hartley, D. (Eds): Inverse Methods in Global Biogeochemical Cycles, Geophysical Monograph Series, Volume 114, American Geophysical Union, Washington DC, 324 pp., 2000.

Keller, C. A., Brunner, D., Henne, S., Vollmer, M. K., O'Doherty, S., and Reimann, S.: Evidence for under-reported Western European emissions of the potent greenhouse gas HFC-23, Geophys. Res. Lett., 38, L15808, https://doi.org/10.1029/2011GL047976, 2011.

Keller, C. A., Hill, M., Vollmer, M. K., Henne, S., Brunner, D., Reimann, S., O'Doherty, S., Arduini, J., Maione, M., Ferenczi, Z., Haszpra, L., Manning, A. J., and Peter, T.: European emissions of halogenated greenhouse gases inferred from atmospheric measurements, Environ. Sci. Technol., 46, 217–225, https://doi.org/10.1021/es202453j, 2012.

Kim, J., Li, S., Kim, K.-R., Stohl, A., Mühle, J., Kim, S.-K., Park, M.-K., Kang, D.-J., Lee, G., Harth, C. M., Salameh, P. K., and Weiss, R. F.: Regional atmospheric emissions determined from measurements at Jeju Island, Korea: Halogenated compounds from China, Geophys. Res. Lett., 37, L12801, https://doi.org/10.1029/2010gl043263, 2010.

Kim, J., Li, S., Mühle, J., Stohl, A., Kim, S.-K., Park, S., Park, M.-K., Weiss, R. F., and Kim, K.-R.: Overview of the findings from measurements of halogenated compounds at Gosan (Jeju Island, Korea) quantifying emissions in East Asia, J. Integr. Environ. Sci., 9, Supplement 1, 71–80, https://doi.org/10.1080/1943815X.2012.696548, 2012.

K im, J., Fraser, P. J., Li, S., Mühle, J., Ganesan, A. L., Krummel, P. B., Steele, L. P., Park, S., Kim, S.-K., Park, M.-K., Arnold, T., Harth, C. M., Salameh, P. K., Prinn, R. G., Weiss, R. F., and Kim, K.-R.: Quantifying aluminum and semiconductor industry perfluorocarbon emissions from atmospheric measurements, Geophys. Res. Lett., 41, 4787–4794, https://doi.org/10.1002/2014GL059783, 2014.

Kirschke, S., Bousquet, P., Ciais, P., Marielle Saunois, M., Canadell, J. G., Dlugokencky, E. J., Bergamaschi, P., Bergmann, D., Blake, D. R., Bruhwiler, L., Cameron-Smith, P., Castaldi, S., Chevallier, F., Feng, L., Fraser, A., Fraser, P. J., Heimann, M., Hodson, E. L., Houweling, S., Josse, B., Krummel, P. B, Lamarque, J.-F., Langenfelds, R. L., Le Quéré, C., Naik, V., O'Doherty, S., Palmer, P. I., Pison, I., Plummer, D., Poulter, B., Prinn, R. G., Rigby, M., Ringeval, B., Santini, M., Schmidt, M., Shindell, D.T., Simpson, I. J., Spahni, R., Steele, L.P., Strode, S. A., Sudo, K., Szopa, S., van der Werf, G. R., Voulgarakis, A., van Weele, M., Weiss, R. F., Williams, J. E., and Zeng, G.: Three decades of global methane sources and sinks, Nat. Geosci., 6, 813–823, https://doi.org/10.1038/NGEO1955, 2013.

Kleiman, G. and Prinn, R. G.: Measurement and deduction of emissions of trichloroethene, tetrachloroethene and trichloromethane (chloroform) in the Northeastern U.S. and Southeastern Canada, J. Geophys. Res., 105, 28875–28893, 2000.

Krol, M. and Lelieveld, J.: Can the variability in tropospheric OH be deduced from measurements of 1,1,1-trichloroethane (methyl chloroform)?, J. Geophys. Res., 108, ACH16, https://doi.org/10.1029/2002JD002423, 2003

Krol, M., Lelieveld, J., Oram, D. E., Sturrock, G. A., Penkett, S. A., Brenninkmeijer, C. A. M., Gros, V., Williams, J., and Scheeren, H. A.: Continuing emissions of methyl chloroform from Europe, Nature, 421, 131–135, 2003.

Lawrence, M. G., Crutzen, P. J., Rasch, P. J., Eaton, B. E., and Mahowald, N. M.: A model for studies of tropospheric photochemistry: description, global distributions and evaluation, J. Geophys. Res., 104, 26245–26277, 1999.

Li, J., Cunnold, D. M., Wang, H.-J., Weiss, R. F., Miller, B. R., Harth, C., Salameh, P., and Harris, J. M.: Halocarbon emissions estimated from Advanced Global Atmospheric Gases Experiment measured pollution events at Trinidad Head, California, J. Geophys. Res., 110, D14308, https://doi.org/10.1029/2004JD005739, 2005.

Li, S., Kim, J., Kim, K.-R., Mühle, J., Kim, S.-K., Park, M.-K., Stohl, A., Kang, D.-J., Arnold, T., Harth, C. M., Salameh, P. K., and Weiss, R.F.: Emissions of halogenated compounds in East Asia determined from measurements at Jeju Island, Korea, Environ. Sci. Technol., 45, 5668–5675, https://doi.org/10.1021/es104124k, 2011.

Li, S., Kim, J., Park, S., Kim, S.-K., Park, M.-K., Mühle, J., Lee, G., Lee, M., Jo, C. O., and Kim, K.-R.: Source Identification and Apportionment of Halogenated Compounds Observed at a Remote Site in East Asia, Environ. Sci. Technol., 48, 491–498, https://doi.org/10.1021/es402776w, 2014.

Liang, Q., Chipperfield, M. P., Fleming, E. L., Abraham, N. L., Braesicke, P., Burkholder, J. B., Daniel, J. S., Dhomse, S., Fraser, P. J., Hardiman, S. C., Jackman, C. H., Kinnison, D.E., Krummel, P. B., Montzka, S. A., Morgenstern, O., McCulloch, A., Mühle, J., Newman, P.A., Orkin, V. L., Pitari, G., Prinn, R. G., Rigby, M., Rozanov, E., Stenke, A., Tummon, F., Velders, G. J. M., Visioni, D., and Weiss, R. F.: Deriving global OH abundance and atmospheric lifetimes for long-lived gases: A search for CH_3CCl_3 alternatives, J. Geophys. Res.-Atmos., 122, 11914–11933, https://doi.org/10.1002/2017JD026926, 2017.

Locatelli, R., Bousquet, P., Chevallier, F., Fortems-Cheney, A., Szopa, S., Saunois, M., Agusti-Panareda, A., Bergmann, D., Bian, H., Cameron-Smith, P., Chipperfield, M. P., Gloor, E., Houweling, S., Kawa, S. R., Krol, M., Patra, P. K., Prinn, R. G., Rigby, M., Saito, R., and Wilson, C.: Impact of transport model errors on the global and regional methane emissions estimated by inverse modelling, Atmos. Chem. Phys., 13, 9917–9937, https://doi.org/10.5194/acp-13-9917-2013, 2013.

Loh, Z. M., Law, R. M., Haynes, K. D., Krummel, P. B., Steele, L. P., Fraser, P. J., Chambers, S. D., and Williams, A. G.: Simulations of atmospheric methane for Cape Grim, Tasmania, to constrain southeastern Australian methane emissions, Atmos. Chem. Phys., 15, 305–317, https://doi.org/10.5194/acp-15-305-2015, 2015.

Lo Vullo, E., Furlani, F., Arduini, J., Giostra, U., Cristofanelli, P., Williams, M. L., and Maione, M.: Non- Methane Volatile Organic Compounds in the Background Atmospheres of a Southern European Mountain Site (Mt. Cimone, Italy): Annual and Seasonal Variability, Aerosol Air Qual. Res., 16, 581–592, https://doi.org/10.4209/aaqr.2015.05.0364, 2016a.

Lo Vullo, E., Furlani, F., Arduini, J., Giostra, U., Graziosi, F., Cristofanelli, P., Williams, M. L., and Maione, M.: Anthropogenic non-methane volatile hydrocarbons at Mt. Cimone (2165 m a.s.l, Italy): Impact of sources and transport on atmospheric composition, Atmos. Environ., 140, 395–440, https://doi.org/10.1016/j.atmosenv.2016.05.060, 2016b.

Lunt, M. F., Rigby, M., Ganesan, A. L., Manning, A. J., Prinn, R. G., O'Doherty, S., Muhle, J., Harth, C. M., Salameh, P. K., Arnold, T., Weiss, R. F., Saito, T., Yokouchi, Y., Krummel, P. B., Steele, L. P., Fraser, P. J., Li, S., Park, S., Reimann, S., Vollmer, M. K., Lunder, C., Hermansen, O., Schmidbauer, N., Malone, M., Young, D., and Simmonds, P. G.: Reconciling reported and unreported HFC emissions with atmospheric observations, P. Natl. Acad. Sci. USA, 112, 5927–5931, https://doi.org/10.1073/pnas.1420247112, 2015.

Lunt, M. F., Rigby, M., Ganesan, A. L., and Manning, A. J.: Estimation of trace gas fluxes with objectively determined basis functions using reversible-jump Markov chain Monte Carlo, Geosci. Model Dev., 9, 3213–3229, https://doi.org/10.5194/gmd-9-3213-2016, 2016.

MacFarling-Meure, C., Etheridge, D., Trudinger, C., Steele, P., Langenfelds, R., van Ommen, T., Smith, A., and Elkins, J.: Law Dome CO_2, CH_4, and N_2O ice core records extended to 2000 years BP, Geophys. Res. Lett., 33, L14810, https://doi.org/10.1029/2006GL026152, 2006.

Mahowald, N. M., Prinn, R. G., and Rasch, P. J.: Deducing CCl_3F emissions using an inverse method and chemical transport models with assimilated winds, J. Geophys. Res., 102, 28153–28168, 1997.

Maione, M., Giostra, U., Arduini, J., Furlani, F., Graziosi, F., Lo Vullo, E., and Bonasoni, P.: Ten years of continuous observations of stratospheric ozone depleting gases at Monte Cimone (Italy) – Comments on the effectiveness of the Montreal Protocol from a regional perspective, Sci. Total Environ., 445–446, 155–164, 2013.

Maione, M., Graziosi, F., Arduini, J., Furlani, F., Giostra, U., Blake, D. R., Bonasoni, P., Fang, X., Montzka, S. A., O'Doherty, S. J., Reimann, S., Stohl, A., and Vollmer, M. K.: Estimates of European emissions of methyl chloroform using a Bayesian inversion method, Atmos. Chem. Phys., 14, 9755–9770, https://doi.org/10.5194/acp-14-9755-2014, 2014.

Manning, A., O'Doherty, S., Jones, A. R., Simmonds, P. G., and Derwent, R. G.: Estimating UK methane and nitrous oxide emissions from 1990 to 2007 using an inversion modeling approach, J. Geophys. Res., 116, D02305, https://doi.org/10.1029/2010JD014763, 2011.

Manning, M. R., Lowe, D. C., Moss, R. C., Bodeker, G. E., and Allan, W.: Short term variations in the oxidizing power of the atmosphere, Nature, 436, 1001–1004, 2005.

McCulloch, A. and Midgley, P.: The history of methyl chloroform emissions: 1951–2000, Atmos. Environ., 35, 5311–5319, 2001.

McNorton, J., Chipperfield, M. P., Gloor, M., Wilson, C., Feng, W., Hayman, G. D., Rigby, M., Krummel, P. B., O'Doherty, S., Prinn, R. G., Weiss, R. F., Young, D., Dlugokencky, E., and Montzka, S. A.: Role of OH variability in the stalling of the global atmospheric CH_4 growth rate from 1999 to 2006, Atmos. Chem. Phys., 16, 7943–7956, https://doi.org/10.5194/acp-16-7943-2016, 2016.

Meinshausen, M., Vogel, E., Nauels, A., Lorbacher, K., Meinshausen, N., Etheridge, D. M., Fraser, P. J., Montzka, S. A., Rayner, P. J., Trudinger, C. M., Krummel, P. B., Beyerle, U., Canadell, J. G., Daniel, J. S., Enting, I. G., Law, R. M., Lunder, C. R., O'Doherty, S., Prinn, R. G., Reimann, S., Rubino, M., Velders, G. J. M., Vollmer, M. K., Wang, R. H. J., and Weiss, R.: Historical greenhouse gas concentrations for climate modelling (CMIP6), Geosci. Model Dev., 10, 2057–2116, https://doi.org/10.5194/gmd-10-2057-2017, 2017.

Meirink, J. F., Bergamaschi, P., and Krol, M. C.: Four-dimensional variational data assimilation for inverse modelling of atmospheric methane emissions: method and comparison with synthesis inversion, Atmos. Chem. Phys., 8, 6341–6353, https://doi.org/10.5194/acp-8-6341-2008, 2008.

Miller, B. R., Weiss, R. F., Prinn, R. G., Huang, J., and Fraser, P. J.: Atmospheric trend and lifetime of chlorodifluoromethane (HCFC-22) and the global tropospheric OH concentration, J. Geophys. Res., 103, 13237–13248, 1998.

Miller, B. R., Weiss, R. F., Salameh, P. K., Tanhua, T., Greally, B. R., Mühle, J., and Simmonds, P.: Medusa: a sample preconcentration and GC-MSD system for in situ measurements of atmospheric trace halocarbons, hydrocarbons and sulfur compounds, Anal. Chem., 80, 1536–1545, 2008.

Miller, B. R., Rigby, M., Kuijpers, L. J. M., Krummel, P. B., Steele, L. P., Leist, M., Fraser, P. J., McCulloch, A., Harth, C., Salameh, P., Mühle, J., Weiss, R. F., Prinn, R. G., Wang, R. H. J., O'Doherty, S., Greally, B. R., and Simmonds, P. G.: HFC-23 (CHF_3) emission trend response to HCFC-22 ($CHClF_2$) production and recent HFC-23 emission abatement measures, Atmos. Chem. Phys., 10, 7875–7890, https://doi.org/10.5194/acp-10-7875-2010, 2010.

Mohn, J., Guggenheim, C., Tuzson, B., Vollmer, M. K., Toyoda, S., Yoshida, N., and Emmenegger, L.: A liquid nitrogen-free preconcentration unit for measurements of ambient N_2O isotopomers by QCLAS, Atmos. Meas. Tech., 3, 609–618, https://doi.org/10.5194/amt-3-609-2010, 2010.

Mohn, J., Wolf, B., Toyoda, S., Lin, C.-T., Liang, M.-C., Brüggemann, N., Wissel, H., Steiker, A. E., Dyckmans, J., Szwec, L., Ostrom, N. E., Casciotti, K. L., Forbes, M., Giesemann, A., Well, R., Doucett, R. R., Yarnes, C. T., Ridley, A. R., Kaiser, J., and Yoshida, N., Interlaboratory assessment of nitrous oxide isotopomer analysis by isotope ratio mass spectrometry and laser spectroscopy: current status and perspectives. Rapid Commun. Mass Spectrom., 28, 1995–2007, https://doi.org/10.1002/rcm.6982, 2014.

Montzka, S. A., Spivakovsky, C. M., Butler, J. H., Elkins, J. W., Lock, L. T., and Mondeel, D. J.: New observational constraints on atmospheric hydroxyl on global and hemispheric scales, Science, 288, 500–503, 2000.

Montzka, S. A., Reimann, S., Engel, A., Krüger, K., O'Doherty, S., Sturges, W. T., Blake, D., Dorf, M., Fraser, P., Froidevaux, L., Jucks, K., Kreher, K., Kurylo, M. J., Mellouki, A., Miller, J., Nielsen, O.-J., Orkin, V. L., Prinn, R. G., Rhew, R., Santee, M. L., Stohl, A., and Verdonik, D.: Ozone-Depleting Substances (ODSs) and Related Chemicals, Chapter 1, in: Scientific Assessment of Ozone Depletion: 2010, World Meteorological Organization, Geneva, Switzerland, 2011a.

Montzka, S., Krol, M., Dlugokencky, E., Hall, B., Jockel, P., and Lelieveld, J.: Small interannual variability of global atmospheric hydroxyl, Science, 331, 67–69, https://doi.org/10.1126/science.1197640, 2011b.

Montzka, S. A., Dutton, G. S., Yu, P., Ray, E., Portmann, R. W., Daniel, J. S., Kuijpers, L., Hall, B. D., Mondeel, D., Siso, C., Nance, D. J., Rigby, M., Manning, A. J., Hu, L., Moore, F., Miller, B. R., and Elkins, J. W.: A persistent and unexpected increase in global emissions of ozone-depleting CFC-11, Nature, 557, 413–417, https://doi.org/10.1038/s41586-018-0106-2, 2018, 2018.

Mühle, J., Huang, J., Weiss, R. F., Prinn, R. G., Miller, B. R., Salameh, P. K., Harth, C. M., Fraser, P. J., Porter, L. W., Greally, B. R., O'Doherty, S., Simmonds, P. G., Krummel, P. B., and Steele, L. P.: Sulfuryl Fluoride in the Global Atmosphere, J. Geophys. Res., 114, D05306, https://doi.org/10.1029/2008JD011162, 2009.

Mühle, J., Ganesan, A. L., Miller, B. R., Salameh, P. K., Harth, C. M., Greally, B. R., Rigby, M., Porter, L. W., Steele, L. P., Trudinger, C. M., Krummel, P. B., O'Doherty, S., Fraser, P. J., Simmonds, P. G., Prinn, R. G., and Weiss, R. F.: Perfluorocarbons in the global atmosphere: tetrafluoromethane, hexafluoroethane, and octafluoropropane, Atmos. Chem. Phys., 10, 5145–5164, https://doi.org/10.5194/acp-10-5145-2010, 2010.

Mulquiney, J. E., Taylor, J. A., Jakeman, A. J., Norton, J. P., and Prinn, R. G.: A new inverse method for trace gas flux estimation, 2. Application to tropospheric $CFCl_3$ fluxes, J. Geophys. Res., 103, 1429–1442, 1998.

Nevison, C. D., Dlugokencky, E., Dutton, G., Elkins, J. W., Fraser, P., Hall, B., Krummel, P. B., Langenfelds, R. L., O'Doherty, S., Prinn, R. G., Steele, L. P., and Weiss, R. F.: Exploring causes of interannual variability in the seasonal cycles of tropospheric nitrous oxide, Atmos. Chem. Phys., 11, 3713–3730, https://doi.org/10.5194/acp-11-3713-2011, 2011.

Obersteiner, F., Bönisch, H., and Engel, A.: An automated gas chromatography time-of-flight mass spectrometry instrument for the quantitative analysis of halocarbons in air, Atmos. Meas. Tech., 9, 179–194, https://doi.org/10.5194/amt-9-179-2016, 2016.

O'Doherty, S., Simmonds, P., Cunnold, D., Wang, R. H. J., Sturrock, G. A., Fraser, P. J., Ryall, D., Derwent, R. G., Weiss, R. F., Salameh, P., Miller, B. R., and Prinn, R. G.: In Situ Chloroform Measurements at AGAGE Atmospheric Research Stations from 1994–1998, J. Geophys. Res., 106, 20429–20444, 2001.

O'Doherty, S., Cunnold, D. M., Manning, A., Miller, B. R., Wang, R. H., Krummel, P. B., Fraser, P. J., Simmonds, P. G., McCulloch, A., Weiss, R. F., Salameh, P. Porter, L. W., Prinn, R. G., Huang, J., Sturrock, G., Ryall, D., Derwent, R. G., and Montzka, S. A.: Rapid growth of hydrofluorocarbon 134a and hydrochlorofluorocarbons 141b, 142b, and 22 from Advanced Global Atmospheric Gases Experiment (AGAGE) observations at Cape Grim Tasmania, and Mace Head, Ireland, J. Geophys. Res.,109, D06310, https://doi.org/10.1029/2003JD004277, 2004.

O'Doherty, S., Cunnold, D. M., Miller, B. R., Mühle, J., McCulloch, A., Simmonds, P. G., Manning, A. J., Reimann, S., Vollmer, M. K., Greally, B. R., Prinn, R. G., Fraser, P. J., Steele, L. P., Krummel, P. B., Dunse, B. L., Porter, L. W., Lunder, C. R., Schmidbauer, N.,. Hermansen, O., Salameh, P. K., Harth, C. M., Wang, R. H. J., and Weiss, R. F.: Global and regional emissions of HFC-125 (CHF_2CF_3) from in situ and air archive atmospheric observations at AGAGE and SOGE observatories, J. Geophys. Res., 114, D23304, https://doi.org/10.1029/2009JD012184, 2009.

O'Doherty, S., Rigby, M., Mühle, J., Ivy, D. J., Miller, B. R., Young, D., Simmonds, P. G., Reimann, S., Vollmer, M. K., Krummel, P. B., Fraser, P. J., Steele, L. P., Dunse, B., Salameh, P. K., Harth, C. M., Arnold, T., Weiss, R. F., Kim, J., Park, S., Li, S., Lunder, C., Hermansen, O., Schmidbauer, N., Zhou, L. X., Yao, B., Wang, R. H. J., Manning, A. J., and Prinn, R. G.: Global emissions of HFC-143a (CH_3CF_3) and HFC-32 (CH_2F_2) from in situ and air archive atmospheric observations, Atmos. Chem. Phys., 14, 9249–9258, https://doi.org/10.5194/acp-14-9249-2014, 2014.

Patra, P. K., Houweling, S., Krol, M., Bousquet, P., Belikov, D., Bergmann, D., Bian, H., Cameron-Smith, P., Chipperfield, M. P., Corbin, K., Fortems-Cheiney, A., Fraser, A., Gloor, E., Hess, P., Ito, A., Kawa, S. R., Law, R. M., Loh, Z., Maksyutov, S., Meng, L., Palmer, P. I., Prinn, R. G., Rigby, M., Saito, R., and Wilson, C.: TransCom model simulations of CH4 and related species: linking transport, surface flux and chemical loss with CH4 variability in the troposphere and lower stratosphere, Atmos. Chem. Phys., 11, 12813–12837, https://doi.org/10.5194/acp-11-12813-2011, 2011.

Patra, P. K., Krol, M. C., Montzka, S. A., Arnold, T., Atlas, E. L., Lintner, B. R., Stephens, B. B., Xiang, B., Elkins, J. W., Fraser, P. J., Ghosh, A., Hintsa, E. J., Hurst, D. F., Ishijima, K., Krummel, P. B., Miller, B. R., Miyazaki, K., Moore, F. L., Mühle, J., O'Doherty, S., Prinn, R. G., Steele, L. P., Takigawa, M., Wang, H. J., Weiss, R. F., Wofsy, S. C., and Young, D.: Observational evidence for interhemispheric hydroxyl-radical parity, Nature, 513, 219–223, https://doi.org/10.1038/nature13721, 2014.

Potter, K. E., Ono, S., and Prinn, R. G.: Fully automated, high-precision instrumentation for the isotopic analysis of tropospheric N_2O using continuous flow isotope ratio mass spectrometry, Rapid Commun. Mass. Sp., 27, 1723–1728, https://doi.org/10.1002/rcm.662, 2013.

Prinn, R. G.: Measurement equation for trace chemicals in fluids and solution of its inverse, in: Inverse Methods in Global Biogeochemical Cycles, edited by: Kasibhatla, P., Heimann, M., Rayner P., Mahowald, N., Prinn, R. G., and Hartley, D. E., Geophysical Monograph, 114, American Geophysical Union, Washington DC, 3–18, 2000.

Prinn, R. G., Simmonds, P. G., Rasmussen, R. A., Rosen, R. D., Alyea, F. N., Cardelino, C. A., Crawford, A. J., Cunnold, D. M., Fraser, P. J., and Lovelock, J. E.: The Atmospheric Lifetime Experiment, I: Introduction, Instrumentation and Overview, J. Geophys. Res., 88, 8353–8367, 1983a.

Prinn, R. G., Rasmussen, R. A., Simmonds, P. G., Alyea, F. N., Cunnold, D. M., Lane, B. C., Cardelino, C. A., and Crawford, A. J.: The Atmospheric Lifetime Experiment, 5: results for CH_3CCl_3 based on three years of data, J. Geophys. Res., 88, 8415–8426, 1983b.

Prinn, R. G., Cunnold, D. M., Rasmussen, R., Simmonds, P. G., Alyea, F. N., Crawford, A., Fraser, P. J., and Rosen, R.: Atmospheric trends in methylchloroform and the global average for the hydroxyl radical, Science, 238, 945–950, 1987.

Prinn, R. G., Cunnold, D. M., Simmonds, P. G., Alyea, F. N., Boldi, R., Crawford, A., Fraser, P. J., Gutzler, D., Hartley, D. E., Rosen, R., and Rasmussen, R.: Global average concentration and trend for hydroxyl radicals deduced from ALE/GAGE trichloroethane (methyl chloroform) data for 1978–1990, J. Geophys. Res., 97, 2445–2461, 1992.

Prinn, R. G., Weiss, R. F., Miller, B. R., Huang, J., Alyea, F. N., Cunnold, D. M., Fraser, P. J., Hartley, D. E., and Simmonds, P. G.: Atmospheric trends and lifetime of CH_3CCl_3 and global OH concentrations, Science, 269, 187–192, 1995.

Prinn, R. G., Weiss, R. F., Fraser, P. J., Simmonds, P. G., Cunnold, D. M., Alyea, F. N., O'Doherty, S., Salameh, P., Miller, B. R., Huang, J., Wang, R. H. J., Hartley, D. E., Harth, C., Steele, L. P., Sturrock, G., Midgley, P. M., and McCulloch, A.: A history of chemically and radiatively important gases in air deduced from ALE/GAGE/AGAGE, J. Geophys. Res., 115, 17751–17792, 2000.

Prinn, R. G., Huang, J., Weiss, R. F., Cunnold, D. M., Fraser, P. J., Simmonds, P. G., McCulloch, A., Harth, C., Salameh, P., O'Doherty, S., Wang, R. H. J., Porter, L., and Miller, B. R.: Evidence for substantial variations of atmospheric hydroxyl radicals in the past two decades, Science, 292, 1882–1888, 2001, Correction, Science, 293, p. 1054, 2001.

Prinn, R. G., Huang, J., Weiss, R. F., Cunnold, D. M., Fraser, P. J., Simmonds, P. G., McCulloch, A., Harth, C., Reimann, S., Salameh, P., O'Doherty, S., Wang, R. H. J., Porter, L., Miller, B.

R., and Krummel, P.: Evidence for variability of atmospheric hydroxyl radicals over the past quarter century, Geophys. Res. Lett., 32, L07809, https://doi.org/10.1029/2004GL022228, 2005.

Prinn, R. G., Weiss, R. F., Arduini, J., Arnold, T., Fraser, P. J., Ganesan, A. L., Gasore, J., Harth, C. M., Hermansen, O., Kim, J., Krummel, P. B., Li, S., Loh, Z. M., Lunder, C. R., Maione, M., Manning, A. J., Miller, B. R., Mitrevski, B., Mühle, J., O'Doherty, S., Park, S., Reimann, S., Rigby, M., Salameh, P. K., Schmidt, R., Simmonds, P. G., Steele, L. P., Vollmer, M. K., Wang, R. H., and Young, D.: The ALE/GAGE/AGAGE Network (DB 1001), http://cdiac.ess-dive.lbl.gov/ndps/alegage.html (https://doi.org/10.3334/CDIAC/atg.db1001), last access: June 2018.

Rasch, P. J., Mahowald, N. M., and Eaton, B. E.: Representations of transport, convection, and the hydrologic cycle in chemical transport models: Implications for the modeling of short-lived and soluble species, J. Geophys. Res., 102, 28127–28138, 1997.

Ravishankara, A. R., Daniel, J. S., and Portmann, R. W.: Nitrous Oxide (N_2O): The Dominant Ozone-Depleting Substance Emitted in the 21st Century, Science, 27, 1–4, https://doi.org/10.1126/science.1176985, 2009.

Reimann, S., Manning, A. J., Simmonds, P. G., Cunnold, D. M., Wang, H. J., Li, J., McCulloch, A., Prinn, R. G., Huang, J., Weiss, R. F., Fraser, P. J., O'Doherty, S., Greally, B. R., Stemmler, K., Hill, M., and Folini, D.: Assessment of European methyl chloroform emissions by analysis of long-term measurements, Nature, 433, 506–508, 2005.

Rhew, R. C. and Happell, J. D.: The atmospheric partial lifetime of carbon tetrachloride with respect to the global soil sink: CCl_4 Soil Sink and Partial Lifetime, Geophys. Res. Lett., 43, 2889–2895, https://doi.org/10.1002/2016GL067839, 2016.

Rigby, M., Prinn, R. G., Fraser, P. J., Simmonds, P. G., Langenfelds, R. L., Huang, J., Cunnold, D. M., Steele, L. P., Krummel, P. B., Weiss, R. F., O'Doherty, S., Salameh, P. K., Wang, H. J., Harth, C. M., Mühle, J., and Porter, L. W.: Renewed growth of atmospheric methane, Geophys. Res. Lett., 35, L22805, https://doi.org/10.1029/2008GL036037, 2008.

Rigby, M., Mühle, J., Miller, B. R., Prinn, R. G., Krummel, P. B., Steele, L. P., Fraser, P. J., Salameh, P. K., Harth, C. M., Weiss, R. F., Greally, B. R., O'Doherty, S., Simmonds, P. G., Vollmer, M. K., Reimann, S., Kim, J., Kim, K.-R., Wang, H. J., Olivier, J. G. J., Dlugokencky, E. J., Dutton, G. S., Hall, B. D., and Elkins, J. W.: History of atmospheric SF_6 from 1973 to 2008, Atmos. Chem. Phys., 10, 10305–10320, https://doi.org/10.5194/acp-10-10305-2010, 2010.

Rigby, M., Manning, A. J., and Prinn, R. G.: Inversion of long-lived trace gas emissions using combined Eulerian and Lagrangian chemical transport models, Atmos. Chem. Phys., 11, 9887–9898, https://doi.org/10.5194/acp-11-9887-2011, 2011.

Rigby, M., Manning, A. J., and Prinn, R. G.: The value of high-frequency, high-precision methane isotopologue measurements for source and sink estimation, J. Geophys. Res.-Atmos., 117, D12312, https://doi.org/10.1029/2011JD017384, 2012.

Rigby, M., Prinn, R. G., O'Doherty, S., Montzka, S. A., McCulloch, A., Harth, C. M., Mühle, J., Salameh, P. K., Weiss, R. F., Young, D., Simmonds, P. G., Hall, B. D., Dutton, G. S., Nance, D., Mondeel, D. J., Elkins, J. W., Krummel, P. B., Steele, L. P., and Fraser, P. J.: Re-evaluation of the lifetimes of the major CFCs and CH_3CCl_3 using atmospheric trends, Atmos. Chem. Phys., 13, 2691–2702, https://doi.org/10.5194/acp-13-2691-2013, 2013.

Rigby, M., Prinn, R. G., O'Doherty, S., Miller, B. R., Ivy, D., Muhle, J., Harth, C. M., Salameh, P. K., Arnold, T., Weiss, R. F., Krummel, P. B., Steele, L. P., Fraser, P. J., Young, D., and Simmonds, P. G.: Recent and future trends in synthetic greenhouse gas radiative forcing, Geophys. Res. Letts., 41, 2623–2630, https://doi.org/10.1002/2013GL059099, 2014.

Rigby, M., Montzka, S. A., Prinn, R. G., White, J. W. C., Young, D., O'Doherty, S, Lunt, M., Ganesan, A. L., Manning, A., Simmonds, P., Salameh, P. K., Harth, C. M., Muhle, J., Weiss, R. F., Fraser, P. J., Steele, L. P., Krummel, P. B., McCulloch, A., and Park, S.: The role of atmospheric oxidation in recent methane growth, P. Natl. Acad. Sci. USA, 114, 5373–5377, https://doi.org/10.1073/pnas.1616426114, 2017.

Rödenbeck, C., Gerbig, C., Trusilova, K., and Heimann, M.: A two-step scheme for high-resolution regional atmospheric trace gas inversions based on independent models, Atmos. Chem. Phys., 9, 5331–5342, https://doi.org/10.5194/acp-9-5331-2009, 2009.

Ryall, D. B., Maryon, R. H., Derwent, R. G., and Simmonds, P. G.: Modelling long-range transport of CFCs to Mace Head, Ireland, Q. J. Roy. Meteor. Soc., 124, 417–446, 1998.

Ryall, D. B., Derwent, R. G., Manning, A. J., Simmonds, P. G., and O'Doherty,S.: Estimating source regions of European emissions of trace gases from observations at Mace Head, Atmos. Environ., 35, 2507–2523, 2001.

Saikawa, E., Rigby, M., Prinn, R. G., Montzka, S. A., Miller, B. R., Kuijpers, L. J. M., Fraser, P. J. B., Vollmer, M. K., Saito, T., Yokouchi, Y., Harth, C. M., Mühle, J., Weiss, R. F., Salameh, P. K., Kim, J., Li, S., Park, S., Kim, K.-R., Young, D., O'Doherty, S., Simmonds, P. G., McCulloch, A., Krummel, P. B., Steele, L. P., Lunder, C., Hermansen, O., Maione, M., Arduini, J., Yao, B., Zhou, L. X., Wang, H. J., Elkins, J. W., and Hall, B.: Global and regional emission estimates for HCFC-22, Atmos. Chem. Phys., 12, 10033–10050, https://doi.org/10.5194/acp-12-10033-2012, 2012.

Saikawa, E., Prinn, R. G., Dlugokencky, E., Ishijima, K., Dutton, G. S., Hall, B. D., Langenfelds, R., Tohjima, Y., Machida, T., Manizza, M., Rigby, M., O'Doherty, S., Patra, P. K., Harth, C. M., Weiss, R. F., Krummel, P. B., van der Schoot, M., Fraser, P. J., Steele, L. P., Aoki, S., Nakazawa, T., and Elkins, J. W.: Global and regional emissions estimates for N_2O, Atmos. Chem. Phys., 14, 4617–4641, https://doi.org/10.5194/acp-14-4617-2014, 2014a.

Saikawa, E., Rigby, M., Prinn, R. G., Montzka, S. A., Miller, B. R., Kuijpers, L. J. M., Fraser, P. J. B., Vollmer, M. K., Saito, T., Yokouchi, Y., Harth, C. M., Mühle, J., Weiss, R. F., Salameh, P. K., Kim, J., Li, S., Park, S., Kim, K.-R., Young, D., O'Doherty, S., Simmonds, P. G., McCulloch, A., Krummel, P. B., Steele, L. P., Lunder, C., Hermansen, O., Maione, M., Arduini, J., Yao, B., Zhou, L. X., Wang, H. J., Elkins, J. W., and Hall, B.: Corrigendum to "Global and regional emission estimates for HCFC-22", Atmos. Chem. Phys., 12, 10033–10050, 2012, Atmos. Chem. Phys., 14, 4857–4858, https://doi.org/10.5194/acp-14-4857-2014, 2014b.

Saito, R., Patra, P. K., Sweeney, C., Machida, T., Krol, M., Houweling, S., Bousquet, P., Agusti-Panareda, A., Belikov, D., Bergmann, D., Bian, H., Cameron-Smith, P., Chipperfield, M. P., Fortems-Cheiney, A., Fraser, A., Gatti, L. V., Gloor, E., Hess, P.,

Kawa, S. R., Law, R. M., Locatelli, R., Loh, Z., Maksyutov, S., Meng, L., Miller, J. B., Palmer, P. I., Prinn, R. G., Rigby, M., and Wilson, C.: TransCom model simulations of methane: Comparison of vertical profiles with aircraft measurements, J. Geophys. Res., 118, 3891–3904, https://doi.org/10.1002/jgrd.50380, 2013.

Saito, T., Fang, X., Stohl, A., Yokouchi, Y., Zeng, J., Fukuyama, Y., and Mukai, H.: Extraordinary halocarbon emissions initiated by the 2011 Tohoku earthquake, Geophys. Res. Lett., 42, 2500–2507, https://doi.org/10.1002/2014gl062814, 2015.

Saunois, M., Bousquet, P., Poulter, B., Peregon, A., Ciais, P., Canadell, J. G., Dlugokencky, E. J., Etiope, G., Bastviken, D., Houweling, S., Janssens-Maenhout, G., Tubiello, F. N., Castaldi, S., Jackson, R. B., Alexe, M., Arora, V. K., Beerling, D. J., Bergamaschi, P., Blake, D. R., Brailsford, G., Brovkin, V., Bruhwiler, L., Crevoisier, C., Crill, P., Covey, K., Curry, C., Frankenberg, C., Gedney, N., Höglund-Isaksson, L., Ishizawa, M., Ito, A., Joos, F., Kim, H.-S., Kleinen, T., Krummel, P., Lamarque, J.-F., Langenfelds, R., Locatelli, R., Machida, T., Maksyutov, S., McDonald, K. C., Marshall, J., Melton, J. R., Morino, I., Naik, V., O'Doherty, S., Parmentier, F.-J. W., Patra, P. K., Peng, C., Peng, S., Peters, G. P., Pison, I., Prigent, C., Prinn, R., Ramonet, M., Riley, W. J., Saito, M., Santini, M., Schroeder, R., Simpson, I. J., Spahni, R., Steele, P., Takizawa, A., Thornton, B. F., Tian, H., Tohjima, Y., Viovy, N., Voulgarakis, A., van Weele, M., van der Werf, G. R., Weiss, R., Wiedinmyer, C., Wilton, D. J., Wiltshire, A., Worthy, D., Wunch, D., Xu, X., Yoshida, Y., Zhang, B., Zhang, Z., and Zhu, Q.: The global methane budget 2000–2012, Earth Syst. Sci. Data, 8, 697–751, https://doi.org/10.5194/essd-8-697-2016, 2016.

Saunois, M., Bousquet, P., Poulter, B., Peregon, A., Ciais, P., Canadell, J. G., Dlugokencky, E. J., Etiope, G., Bastviken, D., Houweling, S., Janssens-Maenhout, G., Tubiello, F. N., Castaldi, S., Jackson, R. B., Alexe, M., Arora, V. K., Beerling, D. J., Bergamaschi, P., Blake, D. R., Brailsford, G., Bruhwiler, L., Crevoisier, C., Crill, P., Covey, K., Frankenberg, C., Gedney, N., Höglund-Isaksson, L., Ishizawa, M., Ito, A., Joos, F., Kim, H.-S., Kleinen, T., Krummel, P., Lamarque, J.-F., Langenfelds, R., Locatelli, R., Machida, T., Maksyutov, S., Melton, J. R., Morino, I., Naik, V., O'Doherty, S., Parmentier, F.-J. W., Patra, P. K., Peng, C., Peng, S., Peters, G. P., Pison, I., Prinn, R., Ramonet, M., Riley, W. J., Saito, M., Santini, M., Schroeder, R., Simpson, I. J., Spahni, R., Takizawa, A., Thornton, B. F., Tian, H., Tohjima, Y., Viovy, N., Voulgarakis, A., Weiss, R., Wilton, D. J., Wiltshire, A., Worthy, D., Wunch, D., Xu, X., Yoshida, Y., Zhang, B., Zhang, Z., and Zhu, Q.: Variability and quasi-decadal changes in the methane budget over the period 2000–2012, Atmos. Chem. Phys., 17, 11135–11161, https://doi.org/10.5194/acp-17-11135-2017, 2017.

Schoenenberger, F., Vollmer, M. K., Rigby, M., Hill, M., Fraser, P. J., Krummel, P. B., Langenfelds, R. L., Rhee, T. S., Peter, T., and Reimann, S.: First observations, trends, and emissions of HCFC-31 (CH_2ClF) in the global atmosphere, Geophys. Res. Lett., 42, 7817–7824, https://doi.org/10.1002/2015gl064709, 2015.

Sherry, D., McCulloch, A., Liang, Q., Reimann, S., and Newman, P. A.: Current sources of carbon tetrachloride (CCl_4) in our atmosphere, Environ. Res. Lett., 13, 024004, https://doi.org/10.1088/1748-9326/aa9c87, 2017.

Simmonds, P. G., O'Doherty, S., Nickless, G., Sturrock, G. A., Swaby, R., Knight, R., Picketts, J., Woffendin, G., and Smith, R.: Automated gas chromatograph/Mass spectrometer for routine atmospheric field measurements of the CFC replacement compounds, the hydrofluorocarbons and hydrochlorofluorocarbons, Anal. Chem., 67, 717–723, https://doi.org/10.1021/Ac00100a005, 1995.

Simmonds, P. G., Manning, A. J., Derwent, R. G., Ciais, P., Ramonet, M., Kazan, V., and Ryall, D.: A burning question. Can recent growth rate anomalies in the greenhouse gases be attributed to large-scale biomass burning events?, Atmos. Environ., 39, 2513–2517, https://doi.org/10.1016/j.atmosenv.2005.02.018, 2005.

Simmonds, P. G., Manning, A. J., Athanassiadou, M., Scaife, A. A., Derwent, R. G., O'Doherty, S., Harth, C. M., Weiss, R. F., G. S., Dutton, Hall, B. D., Sweeney, C., and Elkins, J. W.: Interannual fluctuations in the seasonal cycle of nitrous oxide and chlorofluorocarbons due to the Brewer-Dobson circulation, J. Geophys. Res., 118, 10694–10706, 2013.

Simmonds, P. G., Derwent, R. G., Manning, A. J., McCulloch, A., and O'Doherty, S.: USA emissions estimates of CH_3CHF_2, CH_2FCF_3, CH_3CF_3, and CH_2F_2 based on in situ observations at Mace Head, Atmos. Environ., 104, 27–38, https://doi.org/10.1016/j.atmosenv.2015.01.010, 2015.

Simmonds, P. G., Rigby, M., Manning, A. J., Lunt, M. F., O'Doherty, S., McCulloch, A., Fraser, P. J., Henne, S., Vollmer, M. K., Mühle, J., Weiss, R. F., Salameh, P. K., Young, D., Reimann, S., Wenger, A., Arnold, T., Harth, C. M., Krummel, P. B., Steele, L. P., Dunse, B. L., Miller, B. R., Lunder, C. R., Hermansen, O., Schmidbauer, N., Saito, T., Yokouchi, Y., Park, S., Li, S., Yao, B., Zhou, L. X., Arduini, J., Maione, M., Wang, R. H. J., Ivy, D., and Prinn, R. G.: Global and regional emissions estimates of 1,1-difluoroethane (HFC-152a, CH_3CHF_2) from in situ and air archive observations, Atmos. Chem. Phys., 16, 365–382, https://doi.org/10.5194/acp-16-365-2016, 2016.

Simmonds, P. G., Rigby, M., McCulloch, A., O'Doherty, S., Young, D., Mühle, J., Krummel, P. B., Steele, P., Fraser, P. J., Manning, A. J., Weiss, R. F., Salameh, P. K., Harth, C. M., Wang, R. H. J., and Prinn, R. G.: Changing trends and emissions of hydrochlorofluorocarbons (HCFCs) and their hydrofluorocarbon (HFCs) replacements, Atmos. Chem. Phys., 17, 4641–4655, https://doi.org/10.5194/acp-17-4641-2017, 2017.

Spivakovsky, C. M., Logan, J. A., Montzka, S. A., Balkanski, Y. J., Foreman-Fowler, M., Jones, D. B. A., Horowitz, L. W., Fusco, A. C., Brenninkmeijer, C. A. M., Prather, M. J., Wofsy, S. C., and McElroy, M. B.: Three-dimensional climatological distribution of tropospheric OH: update and evaluation, J. Geophys. Res., 105, 8931–8979, 2000.

Stohl, A., Seibert, P., Arduini, J., Eckhardt, S., Fraser, P., Greally, B. R., Lunder, C., Maione, M., Mühle, J., O'Doherty, S., Prinn, R. G., Reimann, S., Saito, T., Schmidbauer, N., Simmonds, P. G., Vollmer, M. K., Weiss, R. F., and Yokouchi, Y.: An analytical inversion method for determining regional and global emissions of greenhouse gases: Sensitivity studies and application to halocarbons, Atmos. Chem. Phys., 9, 1597–1620, https://doi.org/10.5194/acp-9-1597-2009, 2009.

Stohl, A., Kim, J., Li, S., O'Doherty, S., Mühle, J., Salameh, P. K., Saito, T., Vollmer, M. K., Wan, D., Weiss, R. F., Yao, B., Yokouchi, Y., and Zhou, L. X.: Hydrochlorofluorocarbon and hydrofluorocarbon emissions in East Asia determined by inverse modeling, Atmos. Chem. Phys., 10, 3545–3560, https://doi.org/10.5194/acp-10-3545-2010, 2010.

Tarantola, A.: Inverse problem theory and methods for model parameter estimation, Philadelphia, USA, Society for Industrial and Applied Mathematics, 2005.

Thompson R. L., Dlugokencky, E., Chevallier, F., Ciais, P., Dutton, G., Elkins, J. W., Langenfelds, R. L., Prinn, R. G., Weiss, R. F., Tohjima, Y., Krummel, P. B., Fraser, P., and Steele, L. P.: Interannual variability in tropospheric nitrous oxide, Geophys. Res. Lett., 40, 4426–4431, https://doi.org/10.1002/grl.50721, 2013.

Thompson, R. L., Chevallier, F., Crotwell, A. M., Dutton, G., Langenfelds, R. L., Prinn, R. G., Weiss, R. F., Tohjima, Y., Nakazawa, T., Krummel, P. B., Steele, L. P., Fraser, P., O'Doherty, S., Ishijima, K., and Aoki, S.: Nitrous oxide emissions 1999 to 2009 from a global atmospheric inversion, Atmos. Chem. Phys., 14, 1801–1817, https://doi.org/10.5194/acp-14-1801-2014, 2014a.

Thompson, R. L., Patra, P. K., Ishijima, K., Saikawa, E., Corazza, M., Karstens, U., Wilson, C., Bergamaschi, P., Dlugokencky, E., Sweeney, C., Prinn, R. G., Weiss, R. F., O'Doherty, S., Fraser, P. J., Steele, L. P., Krummel, P. B., Saunois, M., Chipperfield, M., and Bousquet, P.: TransCom N_2O model inter-comparison – Part 1: Assessing the influence of transport and surface fluxes on tropospheric N_2O variability, Atmos. Chem. Phys., 14, 4349–4368, https://doi.org/10.5194/acp-14-4349-2014, 2014b.

Thompson, R. L., Ishijima, K., Saikawa, E., Corazza, M., Karstens, U., Patra, P. K., Bergamaschi, P., Chevallier, F., Dlugokencky, E., Prinn, R. G., Weiss, R. F., O'Doherty, S., Fraser, P. J., Steele, L. P., Krummel, P. B., Vermeulen, A., Tohjima, Y., Jordan, A., Haszpra, L., Steinbacher, M., Van der Laan, S., Aalto, T., Meinhardt, F., Popa, M. E., Moncrieff, J., and Bousquet, P.: TransCom N_2O model inter-comparison – Part 2: Atmospheric inversion estimates of N_2O emissions, Atmos. Chem. Phys., 14, 6177–6194, https://doi.org/10.5194/acp-14-6177-2014, 2014c.

Thompson, R. L., Stohl, A., Zhou, L. X., Dlugokencky, E., Fukuyama, Y., Tohjima, Y., Kim, S.-Y., Lee, H., Nisbet, E. G., Fisher, R. E., Lowry, D., Weiss, R. F., Prinn, R. G., O'Doherty, S., Young, D., and White, J. W. C.: Methane emissions in East Asia for 2000–2011 estimated using an atmospheric Bayesian inversion, J. Geophys. Res.-Atmos., 120, 4352–4369, https://doi.org/10.1002/2014JD022394, 2015.

Trudinger, C. M., Fraser, P. J., Etheridge, D. M., Sturges, W. T., Vollmer, M. K., Rigby, M., Martinerie, P., Mühle, J., Worton, D. R., Krummel, P. B., Steele, L. P., Miller, B. R., Laube, J., Mani, F. S., Rayner, P. J., Harth, C. M., Witrant, E., Blunier, T., Schwander, J., O'Doherty, S., and Battle, M.: Atmospheric abundance and global emissions of perfluorocarbons CF_4, C_2F_6 and C_3F_8 since 1800 inferred from ice core, firn, air archive and in situ measurements, Atmos. Chem. Phys., 16, 11733–11754, https://doi.org/10.5194/acp-16-11733-2016, 2016.

Turner, A. J., Frankenberg, C., Wennberg, P. O., and Jacob, D. J.: Ambiguity in the causes for decadal trends in atmospheric methane and hydroxyl, P. Natl. Acad. Sci. USA, 114, 5367–5372, https://doi.org/10.1073/pnas.1616020114, 2017.

Vollmer, M. K., Miller, B. R., Rigby, M., Reimann, S., Mühle, J., Krummel, P. B., O'Doherty, S., Kim, J., Rhee, T. S., Weiss, R. F., Fraser, P. J., Simmonds, P. G., Salameh, P. K., Harth, C. M., Wang, R. H. J., Steele, L. P., Young, D., Lunder, C. R., Hermansen, O., Ivy, D., Arnold, T., Schmidbauer, N., Kim, K.-R., Greally, B. R., Hill, M., Leist, M., Wenger, A., and Prinn, R.G.: Atmospheric histories and global emissions of the anthropogenic hydrofluorocarbons HFC-365mfc, HFC-245fa, HFC-227ea, and HFC-236fa, J. Geophys. Res., 116, D08304, https://doi.org/10.1029/2010JD015309, 2011.

Vollmer, M. K., Reimann, S., Hill, M., and Brunner, D.: First observations of the fourth generation synthetic halocarbons HFC-1234yf, HFC-1234ze(E) and HCFC-1233zd(E) in the atmosphere, Environ. Sci. Technol., 49, 2703–2708, https://doi.org/10.1021/es505123x, 2015a.

Vollmer, M. K., Rhee, T. S., Rigby, M., Hofstetter, D., Hill, M., Schoenenberger, F., and Reimann, S.: Modern inhalation anesthetics: potent greenhouse gases in the global atmosphere, Geophys. Res. Lett., 42, 1606–1611, https://doi.org/10.1002/2014GL062785, 2015b.

Vollmer, M. K., Rigby, M., Laube, J. C., Henne, S., Rhee, T. S., Gooch, L. J., Wenger, A., Young, D., Steele, L.P., Langenfelds, R. L., Brenninkmeijer, C. A. M., Wang, J.-L., Ou-Yang, C.-F., Wyss, S. A., Hill, M., Oram, D. E., Krummel, P. B., Schoenenberger, F., Zellweger, C., Fraser, P. J., Sturges, W. T., O'Doherty, S., and Reimann, S.: Abrupt reversal in emissions and atmospheric abundance of HCFC-133a (CF_3CH_2Cl), Geophys. Res. Lett., 42, 8702–8710, https://doi.org/10.1002/2015gl065846, 2015c.

Vollmer, M. K., Mühle, J., Trudinger, C., Rigby, M., Montzka, S. A., Harth, C. M., Miller, B. R., Henne, S., Krummel, P. B., Hall, B. D.,Young, D., Kim, J., Arduini, J., Wenger, A., Yao, B., Reimann, S., O'Doherty, S., Maione, M., Etheridge, D. M., Li, S., Verdonik, D. P., Park, S., Dutton, G., Steele, L. P., Lunder, C. R., Rhee, T. S., Hermansen, O., Schmidbauer, N., Wang, R. H. J., Hill, M., Salameh, P. K., Langenfelds, R. L., Zhou, L., Blunier, T., Schwander, J., Elkins, J. W., Butler, J. H., Simmonds, P. G., Weiss, R. F., Prinn, R. G., and Fraser, P. J.: Atmospheric histories and global emissions of halons H-1211 (CBrClF2), H-1301 (CBrF3), and H-2402 (CBrF2CBrF2), J. Geophys. Res., 121, 3663–3686, https://doi.org/10.1002/2015jd024488, 2016.

Vollmer, M. K., Young, D., Trudinger, C. M., Mühle, J., Henne, S., Rigby, M., Park, S., Li, S., Guillevic, M., Mitrevski, B., Harth, C. M., Miller, B. R., Reimann, S., Yao, B., Steele, L. P., Wyss, S. A., Lunder, C. R., Arduini, J., McCulloch, A., Wu, S., Rhee, T. S., Wang, R. H. J., Salameh, P. K., Hermansen, O., Hill, M., Langenfelds, R. L., Ivy, D., O'Doherty, S., Krummel, P. B., Maione, M., Etheridge, D. M., Zhou, L., Fraser, P. J., Prinn, R. G., Weiss, R. F., and Simmonds, P. G.: Atmospheric histories and emissions of chlorofluorocarbons CFC-13 ($CClF_3$), SCFC-114 ($C_2Cl_2F_4$), and CFC-115 (C_2ClF_5), Atmos. Chem. Phys., 18, 979–1002, https://doi.org/10.5194/acp-18-979-2018, 2018.

Waechter, H., Mohn, J., Tuzson, B., Emmenegger, L. and Sigrist, M.: Determination of N_2O isotopomers with quantum cascade laser based absorption spectroscopy, Opt. Express, 16, 9239–9244, 2008.

Weiss, R. F. and Prinn, R. G.: Quantifying greenhouse-gas emissions from atmospheric measurements: a critical reality check for climate legislation, Philos. T. R. Soc. A, 369, 1925–1942, https://doi.org/10.1098/rsta.2011.0006, 2011.

Weiss, R. F., Mühle, J., Salameh, P. K., and Harth, C. M.: Nitrogen trifluoride in the global atmosphere, Geophys. Res. Lett., 35, L20821, https://doi.org/10.1029/2008GL035913, 2008.

Welp, L. R., Keeling, R. F., Weiss, R. F., Paplawsky, W., and Heckman, S.: Design and performance of a Nafion dryer for continuous operation at CO_2 and CH_4 air monitoring sites, Atmos. Meas. Tech., 6, 1217–1226, https://doi.org/10.5194/amt-6-1217-2013, 2013.

Werle, P., Mücke, R., and Slemr, F.: The limits of signal averaging in atmospheric trace-gas monitoring by tunable diode-laser absorption spectroscopy (TDLAS), Appl. Phys. B, 57, 131–139, 1993.

Xiang, B., Patra, P. K., Montzka, S. A., Miller, S. M., Elkins, J. W., Moore, F., Atlas, E. L., Miller, B. R., Weiss, R. F., Prinn, R. G., and Wofsy, S. C.: Global Emissions of Refrigerants HCFC-22 and HFC-134a: Unforeseen Seasonal Contributions, P. Natl. Acad. Sci. USA, 111, 17379–17384, https://doi.org/10.1073/pnas.1417372111, 2014.

Xiao, X., Prinn, R. G., Simmonds, P. G., Steele, L. P., Novelli, P. C., Huang, J., Langenfelds, R. L., O'Doherty, S., Krummel, P. B., Fraser, P. J., Porter, L. W., Weiss, R. F., Salameh, P., and Wang, R. H. J.: Optimal estimation of the soil uptake rate of molecular hydrogen from AGAGE and other measurements, J. Geophys. Res., 112, D07303, https://doi.org/10.1029/2006JD007241, 2007.

Xiao, X., Prinn, R. G., Fraser, P. J., Simmonds, P. G., Weiss, R. F., O'Doherty, S., Miller, B. R., Salameh, P. K., Harth, C. M., Krummel, P. B., Porter, L. W., Mühle, J., Greally, B. R., Cunnold, D., Wang, R., Montzka, S. A., Elkins, J. W., Dutton, G. S., Thompson, T. M., Butler, J. H., Hall, B. D., Reimann, S., Vollmer, M. K., Stordal, F., Lunder, C., Maione, M., Arduini, J., and Yokouchi, Y.: Optimal estimation of the surface fluxes of methyl chloride using a 3-D global chemical transport model, Atmos. Chem. Phys., 10, 5515–5533, https://doi.org/10.5194/acp-10-5515-2010, 2010a.

Xiao, X., Prinn, R. G., Fraser, P. J., Weiss, R. F., Simmonds, P. G., O'Doherty, S., Miller, B. R., Salameh, P. K., Harth, C. M., Krummel, P. B., Golombek, A., Porter, L. W., Butler, J. H., Elkins, J. W., Dutton, G. S., Hall, B. D., Steele, L. P., Wang, R. H. J., and Cunnold, D. M.: Atmospheric three-dimensional inverse modeling of regional industrial emissions and global oceanic uptake of carbon tetrachloride, Atmos. Chem. Phys., 10, 10421–10434, https://doi.org/10.5194/acp-10-10421-2010, 2010b.

Yao, B, Vollmer, M. K., Xia, L., Zhou, L., Simmonds, P. G., Stordal, F., Maione, M., Reimann, S., and O'Doherty, S.: A study of four-year HCFC-22 and HCFC-142b, in-situ measurements at the Shangdianzi regional background station in China, Atmos. Environ. 63, 43–49, https://doi.org/10.1016/j.atmosenv.2012.09.011, 2012a.

Yao, B., Vollmer, M. K., Zhou, L. X., Henne, S., Reimann, S., Li, P. C., Wenger, A., and Hill, M.: In-situ measurements of atmospheric hydrofluorocarbons (HFCs) and perfluorocarbons (PFCs) at the Shangdianzi regional background station, China, Atmos. Chem. Phys., 12, 10181–10193, https://doi.org/10.5194/acp-12-10181-2012, 2012b.

Yates, E. L., Derwent, R. G., Simmonds, P. G., Greally, B. R., O'Doherty, S., and Shallcross, D. E.: The seasonal cycles and photochemistry of C2–C5 alkanes at Mace Head, Atmos. Environ., 44, 2705–2713, https://doi.org/10.1016/j.atmos.env.2010.04.043, 2010.

Yokouchi, Y., Taguchi, S., Saito, T., Tohjima, Y., Tanimoto, H., and Mukai, H.: High frequency measurements of hfcs at a remote site in east asia and their implications for chinese emissions, Geophys. Res. Lett., 33, L21814, https://doi.org/10.1029/2006GL026403, 2006.

Yokouchi, Y., Nojiri, Y., Toom-Sauntry, D., Fraser, P., Inuzuka, Y., Tanimoto, H., Nara, H., Murakami R., and Mukai, H.: Long-term variation of atmospheric methyl iodide and its link to global environmental change, Geophys. Res. Lett., 39, L23805, https://doi.org/10.29/2012GL053695, 2012.

6

Long-term observations of tropospheric particle number size distributions and equivalent black carbon mass concentrations in the German Ultrafine Aerosol Network (GUAN)

Wolfram Birmili[1,a], Kay Weinhold[1], Fabian Rasch[1], André Sonntag[1], Jia Sun[1], Maik Merkel[1], Alfred Wiedensohler[1], Susanne Bastian[2], Alexander Schladitz[2,b], Gunter Löschau[2], Josef Cyrys[3,4], Mike Pitz[3,c], Jianwei Gu[3,4], Thomas Kusch[3,4], Harald Flentje[5], Ulrich Quass[6], Heinz Kaminski[6], Thomas A. J. Kuhlbusch[6,d], Frank Meinhardt[7], Andreas Schwerin[7], Olaf Bath[7], Ludwig Ries[7], Holger Gerwig[7], Klaus Wirtz[7], and Markus Fiebig[8]

[1]Leibniz Institute for Tropospheric Research (TROPOS), Leipzig, Germany
[2]Saxon State Office for Environment, Agriculture and Geology (LfULG), Dresden, Germany
[3]Helmholtz Zentrum Munich (HMGU), Institute of Epidemiology II, Neuherberg, Germany
[4]University of Augsburg (UA), Wissenschaftszentrum Umwelt, Augsburg, Germany
[5]Deutscher Wetterdienst (DWD), Meteorologisches Observatorium Hohenpeißenberg, Hohenpeißenberg, Germany
[6]Institute of Energy and Environmental Technology (IUTA), Duisburg, Germany
[7]German Federal Environment Agency (UBA), Dessau-Rosslau, Germany
[8]Norwegian Institute for Air Research (NILU), Kjeller, Norway
[a]now at: German Federal Environment Agency (UBA), Berlin, Germany
[b]now at: SICK Engineering GmbH, Ottendorf-Okrilla, Germany
[c]now at: Bavarian Environment Agency (LUA), Augsburg, Germany
[d]now at: Federal Institute for Occupational Safety and Health (BAuA), Dortmund, Germany

Correspondence to: Wolfram Birmili (wolfram.birmili@uba.de)

Abstract. The German Ultrafine Aerosol Network (GUAN) is a cooperative atmospheric observation network, which aims at improving the scientific understanding of aerosol-related effects in the troposphere. The network addresses research questions dedicated to both climate- and health-related effects. GUAN's core activity has been the continuous collection of tropospheric particle number size distributions and black carbon mass concentrations at 17 observation sites in Germany. These sites cover various environmental settings including urban traffic, urban background, rural background, and Alpine mountains. In association with partner projects, GUAN has implemented a high degree of harmonisation of instrumentation, operating procedures, and data evaluation procedures. The quality of the measurement data is assured by laboratory intercomparisons as well as on-site comparisons with reference instruments. This paper describes the measurement sites, instrumentation, quality assurance, and data evaluation procedures in the network as well as the EBAS repository, where the data sets can be obtained.

1 Introduction

Atmospheric aerosol particles, or particulate matter (PM), are essential constituents in the atmosphere influencing issues such as atmospheric visibility, global climate, and human health. A climate-relevant effect is their interaction with solar shortwave radiation (Ramanathan et al., 2001). Two major aerosol effects influencing the terrestrial radiation budget have been distinguished: direct radiative forcing – scattering and absorption of upwelling and downwelling radiation in the absence of clouds (Haywood and Boucher, 2000) – and indirect radiative forcing – the modification of cloud radiative properties through the activation of additional particles as cloud condensation nuclei (Lohmann and Feichter, 2005). The magnitude of direct radiative forcing depends, in general, on various properties of the aerosol particles including particle diameter and chemical composition (Bohren and Huffman, 1998) but also shape, state of mixture, and hygroscopicity (Hänel, 1976; Zieger et al., 2013). Black carbon (BC) is among the species contributing to light absorption and, thus, atmospheric warming. According to a recent survey, the radiative forcing due to BC can, at present, only be bound to the uncertainty of a factor of 2 (Bond et al., 2013). The particle number size distribution and the light absorption coefficient are useful parameters to predict the direct radiative forcing on the basis of in situ measurements.

On the other hand, ambient aerosol particles have been recognised to affect human health (e.g. Dockery and Pope, 1994; Pope et al., 2004; Dockery and Stone, 2007). Recent projections of health effects yield drastic numbers of morbidity and premature deaths due to particulate pollution worldwide (Lelieveld et al., 2015). In the European Union, the mass concentration of PM_{10} and $PM_{2.5}$ (particles smaller than 10 and 2.5 µm in aerodynamic diameter, respectively) currently serve as legal metrics to assess a population's exposure to ambient particles (European Council, 2008/50/EC). A rationale for using PM_{10} and $PM_{2.5}$ has been the large body of epidemiological evidence of adverse health effects based on these metrics.

Some studies, however, have suggested that the mass-based metrics might not be the most favourable parameter to characterise PM-induced health effects (HEI Review Panel on Ultrafine Particles, 2013). Some epidemiological studies have associated health endpoints with the number of ultrafine particles or the particle surface area rather than particle mass (Ibald-Mulli et al., 2002; Franck et al., 2011; Rückerl et al., 2011). Ultrafine particles are ubiquitous in urban atmospheres (Kumar et al., 2014), and their ability to penetrate deep into the human body after inhalation has been forwarded as a rationale for their adverse health effects. A recent overview by the world health organisation WHO (Janssen et al., 2012) also counted atmospheric soot particles (BC) among the relevant environmental risk factors for human health. While there seems little doubt about the potential adverse health effects of ultrafine particles and BC, their relatively low mass concentration makes them hardly accessible by total-mass-based measurements. Particle number size distribution and BC mass concentration have consequently been recommended as exposure parameters for future epidemiological studies (HEI Review Panel on Ultrafine Particles, 2013).

International observation networks for in situ atmospheric aerosol measurements include WMO-GAW (World Meteorological Organization Global Atmosphere Watch) and EMEP (European Monitoring and Evaluation Programme). European research infrastructure programmes have contributed to the systematic collection of in situ atmospheric aerosol data as well: EUSAAR (European Supersites for Atmospheric Aerosol Research) and ACTRIS (Aerosols, Clouds, and Trace gases Research InfraStructure Network). In the early 2000s the Nordic Aerosol Network carried out particle number size distribution measurements in a number of rural locations in Scandinavia (Tunved et al., 2003). The nature of most of these networks, however, has been to measure aerosol abundance and characteristics on a continental and global scale. Accordingly, the measurement sites are predominantly located in rural settings where direct anthropogenic influence is weak. Observation networks including urban sites are, for example, the Black Carbon and Particle Numbers and Concentrations Networks in the UK, operated by the National Physical Laboratory (Jones et al., 2012).

Government air quality networks in Europe operate many stations that collect data on PM_{10} and $PM_{2.5}$ mass concentrations, which are relevant to European air quality legislation. In the view of limited financial resources, however, there is usually a limited incentive to measure aerosol and PM metrics that go beyond legal requirements, although such activities might provide enhanced scientific insights into climate-relevant or health-related processes.

In 2008, the German Federal Environment Agency (UBA) and the Leibniz Institute for Tropospheric Research (TROPOS) founded a new Germany-wide network for the characterisation of fine and ultrafine particles in the atmospheric aerosol. Several of UBA's manned background monitoring stations and numerous other legal and research institutions with their personnel and existing infrastructure have been involved. Notable institutions have included the Saxon State Office for Environment, Agriculture and Geology (LfULG), the Helmholtz Zentrum Munich (HMGU), the Institute of Energy and Environmental Technology (IUTA), and the German Meteorological Service (DWD). As a result, continuous measurements of sub-µm particle number size distributions and equivalent BC mass concentrations have been installed and maintained at a total of 17 observation sites. This paper serves to describe in detail the characteristics of the measurement sites, the instrumentation deployed for continuous particle measurements, and the location and properties of the data files.

2 Concept for long-term measurements

To date, there are wide experimental options to characterise atmospheric aerosol particles in much physical and chemical detail (e.g. McMurry, 2000; Baltensperger and Prévôt, 2008; Laj et al., 2009). The requirements of long-term deployment in a network, however, reduce these options to experimental methods that are sufficiently stable and reproducible but also financially viable. When designing the German Ultrafine Aerosol Network (GUAN) in 2008, it was decided to implement a limited number of aerosol parameter measurements only but with enhanced spatial coverage and operational reliability (Birmili et al., 2009a). The measurements include, in particular

- sub-µm particle number size distributions;
- sub-µm particle number size distributions of non-volatile particles;
- equivalent black carbon (eBC) mass concentrations.

Number size distributions of particles in dry conditions are measured by mobility particle size spectrometers. Depending on their individual set-up, these instruments are called scanning mobility particle sizer (SMPS), twin differential mobility particle sizer (TDMPS), or twin scanning mobility particle sizer (TSMPS). Number size distributions of non-volatile particles are measured after passage through a thermodenuder at a temperature of 300 °C. (Thermodenuders remove particulate compounds that are volatile at this temperature.) eBC mass concentrations are mainly measured by multi-angle absorption photometers (MAAPs). During GUAN's first operation phase (2009–2014), these instrumental methods proved robust and yielded reproducible results so that they warrant successful deployment in an observation network.

3 Atmospheric observation sites

Figure 1 illustrates the location of GUAN's 17 ground-based atmospheric measurement sites in Germany. Figure 2 provides maps and illustrations of the immediate surroundings of each measurement. Several Tables supply detailed information: Table 1, the characteristics of the measurement sites; Table 2, instrumental features; Table 3, co-location with other particle and air pollutant measurements; Table 4, associations with other networks, infrastructure, and research projects; Table 5, the list of institutions involved. The following text gives a brief description of each GUAN site, its measurement programme, and references for published results.

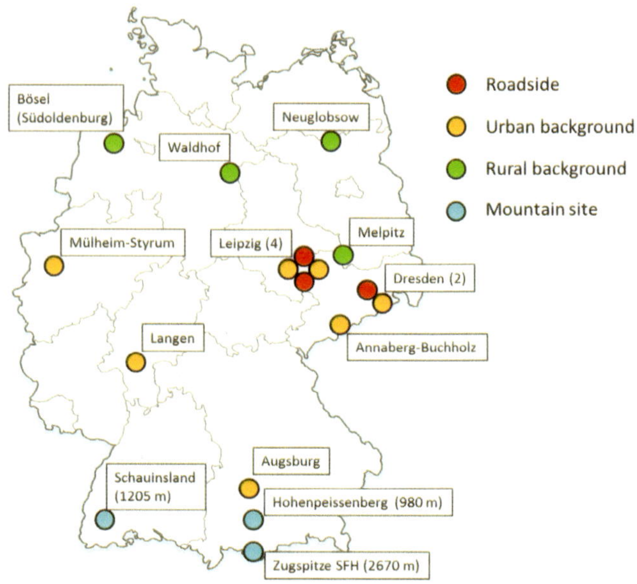

Figure 1. Location of the atmospheric observation sites in the German Ultrafine Aerosol Network (GUAN), currently consisting of 17 sites. See Table 1 for the full names and characteristics of the sites.

3.1 Annaberg-Buchholz

Annaberg-Buchholz is a site of the Air Quality Monitoring Network of Saxony (AQMNS) run by the Saxon State Office for the Environment, Agriculture and Geology (LfULG, Dresden). Like all AQMNS sites, technical operations are conducted by the State Department for Environmental and Agricultural Operations in Saxony (BfUL, Radebeul). The site is located in the city of Annaberg-Buchholz (population ca. 21 000) in the Ore Mountains (Erzgebirge), distant about 10 km from the German–Czech border. Particle number size distributions and eBC mass concentrations have been measured continuously since 2012 in the framework of UltraSchwarz, a research project dedicated to ultrafine particles and health research in the German–Czech border region (Schladitz et al., 2015). The measurements are complemented by a suite of basic particulate and gaseous pollutants (cf. Table 3).

3.2 Augsburg

Augsburg is an urban background monitoring station operated by the Helmholtz Zentrum Munich (HMGU), Institute of Epidemiology II, and the University of Augsburg. The site was established in 2004 in the city of Augsburg (population ca. 270 000) in southern Germany with a main purpose to provide input to epidemiological studies of respiratory and cardiovascular disease (KORA, cooperative health research in the Augsburg region). The station is located on the university premises, about 1 km south-east of the city centre. Particle number size distributions and eBC mass concentrations have been collected continuously since 2004 (Pitz et al., 2008; Birmili et al., 2010a; Gu et al., 2012).

3.3 Bösel (Südoldenburg)

Bösel (Südoldenburg) is a regular site in the government air quality monitoring system of Lower Saxony operated by the Labour Inspectorate of Lower Saxony (Staatliches Gewerbeaufsichtsamt Hildesheim, GAA). To the south, the sampling site borders agricultural areas, while to the north, it is adjacent to residential areas of the village of Bösel (population ca. 7400). The station is situated in an area where livestock production is most intense. The gaseous ammonia and organic emissions related to these activities are anticipated to have an impact on the regional budget of secondary aerosols. Bösel lies at a distance of about 100 km from the North Sea, so that maritime air masses can be sampled with a relatively minor impact of continental sources. Particle number size distributions and eBC mass concentration measurements were carried out on a continuous basis between 2008 and 2015.

3.4 Dresden-Nord

Dresden-Nord is another AQMNS station, located along the roadside in the city of Dresden (population ca. 500 000). Continuous particle number size distribution measurements since 2001 have confirmed a pronounced diurnal cycle of traffic-related pollutants (Löschau et al., 2010). Around 36 000 vehicles pass by the site per day, including 3.5 % heavy-duty vehicles. The horizontal distances to the traffic flows range between 7 m (minor traffic flow; southerly direction) and 80 m (major traffic flow, westerly direction). The site borders the railway station Dresden-Neustadt, with mainly electrified trains passing by at a distance of 200 m to the north. Basic features of the particle number size distributions are presented in Gnauck et al. (2012) and Birmili et al. (2013).

3.5 Dresden-Winckelmannstrasse

Dresden-Winckelmannstrasse is another AQMNS station, located in the urban background of Dresden, about 1.7 km south of the city centre. The next major road passes by the site at a distance of 100 m in easterly direction. Particle number size distributions measurements were established in 2010 and eBC measurements in 2012. A major purpose of the site has been the provision of representative urban background concentrations for particle size distributions and particle mass within the UFIREG (Ultrafine Particles – an evidence based contribution to the development of regional and European environmental and health policy) project (Table 4). The measurements are complemented by a suite of basic particulate and gaseous pollutants (Table 3).

3.6 Hohenpeißenberg

The Meteorological Observatory Hohenpeißenberg (MOHp) has a long tradition of meteorological and climatological observations. It is operated by the German Meteorological Service (DWD) and contributes to WMO-GAW, ACTRIS, and EMEP. MOHp is located on a solitary hill in the rural countryside of southern Bavaria (980 m a.s.l.), approximately 40 km north of the Alpine mountain range. The observatory is located around 300 m above the surrounding rural areas, which are composed mainly of agricultural pasture (70 %) and forests (30 %). MOHp hosts a wide range of atmospheric aerosol and gas phase measurements. Some highlights include continuous observations of volatile organic compounds (VOCs), gaseous sulfuric acid (H_2SO_4), and numerous remote-sensing parameters. MAAP measurements started in December 2003, while SMPS measurements started in 2008. The observatory has delivered atmospheric data as part of GAW since 1994, starting with elemental carbon (EC), total suspended particulate matter (TSP), and total particle number. For the basic characteristics of particle number size distributions and their relation to trace gas and meteorological parameters, see Birmili et al. (2003).

3.7 Langen

Langen is an urban background measurement site located on the premises of the German Federal Environment Agency (UBA) in Langen. The site is located 15 km south of the city of Frankfurt am Main and 5 km south-east of Frankfurt Airport. Aerosol particles are sampled at a height of 14 m on the rooftop of the UBA building. Continuous particle number size distribution measurements started in 2008, complemented since 2009 by measurements of total particle number concentrations (ultrafine condensation particle counter (UCPC), TSI model 3776) and lung-disposable surface area (nanoparticle surface area monitor, TSI model 3550) (Gerwig et al., 2014).

3.8 Leipzig-Eisenbahnstrasse

Leipzig-Eisenbahnstrasse is a roadside observation site in the city of Leipzig (population ca. 500 000), operated by TROPOS since 2002. The site is located in a street canyon within a densely built-up residential area, characterised by multistorey period buildings. The street canyon is regular in that its aspect ratio is close to unity (height: 18 m, width: 20 m), and no building gaps are present. The street experiences traffic of about 12 000 motor vehicles per working day. Ambient aerosol is sampled 6 m above street level on the northern side of the street. Due to the formation of a vortex inside the street canyon, northerly winds have been identified as the condition that favours high particle number concentrations (Voigtländer et al., 2006). For an account of the spatial and temporal variability of particle number size distributions in this area of Leipzig, see Costabile et al. (2009).

3.9 Leipzig-Mitte

Leipzig-Mitte is another AQMNS station, located at the roadside in the city of Leipzig. The site borders the inner-city ring road, in close vicinity to the central train station. Immediately north of the site, three main roads merge at an intersection with daily average traffic volumes around 44 000 vehicles (48 000 on working day). Among all GUAN sites, Leipzig-Mitte exhibits the greatest exposure to traffic-related pollutants. PM_{10} mass and ultrafine particle number concentrations were discussed by Engler et al. (2012) and Ma and Birmili (2015), respectively. Leipzig-Mitte was added to GUAN for the purpose of monitoring possible changes in ultrafine particle number and eBC mass concentrations along with the introduction of the low-emission zone (*Umweltzone*) in Leipzig (Rasch et al., 2013; Löschau et al., 2015). Particle number size distribution and eBC measurements started in 2010. The station's portable cabin borders a tributary road connected to the ring road by traffic lights. Construction activities in the vicinity of the site have occasionally disturbed the measurements between 2010 and 2012. Days on which the impact of construction work was significant were documented in Löschau et al. (2012, 2013).

3.10 Leipzig-Tropos

Leipzig-Tropos (simply called "Leipzig" in some databases) is an atmospheric research station operated by TROPOS since 1997. The station is situated on the roof of the TROPOS institute building. Aerosol particles are sampled at a height of 16 m above the ground. High-traffic roads border the premises at distances of at least 100 m. Comparisons of particle number size distributions at multiple sites in Leipzig have confirmed Leipzig-Tropos as an urban background station (Costabile et al., 2009; Ma and Birmili, 2015). A cross-sectional study suggested a total particle number concentration mean of 9400 cm^{-3}, which proved to be higher than at comparable sites in Helsinki and Copenhagen (von Bismarck-Osten et al., 2013). During the heating season, the site can be influenced by a gas heating stack 50 m south of the aerosol inlet. Screening of the data showed that a perturbation of the measurements is likely under southerly winds and at temperatures below 0 °C. The size distribution profile of the gas heating stack only affects particle number concentration below 30 nm.

3.11 Leipzig-West

Leipzig-West is another AQMNS site in Leipzig, located in the western suburbs of Leipzig. The purpose of the site is to provide a second set of measurements for urban background concentrations in Leipzig. The distance to Leipzig-Mitte is about 6 km and that to Leipzig-Tropos about 10 km. The residential area around the station consists of multistorey apartment blocks that are heated by district heating.[1] The station's cabin is placed in a park area on the premises of a hospital. A minor road passes by the station around 30 m west of the site but has shown to have negligible influence on the measurements. Particle number size distribution and eBC measurements started in 2010 (Rasch et al., 2013).

3.12 Melpitz

Melpitz is an atmospheric research station operated by TROPOS since 1992. The station is located in eastern Germany near the city of Torgau and is ca. 50 km northeast of Leipzig. The site is surrounded by flat and seminatural grasslands without any obstacles in any directions. Besides GUAN, Melpitz contributes to WMO-GAW as a regional background site as well as ACTRIS and EMEP. Measurements at Melpitz can be taken as representative of the central European background atmosphere (Asmi et al., 2011). Melpitz hosts a wide suite of physical and chemical measurements of atmospheric aerosols (Birmili et al., 2008). The site has proved useful to detect long-range pollution transport from continental areas in an easterly direction (Spindler et al., 2010). The distance to the North Sea is about 400 km in a north-westerly direction and that to the Atlantic Ocean about 1000 km in a westerly direction. An account of the basic aerosol characteristics can be found in Engler et al. (2007), Spindler et al. (2010), and Ma et al. (2014). The effects of hygroscopic particle properties on atmospheric light scattering were discussed in Zieger et al. (2014). The formation of new atmospheric particles from gaseous precursors has been analysed here since the 1990s (Birmili and Wiedensohler, 2000; Größ et al., 2015).

3.13 Mülheim-Styrum

Mülheim-Styrum is part of the air quality monitoring network of North Rhine–Westphalia (LUQS). The site is operated by the State Agency for Nature, Environment, and Consumer Protection (LANUV). IUTA Duisburg operates an additional portable cabin for research measurements. The site is situated within a residential area but is also within reach of a motorway (around 250 m to the north), a national road (B223, around 400 m to the west), and industrial premises (around 600 m to the east and south). Overall, the site qualifies as an urban background monitoring station (Quass et al., 2004) and has been used for exposure assessment in health-related studies (ESCAPE, European Study of Cohorts for Air Pollution Effects) and source apportionment studies (Beuck et al., 2011). Mülheim-Styrum was added to GUAN as a representative for the Ruhr district, the largest urban agglomeration in Germany. Continuous particle number size distribution measurements started in 2008. Lung-disposable surface

[1] Hot water is generated in a central power plant 20 km away and delivered to the area through pipelines. Due to this distance, it is expected that any emissions generated at the power plant will have no influence on the local measurements at Leipzig-West.

Site 1: Annaberg-Buchholz (urban background)

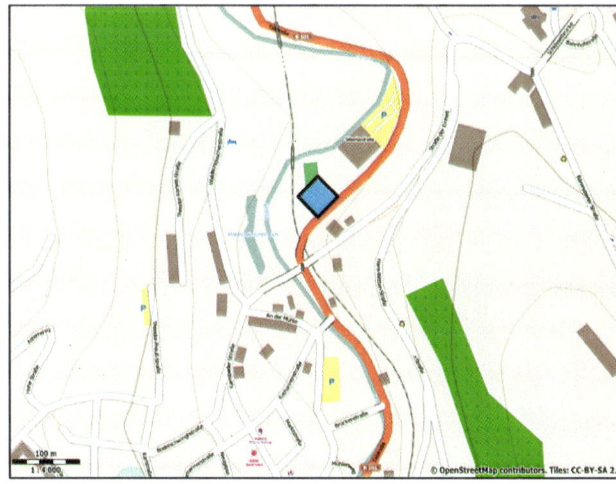

Operator: LfULG

Site 2: Augsburg (urban background)

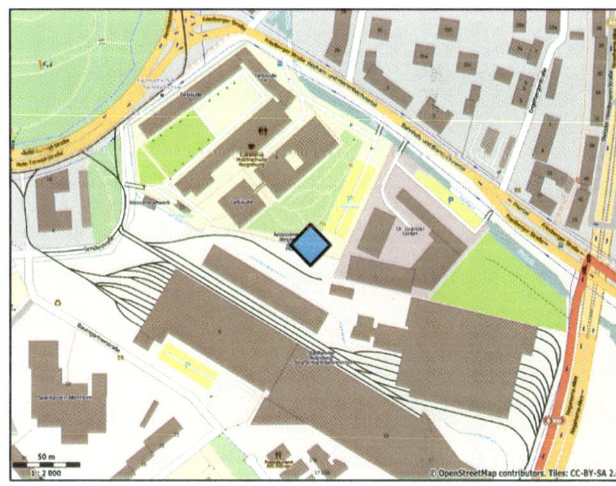

Operator: HMGU/UA

Site 3: Bösel/Südoldenburg (rural)

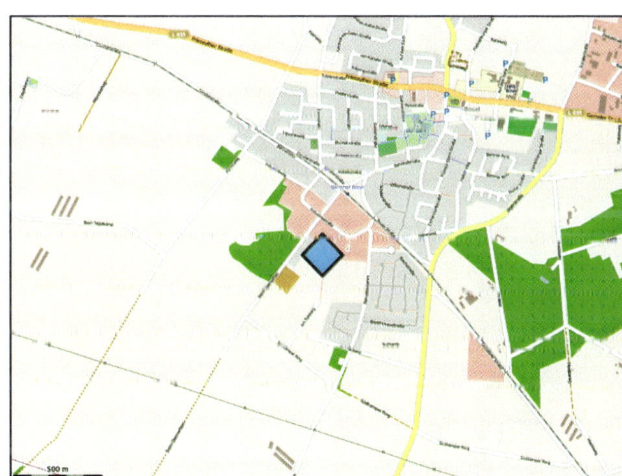

Operator: GAA

Figure 2.

Site 4: Dresden-Nord (roadside)

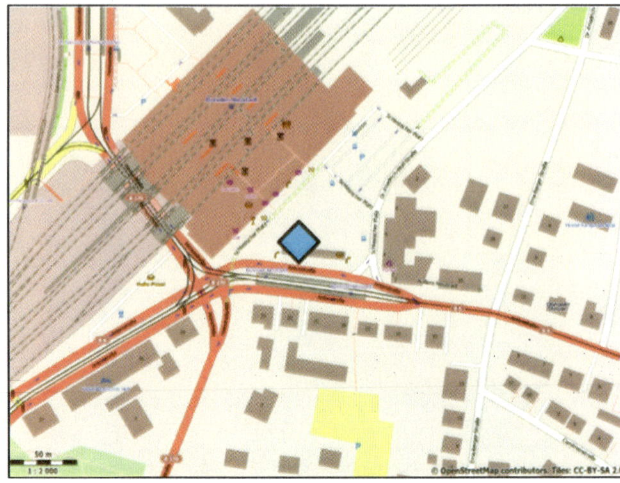

Operator: LfULG

Site 5: Dresden-Winckelmannstr. (urban background)

Operator: LfULG

Site 6: Hohenpeißenberg (rural, mountain)

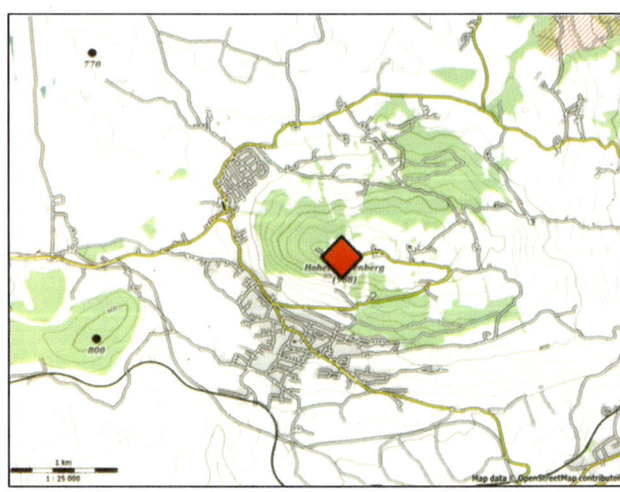

Operator: DWD

Figure 2.

Site 7: Langen (urban background)

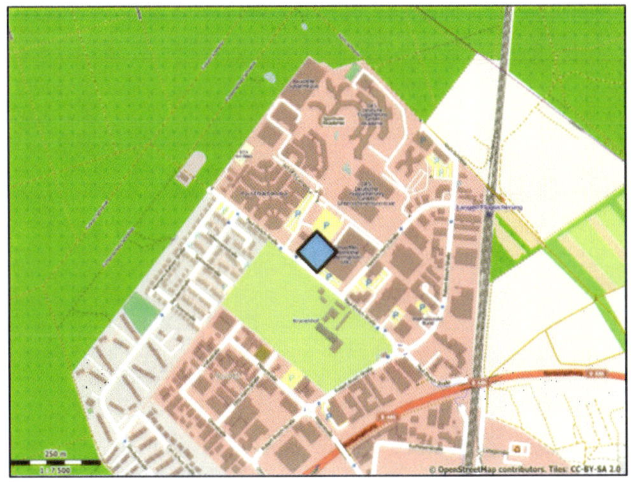

Operator: UBA

Site 8: Leipzig-Eisenbahnstr. (roadside)

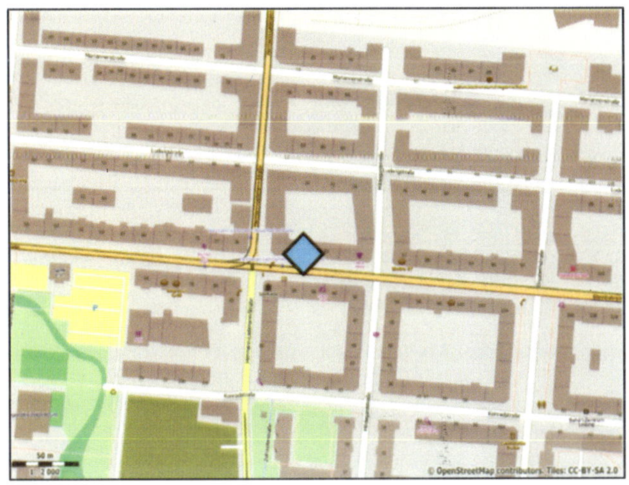

Operator: TROPOS

Site 9: Leipzig-TROPOS (urban background)

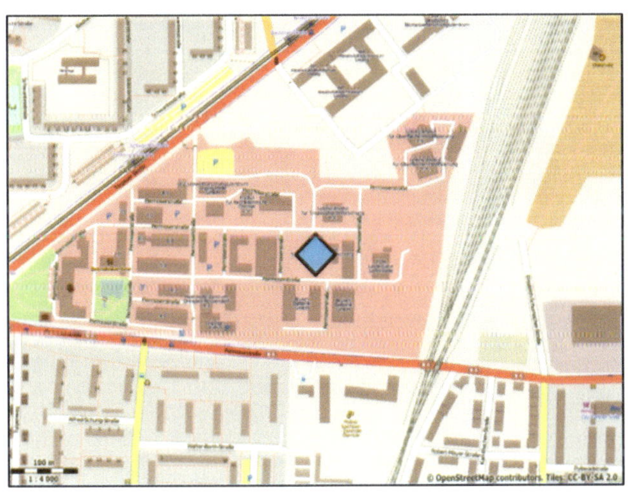

Operator: TROPOS

Figure 2.

Site 10: Leipzig-Mitte (roadside)

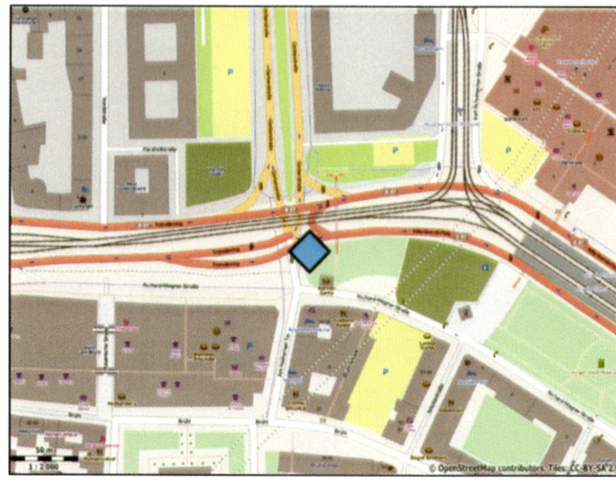

Operator: LfULG

Site 11: Leipzig-West (urban background)

Operator: LfULG

Site 12: Melpitz (rural)

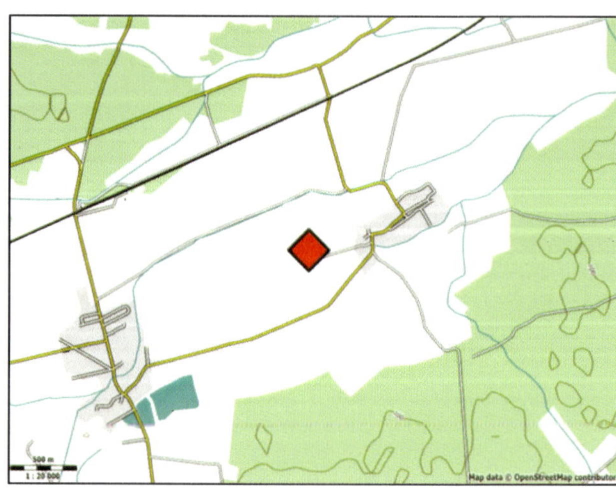

Operator: TROPOS

Figure 2.

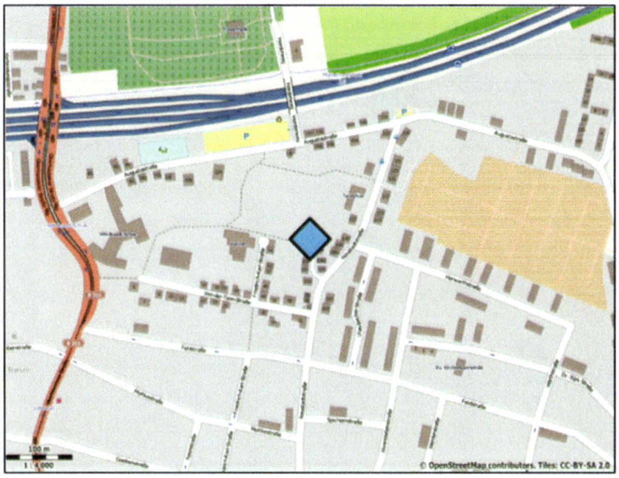

Site 13: Mülheim-Styrum (urban background)

Operator: IUTA/LANUV

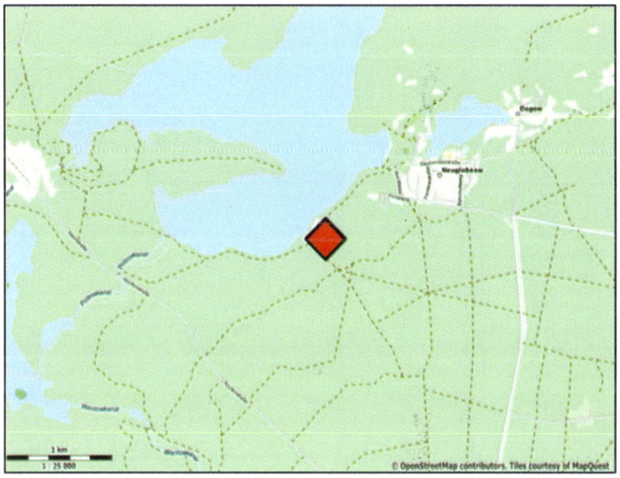

Site 14: Neuglobsow (rural)

Operator: UBA

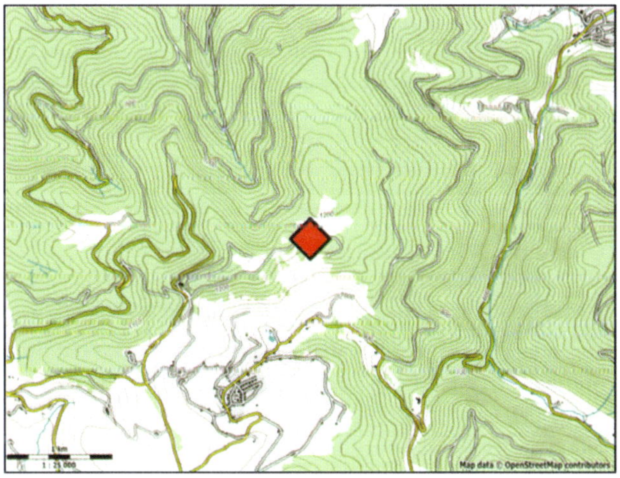

Site 15: Schauinsland (rural, mountain)

Operator: UBA

Figure 2.

Site 16: Waldhof (rural)

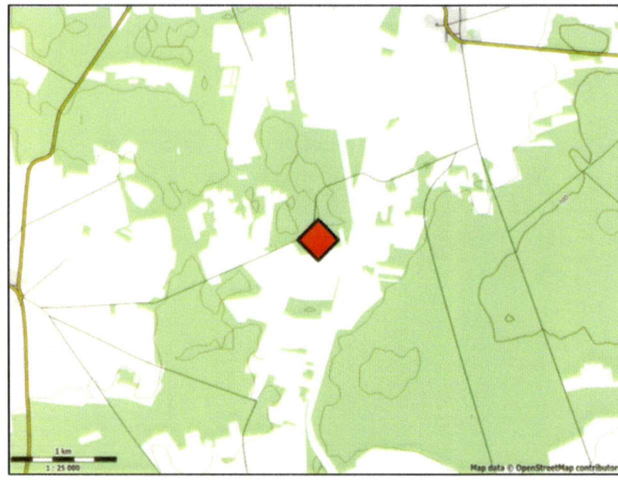

Operator: UBA

Site 17: Zugspitze/Schneefernerhaus (Alpine mountain)

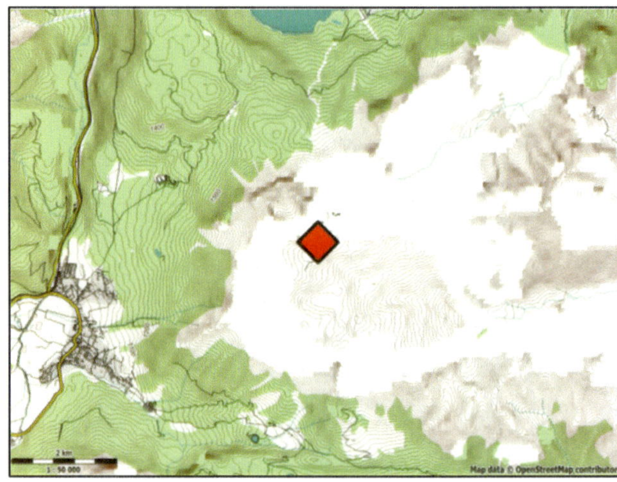

Operator: UBA/DWD

Figure 2. Illustration of the GUAN measurement sites: **(a)** sites 1–3, **(b)** sites 4–6, **(c)** sites 7–9, **(d)** sites 10–12, **(e)** sites 13–15, **(e)** sites 16–17. Pictures illustrate the surroundings of each station. The maps show the geographic area, illustrating major types of land use (Source: OpenStreetMap, processed by Maperitive V.2.3.22). The location of each measurement site is marked by a diamond.

area has been measured by a nanoparticle surface area monitor (TSI model 3550) as an additional measure of quality assurance.

3.14 Neuglobsow

Neuglobsow is one of the permanently manned stations within UBA's continuous observation network (UBA, 2013). The sampling site is surrounded by lakes and forested areas in all directions and is therefore only influenced very little by local sources. Neuglobsow contributes to EMEP. Measurements here can be taken as representative of the atmospheric background in north-eastern Germany. Particle number size distribution and eBC measurements started in 2010.

3.15 Schauinsland

Schauinsland is another of UBA's manned observatories. The station is located at 1205 m a.s.l. near the Schauinsland peak in the Black Forest in south-western Germany. Measurements started as early as 1965 as part of research programmes funded by the German Research Foundation (DFG). The station is well-suited to characterise air masses that approach central Europe from westerly directions. In winter, the site tends to lie in relatively clean air above the Rhine valley's inversion layer. Observations at Schauinsland have focused on the detection of long-term trends. For instance, the site hosts the longest continuous observation of carbon dioxide (CO_2) observations in Europe (since 1972; Schmidt et al., 2003).

Particle number size distribution measurements started in 2005 and eBC measurements in 2008.

3.16 Waldhof

Waldhof is another manned station within UBA's observation network. The sampling site is surrounded by forest in all directions and is therefore only influenced very little by local sources. Measurements here are representative of the background in the north German lowlands. Waldhof is the only German atmospheric station contributing to GMOS (Global Mercury Observation System) (Weigelt et al., 2013). Particle number size distribution and eBC measurements started in 2008.

3.17 Zugspitze (Schneefernerhaus)

Zugspitze (Schneefernerhaus) is part of the global WMO-GAW station Zugspitze/Hohenpeissenberg and jointly operated by the German Federal Environment Agency (UBA) and the German Meteorological Service (DWD). The observatory is located at 2670 m a.s.l., about 300 m below the Zugspitze summit and on the southern slope of the mountain massif. The high altitude leads to a significant annual cycle in observed aerosol particle number and mass concentration, caused by different boundary layer heights in summer and winter (Birmili et al., 2009b) The station's elevated position allows us to sample air masses that had only little contact with the local boundary layer, especially in the cold season from October to March. Zugspitze occasionally receives lofted aerosol layers from remote source regions, such as North America (Birmili et al., 2010b) or the Eyjafjallajökull volcanic eruption (Schäfer et al., 2011). Besides the GUAN measurements, the Karlsruhe Institute of Technology conducts long-term measurements of total particle number concentration (condensation particle counter – CPC), particle number size distribution (optical particle counter – OPC; aerodynamic particle sizer – APS), and primary biological aerosol particles. Particle number size distribution measurements started in 2004 and eBC measurements in 2008.

4 Instrumental set-up

4.1 Particle number size distributions

Particle number size distributions are recorded on a continuous basis using mobility particle size spectrometers. Depending on the specific configuration, these can be TDMPSs (typical particle size range 5–800 nm), TSMPSs (size range 5–800 nm), or SMPSs (typical size range 10–800 nm). The core component of each mobility particle size spectrometer is a differential mobility analyser (DMA).

Note that in GUAN, similar to WMO-GAW, particle number size distributions are measured in a "dry state" i.e. at relative humidities (RHs) below 50 %. The reason is that particle size distributions from different sites should remain comparable. Atmospheric particles are known to grow significantly in diameter above RH 50 % (Swietlicki et al., 2008). To prevent such effects in the instrumentation, aerosol dryers are installed upstream of each mobility particle size spectrometer and also within the sheath flow circuit (Wiedensohler et al., 2012). The actual values of RH and temperature in every instrument are stored at the EBAS database and can be accessed through the database interface there.

4.1.1 SMPS (TROPOS design)

The principle of the custom-built SMPS is described in detail in Wiedensohler et al. (2012). Briefly, the instrument uses a Vienna-type DMA (electrode length 28 cm) with a condensation particle counter (CPC model 3772, TSI Inc.) to measure particles between 10 and 800 nm. Before entering the DMA, ambient aerosol is conveyed to a bipolar charge equilibrium using a ^{85}Kr neutraliser. The sheath flow rate is 5 L min^{-1} at an aerosol flow rate of 1 L min^{-1}. The high-voltage supply provides positive polarity, thus collecting negatively charged particles. The sheath air is generally circulated in a closed loop. Both the aerosol sample flow and the sheath air flow are actively dried in this instrument, thus ensuring a relative humidity during size classification below 40 % at most times. Temperature, relative humidity, and pressure inside the instrument are continuously monitored. The typical time resolution for one combined upscan and downscan is 5 min. TROPOS-designed SMPS instruments are currently deployed at the GUAN stations 1, 3, 5, 6, 14, 15, and 16 (see Table 2).

4.1.2 SMPS (TSI design)

Commercial SMPS instruments (model 3936, TSI Inc., Shoreview, USA) are deployed at the GUAN stations 7, 13, and 17. The sheath air to aerosol flow ratio is 5 : 1 L min^{-1} at stations 7 and 17 – yielding a particle size range of 10–600 nm – and 3 : 0.3 L min^{-1} at station 13 – yielding a particle size range of 14–750 nm. Time resolution is 5 min. The instruments at all stations were upgraded to meet the quality criteria recommended for ambient aerosol measurements by the EUSAAR and ACTRIS initiatives (Wiedensohler et al., 2012). The upgrades concern dryers for the aerosol sample flow and sheath air, as well as additional sensors for temperature, relative humidity, and pressure.

At stations 7 and 13, these upgrades took place prior to the start of measurements. At station 17, the modifications took place at the end of 2008. In addition, the high-voltage supply at station 17 was changed from negative to positive polarity and a range of up to 12.5 kV. This effect was con-

Table 1. Atmospheric measurement sites in GUAN, in alphabetic order.

No.	Name	EBAS Code	Site operator	Type	Altitude	Location	Site and/or data set description
1	Annaberg-Buchholz	DE0061B	LfULG	urban background	545 m	50°34′18″ N, 12°59′56″ E	Schladitz et al. (2015)
2	Augsburg	DE0062B	HMGU/UA	urban background	485 m	48°21′29″ N, 10°54′25″ E	Pitz et al. (2008), Gu et al. (2012)
3	Bösel (Südoldenburg)	DE0056R	GAA	rural	17 m	52°59′53″ N, 07°56′34″ E	Asmi et al. (2011)
4	Dresden-Nord	DE0063K	LfULG	roadside	116 m	51°03′54″ N, 13°44′29″ E	Löschau et al. (2010), Birmili et al. (2013)
5	Dresden-Winckelmannstr.	DE0064B	LfULG	urban background	120 m	51°02′10″ N, 13°43′50″ E	Löschau et al. (2012)
6	Hohenpeißenberg	DE0043G	DWD	rural (mountain)	980 m	47°48′06″ N, 11°00′34″ E	Birmili et al. (2003)
7	Langen	DE0065B	UBA	urban background	130 m	50°00′18″ N, 08°39′05″ E	Gerwig et al. (2014), UBA (2013)
8	Leipzig-Eisenbahnstr.	DE0066K	TROPOS	roadside	120 m	51°20′45″ N, 12°24′23″ E	Voigtländer et al. (2006), Costabile et al. (2009)
9	Leipzig-Mitte	DE0067K	LfULG	roadside	111 m	51°20′39″ N, 12°22′38″ E	Engler et al. (2012), Rasch et al. (2013)
10	Leipzig-Tropos	DE0055B	TROPOS	urban background	126 m	51°21′10″ N, 12°26′03″ E	Costabile et al. (2009), Ma and Birmili (2015)
11	Leipzig-West	DE0068B	LfULG	urban background	122 m	51°19′05″ N, 12°17′51″ E	Löschau et al. (2012), Rasch et al. (2013)
12	Melpitz	DE0044R	TROPOS	rural	84 m	51°31′32″ N, 12°55′40″ E	Engler et al. (2007), Ma et al. (2014)
13	Mülheim-Styrum	DE0069B	LANUV/IUTA	urban background	37 m	51°27′17″ N, 06°51′56″ E	Beuck et al. (2011)
14	Neuglobsow	DE0007R	UBA	rural	70 m	53°08′28″ N, 13°01′52″ E	UBA (2013)
15	Schauinsland	DE0003R	UBA	rural (mountain)	1205 m	47°54′49″ N, 07°54′29″ E	UBA (2013), Asmi et al. (2011)
16	Waldhof	DE0002R	UBA	rural	75 m	52°48′04″ N, 10°45′23″ E	UBA (2013), Asmi et al. (2011)
17	Zugspitze (Schneefernerhaus)	DE0054R	UBA/DWD	Alpine mountain	2650 m	47°25′00″ N, 10°58′47″ E	Birmili et al. (2009b, 2010b)

Table 2. Technical features of GUAN instrumentation. Mobility particle size spectrometers follow the TROPOS design (Sect. 4.1.1) unless stated otherwise. Two types of thermodenuders are used to measure non-volatile size distributions (Sect. 4.2).

No.	Name	Type	Inlet height above ground	Particle mobility size spectrometer type	Size range	Thermodenuder	eBC instrument	eBC cut-off size
1	Annaberg-Buchholz	portable cabin	4 m	SMPS	10–800 nm	–	MAAP	PM_1
2	Augsburg	portable cabin	4 m	TSMPS	5–800 nm	Wehner et al. (2002)	aethalometer	$PM_{2.5}$
3	Bösel (Südoldenburg)	portable cabin	4 m	SMPS	10–800 nm	simplistic type	MAAP	PM_{10}
4	Dresden-Nord	portable cabin	4 m	TSMPS	5–800 nm	–	MAAP	PM_1
5	Dresden-Winckelmannstr.	portable cabin	4 m	SMPS	10–800 nm	–	MAAP	PM_1
6	Hohenpeißenberg	building	12 m	SMPS	10–800 nm	simplistic type	MAAP	PM_{10}
7	Langen	portable cabin	14 m	SMPS (TSI 3936)	10–600 nm	–	–	PM_1
8	Leipzig-Eisenbahnstr.	building	6 m	TDMPS	5–800 nm	Wehner et al. (2002)	MAAP	PM_1
9	Leipzig-Mitte	portable cabin	4 m	TDMPS	5–800 nm	–	MAAP	PM_{10}
10	Leipzig-Tropos	portable cabin	16 m	TDMPS	5–800 nm	Wehner et al. (2002)	MAAP	PM_{10}
11	Leipzig-West	portable cabin	4 m	TDMPS	10–800 nm	–	MAAP	PM_{10}
12	Melpitz	portable cabin	4 m	TSMPS	5–800 nm	Wehner et al. (2002)	MAAP	PM_{10}
13	Mülheim-Styrum	portable cabin	4 m	SMPS (TSI 3936)	14–750 nm	–	–	PM_{10}
14	Neuglobsow	building	6 m	SMPS	10–800 nm	–	MAAP	PM_{10}
15	Schauinsland	building	6 m	SMPS	10–800 nm	simplistic type	MAAP	PM_{10}
16	Waldhof	building	6 m	SMPS	10–800 nm	–	MAAP	PM_{10}
17	Zugspitze (Schneefernerhaus)	building	6 m	SMPS (TSI 3936)	10–600 nm	simplistic type	MAAP	PM_{10}

sidered in the processing of the data by the use of a different corresponding bipolar charge distribution.

4.1.3 TDMPS/TSMPS

Mechanically, the TDMPS and TSMPS are dual DMA versions of the SMPS described in Sect. 4.1.1 (Birmili et al., 1999). The first subsystem combines an ultrafine Vienna-type DMA (electrode length 11 cm) with a UCPC (model 3025, TSI Inc., Shoreview (MN), USA) to measure particles across the range 3–80 nm. The second subsystem combines another DMA (electrode length 28 cm) with a condensation particle counter (CPC model 3010 or 3772, TSI Inc.) to measure particles between 10 and 800 nm. Due to enhanced measurement uncertainties below 5 nm, only the diameter range 5–800 nm is further analysed and fed into the EBAS database. Like with the SMPS, sheath air is circulated in a closed loop at relative humidities ranging mostly between 10 and 40 %. The typi-

Table 3. Co-location of GUAN measurements with other continuous aerosol and air pollutant measurements. The core of GUAN measurements include sub-µm particle number size distributions (PNSD), sub-µm particle number size distributions of non-volatile particles (NV-PNSD), and equivalent black carbon mass concentrations (eBC). Additional continuous measurements may include total particle number concentration (TNC) – measured by condensation particle counters; coarse particle number size distribution CPNSD – using an aerodynamic particle sizer (APS) or an optical particle counter (OPC); nanoparticle surface area (NSA) – using an NSAM monitor; aerosol scattering coefficient (σ_p) – using a nephelometer; PM_x particle mass concentrations; and basic meteorological parameters (Meteo) including T, RH, wind speed, wind direction, global radiation, and precipitation. These additional data need to be obtained directly from the operator of the respective measurements.

No.	Name	PNSD	NV-PNSD	eBC	TNC	CPNSD	NSA	σ_p	PM_{10}	$PM_{2.5}$	NO_x	SO_2	CO	O_3	CO_2	Meteo
1	Annaberg-Buchholz	•		•					•	•	•	•		•		•
2	Augsburg	•	•	•	•	•			•	•						•
3	Bösel (Südoldenburg)	•	•	•					•		•	•	•	•		•
4	Dresden-Nord	•		•					•		•			•		•
5	Dresden-Winckelmannstr.	•		•					•	•	•	•		•		•
6	Hohenpeißenberg	•	•	•	•	•		•	•	•	•	•	•	•	•	•
7	Langen	•			•		•									•
8	Leipzig-Eisenbahnstr.	•	•	•												•
9	Leipzig-Mitte	•		•		•			•	•	•	•				•
10	Leipzig-Tropos	•	•	•							•	•		•		•
11	Leipzig-West	•		•					•	•	•					•
12	Melpitz	•	•	•		•		•	•	•	•	•		•		•
13	Mülheim-Styrum	•					•		•	•	•	•	•	•		•
14	Neuglobsow	•		•					•		•	•	•	•		•
15	Schauinsland	•	•	•		•			•		•	•	•	•	•	•
16	Waldhof	•		•					•		•	•	•	•	•	•
17	Zugspitze (Schneefernerhaus)	•	•	•	•	•		•	•	•	•	•	•	•	•	•

Table 4. Associations of GUAN stations with other air quality networks, infrastructure, and research projects. For further abbreviations, see Table 5.

No.	Name	WMO-GAW[a]	EMEP[b]	UBA	ACTRIS[c]	AQMNS[d]	UFIREG[e]	LLEZ[f]	Others
1	Annaberg-Buchholz					•		UltraSchwarz[g]	
2	Augsburg						•		KORA[h]
3	Bösel (Südoldenburg)	regional station							LÜN[i]
4	Dresden-Nord					•		•	
5	Dresden-Winckelmannstr.					•	•	•	
6	Hohenpeißenberg	global station	•		•				DWD, VAO[j]
7	Langen			•					
8	Leipzig-Eisenbahnstr.							•	
9	Leipzig-Mitte					•		•	
10	Leipzig-Tropos	regional station						•	
11	Leipzig-West					•		•	
12	Melpitz	regional station	•		•			•	
13	Mülheim-Styrum								LUQS[k]
14	Neuglobsow	regional station	•	•					
15	Schauinsland	regional station	•	•					
16	Waldhof	regional station	•	•					GMOS[l]
17	Zugspitze (Schneefernerhaus)	global station		•					VAO[j]

[a] WMO-GAW: World Meteorological Organization/Global Atmosphere Watch (http://www.wmo.int/pages/prog/arep/gaw/gaw_home_en.html, http://gaw.empa.ch/gawsis).
[b] EMEP: European Monitoring and Evaluation Programme (http://www.emep.int).
[c] ACTRIS: Aerosols, Clouds, and Trace gases Research InfraStructure Network (http://www.actris.eu).
[d] AQMNS: Air quality monitoring network of Saxony (Luftgütemessnetz Sachsen), coordinated by LfULG, operated by BfUL.
[e] UFIREG: Ultrafine particles – an evidence based contribution to the development of regional and European environmental and health policy (http://www.ufireg-central.eu).
[f] LLEZ: Leipzig low-emission zone studies (Löschau et al., 2012, 2013, 2014, 2015; Rasch et al., 2013).
[g] UltraSchwarz: Ultrafine particles and health in the Ore Mountains district and the region of Usti (*Ultrafeinstaub und Gesundheit im Erzgebirgskreis und Region Usti*) (http://www.ultraschwarz-ziel3.de).
[h] KORA: Cooperative health research in the Augsburg region (http://www.helmholtz-muenchen.de/kora).
[i] LÜN: Air quality monitoring system in Lower Saxony (Luftüberwachungssystem Niedersachsen), operated by GAA.
[j] VAO: Virtual Alpine Observatory, coordinated by the Bavarian Research Alliance (http://www.bayfor.org).
[k] LUQS: Air quality monitoring network of North Rhine–Westphalia (Kontinuierliches Luftmessnetz), operated by LANUV.
[l] GMOS: Global Mercury Observation System (http://www.gmos.eu).

Table 5. Institutions involved in GUAN, in alphabetic order.

Acronym	Name	Location (all in Germany)
BfUL	State Dept. for Environmental and Agricultural Operations in Saxony (Betriebsgesellschaft für Umwelt und Landwirtschaft)	Radebeul
DWD	German Meteorological Service (Deutscher Wetterdienst)	Hohenpeißenberg
GAA	Labour Inspectorate of Lower Saxony (Staatliches Gewerbeaufsichtsamt)	Hildesheim
HMGU	Helmholtz Zentrum Munich, Institute of Epidemiology II	Neuherberg
IUTA	Institute of Energy and Environmental Technology e.V.	Duisburg
LANUV	State Agency for Nature, Environment and Consumer Protection in North Rhine–Westphalia (Landesamt für Natur, Umwelt und Verbraucherschutz Nordrhein-Westfalen)	Essen
LfULG	Saxon State Agency for the Environment, Agriculture and Geology (Sächsisches Landesamt für Umwelt, Landwirtschaft und Geologie)	Dresden
TROPOS	Leibniz Institute for Tropospheric Research	Leipzig
UA	University of Augsburg	Augsburg
UBA	German Federal Environment Agency (Umweltbundesamt)	Dessau-Rosslau

cal time resolution of the instrument is 10 min. TDMPS instruments are deployed at sites 2, 4, and 12, while TDMPS instruments are operated at sites 8–11. TSMPS and TSMPS have, in principle, no major differences in hardware. In software, TSMPS use a continuous ramping of the high voltage, like in the SMPS, rather than the stepwise change used in the TDMPS.

4.2 Non-volatile size distributions (thermodenuder)

Upstream of some mobility particle size spectrometers, a thermodenuder (TD) is deployed as a way of removing volatile aerosol components. The standard operation mode for these extended instruments is to record size distributions upstream and downstream of the TD in alternating sampling intervals (Birmili et al., 2010a). This procedure provides a steady flow of size distributions both with and without the TD and an effective time resolution of half the original instrumental time resolution.

One thermodenuder type follows the design of Wehner et al. (2002). Volatile particle material is evaporated at a temperature of 300 °C and subsequently removed with the assistance of active carbon in a cooling section. The temperature of 300 °C was selected with the aim of evaporating the overwhelming mass of volatile and semi-volatile material, particularly ammonium nitrate, ammonium sulfate, and most organic carbonaceous compounds. Major compounds not removed include elemental carbon, crustal material, and sodium chloride (cf. Engler et al., 2007). Meanwhile, 300 °C is a temperature at which charring (i.e. the incomplete combustion of oxygenated hydrocarbons) of organic compounds is avoided. A mass closure for the non-volatile particle fraction at the research station Melpitz suggested that the non-volatile fraction at 300 °C contains not only refractory black carbon but also a comparable share of low-volatility organic aerosol compounds (LVOCs; Poulain et al., 2014).

The second, simplistic thermodenuder type consists of a simple steel tube that is heated to 300 °C by a laboratory furnace. Excess vapours are only adsorbed by the tube walls. Such a simplistic TD has been used, for example, by Ehn et al. (2007). Experiments with ambient aerosol in Leipzig suggested that both thermodenuder designs produced equivalent results for non-volatile particle size distributions across the diameter range 10–800 nm. The tubing acting as a TD is usually cleaned once per year.

In the TD, the aerosol sample is subject to enhanced particle losses. One can think of, e.g., diffusional as well as thermophoretic losses, which are both a function of particle size. In the Wehner-type thermodenuder, part of the sample flow passes through the active carbon filter. Therefore, this instrument was calibrated for particle losses by using spherical silver particles (solid at 300 °C) and by measuring the particle counts upstream and downstream of the TD. The penetration of solid particles through the Wehner-type thermodenuder was about 0.34 at 3 nm, 0.66 at 10 nm and 0.85 for particles bigger than 100 nm at a flow rate of 2.5 L min^{-1}. All data collected from this thermodenuder type were corrected for these losses.

4.3 Aerosol absorption and eBC

The aerosol absorption coefficient is measured by MAAPs. The MAAP converts the light attenuation and reflection by a particle-laden quartz fibre filter into an absorption coefficient by calculating the radiative transfer through this two-layer system (Petzold and Schönlinner, 2004). The wavelength ($\lambda = 637$ nm) corresponds to the region of the solar spectrum where BC is the prime absorber, thus minimising interferences with "brown carbon" and mineral dust. Brown carbon and mineral dust tend, in fact, to absorb light more efficiently towards the ultraviolet wavelengths (Sun et al., 2007; Müller et al., 2009).

Because there is no unique relationship between aerosol absorption and black carbon mass concentration, black carbon is typically reported as eBC (Petzold et al., 2013). For this purpose, the aerosol absorption coefficient σ_{abs} is converted into an eBC mass concentration using an experimentally determined mass absorption cross section (MAC). The manufacturer of the MAAP instrument reports a MAC value of $6.6\,m^2\,g^{-1}$, which is automatically applied to all data records. An assessment of aerosol absorption using Raman spectroscopy and EC measurements as reference methods yielded a mean MAC value of $5.3\,m^2\,g^{-1}$, with a range of variability between 3.9 and $7.4\,m^2\,g^{-1}$ across a selection of seven GUAN sites (Nordmann et al., 2009, 2013). An intercomparison of multiple instruments showed that different MAAP instruments produce comparable results with less than 5 % inter-device variability (Müller et al., 2011). Besides eBC mass concentrations, the sample flow rate, temperature, and pressure, the MAAP yields the raw signals of loaded and blank filter material at scattering angles of 0, 135, and 165°. The latter parameters are stored internally in a so-called "extended data format".[2] To provide accurate and comparable measurements under dry sample conditions, the MAAP aerosol flow is usually conditioned by a membrane dryer.

At Augsburg, an aethalometer (Type 8100, Thermo Fisher Scientific Inc.) is deployed using a cut-off of 2.5 µm. This instrument yields mean eBC mass concentrations that are comparable to those from a MAAP instrument. Details on this instrument can be seen in Appendix A.

Unfortunately, not all eBC measurements in GUAN use the same inlet configuration. Inlets for PM_{10}, $PM_{2.5}$ and PM_1 are used throughout the network (Table 2). In order to harmonise the eBC values, the data recorded downstream of the PM_1 cyclone inlets may be adjusted to the corresponding level of a PM_{10} inlet using suitable correction factors. The multiplication factors recommended here are 1.10 for rural sites, 1.08 for urban background sites, and 1.05 for roadside sites. These values were determined by a direct intercomparison of the readings of two MAAP instruments using a PM_{10} and a PM_1 inlet at the sites Leipzig-Eisenbahnstrasse (roadside), Leipzig Tropos (urban background), and Melpitz (rural background). Because very high correlations were found during those intercomparison experiments ($R^2 = 0.99$), a post correction appears justified (Löschau et al., 2012).

[2] The extended version of the data, i.e. the raw signals at various scattering angles could come to use if at some point in the future, the radiative transfer algorithm to retrieve eBC from the MAAP measurement might be improved. The raw signals would give the possibility to recalculate eBC according to such a different scheme.

5 Quality assurance

Quality assurance (QA) in GUAN consists of procedures to ensure that measurements remain stable both instrument to instrument (or site to site) and instrument to standard. Temporal or spatial deviations in atmospheric pollutants often occur in the range of a few per cent of the absolute concentration of a specific parameter. The issue of QA is therefore vital to successfully address scientific questions on the basis of such data. It has been an aim in GUAN to ensure an accuracy of within a few per cent for the eBC mass concentration measurements, of a few per cent for the particle sizing accuracy in number size distributions, and of ±10 % for particle number concentrations over the entire measurement period.

The following paragraphs describe the state-of-the-art QA procedures for mobility particle size spectrometers, which were followed at the majority of the stations. Due to the different degree of access and availability of manpower, they may not be valid in all details at every single station and for all historical parts of the data collection. Most of these measures were co-developed within the framework of previous infrastructure projects, such as WMO-GAW, EUSAAR, ACTRIS, and research projects initiated by the Saxon State Office for Environment, Agriculture and Geology (LfULG).

5.1 Maintenance

5.1.1 Weekly or biweekly inspection

At unmanned GUAN stations, the mobility particle spectrometers and MAAPs are inspected in person at least every 2 weeks, preferably every week. At the manned GUAN stations (sites 6 and 14–17), the instrumentation is usually inspected more often, up to once per working day. At unmanned sites, remote data access has helped to check instrumental performance. Unfortunately, remote data access to unmanned stations is currently available only for a few stations (e.g. 1, 2, 4, 5, 12). The weekly or biweekly inspection of mobility particle size spectrometers includes visual checks whether all instrumental components are switched on and working correctly: high-voltage power supply, sheath air flow, and aerosol flow; CPCs; supply of CPC working liquid (butanol or water); data acquisition program; flow status and operation of the MAAP.

5.1.2 Monthly maintenance

A full maintenance check is typically carried out with mobility particle size spectrometers every 4 weeks. Here, the following instrumental flow rates are verified using an external reference flow meter, usually a bubble flow meter: aerosol flow rate, sheath air flow rate, flow rate of the aerosol dryer's counter flow, flow rate of the sheath air dryer counter flow. Aerosol and sheath air flow meters in the mobility particle size spectrometers are recalibrated if they deviate by more than 5 and 2 % from their set point values, respectively. Instruments are also checked for leaks using a total particle filter. The performance of the mobility particle size spectrometers is deemed satisfactory if the total particle number collected by the instrument after a waiting time of 15 min does not exceed 10 particles cm^{-3}.

The high-voltage supply for each DMA is checked with a digital multimeter, involving a verification of the high voltage between 0 and 1000 V. Recalibrations are made if the voltages exceed defined thresholds at 0 V (±3 V), 6.25 V (±25 %), 100 V (±10 %), and 1000 V (±1 %). Frequently checking the high voltage is essential to provide a correct sizing of the DMA, particularly at the lower end of the particle size distribution. A NIST (National Institute of Standards and Technology, US Department of Commerce) certified particle size standard is used to verify the exact sizing of the instrument. Currently, the most popular standard is 203 nm Polystyrene latex (PSL) spheres, which are certified within 2.5 % of the nominal particle diameter. The particles are nebulised from aqueous suspension using a jet nebulizer (e.g. PariBoy, Pari GmbH, Starnberg, Germany). If the sizing of the mobility particle size spectrometer deviates more than 2 % from the standard size (i.e. is outside the interval 200–206 nm), the sheath air flow is adjusted until the DMA matches the certified diameter of the PSL particles.

5.1.3 Annual maintenance

The annual maintenance event includes more extensive instrumental checks and care. It is preferentially performed in a central laboratory. Many GUAN instruments are returned to the World Calibration Centre for Aerosol Physics (WCCAP; Wiedensohler et al., 2012) at TROPOS in Leipzig once a year. The annual maintenance includes check and calibration of the humidity sensors; calibration of the sheath air flow rate zero offset; check and calibration of the pressure transducer; disassembly and cleaning of the DMA(s); check of the saturator sponge inside the CPC; check and calibration of the CPC(s) against a particle number concentration standard. It has also been found important to check the activity of the charge neutraliser. A ^{85}Kr beta source, for example, degrades substantially in its ion production rate after about 10 years. The annual maintenance event is also used to perform hardware improvements and software updates.

5.2 Comparison to reference instruments

Intercomparisons with reference instruments are essential in establishing a relationship to a defined standard. In the case of mobility particle size spectrometers, we use CPCs of a specific type (model 3010 and 3772, TSI Inc., Shoreview, USA) and an electrometer (model 3068B, TSI Inc.) as an intermediate standard for particle number concentration. It is also a standard procedure to compare a mobility particle size spectrometer once per year to a reference instrument, which will reveal possible deviations with respect to the size-dependent instrument response. In GUAN, these intercomparisons are made in the central laboratory, at calibration workshops, or in the field as part of a "round-robin test".

5.2.1 Laboratory intercomparisons

Intercomparisons with the TROPOS reference mobility particle size spectrometers can be made at the WCCAP "on the fly", i.e. at short notice at most times. However, it is common practice to collect a pool of instruments, which are then examined collectively against one or two reference instruments (Wiedensohler et al., 2012). Because of their state close to the bipolar charge equilibrium, we generally prefer ambient aerosols for all instrumental intercomparisons, even if the experiments are conducted in the central laboratory. Three reference instruments are provided by WCCAP for quality assurance, which are checked regularly against each other and against total particle counters. (Reference instruments need to agree within ±5 % for all particle sizes between 20 and 300 nm.) Laboratory intercomparisons are preferentially accompanied by a total particle counter (CPC). The deviation between the total particle number concentration derived from any mobility particle size spectrometer and the total particle counter must not exceed ±10 % for ambient aerosols. On the basis of such laboratory intercomparisons, we estimate the accuracy of the particle number concentration measured by mobility particle size spectrometers to be ±10 % for the diameter range 20–300 nm. Outside this range, larger uncertainties are possible. Below 20 nm, deviations between SMPS instruments may amount up to 50 % and above 300 nm up to 30 %. TDMPS and TSMPS instruments exhibit a higher accuracy at the lower particle size end: ±10 % down to 10 nm and ca. ±30 % at 5 nm.

5.2.2 Round-robin intercomparisons

Ideally, every mobility particle size spectrometer undergoes at least one calibration experiment per year. To bridge gaps between the scheduled WCCAP calibration workshops, the QA measures may include a round-robin test on its measurement site using a reference mobility particle size spectrometer. In practice, such field intercomparisons usually involve (a) setting up a reference instrument in parallel to the mobility particle size spectrometers on site, (b) performing parallel size distribution measurements of ambient aerosol with the reference instrument during one night (introductory performance test), (c) maintaining and/or improving the instrument in case of problems, and (d) repeating the ambient aerosol comparison for several days as a final performance test. The round-robin test usually involves checks of particle sizing and parallel measurements by a total particle counter (CPC) as well. As in the case of the laboratory intercomparisons, an agreement between the total number concentration of a mobility particle size spectrometer and the reading of a total particle counter of ±10 % is required. In most practical cases, we also found a good agreement between the test and reference instruments within ±10 % for the diameter range 20–300 nm.

5.2.3 Enhanced quality assurance

For three stations in Saxony (sites 1, 4, and 5), enhanced quality assurance measures have been developed (Schladitz et al., 2014). At these sites, the mobility particle size spectrometers are equipped with an automatic function control unit that performs unattended instrumental comparisons. First, a leak check using a total particle filter is performed every 3 days. A dedicated total particle counter (TSI model 3772, "transfer CPC") is moved to each of the three stations every 8 weeks. During a 2-week presence of the transfer CPC, the control unit will perform comparison measurements between this CPC and the particle size spectrometer every 23 h. To avoid uncertainties due to nucleation mode particles present in the lower cut-off region of the CPC, particles at the lower end of the number size distribution are removed by a diffusion screen. For traceability, the transfer CPC is checked every 8 weeks against a dedicated reference CPC at WCCAP and once a year against a calibrated electrometer. This procedure allows an even closer tracking of the actual performance of the mobility particle size spectrometer than under less infrequent intercomparisons. In our experience, the procedure can narrow down the uncertainty with respect to particle number concentration across the diameter range of 20–300 nm from ±10 to ±5 %.

At several sites, additional instrumentation has been used as a support for quality assurance and control: at sites 2, 6, and 17, CPCs have been deployed on a continuous basis to measure total particle number concentration. As total particle concentration can also be computed from the particle number size distribution, a divergence of the two readings can help identify instrumental problems. At sites 7 and 13, nanoparticle surface area monitors (NSAM, TSI Inc.) have been deployed as a QA tool because the lung-deposited particle surface area (LDSA) can be computed from the particle number size distribution as well. At site 13, significant deviations between the measured and calculated LDSA over several consecutive days indicated the failure of one of the instruments which, in practice, almost always turned out to be the SMPS or its CPC.

6 Data processing and validation

6.1 Particle number size distribution processing

Particle number size distributions from all stations except station 13 were processed using the TropINV software package, written in LabVIEW (Version 8.5, National Instrument, Austin, USA). This program evokes a linear multiple charge inversion algorithm (Muchain, Multiple Charge Inversion; Pfeifer et al., 2014) and performs various additional corrections before yielding the final particle number size distributions. TropINV performs the following steps:

1. reformatting the raw concentration and instrument diagnostics data (optional);
2. condensing the number of size bins of the raw data to a fraction of its original number (optional);
3. automatic flagging of periods exhibiting diagnostics data outside the nominal range;
4. assimilation of the two DMA branches (only for TDMPS and TSMPS instruments);
5. multiple charge inversion (Pfeifer et al., 2014);
6. adjustment of particle number concentrations for non-ideal aerosol flow rates (optional);
7. size-dependent corrections of particle number concentration (Wiedensohler et al., 2012):
   - CPC counting efficiency
   - DMA transfer function
   - bipolar charger
   - aerosol dryer.
   - connecting tubes inside and outside of the instrument
8. normalisation of ambient concentrations to 0 °C, 1013 hPa (optional);
9. re-binning the size channels to a predefined standard set of standard channels.

Step 7 yields technically correct data that are ready for scientific use (dubbed "Level-1" in Sect. 7.2). Step 8 yields data that, after averaging to 1 h resolution, form "Level-2" data (Sect. 7.2). GUAN has followed a practice that during step 9, all particle number size distributions are re-binned to a uniform set of 40 channels between 10 and 800 nm in the case of the SMPS instrument and 46 channels between 5.1 and 800 nm by linear interpolation. It is expected that in this way, data processing will be facilitated for the scientific end users. The data at station 13 were processed using the Aerosol Instrument Manager software (TSI Inc., Revision G, October 2006), which performs all the necessary steps (5–7) in a manner equivalent to the TropINV software (Pfeifer et al., 2014).

6.2 Technical and manual data validation

The first step of quality control of mobility particle size spectrometer data involves the automatic flagging of data records that are associated with invalid instrument diagnostic data. For example, the relative humidity in the instrument needs to remain below 40 %. Instrumental temperature needs to be within a range of +10 to 30 °C. Aerosol sampling flows are not permitted to deviate from their set point by more than ±10 %, and the sheath air flows must not diverge more than ±5 % from their set point. These thresholds are motivated in Wiedensohler et al. (2012).

Further quality control involves additional visual checks by the scientists processing the data. One procedure is to carefully inspect contour diagrams of the particle number size distribution. An example for such contour diagrams at the highest available time resolution is given in Fig. 3. The human eye is rather sensitive to unusual features in such two-dimensional structures, and in our experience, this visual inspection has proved very effective in detecting irregularities in the data, such as

- flash-over of high voltage between the two DMA electrodes (this unwanted effect tends to produce artificial particles at the upper end of the size distribution);
- decrease in the counting efficiency of a CPC, as a result of laser diode power degradation;
- contaminations, such as room air leaking into the instrument
- disturbing particle sources inside and outside of the measurement station.

A second tool is the screening of the time series, spanning half a year or more for the all-time maxima and minima. We look, for example, at the total particle number and volume concentrations calculated from number size distributions and review their time series and histograms for unusual outliers. Individual values far outside the main frequency distribution are then inspected more closely and, if judged as the result of technical faults, deleted from the subsequent processing.

7 Data repository and format

7.1 EBAS (World Data Center for Aerosols)

Particle number size distributions and aerosol absorption coefficients from all GUAN stations are stored at the World Data Center for Aerosols EBAS database, where they are publicly available and free of charge (Birmili et al., 2015). The GUAN data can be accessed by following the persistent identifier doi:10.5072/guan or by visiting the EBAS user interface directly (http://ebas.nilu.no). EBAS is a database infrastructure operated by NILU (Norwegian Institute for Air Research), with basic funding secured through the EMEP protocol of the Convention on Long-range Transboundary Air Pollution. EBAS handles, stores, and disseminates atmospheric composition data generated by international and national frameworks in long-term monitoring programmes and research projects and has, for instance, been used by more than 50 European Union research projects. Dedicated staff (ca. 10 persons) secure the reliability for long-term data access.

Data integrity, interpreted as quality assurance and protection of the data holdings, is secured by the following measures: each data submission to EBAS undergoes manual quality assurance for the correctness of the accompanying metadata, as well as an extra check of the data records for outliers and implausible values. Errors are iterated with the data originators until the submission passes quality assurance. Once inserted into the database, the data are backed up together with the whole database at least twice per week (two full and two incremental back-ups per week) at an off-site location.

With regard to copyright, any scientific user is free to download, copy, distribute, transmit, and adapt the data sets as long as he/she gives credit to the original authors (equivalent to the Creative Commons Attribution License). Data users accept that they should make an offer of co-authorship through personally contacting the data providers or owners whenever substantial use is made of their data. It is desired that the data users contact the data originators in order to create positive feedback and foster scientific communication. In all cases, acknowledgement must be made of the data originators and of the project name when these data are used within a publication. Tørseth et al. (2012) illustrate the possibilities of a cross-sectional study that makes extensive use of data stored at EBAS.

7.2 Data format

At EBAS, aerosol data are stored in three levels, reflecting an enhanced data submission protocol for particle number size distribution and BC mass concentration data. The objective of a three-level protocol is to enhance the traceability of measurements and data processing, ranging from instrumental raw data up to final hourly averages.

Level-0 data contain raw data, i.e. electrical particle mobility distribution data and instrumental diagnostic parameters, as they are measured directly by the instrument. The electrical particle mobility distribution is provided as an array of measured particle number concentrations vs. particle diameters (electrical particle mobility for singly charged particles).

Level-1 data are scientifically correct particle number size distributions after multiple-charge inversion and the correction for particle losses. In the case of particle number size distributions, the data refer to the original conditions during the measurement, i.e. ambient pressure at the station, and typically 20 °C laboratory temperature. In the case of the aerosol absorption coefficient, the data refer to standard temperature

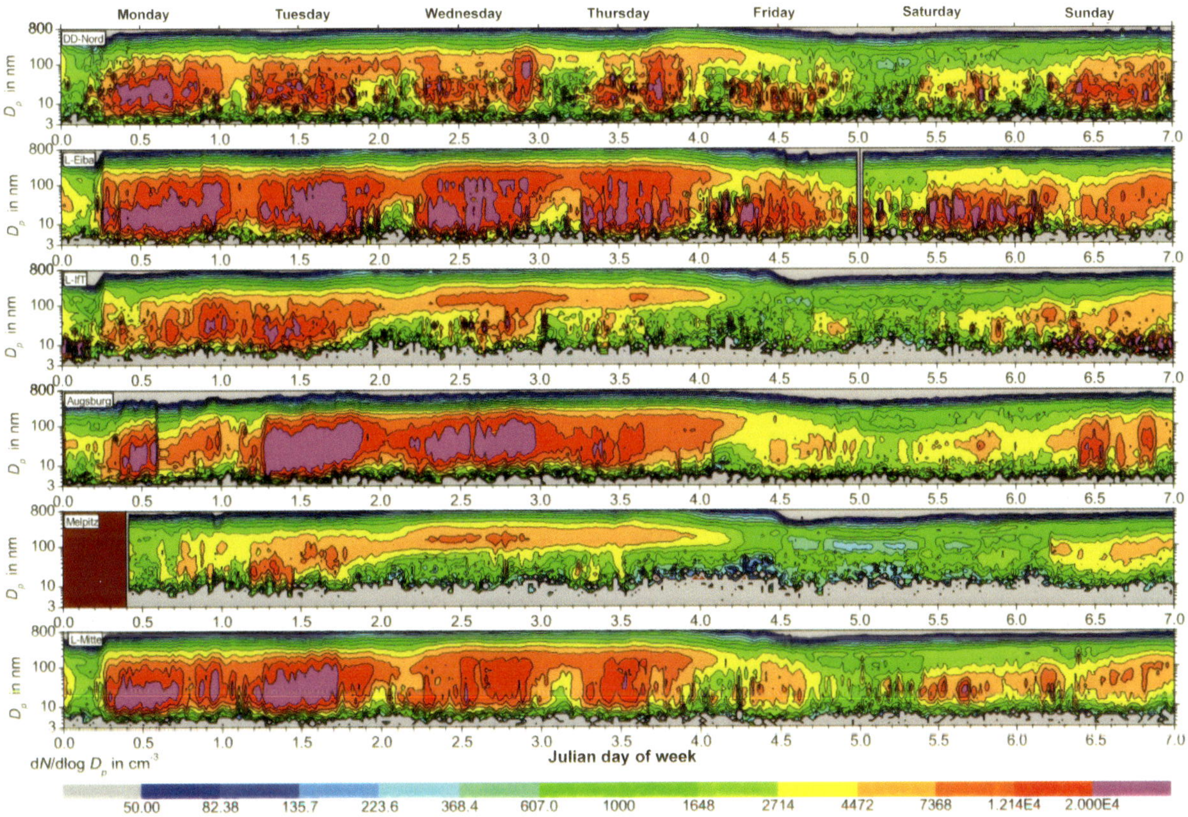

Figure 3. Exemplary contour diagrams of particle number size distributions at six mainly urban GUAN sites (20–26 December 2009). Such diagrams assist the visual quality control of the data.

(273.15 K) and pressure (1013.25 hPa) already. In any case, the time resolution refers to the original instrumental time resolution.

Level-2 data are given as hourly averages and adjusted to a uniform standard temperature (273.15 K) and pressure (1013.25 hPa). Moreover, most of the particle number size distribution data sets have been re-binned to the same standard set of 40 or 46 diameter channels (cf. Sect. 6.1). Level-2 data are presumed to be the main point of interest to most data users. Level-2 data are available through the EBAS user interface (see Supplement), while Level-0 and Level-1 data are available on request from the administrators of the database.

Data are stored at EBAS in the NASA-Ames format, which is based on the ASCII text NASA Ames 1001 format but contains additional metadata specifications ensuring instrumental documentation. For more information on the format, we refer the reader to Wiedensohler et al. (2012) and the EBAS website. An extract of NASA-Ames 1001 text is provided in the Supplement.

7.3 State of data submission

Although data are collected continuously, the body of GUAN data currently requires some degree of manual quality control and semi-manual data post processing. Data parcels encompassing half a calendar year are currently submitted to EBAS at regular intervals. Feeding the data from all 17 GUAN stations and also non-volatile particle number size distributions into EBAS has not been completed and will require, especially in the case of some retrospective data sets, additional time. It has been our first priority to transfer all recent GUAN data spanning the years 2009–2014 into EBAS. At the time of writing, data from 2015 have been published as well. The Tables provided on GUAN's download page doi:10.5072/guan present the current state of the data submission and availability. (A snapshot of that page is also shown in the Supplement to this article.) The most complete year of data availability is 2012, with particle number size distributions available at 14 sites. The longest continuous time series are available for particle number size distributions in Dresden-Nord and Melpitz.

It needs to be mentioned that certain data sets prior to 2009 were still processed under previous data processing standards. This is a result of the EBAS repository emerging from projects that go back to the early 2000s. These older particle number size distribution data sets lack, particularly, the corrections for the losses in the internal tubing of the mobility particle size spectrometers. These losses come into ef-

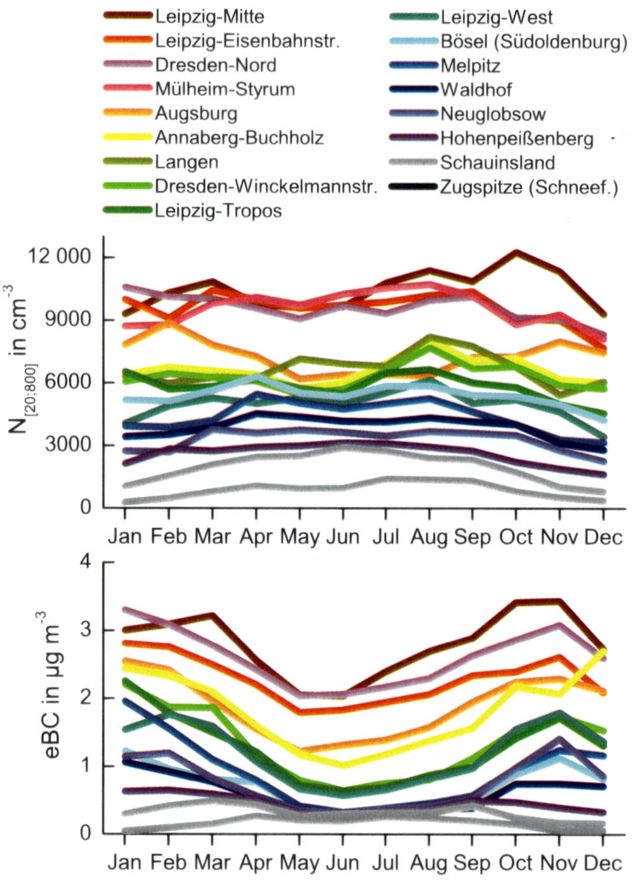

Figure 4. Preview of the mean annual cycles of total particle number concentration ($N_{[20;800]}$) and eBC mass concentration. The data coverage ranges between 2 and 6 years (see text and Table 6).

fect mainly for particle diameters below 30 nm, and inconsistencies may occur when older and modern data sets are directly compared. The older data sets are clearly marked on the download page. It is planned that the old data sets will be updated to modern standards.

8 Preliminary results

This section provides a brief overview of the data collected in GUAN so far. Figure 4 displays the total particle number concentration (diameter range 20–800 nm) and eBC mass concentration at the various GUAN sites as a function of the month of the year, while Table 6 indicates a brief overall statistics of the two parameters based on hourly average values. The data coverage used in this compilation ranged between 2 and 6 years: Annaberg-Buchholz – 3 years (2012–2014); Langen – 3 years (2011–2013); Dresden-Winckelmannstr. and Neuglobsow – 4 years (2011–2014); Leipzig-Mitte and Leipzig-West – 5 years (2010–2014); and Augsburg, Bösel (Südoldenburg), Dresden-Nord, Hohenpeißenberg, Leipzig-Eisenbahnstr., Leipzig-Tropos, Melpitz, Mülheim-Styrum, Schauinsland, Waldhof, and Zugspitze (Schneefernerhaus) – 6 years (2009–2014). The different periods of data coverage reflect, above all, different operation periods at some stations.

Figure 4 illustrates the mean concentrations at the various sites but also their annual course. To facilitate the reading of the figure, the legend has been arranged in descending order of mean concentration. Observations at the roadside are distinct in that they yield an average $N_{[20;800]}$ above 9000 cm^{-3} most of the year, while urban background and rural observations sites form a continuum of concentrations below that region. Zugspitze (Schneefernerhaus) shows the lowest particle concentrations, due to its altitude above 2500 m.

These findings are also reflected in Table 6, where the sites and values have been sorted according to descending median concentrations of $N_{[20;800]}$. One can clearly see how the stations cluster according to their station type. The concentrations at the stations can therefore be arranged almost completely in the order roadside > urban background > rural > rural (mountain) > Alpine mountain. Overall, the GUAN sites provide a continuum of atmospheric observations over 1 magnitude of median values in total particle number concentration and eBC mass concentration. $N_{[20;800]}$ spans a range between 900 and 9000 cm^{-3}, while eBC ranges between 0.15 and 2.3 µg cm^{-3}. One notable exception in the sequence of stations is Bösel (Südoldenburg) — classified as "rural", which seems, however, influenced by sources of particle number that are likely to be found in the adjacent village (cf. location of the site in Fig. 2).

The annual cycle of total particle number concentration yields only a weak annual cycle, whereas this is not the case for the eBC mass concentrations (Fig. 4). For eBC, the monthly maximum concentration occurs between October and March at the urban and rural lowland sites. The summer values (May–September), in contrast, are significantly lower. One explanation is that the atmosphere tends to be less vertically mixed in the winter, leading to the trapping of pollution at the urban and rural lowland sites. Second, eBC emissions are enhanced in wintertime due to the more intensive power and heat generation. In summer, vertical dilution is a common atmospheric phenomenon that will decrease the corresponding mean values. It is worth noting that in summer, the particle mass concentrations of eBC at rural sites are very close to each other, suggesting rather uniform eBC concentrations in the rural atmosphere.

A more elaborate interpretation of the GUAN data, notably the 2009–2014 section, is currently underway and in preparation for publication in a separate paper.

9 Conclusions

The co-operative German Ultrafine Aerosol Network (GUAN) delivers atmospheric particle number size distributions and equivalent black carbon concentrations. Data are

Table 6. Mean values of total particle number concentration ($N_{[20;800]}$) and eBC mass concentration at 17 GUAN sites. The table indicates the number of hourly samples n, mean value μ, standard deviation σ, median ("med"), and the 95th percentile of the hourly statistics ("p95"). Numbers are indicated at a precision of three significant digits. The stations are sorted according the decreasing median of $N_{[20;800]}$.

No.	Site	Type	Period	$N_{[20;800]}$ in cm^{-3}					eBC in $\mu g\, m^{-3}$				
				n	μ_N	σ_N	med	p95	n	μ_{eBC}	σ_{eBC}	med	p95
9	Leipzig-Mitte	roadside	2010–2014	35 400	10 500	6500	9000	22 400	37 800	2.80	2.03	2.30	6.57
8	Leipzig-Eisenbahnstr.	roadside	2009–2014	43 000	9610	5600	8460	20 000	48 400	2.15	1.65	1.71	5.28
4	Dresden-Nord	roadside	2009–2014	33 900	9570	5440	8400	19 900	46 900	2.48	1.61	2.10	5.56
13	Mülheim-Styrum	urban backg.	2009–2014	45 100	9570	5450	8330	20 000	–	–	–	–	–
7	Langen	urban backg.	2011–2013	24 800	6490	3840	5740	13 500	–	–	–	–	–
2	Augsburg	urban backg.	2009–2014	41 900	7260	5780	5630	17 600	42 700	1.88	1.52	1.41	4.58
1	Annaberg-Buchholz	urban backg.	2012–2014	22 400	6620	6450	4950	16 800	23 300	1.65	1.75	1.12	4.99
10	Leipzig-Tropos	urban backg.	2009–2014	49 200	5840	4110	4840	13 000	51 100	1.26	1.33	0.82	3.76
3	Bösel (Südoldenburg)	rural	2009–2014	46 600	5380	3120	4810	10 900	44 100	0.76	0.80	0.51	2.25
5	Dresden-Winckelmannstr.	urban backg.	2011–2014	22 400	6290	6130	4710	15 900	23 200	1.21	1.14	0.86	3.44
11	Leipzig-West	urban backg.	2010–2014	28 600	4960	3210	4210	10 600	36 000	1.21	1.23	0.79	3.64
12	Melpitz	rural	2009–2014	47 400	4380	3000	3710	9620	51 400	0.90	1.16	0.52	3.03
16	Waldhof	rural	2009–2014	49 700	3860	2350	3390	8050	39 000	0.60	0.68	0.39	1.92
14	Neuglobsow	rural	2011–2014	27 000	3270	1840	2940	6590	41 000	0.74	0.91	0.40	2.50
6	Hohenpeißenberg	rural (mountain)	2009–2014	48 800	2600	1470	2360	5280	52 500	0.44	0.43	0.33	1.27
15	Schauinsland	rural (mountain)	2009–2014	44 100	1950	1430	1640	4590	43 000	0.32	0.36	0.22	0.96
17	Zugspitze (Schneef.)	Alpine mount.	2013–2014	39 000	1120	874	886	2780	46 900	0.17	0.10	0.15	0.37

transferred to the World Data Center for Aerosols repository EBAS (doi:10.5072/guan) at regular intervals. These continuously measured data are expected to provide the basis for a better scientific understanding of sub-μm aerosol processes in the troposphere, addressing questions related to both human particle exposure and climate-relevant effects. Particle number size distributions down to 5 nm allow us to study the emission and formation processes of atmospheric ultrafine particles (e.g. Ma and Birmili, 2015). Black carbon, on the other hand, is an essential parameter that is linked to atmospheric particulate light absorption (Bond et al., 2013; Nordmann et al., 2014). Number size distributions of refractory particles (300 °C) represent particle cores that are of likely relevance for health studies and which have been associated with harmful soot particles (Nordmann et al., 2009; Poulain et al., 2014).

The selection of GUAN's measurement sites provides a continuum of exposure situations from roadside sites, urban background sites, and rural background sites to a high Alpine mountain site that spans approximately 1 order in mean concentrations. In association with partner projects, GUAN has implemented a high degree of harmonisation of instrumentation, operating procedures, and data evaluation procedures. The GUAN data have already proved suitable for the validation of atmospheric dispersion and process simulations involving atmospheric aerosols. Examples for their successful use include the validation of a global chemical transport model (Reddington et al., 2011), a regional-scale radiative transfer model (Nordmann et al., 2014; Chen et al., 2016), a street canyon aerosol dynamics model (Toenges-Schuller et al., 2015), and statistical prediction tools for particle number size distributions in rural and urban environments (Mølgaard et al., 2013; von Bismarck-Osten et al., 2015).

The high standards of the GUAN data make them suitable for use in cross-sectional air quality and health studies, particularly alleviating the lack of health studies using BC and ultrafine particles as exposure variables. In the future, the data set may also serve as an excellent basis for the future discussion of legislative regulation of (ultrafine) particle number and/or BC mass concentrations in ambient air.

10 Data availability

This manuscript refers to data from the German Ultrafine Aerosol Network (GUAN) that are stored at the World Data Center for Aerosols EBAS database (Birmili et al., 2015). Data access is possible through the persistent identifier doi:10.5072/guan or by visiting the EBAS user interface directly (http://ebas.nilu.no). When accessing the data set through the persistent identifier (doi:10.5072/guan), visitors will find detailed information on data availability, data acquisition and processing.

Scientific users are free to download, copy, distribute, transmit, and adapt the data sets as long as they adhere to EBAS' data policy, which involves, amongst other things, giving credit to the original authors (cf. Sect. 7). The GUAN data set is not static in that the length of the data records is augmented at regular intervals by fresh, quality-controlled measurements – currently every 6 months (cf. Sect. 7.3). This will lead to a considerably longer, continuous data set in the foreseeable future.

Acknowledgements. The establishment of GUAN was made possible by the German Federal Environment Ministry (BMU) grants F&E 370343200 (German title: "Erfassung der Zahl feiner und ultrafeiner Partikel in der Außenluft"), 2008–2010, and F&E 371143232 (German title: "Trendanalysen gesundheitsgefährdender Fein- und Ultrafeinstaubfraktionen unter Nutzung der im German Ultrafine Aerosol Network (GUAN) ermittelten Immissionsdaten durch Fortführung und Interpretation der Messreihen"), 2012–2014. We acknowledge additional funding within the infrastructure projects EUSAAR (European Supersites for Atmospheric Aerosol Research; EU FP6 contract RII3-CT-2006-026140), and ACTRIS (Aerosols, Clouds, and Trace gases Research InfraStructure Network; EU FP7 grant 262254). WCCAP was funded by the German Federal Environment Agency (UBA) grant FKZ 35101086 (The World Calibration Centre for Aerosol Physics within WMO-GAW), 2012–2014. We acknowledge funding by the LfULG grants UFP 2011 (German title: "Qualitätssicherung der Messung ultrafeiner Partikel in Außenluft 2011") and UFP 2012 ("Aufwandsreduzierung in der Qualitätskontrolle der Messung ultrafeiner Partikel in der Außenluft im Luftgütemessnetz Sachsen"). Measurements at Annaberg-Buchholz were supported by the EU-Ziel3 project UltraSchwarz (German title: "Ultrafeinstaub und Gesundheit im Erzgebirgskreis und Region Usti"), grant 100083657. Measurements at Dresden-Winckelmannstrasse were co-funded by the European Regional Development Fund Financing Programme Central Europe, grant No. 3CE288P (UFIREG). Measurements in Augsburg were financed by the Helmholtz Zentrum Munich and partly by UFIREG (see above). IUTA e.V. acknowledges funding by the State Agency for Nature, Environment, and Consumer Protection North Rhine–Westphalia (LANUV), grant title: "Untersuchung ultrafeiner Partikel zur räumlichen und zeitlichen Verteilung im Ruhrgebiet". Individual credits for TROPOS personnel: Thomas Tuch and Achim Grüner (TROPOS station Melpitz), Stephan Nordmann (MAAP measurements), and Johanna Rehn (map generation). We thank Horst-Günther Kath (State Dept. for Environmental and Agricultural Operations in Saxony, Betriebsgesellschaft für Umwelt und Landwirtschaft – BfUL), Andreas Hainsch and Dirk Haase (Labour Inspectorate of Lower Saxony, Staatliches Gewerbeaufsichtsamt Hildesheim – GAA), and Dieter Gladtke (LANUV), who made the GUAN measurements possible at the observations sites of their air quality networks.

Edited by: A. Kokhanovsky

References

Asmi, A., Wiedensohler, A., Laj, P., Fjaeraa, A.-M., Sellegri, K., Birmili, W., Weingartner, E., Baltensperger, U., Zdimal, V., Zikova, N., Putaud, J.-P., Marinoni, A., Tunved, P., Hansson, H.-C., Fiebig, M., Kivekäs, N., Lihavainen, H., Asmi, E., Ulevicius, V., Aalto, P. P., Swietlicki, E., Kristensson, A., Mihalopoulos, N., Kalivitis, N., Kalapov, I., Kiss, G., de Leeuw, G., Henzing, B., Harrison, R. M., Beddows, D., O'Dowd, C., Jennings, S. G., Flentje, H., Weinhold, K., Meinhardt, F., Ries, L., and Kulmala, M.: Number size distributions and seasonality of submicron particles in Europe 2008–2009, Atmos. Chem. Phys., 11, 5505–5538, doi:10.5194/acp-11-5505-2011, 2011.

Baltensperger, U. and Prévôt, A. S. H.: Chemical analysis of atmospheric aerosols, Anal. Bioanal. Chem., 390, 277–280, doi:10.1007/s00216-007-1687-z, 2008.

Beuck, H., Quass, U., Klemm, O., and Kuhlbusch, T. A. J.: Assessment of sea salt and mineral dust contributions to PM_{10} in NW Germany using tracer models and positive matrix factorization, Atmos. Environ., 45, 5813–5821, doi:10.1016/j.atmosenv.2011.07.010, 2011.

Birmili, W. and Wiedensohler, A.: New particle formation in the continental boundary laxer: meteorological and gas phase parameter influence, Geophys. Res. Lett., 27, 3325–3328, doi:10.1029/1999GL011221, 2000.

Birmili, W., Stratmann, F., and Wiedensohler, A.: Design of a DMA-based size spectrometer for a large particle size range and stable operation, J. Aerosol Sci., 30, 549–553, doi:10.1016/S0021-8502(98)00047-0, 1999.

Birmili, W., Berresheim, H., Plass-Dülmer, C., Elste, T., Gilge, S., Wiedensohler, A., and Uhrner, U.: The Hohenpeissenberg aerosol formation experiment (HAFEX): a long-term study including size-resolved aerosol, H_2SO_4, OH, and monoterpenes measurements, Atmos. Chem. Phys., 3, 361–376, doi:10.5194/acp-3-361-2003, 2003.

Birmili, W., Schepanski, K., Ansmann, A., Spindler, G., Tegen, I., Wehner, B., Nowak, A., Reimer, E., Mattis, I., Müller, K., Brüggemann, E., Gnauk, T., Herrmann, H., Wiedensohler, A., Althausen, D., Schladitz, A., Tuch, T., and Löschau, G.: A case of extreme particulate matter concentrations over Central Europe caused by dust emitted over the southern Ukraine, Atmos. Chem. Phys., 8, 997–1016, doi:10.5194/acp-8-997-2008, 2008.

Birmili, W., Weinhold, K., Nordmann, S., Wiedensohler, A., Spindler, G., Müller, K., Herrmann, H., Gnauk, T., Pitz, M., Cyrys, J., Flentje, H., Nickel, C., Kuhlbusch, T. A. J., Löschau, G., Haase, D., Meinhardt, F., Schwerin, A., Ries, L., and Wirtz, K.: Atmospheric aerosol measurements in the German Ultrafine Aerosol Network (GUAN): Part 1 – soot and particle number size distributions, Gefahrst. Reinhalt. L., 69, 137–145, 2009a.

Birmili, W., Ries, L., Sohmer, R., Anastou, A., Sonntag, A., König, K., and Levin, I.: Fine and ultrafine aerosol particles at the GAW station Schneefernerhaus/Zugspitze, Gefahrst. Reinhalt. L., 69, 31–35, 2009b.

Birmili, W., Heinke, K., Pitz, M., Matschullat, J., Wiedensohler, A., Cyrys, J., Wichmann, H.-E., and Peters, A.: Particle number size distributions in urban air before and after volatilisation, Atmos. Chem. Phys., 10, 4643–4660, doi:10.5194/acp-10-4643-2010, 2010a.

Birmili, W., Göbel, T., Sonntag, A., Ries, L., Sohmer, R., Gilge, S., Levin, I., and Stohl, A.: A case of transatlantic aerosol transport detected at the Schneefernerhaus Observatory (2650 m) on the northern edge of the Alps, Meteorol. Z., 19, 591–600, doi:10.1127/0941-2948/2010/0465, 2010b.

Birmili, W., Tomsche, L., Sonntag, A., Opelt, C., Weinhold, K., Nordmann, S., and Schmidt, W.: Variability of aerosol particles in the urban atmosphere of Dresden (Germany): effects of spatial scale and particle size, Meteorol. Z., 22, 195–211, doi:10.1127/0941-2948/2013/0395, 2013.

Birmili, W., Weinhold, K., Merkel, M., Rasch, F., Sonntag, A., Wiedensohler, A., Bastian, S., Schladitz, A., Löschau, G., Cyrys, J., Pitz, M., Gu, J., Kusch, T., Flentje, H., Quass, U., Kaminski, H., Kuhlbusch, T. A. J., Meinhardt, F., Schwerin, A., Bath, O., Ries, L., and Wirtz, K.: The German Ultrafine Aerosol

Network (GUAN): Atmospheric particle number size distributions and equivalent black carbon mass concentrations, 2009–2014, doi:10.5072/guan, 2015.

Bohren, C. F. and Huffman, D. R.: Absorption and Scattering of Light by Small Particles, John Wiley & Sons, doi:10.1002/9783527618156, 544 pp., Weinheim, Germany, 1998.

Bond, T. C., Doherty, S. J., Fahey, D. W., Forster, P. M., Berntsen, T., DeAngelo, B. J., Flanner, M. G., Ghan, S., Kärcher, B., Koch, D., Kinne, S., Kondo, Y., Quinn, P. K., Sarofim, M. C., Schultz, M. G., Schulz, M., Venkataraman, C., Zhang, H., Zhang, S., Bellouin, N., Guttikunda, S. K., Hopke, P. H., Jacobson, M. Z., Kaiser, J. W., Klimont, Z., Lohmann, U., Schwarz, J. P., Shindell, D., Storelvmo, T., Warren, S. G., and Zender, C. S.: Bounding the role of black carbon in the climate system: a scientific assessment, J. Geophys. Res.-Atmos., 118, 5380–5552, doi:10.1002/jgrd.50171, 2013.

Chen, Y., Cheng, Y. F., Nordmann, S., Birmili, W., Denier van der Gon, H. A. C., Ma, N., Wolke, R., Wehner, B., Sun, J., Spindler, G., Mu, Q., Pöschl, U., Su, H., and Wiedensohler, A.: Evaluation of the size segregation of elemental carbon (EC) emission in Europe: influence on the simulation of EC long-range transport, Atmos. Chem. Phys., 16, 1823–1835, doi:10.5194/acp-16-1823-2016, 2016.

Costabile, F., Birmili, W., Klose, S., Tuch, T., Wehner, B., Wiedensohler, A., Franck, U., König, K., and Sonntag, A.: Spatio-temporal variability and principal components of the particle number size distribution in an urban atmosphere, Atmos. Chem. Phys., 9, 3163–3195, doi:10.5194/acp-9-3163-2009, 2009.

Dockery, D. W. and Pope, C. A.: Acute respiratory effects of particulate air pollution, Annu. Rev. Publ. Health, 15, 107–132, doi:10.1146/annurev.pu.15.050194.000543, 1994.

Dockery, D. W. and Stone, P. H.: Cardiovascular risks from fine particulate air pollution, New Engl. J. Med., 356, 511–513, doi:10.1056/NEJMe068274, 2007.

Ehn, M., Petäjä, T., Birmili, W., Junninen, H., Aalto, P., and Kulmala, M.: Non-volatile residuals of newly formed atmospheric particles in the boreal forest, Atmos. Chem. Phys., 7, 677–684, doi:10.5194/acp-7-677-2007, 2007.

Engler, C., Rose, D., Wehner, B., Wiedensohler, A., Brüggemann, E., Gnauk, T., Spindler, G., Tuch, T., and Birmili, W.: Size distributions of non-volatile particle residuals ($D_p < 800$ nm) at a rural site in Germany and relation to air mass origin, Atmos. Chem. Phys., 7, 5785–5802, doi:10.5194/acp-7-5785-2007, 2007.

Engler, C., Birmili, W., Spindler, G., and Wiedensohler, A.: Analysis of exceedances in the daily PM_{10} mass concentration ($50\,\mu g\,m^{-3}$) at a roadside station in Leipzig, Germany, Atmos. Chem. Phys., 12, 10107–10123, doi:10.5194/acp-12-10107-2012, 2012.

Franck, U., Odeh, S., Wiedensohler, A., Wehner, B., and Herbarth, O.: The effect of particle size on cardiovascular disorders – the smaller the worse, Sci. Total Environ., 409, 4217–4221, doi:10.1016/j.scitotenv.2011.05.049, 2011.

Gerwig, H., Pecher, W., and Wirtz, K.: Ursache für maximale Partikeloberflächenkonzentrationen in Langen bei Frankfurt, oral presentation at: 49. Messtechnisches Kolloquium, Dresden, 26–28 May 2014.

Gnauk, T., Müller, K., Brüggemann, E., Birmili, W., Weinhold, K., van Pinxteren, D., Löschau, G., Spindler, G., and Herrmann, H.: A study to discriminate local, urban and regional source contributions to the particulate matter concentrations in the city of Dresden, Germany, J. Atmos. Chem., 68, 199–231, doi:10.1007/s10874-012-9216-7, 2012.

Größ, J., Birmili, W., Hamed, A., Sonntag, A., Wiedensohler, A., Spindler, G., Maninnen, H. E., Nieminen, T., Kulmala, M., Hõrrak, U., and Plass-Dülmer, C.: Evolution of gaseous precursors and meteorological parameters during new particle formation events in the Central European boundary layer, Atmos. Chem. Phys. Discuss., 15, 2305–2353, doi:10.5194/acpd-15-2305-2015, 2015.

Gu, J., Pitz, M., Breitner, S., Birmili, W., von Klot, S., Schneider, A., Soentgen, J., Reller, A., Peters, A., and Cyrys, J.: Selection of key ambient particulate variables for epidemiological studies – applying cluster and heatmap analyses as tools for data reduction, Sci. Total Environ., 435/436, 541–550, doi:10.1016/j.scitotenv.2012.07.040, 2012.

Hänel, G.: The properties of atmospheric aerosol particles as functions of the relative humidity at thermodynamic equilibrium with the surrounding moist air, Adv. Geophys., 19, 74–188, doi:10.1016/S0065-2687(08)60142-9, 1976.

Haywood, J. and Boucher, O.: Estimates of the direct and indirect radiative forcing due to tropospheric aerosols: a review, Rev. Geophys., 38, 513–543, doi:10.1029/1999RG000078, 2000.

HEI Review Panel on Ultrafine Particles: Understanding the Health Effects of Ambient Ultrafine Particles. HEI Perspectives 3, Health Effects Institute, Boston, MA, January 2013.

Janssen, N. A. H., Gerlofs-Nijland, M. E., Lanki, T., Salonen, R. O., Cassee, F., Hoek, G., Fischer, P., Brunekreef, B., and Krzyzanowski, M.: Health effects of black carbon, World Health Organisation (WHO), Regional Offce for Europe, Copenhagen, Denmark, 86 pp., 2012.

Jones, A. M., Harrison, R. M., Barratt, B., and Fuller, G.: A large reduction in airborne particle number concentrations at the time of the introduction of "sulphur free" diesel and the London Low Emission Zone, Atmos. Environ., 50, 129–138, doi:10.1016/j.atmosenv.2011.12.050, 2012.

Kumar, P., Morawska, L., Birmili, W., Paasonen, P., Hu, M., Kulmala, M., Harrison, R. M., Norford, L., and Britter, R.: Ultrafine particles in cities, Environ. Int., 66, 1–10, doi:10.1016/j.envint.2014.01.013, 2014.

Laj, P., Klausen, J., Bilde, M., Plass-Duelmer, C., Pappalardo, G., Clerbaux, C., Baltensperger, U., Hjorth, J., Simpson, D., Reimann, S., Coheur, P.-F., Richter, A., De Maziere, M., Rudich, Y., McFiggans, G., Torseth, K., Wiedensohler, A., Morin, S., Schulz, M., Allan, J. D., Attie, J.-L., Barnes, I., Birmili, W., Cammas, J. P., Dommen, J., Dorn, H.-P., Fowler, D., Fuzzi, S., Glasius, M., Granier, C., Hermann, M., Isaksen, I. S. A., Kinne, S., Koren, I., Madonna, F., Maione, M., Massling, A., Moehler, O., Mona, L., Monks, P. S., Müller, D., Müller, T., Orphal, J., Peuch, V. H., Stratmann, F., Tanre, D., Tyndall, D., Abo Riziq, A., Van Roozendael, M., Villani, P., Wehner, B., Wex, H., and Zardini, A. A.: Measuring atmospheric composition change, Atmos. Environ., 43, 5351–5414, doi:10.1016/j.atmosenv.2009.08.020, 2009.

Lelieveld, J., Evans, J. S., Fnais, M., Giannadaki, D., and Pozzer, A.: The contribution of outdoor air pollution sources to premature mortality on a global scale, Nature, 525, 367–371, doi:10.1038/nature15371, 2015.

Lohmann, U. and Feichter, J.: Global indirect aerosol effects: a review, Atmos. Chem. Phys., 5, 715–737, doi:10.5194/acp-5-715-2005, 2005.

Löschau, G., Wiedensohler, A., Wehner, B., Birmili, W., and Gerwig, H.: Measurement of the number concentration of ultrafine particles in ambient air in an air quality monitoring network – Part 2: Results of a traffic-orientated long term measurement, Gefahrst. Reinhalt. L., 70, 183–187, 2010.

Löschau, G., Wiedensohler, A., Birmili, W., Rasch, F., Spindler, G., Müller, K., Wolf, U., Hausmann, A., and Herrmann, H.: The low-emission zone in Leipzig, part 1: initial state (Umweltzone Leipzig, Teil 1: Ausgangsbeurteilung), Landesamt für Umwelt, Landwirtschaft und Geologie, 83 pp., Dresden, Germany, 2012 (in German).

Löschau, G., Wiedensohler, A., Birmili, W., Rasch, F., Spindler, G., Müller, K., Wolf, U., Hausmann, A., Böttger, M., Anhalt, M., and Herrmann, H.: The low-emission zone in Leipzig, part 2: state of air quality in 2011 (Umweltzone Leipzig, Teil 2: Immissionssituation 2011), Landesamt für Umwelt, Landwirtschaft und Geologie, 89 pp., Dresden, Germany, 2013 (in German).

Löschau, G., Wiedensohler, A., Birmili, W., Rasch, F., Spindler, G., Müller, K., Wolf, U., Hausmann, A., Böttger, M., Anhalt, M., Dietz, V., Herrmann, H., and Böhme, U.: The low-emission zone in Leipzig, part 3: state of air quality 2010–2013 (Umweltzone Leipzig, Teil 3: Immissionssituation 2010–2013), Landesamt für Umwelt, Landwirtschaft und Geologie, 73 pp., Dresden, Germany, 2014 (in German).

Löschau, G., Wiedensohler, A., Birmili, W., Rasch, F., Spindler, G., Müller, K., Wolf, U., Hausmann, A., Böttger, M., Bastian, S., Anhalt, M., Dietz, V., Herrmann, H., and Böhme, U.: The low-emission zone in Leipzig, part 4: state of air quality 2010–2014 (Umweltzone Leipzig, Teil 4: Immissionssituation 2010–2014), Landesamt für Umwelt, Landwirtschaft und Geologie, 75 pp., Dresden, Germany, 2015 (in German).

Ma, N. and Birmili, W.: Estimating the contribution of photochemical particle formation to ultrafine particle number averages in an urban atmosphere, Sci. Total Environ., 512/513, 154–166, doi:10.1016/j.scitotenv.2015.01.009, 2015.

Ma, N., Birmili, W., Müller, T., Tuch, T., Cheng, Y. F., Xu, W. Y., Zhao, C. S., and Wiedensohler, A.: Tropospheric aerosol scattering and absorption over central Europe: a closure study for the dry particle state, Atmos. Chem. Phys., 14, 6241–6259, doi:10.5194/acp-14-6241-2014, 2014.

McMurry, P. H.: A review of atmospheric aerosol measurements, Atmos. Environ., 34, 1959–1999, doi:10.1016/S1352-2310(99)00455-0, 2000.

Mølgaard, B., Birmili, W., Clifford, S., Massling, A., Eleftheriadis, K., Norman, M., Vratolis, S., Wehner, B., Corander, J., Hämeri, K., and Hussein, T.: Evaluation of a statistical forecast model for size-fractionated urban particle number concentrations using data from five European cities, J. Aerosol Sci., 66, 96–110, doi:10.1016/j.jaerosci.2013.08.012, 2013.

Müller, T., Schladitz, A., Massling, A., Kaaden, N., Kandler, K., and Wiedensohler, A.: Spectral absorption coefficients and imaginary parts of refractive indices of Saharan dust during SAMUM-1, Tellus B, 61, 79–95, doi:10.3402/tellusb.v61i1.16816, 2009.

Müller, T., Henzing, J. S., de Leeuw, G., Wiedensohler, A., Alastuey, A., Angelov, H., Bizjak, M., Collaud Coen, M., Engström, J. E., Gruening, C., Hillamo, R., Hoffer, A., Imre, K., Ivanow, P., Jennings, G., Sun, J. Y., Kalivitis, N., Karlsson, H., Komppula, M., Laj, P., Li, S.-M., Lunder, C., Marinoni, A., Martins dos Santos, S., Moerman, M., Nowak, A., Ogren, J. A., Petzold, A., Pichon, J. M., Rodriquez, S., Sharma, S., Sheridan, P. J., Teinilä, K., Tuch, T., Viana, M., Virkkula, A., Weingartner, E., Wilhelm, R., and Wang, Y. Q.: Characterization and intercomparison of aerosol absorption photometers: result of two intercomparison workshops, Atmos. Meas. Tech., 4, 245–268, doi:10.5194/amt-4-245-2011, 2011.

Nordmann, S., Birmili, W., Weinhold, K., Wiedensohler, A., Mertes, S., Müller, K., Gnauk, T., Herrmann, H., Pitz, M., Cyrys, J., Flentje, H., Ries, L., and Wirtz, K.: Atmospheric aerosol measurements in the German Ultrafine Aerosol Network (GUAN) – Part 2: Comparison of measurements techniques for graphitic, light-absorbing, and elemental carbon, and the non-volatile particle volume under field conditions, Gefahrst. Reinhalt. L., 69, 469–474, 2009.

Nordmann, S., Birmili, W., Weinhold, K., Müller, K., Spindler, G., and Wiedensohler, A.: Measurements of the mass absorption cross section of atmospheric soot particles using Raman spectroscopy, J. Geophys. Res.-Atmos., 118, 12075–12085, doi:10.1002/2013JD020021, 2013.

Nordmann, S., Cheng, Y. F., Carmichael, G. R., Yu, M., Denier van der Gon, H. A. C., Zhang, Q., Saide, P. E., Pöschl, U., Su, H., Birmili, W., and Wiedensohler, A.: Atmospheric black carbon and warming effects influenced by the source and absorption enhancement in central Europe, Atmos. Chem. Phys., 14, 12683–12699, doi:10.5194/acp-14-12683-2014, 2014.

Ibald-Mulli, A., Wichmann, H.-E., Kreyling, W., and Peters, A.: Epidemiological evidence on health effects of ultrafine particles, J. Aerosol Med., 15, 189–201, doi:10.1089/089426802320282310, 2002.

Petzold, A. and Schönlinner, M.: Multi-angle absorption photometry – a new method for the measurement of aerosol light absorption and atmospheric black carbon, J. Aerosol Sci., 35, 421–441, doi:10.1016/j.jaerosci.2003.09.005, 2004.

Petzold, A., Ogren, J. A., Fiebig, M., Laj, P., Li, S.-M., Baltensperger, U., Holzer-Popp, T., Kinne, S., Pappalardo, G., Sugimoto, N., Wehrli, C., Wiedensohler, A., and Zhang, X.-Y.: Recommendations for reporting "black carbon" measurements, Atmos. Chem. Phys., 13, 8365–8379, doi:10.5194/acp-13-8365-2013, 2013.

Pfeifer, S., Birmili, W., Schladitz, A., Müller, T., Nowak, A., and Wiedensohler, A.: A fast and easy-to-implement inversion algorithm for mobility particle size spectrometers considering particle number size distribution information outside of the detection range, Atmos. Meas. Tech., 7, 95–105, doi:10.5194/amt-7-95-2014, 2014.

Pitz, M., Birmili, W., Schmid, O., Peters, A., Wichmann, H.-E., and Cyrys, J.: Quality control and quality assurance for particle size distribution measurements at an urban monitoring station in Augsburg, Germany, J. Environ. Monitor., 10, 1017–1024, doi:10.1039/b807264g, 2008.

Pope, C. A., Burnett, R. T., Thurston, G. D., Thun, M. J., Calle, E. E., Krewski, D., and Godleski, J. J.: Cardiovascular mortality and long-term exposure to particulate air pollution, Circulation, 109, 71–77, doi:10.1161/01.CIR.0000108927.80044.7F, 2004.

Pöschl, U.: Atmospheric aerosols: composition, transformation, climate and health effects, Angew. Chem. Int. Edit., 44, 7520–7540, doi:10.1002/anie.200501122, 2005.

Poulain, L., Birmili, W., Canonaco, F., Crippa, M., Wu, Z. J., Nordmann, S., Spindler, G., Prévôt, A. S. H., Wiedensohler, A., and Herrmann, H.: Chemical mass balance of 300 °C non-volatile particles at the tropospheric research site Melpitz, Germany,

Atmos. Chem. Phys., 14, 10145–10162, doi:10.5194/acp-14-10145-2014, 2014.

Quass, U., Kuhlbusch, T., Koch, M., Fissan, H., Schmidt, K.-G., Bruckmann, P., Pfeffer, U., Gladtke, D., and Zang, T.: Identification of source groups for fine dust (in German, Identifizierung von Quellgruppen für die Feinstaubfraktion), Public report to the Ministry of the Environment of North Rhine-Westphalia, Germany, 2004.

Ramanathan, V., Crutzen, P. J., Lelieveld, J., Mitra, A. P., Althausen, D., Anderson, J., Andreae, M. O., Cantrell, W., Cass, G. R., Chung, C. E., Clarke, A. D., Coakley, J. A., Collins, W. D., Conant, W. C., Dulac, F., Heintzenberg, J., Heymsfield, A. J., Holben, B., Howell, S., Hudson, J., Jayaraman, A., Kiehl, J. T., Krishnamurti, T. N., Lubin, D., McFarquhar, G., Novakov, T., Ogren, J. A., Podgorny, I. A., Prather, K., Priestley, K., Prospero, J. M., Quinn, P. K., Rajeev, K., Rasch, P., Rupert, S., Sadourny, R., Satheesh, S. K., Shaw, G. E., Sheridan, P., and Valero, F. P. J.: Indian Ocean Experiment: an integrated analysis of the climate forcing and effects of the great Indo-Asian haze, J. Geophys. Res., 106, 28371–28398, doi:10.1029/2001JD900133, 2001.

Rasch, F., Birmili, W., Weinhold, K., Nordmann, S., Sonntag, A., Spindler, G., Herrmann, H., Wiedensohler, A., and Löschau, G.: Significant reduction of ambient black carbon and particle number in Leipzig as a result of the low emission zone, Gefahrst. Reinhalt. L., 73, 483–489, 2013.

Reddington, C. L., Carslaw, K. S., Spracklen, D. V., Frontoso, M. G., Collins, L., Merikanto, J., Minikin, A., Hamburger, T., Coe, H., Kulmala, M., Aalto, P., Flentje, H., Plass-Dülmer, C., Birmili, W., Wiedensohler, A., Wehner, B., Tuch, T., Sonntag, A., O'Dowd, C. D., Jennings, S. G., Dupuy, R., Baltensperger, U., Weingartner, E., Hansson, H.-C., Tunved, P., Laj, P., Sellegri, K., Boulon, J., Putaud, J.-P., Gruening, C., Swietlicki, E., Roldin, P., Henzing, J. S., Moerman, M., Mihalopoulos, N., Kouvarakis, G., Ždímal, V., Zíková, N., Marinoni, A., Bonasoni, P., and Duchi, R.: Primary versus secondary contributions to particle number concentrations in the European boundary layer, Atmos. Chem. Phys., 11, 12007–12036, doi:10.5194/acp-11-12007-2011, 2011.

Rückerl, R., Schneider, A., Breitner, S., Cyrys, J., and Peters, A.: Health effects of particulate air pollution: a review of epidemiological evidence, Inhal. Toxicol., 23, 555–592, doi:10.3109/08958378.2011.593587, 2011.

Schäfer, K., Thomas, W., Peters, A., Ries, L., Obleitner, F., Schnelle-Kreis, J., Birmili, W., Diemer, J., Fricke, W., Junkermann, W., Pitz, M., Emeis, S., Forkel, R., Suppan, P., Flentje, H., Gilge, S., Wichmann, H. E., Meinhardt, F., Zimmermann, R., Weinhold, K., Soentgen, J., Münkel, C., Freuer, C., and Cyrys, J.: Influences of the 2010 Eyjafjallajökull volcanic plume on air quality in the northern Alpine region, Atmos. Chem. Phys., 11, 8555–8575, doi:10.5194/acp-11-8555-2011, 2011.

Schmidt, M., Graul, R., Sartorius, H., and Levin, I.: The Schauinsland CO_2 record: 30 years of continental observations and their implications for the variability of the European CO_2 budget, J. Geophys. Res., 108, 4619, doi:10.1029/2002JD003085, 2003.

Schladitz, A., Merkel, M., Bastian, S., Birmili, W., Weinhold, K., Löschau, G., and Wiedensohler, A.: A concept of an automated function control for ambient aerosol measurements using mobility particle size spectrometers, Atmos. Meas. Tech., 7, 1065–1073, doi:10.5194/amt-7-1065-2014, 2014.

Schladitz, A., Lenícek, J., Beneš, I., Kovác, M., Skorkovský, J., Soukup, A., Jandlová, J., Poulain, L., Plachá, H., Löschau, G., and Wiedensohler, A.: Air quality in the German-Czech border region: a focus on harmful fractions of PM and ultrafine particles, Atmos. Environ., 122, 236–249, doi:10.1016/j.atmosenv.2015.09.044, 2015.

Spindler, G., Brüggemann, E., Gnauk, T., Grüner, A., Müller, K., and Herrmann, H.: A four-year size-segregated characterization study of particles PM_{10}, $PM_{2.5}$ and PM_1 depending on air mass origin at Melpitz, Atmos. Environ., 44, 164–173, doi:10.1016/j.atmosenv.2009.10.015, 2010.

Sun, H., Biedermann, L., and Bond, T.: Color of brown carbon: a model for ultraviolet and visible light absorption by organic carbon aerosol, Geophys. Res. Lett., 34, L17813, doi:10.1029/2007GL029797, 2007.

Swietlicki, E., Hansson, H.-C., Hämeri, K., Svenningsson, B., Massling, A. McFiggans, G., McMurry, P. H., Petäjä, T., Tunved, P., Gysel, M., Topping, D., Weingartner, E., Baltensperger, U., Rissler, J., Wiedensohler, A., and Kulmala, M.: Hygroscopic properties of submicrometer atmospheric aerosol particles measured with H-TDMA instruments in various environments – a review, Tellus, 60B, 432–469, doi:10.1111/j.1600-0889.2008.00350.x, 2008.

Toenges-Schuller, N., Schneider, C., Niederau, A., Vogt, R., and Birmili, W.: Modelling particle number concentrations in a typical street canyon in Germany and analysis of future trends, Atmos. Environ., 111, 127–135, doi:10.1016/j.atmosenv.2015.04.006, 2015.

Tørseth, K., Aas, W., Breivik, K., Fjæraa, A. M., Fiebig, M., Hjellbrekke, A. G., Lund Myhre, C., Solberg, S., and Yttri, K. E.: Introduction to the European Monitoring and Evaluation Programme (EMEP) and observed atmospheric composition change during 1972–2009, Atmos. Chem. Phys., 12, 5447–5481, doi:10.5194/acp-12-5447-2012, 2012.

Tunved, P., Hansson, H.-C., Kulmala, M., Aalto, P., Viisanen, Y., Karlsson, H., Kristensson, A., Swietlicki, E., Dal Maso, M., Ström, J., and Komppula, M.: One year boundary layer aerosol size distribution data from five nordic background stations, Atmos. Chem. Phys., 3, 2183–2205, doi:10.5194/acp-3-2183-2003, 2003.

UBA: Das Luftmessnetz des Umweltbundesamtes: Langzeitmessungen, Prozessverständnis und Wirkungen ferntransportierter Luftverunreinigungen, Tech. Rep., edited by: Schleyer, R., Bieber, E., and Wallasch, M., Umweltbundesamt (UBA), Dessau-Rosslau, 112 pp., 2013.

Voigtländer, J., Tuch, T., Birmili, W., and Wiedensohler, A.: Correlation between traffic density and particle size distribution in a street canyon and the dependence on wind direction, Atmos. Chem. Phys., 6, 4275–4286, doi:10.5194/acp-6-4275-2006, 2006.

von Bismarck-Osten, C., Birmili, W., Ketzel, M., Massling, A., Petäjä, T., and Weber, S.: Characterization of parameters influencing the spatio-temporal variability of urban aerosol particle number size distributions in four European cities, Atmos. Environ., 77, 415–429, doi:10.1016/j.atmosenv.2013.05.029, 2013.

von Bismarck-Osten, C., Birmili, W., Ketzel, M., and Weber, S.: Statistical modelling of aerosol particle number size distributions in urban and rural environments – a multi-site study, Urban Climate, 11, 51–66, doi:10.1016/j.uclim.2014.11.004, 2015.

Weigelt, A., Temme, C., Bieber, E., Schwerin, A., Schuetze, M., Ebinghaus, R., and Kock, H. H.: Measurements of atmospheric mercury species at a German rural background site from 2009 to 2011 – methods and results, Environ. Chem., 10, 102–110, doi:10.1071/EN12107, 2013.

Wehner, B., Philippin, S., Wiedensohler, A., and Haudek, A.: Design and calibration of an improved thermodenuder to study the volatility fraction of aerosol particles, J. Aerosol Sci., 33, 1087–1093, doi:10.1016/S0021-8502(02)00056-3, 2002.

Wiedensohler, A., Birmili, W., Nowak, A., Sonntag, A., Weinhold, K., Merkel, M., Wehner, B., Tuch, T., Pfeifer, S., Fiebig, M., Fjäraa, A. M., Asmi, E., Sellegri, K., Depuy, R., Venzac, H., Villani, P., Laj, P., Aalto, P., Ogren, J. A., Swietlicki, E., Williams, P., Roldin, P., Quincey, P., Hüglin, C., Fierz-Schmidhauser, R., Gysel, M., Weingartner, E., Riccobono, F., Santos, S., Grüning, C., Faloon, K., Beddows, D., Harrison, R., Monahan, C., Jennings, S. G., O'Dowd, C. D., Marinoni, A., Horn, H.-G., Keck, L., Jiang, J., Scheckman, J., McMurry, P. H., Deng, Z., Zhao, C. S., Moerman, M., Henzing, B., de Leeuw, G., Löschau, G., and Bastian, S.: Mobility particle size spectrometers: harmonization of technical standards and data structure to facilitate high quality long-term observations of atmospheric particle number size distributions, Atmos. Meas. Tech., 5, 657–685, doi:10.5194/amt-5-657-2012, 2012.

Zieger, P., Fierz-Schmidhauser, R., Weingartner, E., and Baltensperger, U.: Effects of relative humidity on aerosol light scattering: results from different European sites, Atmos. Chem. Phys., 13, 10609–10631, doi:10.5194/acp-13-10609-2013, 2013.

Zieger, P., Fierz-Schmidhauser, R., Poulain, L., Müller, T., Birmili, W., Spindler, G., Wiedensohler, A., Baltensperger, U., and Weingartner, E.: Influence of water uptake on the aerosol particle light scattering coefficients of the Central European aerosol, Tellus B, 66, 22716, doi:10.3402/tellusb.v66.22716, 2014.

7

A general database of hydrometeor single scattering properties at microwave and sub-millimetre wavelengths

Patrick Eriksson[1], Robin Ekelund[1], Jana Mendrok[1], Manfred Brath[2], Oliver Lemke[2], and Stefan A. Buehler[2]

[1]Department of Space, Earth and Environment, Chalmers University of Technology, Gothenburg, Sweden
[2]Meteorological Institute, Department of Earth Sciences, Center for Earth System Research and Sustainability, Universität Hamburg, Hamburg, Germany

Correspondence: Patrick Eriksson (patrick.eriksson@chalmers.se)

Abstract. A main limitation today in simulations and inversions of microwave observations of ice hydrometeors (cloud ice, snow, hail, etc.) is the lack of data describing the interaction between electromagnetic waves and the particles. To improve the situation, the development of a comprehensive dataset of such "scattering properties" has been started. The database aims at giving a broad coverage in both frequency (1 to 886 GHz) and temperature (190 to 270 K), to support both passive and active current and planned measurements, and to provide data corresponding to the full Stokes vector. This first version of the database is restricted to totally random particle orientation. Data for 34 particle sets, i.e. habits, have been generated. About 17 of the habits can be classified as single crystals, three habits can be seen as heavily rimed particles, and the remaining habits are aggregates of different types, e.g. snow and hail. The particle sizes considered vary between the habits, but maximum diameters of 10 and 20 mm are typical values for the largest single crystal and aggregate particles, respectively, and the number of sizes per habit is at least 30. Particles containing liquid water are also inside the scope of the database, but this phase of water is so far only represented by a liquid sphere habit. The database is built upon the netCDF4 file format. Interfaces to browse, extract and convert data for selected radiative transfer models are provided in MATLAB and Python. The database and associated tools are publicly available from Zenodo (https://doi.org/10.5281/zenodo.1175572, Ekelund et al., 2018b, and https://doi.org/10.5281/zenodo.1175588, Mendrok et al., 2018, respectively). Planned extensions include non-spherical raindrops, melting particles and a second orientation case that can be denoted as azimuthally random.

1 Introduction

1.1 Motivation

Atmospheric particles absorb, emit and scatter electromagnetic radiation. The optical properties governing these processes are not easily determined for an arbitrarily shaped particle (e.g. Min et al., 2006). Accordingly, these properties can not be calculated "on-the-fly" and radiative transfer simulations must frequently make use of tabulated values. At microwave wavelengths, the targets of a database of particle optical properties are hydrometeors, i.e. cloud and precipitating particles that consist of condensed water. The need to consider particle shape in detail increases with frequency and is the highest for hydrometeors containing ice. Aerosols can in general be neglected for microwave radiative transfer, but measurements of volcanic ash are another potential application of sub-millimetre radiometry (Baran, 2012a).

Microwave particle optical properties, also often denoted as scattering data, are most obviously needed for measurements of precipitation and cloud properties. Active instruments, i.e. radars, perform such measurements by detecting backscattering of hydrometeors, but extinction of the signal must also be considered. Radars targeting rain operate at relatively low frequencies and are well established

(Doviak and Zrnic, 2014). For example, ground-based networks of precipitation radars are today found in many countries. The sensitivity to cloud particles and snow increases with frequency. The highest standard radar frequency today is 94 GHz (Kollias et al., 2007), used from both ground (e.g Hogan et al., 2000) and aircraft (e.g. Bouniol et al., 2010) and by the first space-based cloud radar onboard CloudSat launched in 2006 (Stephens et al., 2008) as well as the CPR instrument onboard ESA's EarthCARE mission (Illingworth et al., 2015).

Rain and ice particles affect passive microwave radiances by scattering as well as by contributing to absorption and emission (e.g. Liou, 2002). The relative importance of absorption and emission is the highest for particles that are small compared to the wavelength, and also higher for liquid particles, compared to ice ones, due to differences in refractive index. The present set of operational microwave sensors covers frequencies of up to about 190 GHz (e.g. Weng et al., 2012). The main hydrometeor products of these measurements today are cloud liquid water and rain (Boukabara et al., 2011). Ice hydrometeor retrievals exist (e.g. Boukabara et al., 2013), but are less established and are often purely empirical using satellite radar data for "training" (Holl et al., 2014; Piyush et al., 2017).

At present, the sub-millimetre part of the microwave region is primarily used for limb sounding. These instruments target mainly the stratosphere and mesosphere, but can provide data on humidity and cloud ice down to about 10 km (Wu et al., 2008; Eriksson et al., 2014). Sub-millimetre measurements dedicated to the troposphere are so far only available for some aircraft instruments (Evans et al., 2012; Fox et al., 2017; Brath et al., 2018). Operational sensors, such as used for weather forecasting, making use of sub-millimetre wavelengths will be available from about 2022 (Kangas et al., 2012). The addition of this wavelength region will improve the general sensitivity of passive microwave measurements, particularly regarding ice hydrometeors. Adding this wavelength region should provide a better constraint on particle size of frozen hydrometeors and, not only the sensitivity, but also the accuracy of ice hydrometeor mass retrievals should be improved (Evans and Stephens, 1995; Buehler et al., 2007a).

Scattering data are not only important for the direct measurement of hydrometeors, but also essential to characterise the interference of clouds and precipitation on measurements targeting other quantities. One such example is the "cloud screening", applied in humidity retrievals, based on channels around the 183 GHz water vapour spectral line (Buehler et al., 2007b). The addition of sub-millimetre channels should facilitate this cloud screening. Passive microwave data, particular using channels on the low-frequency end, are also used to estimate ocean surface variables, and even these retrievals are affected by hydrometeors. The characterisation of this interference can even involve polarisation effects (Adams et al., 2008).

In summary, microwave sensors offer information on cloud and precipitation, as well as surface variables, humidity and other gases with relatively low cloud interference (Kunzi et al., 2011). These advantageous aspects are today heavily used for both climate studies and weather prediction. However, the full potential of existing data is far from fully exploited (e.g. Guerbette et al., 2016), and preparations for future measurements are needed (e.g. Buehler et al., 2012; Birman et al., 2017). In both cases, progress is today strongly limited by a lack of hydrometeor scattering data (Geer and Baordo, 2014). Liquid particles can to a large extent be assumed to be spherical, and their scattering properties can be calculated efficiently by Mie theory. On the other hand, both cloud and snow ice hydrometeors exhibit a huge variability in shape (e.g. Lynch et al., 2002; Garrett et al., 2015; O'Shea et al., 2016), and to derive their scattering properties is a demanding calculation task. These complications have resulted in patchy coverage of ice hydrometeor microwave scattering properties.

1.2 Single scattering data

To be of a general character, scattering data must cover a number of attributes, as discussed in later sections. For a given particle, the scattering properties vary with frequency, polarisation, temperature and orientation. Furthermore, particle size and shape play a significant role where the comprehensive description of particle shape is less clear. Several measures on particle size are being used, but those represent a small complication in comparison to morphology classification. Traditional schemes, such as Magono and Lee (1966), offer a good starting point regarding the classification of pristine crystals. Ice hydrometeors above about 200 μm in maximum dimension tend to be aggregates of crystals (Schmitt and Heymsfield, 2014); hence, aggregates should in general dominate the scattering signatures in microwave data, which is mainly determined by particles larger than 100 μm (Eriksson et al., 2008). However, there is no established manner to categorise particle classes. So far aggregates are described in relatively crude terms, by, for example, effective density and degree of riming. A further complication is that both liquid and ice water can be found in a particle simultaneously. The so-called radar bright band is normally taken as a manifestation of the presence of melting ice hydrometeors (Sassen et al., 2007).

A review of existing microwave scattering data was done by Eriksson et al. (2015). At that point, the most general collection was provided by Liu (2008) despite it having been limited to totally random particle orientation, frequencies below 340 GHz and low complexity in particle morphology. The only source of data covering sub-millimetre wavelengths was Hong et al. (2009), but it was found to be based on an outdated ice refractive index parameterisation. More complex ice particles, i.e. particles of aggregate type, had been considered (Tyynelä and Chandrasekar, 2014), but then providing data with limited frequency coverage and mainly targeting radar applications. More recently, Kuo et al. (2016) presented data for a large number of habits, including many aggregate realisations, but the publicly available data lack detailed information and provided only parameters applicable in simplified radiative transfer. The data of Lu et al. (2016) have interesting features regarding particle shapes and orientation, but are not well suited for passive microwave applications since they are based on an obsolete, and inaccurate, refractive index parameterisation (see Sect. 4.1 for details). The data of Ding et al. (2017), which are an updated and extended version of the database by Hong et al. (2009), cover 1 to 874 GHz, at multiple temperatures, and can thus be considered to be the first to provide full coverage of the microwave region, but are still limited to totally random orientation and relatively simple habits.

Besides the production of databases, there are a number of investigations of relationships between particle properties and optical properties. For example, the data of Ding et al. (2017) were compared to the scattering properties of more complex, snow-like, particles by Baran et al. (2018). The impact of riming on scattering properties was studied by Leinonen and Szyrmer (2015), using a triple-frequency approach. A clear shift in radar signature was observed as riming was gradually applied to pristine aggregates. Also, data for melting particles have been reported (Ori et al., 2014; Johnson et al., 2016), showing an increase in extinction and backscattering at increasing melting fractions.

In summary, important progress has been made, but there are still important gaps in available data. Based on the findings in Eriksson et al. (2015), a decision was taken to gradually build up a database of microwave scattering properties, with the long-term goal of providing a comprehensive source of such data. The ambition is to provide data of a general character supporting both passive and active observations involving polarimetric information in a consistent manner. A secondary objective is to ensure that advanced scattering solvers (e.g. RT4 by Evans and Stephens, 1995 and MC by Davis et al., 2005) as included in the ARTS-2 forward model (Eriksson et al., 2011a; Buehler et al., 2018) can be fully utilised. In the following, we will hence refer to this data compilation as the "ARTS scattering database".

The first version of the database is presented below. The database is not complete. This versions lacks at least features found in some databases targeting radar applications, but this version is already more extensive than comparable efforts with a broad scope. The frequency coverage is high, very similar to the one of Ding et al. (2017), and multiple temperatures are covered. Full polarimetric data are given. A special emphasis has been given to improvements in the representation of aggregates; 34 ice hydrometeor habits of totally random orientation are included, with 16 of these being of aggregate type. Pure liquid particles are also covered, in the form of liquid spheres. Data for more specific orientations (such as particles with a preference for the horizontal plane) are left for the next database version. Non-spherical liquid particles and melting particles are other planned database extensions.

An overview of the database content is found in Sect. 2. A number of software tools have been used to generate the database, and those are presented in Sect. 3. The various microphysical assumptions are summarised in Sect. 4. Example results are found in Sect. 5. Data format, interfaces and availability are discussed in Sects. 6 and . Finally, conclusions are found in Sect. 8.

2 Database content

This section gives an overview of the optical properties provided, and the coverage of the database in terms of frequency, size, temperature and habits.

2.1 Radiative properties

The optical properties of a particle can be reported in several ways. The ARTS database assumes that electromagnetic radiation is described using the Stokes formalism and provides the extinction matrix, the absorption vector and the phase matrix. These quantities are most easily introduced by considering the basic radiative transfer equation:

$$\frac{\mathrm{d}\boldsymbol{I}(\nu,\boldsymbol{r},\boldsymbol{n})}{\mathrm{d}s} = -\mathbf{K}(\nu,\boldsymbol{r},\boldsymbol{n})\boldsymbol{I}(\nu,\boldsymbol{r},\boldsymbol{n}) + \boldsymbol{a}(\nu,\boldsymbol{r},\boldsymbol{n})B(\nu,\boldsymbol{r}) + \int_{4\pi} \mathbf{Z}(\nu,\boldsymbol{r},\boldsymbol{n}',\boldsymbol{n})\boldsymbol{I}(\nu,\boldsymbol{r},\boldsymbol{n}')\mathrm{d}\boldsymbol{n}', \quad (1)$$

where $\boldsymbol{I}$ is the Stokes vector of the radiance, ν is the frequency, $\boldsymbol{r}$ is the position, $\boldsymbol{n}$ is the propagation direction, s is the distance along $\boldsymbol{n}$, $\mathbf{K}$ is the extinction matrix, $\boldsymbol{a}$ is the absorption vector, B is the Planck function, $\mathbf{Z}$ is the phase matrix and $\boldsymbol{n}'$ represents the propagation direction of radiation scattered towards $\boldsymbol{n}$. See Mishchenko et al. (2002) for an exact definition of the Stokes vector assumed, as well as for details regarding assumptions and relationships implied by the equation. Note that $\boldsymbol{a}$ can be derived from $\mathbf{K}$ and $\mathbf{Z}$, but is included in the database anyways in order to avoid numerical issues and ensure good accuracy of all three main quantities.

To be clear, $\mathbf{K}$, $\boldsymbol{a}$ and $\mathbf{Z}$ in Eq. (1) represent bulk properties, while the database provides these quantities on a particle basis, which are dependent on particle size and dielectric properties. Furthermore, the scattering is assumed to be incoherent between particles, and the bulk properties are simply obtained by summing up $\mathbf{K}$, $\boldsymbol{a}$ and $\mathbf{Z}$ of all particles inside the considered atmospheric volume. Information allowing radiative transfer with the full Stokes vector (length 4) is provided. However, transformations are used in order to save storage space, and some processing of the data is needed (Sect. 6.1) before they can be applied in Eq. (1).

The radiative transfer equation is not always expressed as in Eq. (1), but it should be possible to derive the quantities found in other versions of the equation from the optical properties provided. For example, some "scattering solvers" operate with the single scattering albedo and the asymmetry parameter. These scalar quantities can be calculated from the elements in $\mathbf{K}$, $\boldsymbol{a}$ and $\mathbf{Z}$ matching the first Stokes component. The main quantity for simulations of radar measurements is backscattering, which essentially is the value of $\mathbf{Z}$ representing the backward direction. For further details around these derived properties, see the database technical report (Ekelund et al., 2018a).

2.2 Particle orientation

The database considers orientation-averaged quantities only. In general, using the scattering matrix $\mathbf{Z}$ as an example, orientation-averaged quantities are calculated as

$$\mathbf{Z}_\mathrm{o}\left(\boldsymbol{n}',\boldsymbol{n}\right)=\int_0^{2\pi}\int_0^{\pi}\int_0^{2\pi} p_\alpha(\alpha)\,p_\beta(\beta)\,p_\gamma(\gamma)\,\mathbf{Z}\left(\boldsymbol{n}',\boldsymbol{n},\alpha,\beta,\gamma\right)\sin\beta\,\mathrm{d}\alpha\,\mathrm{d}\beta\,\mathrm{d}\gamma, \quad (2)$$

where $\mathbf{Z}_\mathrm{o}$ is the orientation-averaged scattering matrix, α, β and γ are the three Euler angles describing the orientation of particles, and p_j are probability density functions describing the distribution of particle orientations.

The database currently only covers particles in one orientation mode: totally random, defined as the distribution where all particle orientations are equally probable. That is, p_α, p_β and p_γ are all uniform distributions. The database is designed and intended to also include what can be denoted as azimuthally random orientation. In this orientation case, the angles α and γ are fully random as for the totally random case, while the data are evaluated at some discrete tilt angle β (i.e. p_β is a Dirac delta function). By integrating over β, arbitrary fluttering distributions can then be determined. It turns out that azimuthally random orientation is much more costly with respect to both calculation burden and data storage, and this orientation mode is left for the next database version.

Also, a reference particle alignment must be defined, as it affects both our definition of aspect ratio (Sect. 2.3) and data for azimuthally random orientation. The reference alignment should be linked to how ice particles on average are oriented in the atmosphere. Accordingly, we align the particle with its maximum principal moment of inertia axis parallel to the z-axis. Furthermore, the minimum principal axis is aligned along the x-axis. This approximates the mean alignment of falling particles without performing extensive aerodynamics simulations. The definition is also stable in the sense that particular features, such as a narrow arm of a particle, will have less impact than the overall mass distribution of the particle.

2.3 Size and shape descriptors

Quantitatively characterising size and shape of irregularly shaped particles is a non-trivial task. We provide several descriptive values on particle size. First of all, particle mass m is reported. Secondly, the volume equivalent diameter is given, which is related to mass by

$$D_\mathrm{veq}=\left(\frac{6m}{\pi\rho}\right)^{1/3}, \quad (3)$$

where ρ is the actual density of the material found in the particle, in the case of ice set to 916.7 kg m^{-3}. Note that D_veq relates to the diameter of the volume-equivalent ice sphere, not of a mass equivalent melted sphere. The size parameter applied, x, is defined with respect to D_veq:

$$x=\pi D_\mathrm{veq}/\lambda, \quad (4)$$

where λ is the wavelength.

Another common parameter is the maximum dimension D_max (sometimes used interchangeably with the maximum diameter), for which no standard convention exists. For instance, photography-based in situ measurements often derive the maximum dimension by the largest distance between two pixels in a 2-D image. Such a definition is highly dependent on the particle orientation with respect to the camera. In the ARTS database D_max is defined as the diameter of the minimum circumscribing sphere, which takes the whole 3-D structure into account. However, ice particles are often highly irregular, and since D_max is a measure of a particle's extreme points, it is an inherently ambiguous and ill-defined parameter.

Similarly, the aspect ratio (AR) is defined differently by different authors. In general it provides a measure of the non-sphericity of the particle of concern, thus affecting the particle's preferred orientation and its scattering properties (polarimetric quantities in particular). The definition of AR is ambiguous and ill defined, similar to D_max. Our definition of AR is linked to how the particle is oriented with respect to the horizontal plane. Based on the particles' reference orientation outlined in Sect. 2.2, we define AR as the ratio of the maximum particle span found in the XY-plane to the span along the z-axis.

Axial ratio is sometimes used interchangeably with aspect ratio. We restrict the use of this term to describe the shape of hexagonal prism columns and plates, with axial ratio defined as $L/(2a)$ where L is the height of the prism and $2a$ its diameter. With this definition, columns/plates have an axial ratio above/below unity.

The database also contains a simplistic measure of the aerodynamic cross-sectional area for each particle. The variable stored is the diameter of a circle with an area equal to the cross-sectional area when viewing along the zenith angle. For this calculation the particle orientation is assumed fixed in its reference position.

2.4 Habits

The term "habit" is frequently used in the context of particle shape. We define habit as a set of particles with a common ba-

Table 1. Habits included in the first version of the database. Habits marked with * have scattering properties calculated using Mie theory. The last column displays the software or source used to create the shape data of the given habit, with abbreviations being SFTK (SnowFlake ToolKit, Sect. 3.1.1), RC (RimeCraft, Sect. 3.1.2), RSP (Recreated Shape Data, Sect. 4.2) and ESP (External Shape Data, Sect. 4.3.1). See the text for further details.

Habits	Id	D_{max} (µm)	D_{veq} (µm)	No. of sizes	a	b	Software used
Ice:							
Single crystals:							
Pristine:							
Plate type 1	9	13–10 000	10–2596	45	0.76	2.48	RSP
Column type 1	7	14–10 000	10–1815	45	0.037	2.05	RSP
Thin plate	16	25–5059	10–2000	35	30	3.00	RSP
Thick plate	15	16–3246	10–2000	35	110	3.00	RSP
Block column	12	13–2632	10–2000	35	210	3.00	RSP
Short column	13	17–3303	10–2000	34	110	3.00	RSP
Long column	14	24–4835	10–2000	35	34	3.00	RSP
Sector snowflake	3	20–12 000	20–1415	34	0.00081	1.44	RSP
Ice sphere*	24	1–50 000	1–50 000	200	480	3.00	Mie
ICON cloud ice	27	13–10 000	10–2929	45	1.6	2.56	SFTK
GEM cloud ice	31	10–3088	10–3000	45	440	3.00	SFTK
6-bullet rosette	6	16–10 000	10–2371	45	0.48	2.42	RSP
5-bullet rosette	2	17–10 000	10–2231	45	0.4	2.43	SFTK
Perpendicular 4-bullet rosette	10	18–10 000	10–2071	45	0.32	2.43	SFTK
Flat 4-bullet rosette	11	18–10 000	10–2071	45	0.32	2.43	SFTK
Perpendicular 3-bullet rosette	4	19–10 000	10–2137	45	0.44	2.47	SFTK
Flat 3-bullet rosette	5	20–10 000	10–1882	45	0.2	2.43	SFTK
Aggregates:							
Pristine:							
Evans snow aggregate	1	32–11 755	50–2506	35	0.20	2.39	ESP
Tyynelä dendrite aggregate	26	595–20 826	228–3328	35	0.10	2.25	ESP
8-column aggregate	8	19–9714	10–5000	39	65	3.00	RSP
Small column aggregate	17	105–3855	37–738	35	0.14	2.45	SFTK
Large column aggregate	18	368–19 981	128–3021	35	0.25	2.43	SFTK
Small block aggregate	21	100–7328	72–1665	35	0.21	2.33	SFTK
Large block aggregate	22	349–21 875	253–4607	35	0.35	2.27	SFTK
Small plate aggregate	19	99–7054	53–1376	35	0.077	2.25	SFTK
Large plate aggregate	20	349–22 860	197–4563	34	0.21	2.26	SFTK
ICON hail	30	120–5349	94–5000	35	380	2.99	RC
ICON snow	28	120–20 000	94–3219	35	0.031	1.95	RC
GEM hail	29	120–5031	94–5000	35	540	3.02	RC
GEM snow	32	170–10 459	94–5000	35	24	2.86	RC
Rimed:							
Spherical graupel	23	622–9744	454–5293	30	13	2.69	SFTK
ICON graupel	29	170–6658	94–5000	35	390	3.13	RC
GEM graupel	33	120–6597	94–5000	35	170	2.96	RC
Liquid:							
Liquid sphere*	25	1–50 000	1–50 000	200	523	3.00	Mie

sic morphology and following a rule on how the morphology varies with size. As an example, a common habit assumption is hexagonal columns having an aspect ratio that depends on size according to an analytic expression. Furthermore, the relationship between maximum diameter and mass of a habit is commonly parameterised by two constants, a and b, as

$$m = a D_{\mathrm{max}}^{b}. \tag{5}$$

The a and b values reported below assume SI units. As a reference, a habit consisting solely of ice spheres has $a = 480$ and $b = 3$. The goal is that the database habits will follow

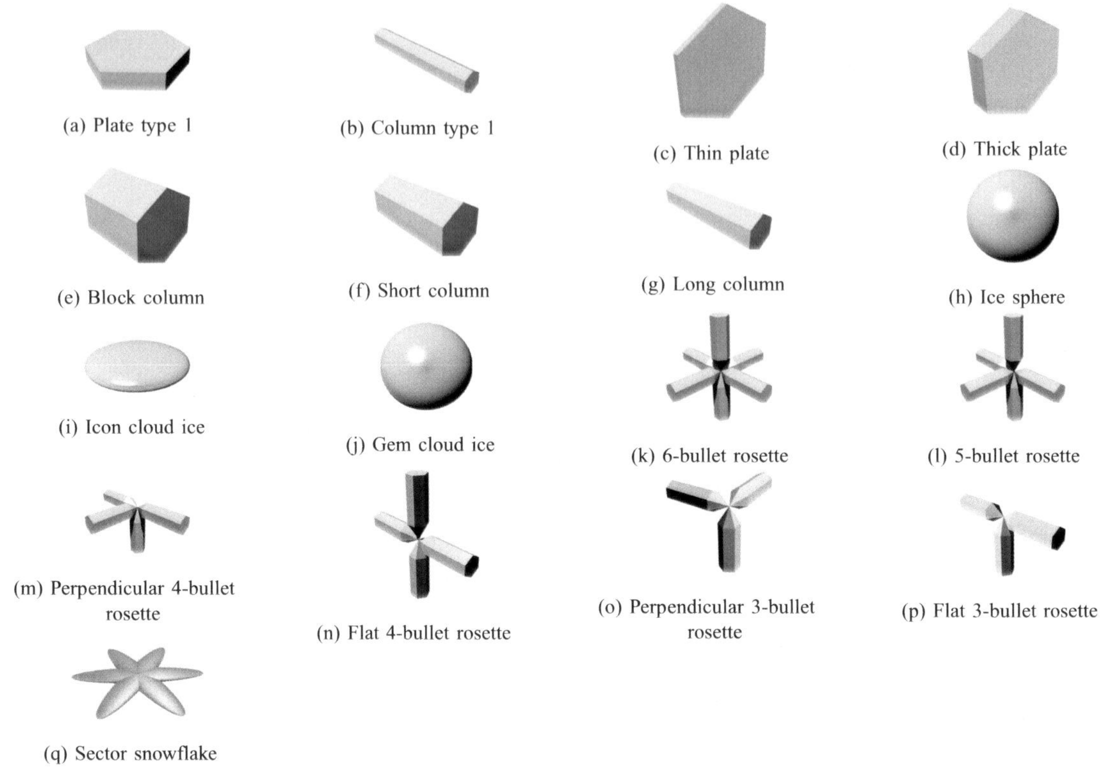

Figure 1. Single crystal habits included in the first database version. Shown particle orientation varies between the habits. Despite images **(h)** and **(j)** looking identical, they depict different particles; gem cloud ice **(j)** is a habit consisting of oblate spheroids with aspect ratio ≈ 0.92.

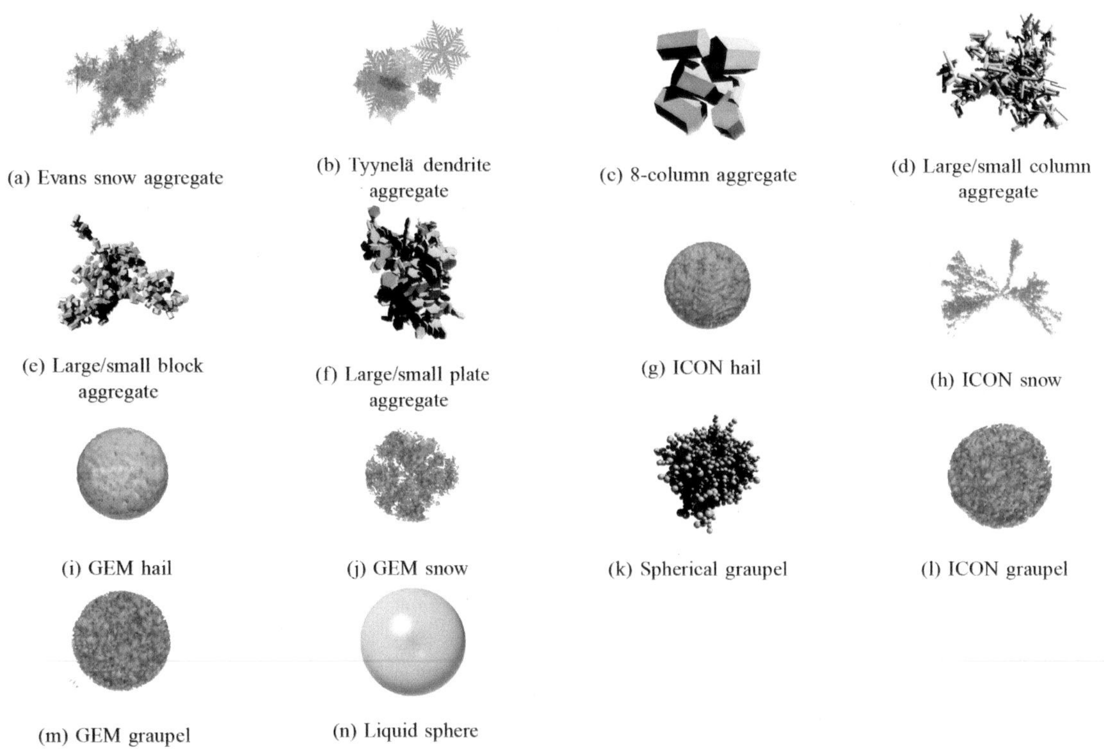

Figure 2. Aggregate and liquid habits included in the first database version. Shown particle orientation varies between the habits.

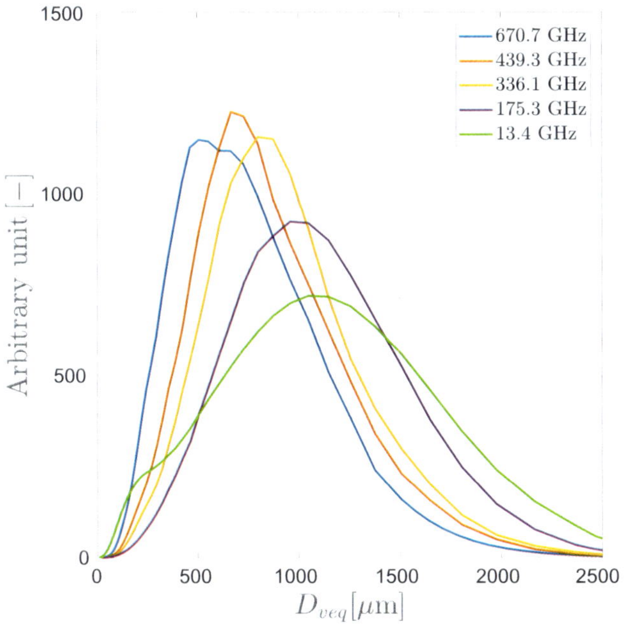

Figure 3. Contribution of differently sized particles to bulk extinction at a number of frequencies. The area under each curve has been normalised to unity. Calculated using the 6-bullet rosette habit and the tropical version of the Field et al. (2007) particle size distribution, assuming 270 K and an ice water content of 1 g m^{-3}.

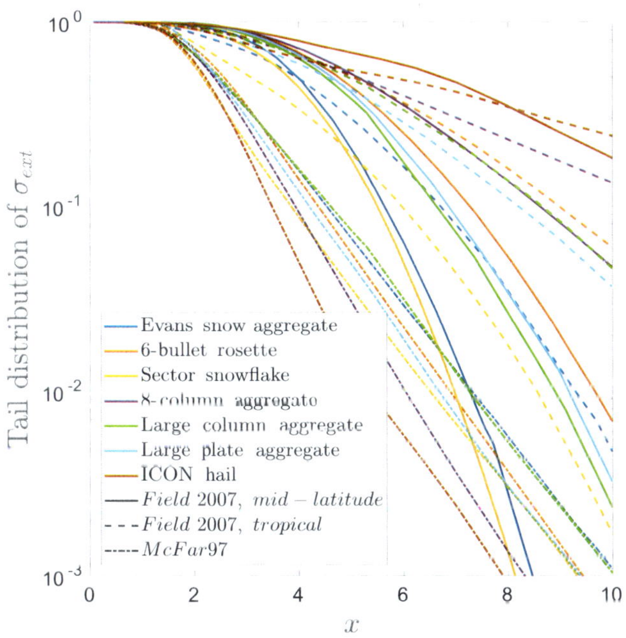

Figure 4. Tail distribution of extinction (see text for details), at 670.7 GHz and assuming 270 K and an ice water content of 1 g m^{-3}, for some combinations of habit and particle size distribution (PSD). Solid and dashed lines are results using the tropical and mid-latitude versions of Field et al. (2007) as the PSD, while dashed–dotted lines show results using the PSD of McFarquhar and Heymsfield (1997).

Table 2. Database habit classification and structure.

Habit classification	
Attribute	Possible cases
Orientation	Totally random Azimuthally random
Phase	Ice Melting Liquid
Aggregation	Single crystal Aggregate
Riming	Pristine Rimed

Eq. (5) as far as possible, but an exact fit is normally not possible, for example, when using third-party generated shape data. For $b < 3$, Eq. (5) can only be considered valid above some threshold size, normally of the order of 100 μm, because Eq. (5) otherwise requires unphysically high densities ("superdensity") for $D_{\max}$ below the threshold size.

In total, 34 habits divided into several subgroups are provided (Table 1). Each habit is assigned a unique Id number. Visualisations of the single crystal and aggregate habits are displayed in Figs. 1 and 2, respectively. The a and b parameters given in Table 1 were calculated by a least squares fit between $\log(m)$ and $\log(D_{\max})$, including only particles having a $D_{\max} \geq 200$ μm.

As mentioned, no totally general classification of particle shape exists, but for user friendliness the database is structured according to a rough categorisation of the habit type. The top category is "phase": liquid, ice or melting. Melting is not covered by the current database version, but such habits are intended to be included in the future and the database structure is set up accordingly. The ice and melting habits are further divided between single crystals and aggregates. This classification level is not applied for liquid particles. Both single crystals and aggregates are classified as either pristine or rimed. Table 2 gives an overview of this categorisation. Further information on how the particles were generated is provided in Sect. 3.1.

2.5 Size grid

The number and range of sizes are not identical between the habits for a number of reasons. For some habits, mainly aggregates, it is not possible to generate shape data on a predefined size grid. Certain habits have been provided by third-party sources, where control of the size grid is limited. Furthermore, the methodologies of designing habits have evolved over the time of the database creation. Lastly, it makes sense to target different size ranges with single crystal and aggregate habits. However, the general strategies to set the size grids are similar.

The minimum number of sizes was set to 30, but single crystal data are still provided for 35 to 45 sizes. Aggregates have mainly 35 sizes. The size ranges of the habits (see Table 1) are not consistent, due to the diversity in methods to generate the shape data. However, target ranges were set to give a reasonable coverage of particle sizes reported in the literature. The single crystals' habits have sizes reaching either 2 mm in D_{veq} or 10 mm in D_{max}. The aggregates go up to 5 mm in D_{veq} or 20 mm in D_{max}. Exceptions are the aggregates consisting of relatively small crystals, where it was computationally too costly to reach the upper limit of any of the target ranges. The smallest single crystal particles are mostly $D_{\text{veq}} = 10\,\mu\text{m}$. The smallest aggregates mainly have $D_{\text{veq}} < 100\,\mu\text{m}$, but this criterion is not met when the constituting crystals are larger than this limit. Habits consisting of solid spheres (handled by Mie code) were allowed both to have more sizes (200) and cover larger size ranges, in order to allow detailed reference calculations using these habits. The final size range of all habits is found in Table 1. The general strategy to select intermediate sizes is to aim for an approximately equidistant linear spacing of up to $D_{\text{veq}} = 100\,\mu\text{m}$ and equidistant spacing in the logarithm of D_{veq} above.

Additionally, an upper limit with respect to size parameter (x) was introduced for practical reasons since the computational burden to calculate the scattering properties increases strongly with x. That it is possible to apply a limit on x is due to the fact that the size range that contributes the most to bulk scattering properties is shifted towards smaller sizes when going up in frequency. That is, for a given (large) particle, x itself increases with frequency, but the particle becomes less influential for the scattering properties of the overall local particle mix. This effect is illustrated in Fig. 3 for extinction. An analysis of backscattering would give a similar result as extinction and backscattering are highly correlated (both increase with D^6 in the Rayleigh regime, etc.).

Figure 4 demonstrates that $x = 10$ is a reasonable limit. It shows the estimated relative error in the calculation of bulk extinction considering data to a given limit in x. Figure 4 presents a worst case scenario as a relatively high ice water content ($1\,\text{g}\,\text{m}^{-3}$) and a frequency in the upper end of the database's coverage (671 GHz) are considered. Particles with $x < 1$ contribute marginally to the bulk extinction here indicated by the tail distribution staying close to 1 up to about $x = 1$. There are three cases in Fig. 4 where the tail distribution is well above 10 % even at $x = 10$. These cases are all combinations with the PSD of Field et al. (2007) and habits having $b = 3$. This PSD is intended for snow-type hydrometeors, where b around 2 is expected (Baran, 2012b). That is, these three cases of concern represent fairly unrealistic, or extreme, assumptions.

It is not unlikely that further problematic situations with respect to the PSD exist apart from those three cases presented above, and occasional errors above 10 % can not be ruled out. On the other hand, it should be remembered that the water vapour absorption at 671 GHz is high, giving a decreased sensitivity to lower altitudes, where most large particles are found most frequently, when observed by satellite instruments. In summary, our interpretation of Fig. 4 is that the error by applying a limit at $x = 10$ should in general be well below 10 %.

The size limit with respect to x is evaluated separately for each frequency. This means that, for example, at typical radar frequencies data are at hand for all sizes, while data can be lacking for large particles at higher frequencies. The limit was applied in such a way that, if possible, coverage up to $x = 10$ is maintained. That is, assuming that particles matching $x \geq 10$ are available, the smallest one exceeding the limit still got included.

2.6 Frequencies and temperatures

The overall frequency coverage was set to be 1–886.4 GHz. Inside this range, optical properties are given for 34 frequencies (Table 3). A majority of the frequencies were selected to bracket channel sets found on both existing and planned operational microwave sensors. These are in the following termed "channels" with arbitrary channel numbers assigned (see Table 3). For example, data are provided for 50.1 and 57.6 GHz, forming channel 3, which provides coverage of the standard set of microwave temperature sounding channels. Additional frequencies were picked to cover standard radar wavelengths, as well as to obtain a general coverage between 1 and 15 GHz. Above 200 GHz, the frequencies are mainly governed by the channels of the planned Ice Cloud Imager (ICI) mission (Kangas et al., 2012). The upper end at 886.4 GHz was set in order to support the ISMAR (Fox et al., 2017) and IceCube (Wu et al., 2015) instruments.

We stress that the database should be used with great care for frequencies outside the defined channel ranges. Whether interpolation between the channels can be applied must be judged from case to case. For example, interpolation to 170 GHz should not be problematic as the distance between channels 6 and 7 is small (in fact smaller than the width of channel 7 itself), while frequencies around, for example, 550 and 770 GHz can not be claimed to be properly covered by the database. The relevant measure to judge whether an interpolation can be performed should be the ratio between spacing and frequency, not the absolute size of the frequency spacing. As a general rule, only the closest database frequencies should be used when performing interpolation.

Optical properties are given for multiple temperatures, rather than a set of real and imaginary refractive index values. The reason for this is that the used parameterisations, Mätzler (2006) and Ellison (2007) for ice and water, respectively, should predict the refractive index with sufficiently high accuracy (Sect. 4.1). The selected set of temperatures contains 190, 230 and 270 K for hydrometeors consisting solely of ice. For pure liquid droplets, the temperatures are 230, 250, 270, 290 and 310 K. See Sect. 4.1 for further details regarding is-

Table 3. Included database frequencies. Frequencies above 18.6 GHz organised into channels (see text).

Channel	1	2	3	4	5	6	7	8	9	10	11	12
Freq.	18.6	31.3	50.1	88.8	115.3	164.1	175.3	228.0	314.2	439.3	657.3	862.4
(GHz)	24.0	31.5	57.6	94.1	122.2	166.9	191.3	247.2	336.1	456.7	670.7	886.4
Other frequencies: 1, 1.4, 3, 5, 7, 8, 9, 10, 13.4, 15												

sues related to refractive index and to temperature interpolation.

3 Software and processing

This section provides an overview of the software used for the generation of the data. The generation of the database consists of two main steps. First, a description of the particle shape must be acquired, either from some analytic expression or by physical models of varying complexity (Sect. 3.1). Second, the actual single scattering properties are calculated (Sect. 3.2), mainly by the discrete dipole approximation (DDA) method (Sect. 3.2.1). The DDA calculations are complemented with a set of quality checks and modifications, with the purpose of ensuring high quality and consistency.

3.1 Shape data generation

3.1.1 Snowflake Toolkit

Particle shape data have mainly been generated using internally developed software (Rathsman, 2016)[1], called the *Snowflake Toolkit* (SFTK). The toolkit's main features are a format for representing crystal shapes in an analytic way, an algorithm for simulation of aggregation, and sampling routines for generating DDA input shape data.

The toolkit's shape format is used both internally in the simulation software and to specify output data. The format represents particles as polygon meshes. This makes it straightforward to not only represent simpler shapes, such as hexagons and bullets, but also aggregates of said crystals. Further, the same shape data file can be used both as input for dipole sampling or aggregate simulation. The dipole sampling routines can take both crystals or aggregates as input. An advantage of using a polygon representation along with a sampling routine is that the grid resolution can be set arbitrarily, depending on application. The grid accuracy can thus be adapted to frequency.

As mentioned, the software includes capabilities to simulate aggregation. A semi-physical model, inspired by Maruyama and Fujiyoshi (2005), is applied. The model can be considered to describe a small section in a cloud, with a number of ice particles inside the volume. Several events can occur at each iteration: the aggregation of two particles, growth of a new crystal, particle melting or fallout of the cloud. The likelihood of these events depends on a number of factors such as size, estimated fall speed and control parameters. Contrary to most aggregate models used in the field, this model allows for the aggregation of two aggregates; the algorithm can consider aggregation between any particles that can be represented by the software's shape format.

Aggregation is modelled as if the two particles collide with each other at random angles and centre offsets. Aggregate sticking is only allowed to occur at particle faces, where the two involved surface normals are forced to be parallel (i.e. face-to-face sticking is ensured). This is achieved through rotation of the involved particles. For each candidate aggregation, it is checked if it has created any overlap between any combination of building blocks. If any overlap is found, the aggregation is discarded or a new try is made depending on the exact settings. The no-overlap condition is motivated by the fact that for pristine and compact crystals, little or no deformation should result from collisions and aggregation (by surface sticking). Also, volume calculations become significantly cheaper. There are a number of available control parameters. One such parameter is the number of particles allowed to populate the cloud at a given iteration. The available memory has to be taken into consideration here, but at least 5000 particles can be handled by a standard desktop computer. Growth, melting and drop-out ratio parameters are used to determine the relative likelihood of the different possible events.

As a complement, the aggregation of spheres is also handled. This version is more straightforward, but is less based on physics. The main application is for generating heavily rimed particles. The model describes a single aggregate particle, originally only a single sphere, onto which new spheres are added. Each new sphere is inserted from a random angle. It is allowed to let the spheres "tunnel" and reach the interior of the particle. The tunnelling feature is required to reach higher effective densities, but is not obligatory. It is also possible to set the angular distribution of the incoming spheres to be non-uniform, as a way of influencing the shape of the resulting aggregate. Other options and features available include maximum diameter (constraining the growth of the particle) and fill ratio (ensuring that a minimum effective density is achieved).

[1] Code available at https://github.com/milasudril/snowflake-toolkit

3.1.2 RimeCraft

There is frequently a desire to generate particles fulfilling Eq. (5) for some predefined a and b. This task is hard to tackle with the Snowflake Toolkit, and, as an alternative, a small set of MATLAB functions was developed for this task (code can by obtained by contacting the authors). As the function set should also be useful for adding different degrees of riming, it was named *RimeCraft* (RC). It works by adding 3-D cubic blocks directly on the shape grid. Different rules can be defined for how to add blocks. Equation (5) is fulfilled by growing the particle inside subsequent thin layers, and calculating a fill ratio for each layer from the target a and b. The position of a new dipole is selected randomly, but weights can be specified to tune how a new dipole prefers to have "neighbours" or not. This controls the overall granularity of the particle. Finer details of the structure can be modified by setting different weights (that can be both positive and negative) with respect to face, edge and vertex neighbours. A complete habit can be created either by taking a snapshot at each intermediate size, as a single particle grows, or by re-running RimeCraft for each target D_{max}.

An additional feature (not yet used for the database) is to simulate the melting of a particle. This feature largely follows the algorithm of Johnson et al. (2016). Here, an existing, gridded particle is used as starting point. The algorithm consists of two main steps at each iteration. First, all ice grid points are assigned weights based on the radius from the centre and their number of icy neighbours. A portion of the grid points with the highest weights are selected and converted to liquid. The liquid grid points are then allowed to move in order to mimic surface tension. The iterations continue until the desired melting fraction is achieved. A column particle, for example, would begin to melt at its tips, successively melt inwards and be converted to a drop.

3.2 Calculation of optical properties

3.2.1 ADDA

A vast majority of the scattering data have been produced by the Amsterdam DDA (ADDA), a DDA implementation developed by Yurkin and Hoekstra (2011). The basic idea of the DDA method is to represent the particle by a discretised equidistant Cartesian grid, where each grid point represents an electric dipole with polarisation $\mathbf{P}$. The DDA solution can be formulated as

$$\alpha_i^{-1}\mathbf{P}_i - \sum_{j\neq i}\overline{\mathbf{H}}_{ij}\mathbf{P}_j = \mathbf{E}_i^{\mathrm{inc}}, \tag{6}$$

where α is the dipole polarisability, i and j are grid point indices, $\overline{\mathbf{H}}$ the total interaction term, and $\mathbf{E}^{\mathrm{inc}}$ the incoming electric field (Yurkin and Hoekstra, 2011). The equation is solved iteratively, and output quantities are derived via the resulting total electric field $\mathbf{E}_i$. This field can be derived when the polarisation field is known, using $\mathbf{P}_i = V_{\mathrm{dp}}\chi_i\mathbf{E}_i$, where χ_i is the susceptibility of the medium and V_{dp} the volume of a dipole. Without going into details, there are different formulations known for the susceptibility and the interaction terms. For this database, the "lattice dispersion relation" was used for the susceptibility and "interaction of point dipoles" for the interaction term (Draine and Flatau, 1994). The ADDA calculations are stopped when the relative norm of the residual, essentially the relative difference between the right and left parts of Eq. (6), reaches a user specified limit, denoted as ϵ_{DDA}. The selected value of ϵ_{DDA} thus determines the accuracy of the output and we have applied

$$\epsilon_{\mathrm{DDA}} = 10^{-2}, \tag{7}$$

aiming for an accuracy of the order of a few percent.

The shape data, in the form of shape files, are the most important input required by ADDA. The discretised grid of a particle is defined in the shape file as three columns of grid coordinates. The grid is scaled such that only integer values are used as coordinates. An extra column can be added to index points with different refractive indices, which is only required for heterogeneous particles. Other essential input parameters are the particle volume equivalent radius (r_{veq}), radiation wavelength (λ), number of dipoles per wavelength (dpl), propagation vector of incoming radiation, the scattering grid, and orientation of the particle (Euler angles using zyz notation). The variable dpl is critical as it must be sufficiently large. We determine dpl such that we fulfil the standard criterion:

$$|m|kd < 0.5, \tag{8}$$

where m is the complex refractive index, k the wavenumber, and d the dipole size. The microwave refractive index of ice means that dpl should be ≈ 22 or higher. As an additional rule, the minimum number of dipoles was set to 1000, to enable the shape of particles of small x to also be resolved in a reasonable manner.

ADDA handles two orientation cases, either fixed or totally random. The later case is handled by Romberg integration (Davis and Rabinowitz, 1984). This method starts out with two scattering calculations, at orientation angles spaced 180° apart. The integrated average is then updated iteratively by adding scattering calculations at new angles, until convergence is achieved. This tends to happen faster for calculations of small size parameter x, whose scattering functions are less varying compared to large x. Note that while this averaging is expressed in Eq. (2) by three angles, only β and γ are varied in ADDA, as rotation over α is equivalent to rotation of the polarisation plane, which can be performed analytically. By effectively removing one dimension from the orientation "grid", significant calculation time is saved. The convergence criterion refers here to maximum residual error of the averaging over each Euler angle, ϵ_{avg}. We have used

$$\epsilon_{\mathrm{avg}} = 10^{-2} \tag{9}$$

for both Euler angles. All ADDA calculations found in the database so far have been calculated using ADDA's orientation averaging.

ADDA outputs the absorption cross section σ_a, extinction cross section σ_e, Mueller matrix **M**, and amplitude scattering matrix **S**. The Mueller matrix is a 4-by-4 matrix that relates the Stokes vector of incoming and scattered radiation, and is defined with respect to the scattering reference frame. The scattering cross section is related to the Mueller matrix by

$$\sigma_s = \frac{1}{k^2} \int_{4\pi} M_{11}\,\mathrm{d}\boldsymbol{n}', \tag{10}$$

where k is the wavenumber. The amplitude scattering matrix relates also incoming and scattered radiation, but operates on the electric field. Both the Mueller and amplitude matrices have incoming and scattered angular dependence. Note that it is not possible to compute the amplitude matrix in conjunction with ADDA's orientation averaging option, as the amplitude matrix refers to coherent radiation and the averaging assumes incoherent interaction between the particles.

Some sanity checks, and possibly also some smaller corrections, are applied to the ADDA output.

1. Positivity. The cross sections σ_e and σ_a must be positive (≥ 0). The same applies to the M_{11} element at all angles. However, a violation of 1 % relative to the total extinction is allowed for σ_a, in which case σ_a is re-set to

$$\sigma_a = \sigma_e - \sigma_s. \tag{11}$$

2. Consistency. The extinction must be larger than both the scattering and absorption separately, i.e. $\sigma_e \geq \sigma_a$ and $\sigma_e \geq \sigma_s$. A violation of 1 % relative to σ_e itself is allowed. Furthermore, it is ensured that equivalent **M** elements, i.e. M_{12} and M_{21}, and M_{34} and $-M_{43}$, do not deviate by more than 10 % from each other. If fulfilled, these elements are then re-set to the means of respective values for the purpose of consistency. For example, M_{34} is re-set to

$$M'_{34} = \frac{(M_{34} - M_{43})}{2}. \tag{12}$$

3. Energy conservation. The extinction must equal the sum of the absorption and scattering cross sections, i.e. $\sigma_e = \sigma_a + \sigma_s$. At this point, differences of up to 30 % have been allowed. This is an unsatisfactory situation, but issues around the sector snowflake habit forced us to apply this high value. These issues are likely associated with the extreme aspect ratios of large sector snowflakes, and we aim for having a more demanding limit in the next database version. The maximum deviation found among other habits is < 5 %.

DDA calculations that do not fulfil all of above criteria are recalculated with higher accuracy criteria.

Finally, the database quantities **K**, $\boldsymbol{a}$ and **Z** are set as

$$\boldsymbol{a}_1 = \sigma_a, \tag{13}$$

$$\mathbf{Z} = \frac{1}{k^2}\mathbf{M}, \tag{14}$$

$$\mathbf{K}_{ii} = \sigma_e = \sigma_a + \sigma_s. \tag{15}$$

Other elements in $\boldsymbol{a}$ and **K**, i.e. $i \neq 1$ in $\boldsymbol{a}_i$ and off-diagonal elements in **K**, are set to zero according to the theory for totally randomly oriented particles (Mishchenko et al., 2002).

3.2.2 Mie theory

Two habits represented by spheres are included in the database, liquid and solid ice spheres. For the derivation of these scattering properties it is appropriate to make use of the well-established Mie theory, which provides an exact solution and is much more computationally efficient than DDA. The implementation in MATLAB by Mätzler (2002) was used. Two functions are used, `mie` and `mie_12`, where the former calculates scalar quantities, such as scattering efficiencies. The efficiency parameters are normalised scattering quantities with respect to the particle cross-sectional area, for example in the case of extinction

$$\sigma_e = \pi r^2 Q_e, \tag{16}$$

where r is the sphere radius. `mie_12` calculates the amplitude scattering matrix elements S_{11} and S_{22}. These elements describe the angular dependence of the plane-perpendicular and plane-parallel components of the scattered light relative to the incoming light. The S_{12} elements are zero since spheres cause no depolarisation. The amplitude scattering matrix is converted to the scattering matrix **Z** using the wrapper code provided by the Atmlab[2] package.

4 Microphysical and dielectric data

This section deals with the microphysical and dielectric data employed for the database. Specifically, Sect. 4.1 discusses the refractive index models used, Sect. 4.2 and 4.3 describe single crystal and aggregate shape data, respectively, Sect. 4.4 explains how rime is included, and the subject of Sect. 4.5 is liquid particles.

4.1 Refractive index

Beside sizes and shapes of particles, single scattering properties are determined by the dielectric properties of the particle material. These are commonly described by the complex refractive index, where the real part mainly affects the scattering properties while the imaginary part primarily determines the absorption. The refractive index of liquid and frozen water at microwave frequencies varies significantly with frequency as well as with temperature.

A review of (water) ice refractive index models is shown in Eriksson et al. (2015), and a comparison between some models is also found in Fig. 5. The refractive index model by Ray (1972) is included in Fig. 5 as it strongly deviates from more recent parameterisations, based on a wider set of laboratory measurements, but it has still been applied in some recent studies (Lu et al., 2016; Gong and Wu, 2017). The most significant deviations are found for the imaginary part. This causes concern for at least passive measurements, as ice absorption has been shown to play a non-negligible role for such observations. (e.g. Eriksson et al., 2011b, 2015; Ding et al., 2017).

In Eriksson et al. (2015), the ice refractive index model by Mätzler (2006) was suggested as a good choice for the full microwave region (up to about 1 THz) and Earth atmospheric temperatures. Newer models (e.g. Warren and Brandt, 2008; Iwabuchi and Yang, 2011) refer or are essentially identical to Mätzler (2006). Hence, for our database, we set the refractive index of ice according to Mätzler (2006).

For liquid water, the model of Ellison (2007) is applied, which provides refractive indices for frequencies of up to 25 THz and temperatures between 0 and 100 °C based on empirical fits to an extensive collection of water refractive index data (Ellison et al., 1996). Deviations to the model by Turner et al. (2016), which focuses on super-cooled liquid water but does not cover sub-millimetre frequencies, remain smaller than 30 % even at the lowest temperatures of concern (−40 °C). Hence, although Ellison (2007) is, in the strict sense, not recommended for super-cooled liquid water, it is applied here at all temperatures.

Finally, a smaller study on the required temperature grid for ice habits was performed. Assuming optical properties of solid Mie spheres, a number of forward calculations with varying temperature grid resolution were compared. Forward calculations using a 2.5 K spaced temperature grid covering 150 to 270 K were used as a reference. A number of coarser grids, all equally spaced between 190 and 270 K and applying linear interpolation and extrapolation in temperature, were compared to the reference set. Simulations were performed for all channels employed by the ISMAR airborne radiometer (Fox et al., 2017) and 5000 atmospheric scenarios contained in Eresmaa and McNally (2014), in order to maximise the temperature profile diversity. Deviation statistics over these 5000 cases and the 13 channels were derived. Figure 6 summarises the results, showing that for a setup with three temperature grid points, maximum deviations to the reference simulations over all scenarios and channels are < 2 K, and the 99.7 percentiles of differences are < 0.2 K. For five grid points, maximum deviations fall below 0.3 K and the 99.7 percentile below 0.05 K. The accuracy can be improved further by applying higher-order polynomial temperature interpolation. A three-point grid was judged as a good compromise between attainable simulation accuracy and the calculation burden required to generate the database.

4.2 Ice crystal shapes

The single crystal habits provided by the database are summarised in the upper part of Table 1. Example visualisations are found in Fig. 1. The crystal shapes are mainly based on parameterisations taken from the literature (Liu, 2008; Hong et al., 2009), but some new crystal habits have also been defined. For completeness and as a reference, the ice sphere habit was also included.

For crystal habits, external data are restricted to numerical parameterisations of shape properties, while the actual shape data matching those parameterisations were generated by the Snowflake Toolkit. The long column, short column, block column, thick plate and thin plate, taken from Liu (2008), are all hexagonal and have axial ratios of 4, 2, 1, 0.2 and 0.05, respectively. These axial ratios were defined by Liu (2008) to be constant over all sizes. As a consequence, they all obtain $b = 3$. In contrast, plate and column type 1, with dimensions taken from Hong (2007), are defined to have a b of 2.48 and 2.05, respectively, implying an axial ratio that grows with mass. These definitions were based on measurements of ice crystal dimensions (Auer Jr. and Veal, 1970; Mitchell and Arnott, 1994; Yang et al., 2000). The sector snowflake, defined in Liu (2008), is an idealised representation of a snowflake, consisting of three intersecting ellipsoids with their mass centres positioned at the origin. This parameterisation was based on surface measurements of aggregates (Kajikawa, 1982).

The different bullet rosette habits all consist of hexagonal bullets connected at the tips. The crystal dimensions are taken from Hong (2007), based on measurements listed in that paper. All bullets grow in a similar fashion, in the sense that all bullets with equal mass will have the same dimensions regardless of the considered habit. Furthermore, in all versions, the bullets are aligned perpendicularly to each other. The flat 3- and 4-bullet rosettes have all their arms lying in a common plane. The perpendicular versions have one arm rotated 90°, resulting in more compact particles.

Two crystal habits were generated specifically to match the a and b assumed for cloud ice in two atmospheric models: GEM (Global Environmental Multiscale Model) and ICON (Icosahedral non-hydrostatic general circulation model). The GEM model was prepared for the EarthCARE mission (Côté et al., 1998) and the ICON model is a part of the German HD(CP)2 project. The GEM values ($a = 440$ and $b = 3$) imply a habit with low and constant aspect ratio, and are in fact very close to the ones of solid spheres. Accordingly, it was decided to apply spheroids for GEM cloud ice. For simplicity and to have some additional reference spheroidal reference data, the same basic shape was selected for ICON cloud

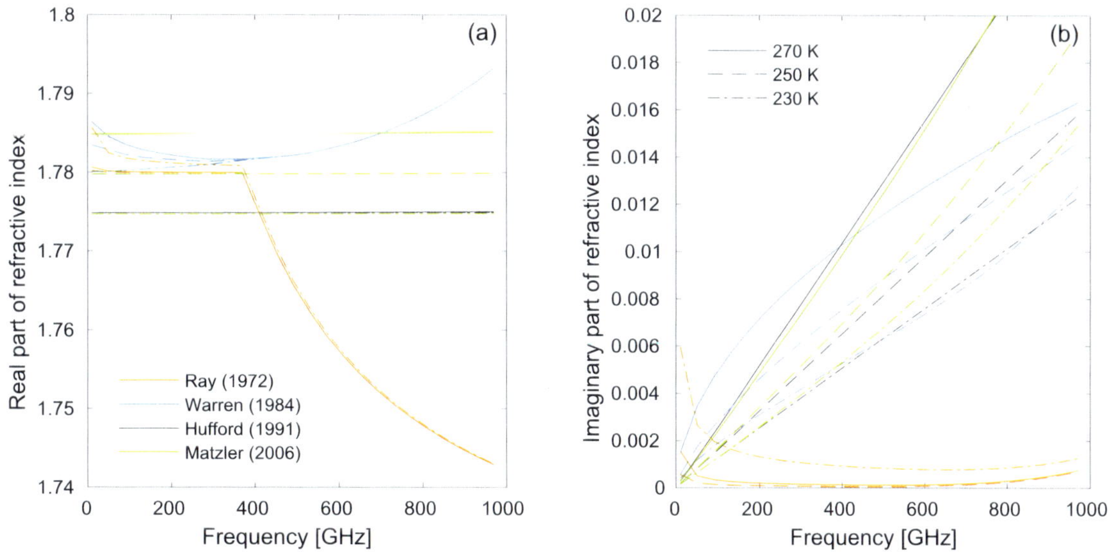

Figure 5. Real **(a)** and imaginary **(b)** parts of the refractive index of pure ice, for selected temperatures as a function of frequency, according to Ray (1972), Warren (1984), Hufford (1991), and Mätzler (2006).

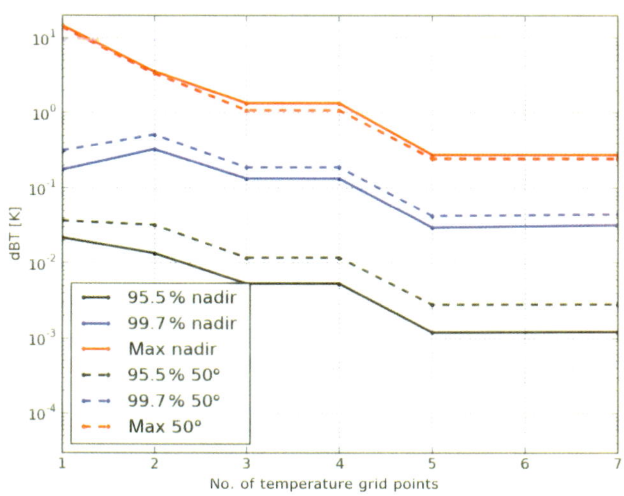

Figure 6. Error statistics of ISMAR observation simulations, given as a function of the temperature grid length used to represent ice hydrometeor scattering properties. The error (dBT) is estimated by comparison to reference calculations. The lines refer to the 95.5 % percentile, the 99.7 % percentile and the maximum value of the errors, for two different observation directions.

ice. The scattering properties of spheroids can be calculated efficiently by the T-matrix method, but for consistency reasons and to avoid possible numerical problems of T-matrix at higher size parameters (Mishchenko and Travis, 1998), also these habits were processed using the Snowflake Toolkit and ADDA.

4.3 Ice aggregate shapes

The ARTS database includes habits developed specifically both for this database and from third-party sources. For calculations at higher frequencies, aggregates are arguably lacking in the current scattering databases and the aim has been to include a relatively diverse set of aggregate habits.

4.3.1 Third-party data

Third-party aggregate data come in two forms, either in the form of shape data ready to be used in the DDA calculations, or as shape descriptions taken from the literature. The 8-column aggregate, originally defined by Yang and Liou (1998) and also represented in Hong et al. (2009), is of the latter case. It is constructed of eight columns of varying sizes. The aspect ratios and relative positions of the columns do not change with mass, and the habit has a $b = 3$. The a-value corresponds to an effective density of 13.5 %.

The Evans snow aggregate (Evans et al., 2012) and Tyynelä snowflake aggregate (Tyynelä, 2011) are both habits where explicit third-party shape data have been used. They both have features in common with the aggregates generated by the Snowflake Toolkit, in that they are generated in a stochastic manner, with multiple particles tracked at the same time. The algorithms used in Tyynelä (2011) and Evans et al. (2012) for generating the aggregates were both inspired by Maruyama and Fujiyoshi (2005), which is the case for the Snowflake Toolkit as well. The dendrites by Evans et al. (2012) were randomly oriented during aggregation events, and they were connected with an overlap of 5 % in volume. The Tyynelä aggregates were generated assuming a quasi-horizontal alignment for the dendrites, more precisely a Gaussian orientation distribution with a standard deviation of 2°. The publicly available shape data are of too low a resolution for large parts of the database. Higher-resolution data were obtained by correspondence to J. Tyynelä and applied here for these aggregates. The shape data by Tyynelä and Evans have a somewhat poor resolution (dpl = 15 at 664 GHz) and corresponding scattering data have likely limited accuracy above roughly 460 GHz.

4.3.2 Database specific

The aggregate simulator of the Snowflake Toolkit (Sect. 3.1.1) was used to create a number of aggregate habits. Throughout, hexagons have been used as constituent crystals. Separate simulations were performed using mean axis ratios of 5 for columns, 1.25 for blocks (short columns), and 1/6 for plates. Furthermore, dimensions of each constituent crystal, length L and and side length a (Sect. 2.3) are not set with an exact fixed value, but are each selected randomly from Gamma distributions, done in such a way that the standard deviations become 25 % of the corresponding mean value. Since L and a are selected independently, the aspect ratios are randomly distributed as well, around the specified mean value. Separate simulations were also run for different constituent crystal mean maximum dimensions D_{max}, for which two values, 100 and 350 µm, were chosen. The motivation for having two constituent crystal sizes is that the simulations could not provide large enough particles when using 100 µm crystals. Hence, habit versions with 350 µm crystals were created as well. In summary, six simulation runs in total were performed, using three aspect ratios and two crystal maximum dimensions. As described in Sect. 3.1.1, a high number of particles are generated in each run, with varying sizes. Accordingly, each run gives sufficient data to produce a full habit set.

The details of this set are summarised in Table 4, and example figures of the particles are shown in Fig. 2. A difference to the Evans aggregate is that no overlap between crystals is allowed: attaching crystals are connected with their faces against each other. Since each simulation generates thousands of particles, a filtering had to be applied. In line with the statement in Sect. 2.4 that habits should follow a single mass-size power law, we fit Eq. (5) to the set of particles of each run and make a somewhat ad hoc selection among the particles that best match the derived a and b to obtain a logarithmically spaced grid in D_{veq}.

Furthermore, habits matching snow and hail hydrometeor assumptions in the GEM and ICON models were generated using the RimeCraft function set. The hail habits were initiated with a single cubic "block" and were allowed to grow inside a spherical volume. Together with the a and b of the GEM and ICON hail habits, this resulted in (macroscopically) quasi-spherical particles exhibiting some surface roughness and internal hollows. For the snow habits, some initial structure was needed to create a visual resemblance

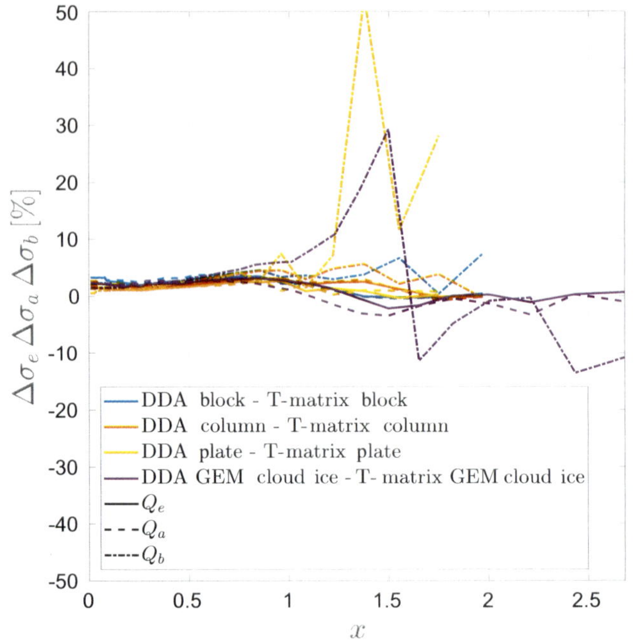

Figure 7. Comparison between the extinction, absorption and backscattering of short columns, block columns and thick plates in the database, and T-matrix results using cylinders having the same length and mass. Shown as a function of size parameter x, with relative difference calculated as $\Delta\sigma = 100\,\% \cdot (\sigma - \sigma_{\mathrm{Tmatrix}})/\sigma_{\mathrm{Tmatrix}}$. Temperature is 230 K and frequency is 94.1 GHz.

of snowflakes. Small 3-D and 2-D crosses with one-block thick arms were used for GEM and ICON habits, respectively. With $b \approx 3$, GEM snow-like hail was allowed to grow in a spherical volume, while for the ICON snow habit with $b \approx 2$ the allowed volume for growth was set to be plate-shaped.

4.4 Rimed particles

Three different habits designed to represent graupel are included in the database. All three can be considered heavily rimed particles, i.e. only composed of rime. The spherical graupel was generated using the Snowflake Toolkit module treating aggregation of spheres. The angle distribution of the incoming spheres was set to be isotropic, resulting in an overall spherical shape of these graupel particles. Furthermore, the constituting spheres have randomly distributed diameters, selected from a gamma distribution with a mean radius of 300 µm and a standard deviation of 75 µm. Furthermore, the habit was designed to have an effective density of approximately 140 kg m^{-3} in order to roughly match observations (Mitchell, 1996).

The ICON and GEM graupel habits were generated in the same general way as the ICON and GEM hail habits, but the settings were tuned to create a granular pattern when examining slices through the resulting graupels. Particles that can be

Table 4. Overview of aggregates produced using the Snowflake Toolkit. The columns axial ratio and D_{max} refer to mean values of constituting crystals, while the last two columns refer to complete aggregates.

Habit name	Id	Axial ratio $L/(2a)$	Crystal D_{max} (μm)	a	b
Small column aggregate	17	5	100	0.14	2.45
Large column aggregate	18	5	350	0.25	2.43
Small block aggregate	21	1.25	100	0.21	2.33
Large block aggregate	22	1.25	350	0.35	2.27
Small plate aggregate	19	1/6	100	0.077	2.25
Large plate aggregate	20	1/6	350	0.21	2.26

classified as lightly or moderately rimed can be generated by RimeCraft, and are intended to be included in a next database version.

4.5 Raindrops and liquid cloud drops

The liquid phase is so far limited to spherical drops, with scattering properties calculated by Mie theory (Sect. 3.2.2). The number of sizes was set to 200, and, as mentioned, data are provided for five temperatures between 230 and 310 K. Spherical drops should be an acceptable approximation for liquid cloud and drizzle particles, but not for larger raindrops, and liquid drops of non-spherical shape will be part of the next database version.

5 Results

This section presents example characteristics and results of the database. The ambition is to give a basic overview of the database. Exploration of all details of the scattering data provided is left for other publications and the users of the database.

5.1 Comparisons

A number of comparisons have been made in order to obtain a rough assessment of the accuracy of the data. At an early stage, dedicated tests were made to ensure that there is an acceptable agreement between our usage of ADDA and equivalent results obtained by the T-matrix implementation of Mishchenko and Travis (1998). Some deviations were unavoidable as T-matrix is a fully analytic method, while DDA is numerical and operates with discretised shape data. This is of special concern for particles where the properties vary quickly as a function of size (due to "Mie resonance features"). However, this should not be critical as these details are largely removed when averaging over size to obtain bulk properties.

Further, Fig. 7 compares the extinction of some hexagonally shaped particles with T-matrix calculations for cylinders. The T-matrix cylinders were given the same length or height and the same cross-sectional area; i.e. volume (and hence also mass) is maintained. The ADDA results for backscattering mainly agree with the T-matrix, but also show indications of some instability. A closer examination of the poorer cases revealed that high deviation in backscattering is associated with low number of orientations used by ADDA, indicating that its iteration criterion is not optimal with respect to this quantity. That is, the Romberg scheme has stopped prematurely with respect to backscattering. On the other hand, an acceptable accuracy appears to have been reached for extinction. As expected, the extinction deviates between ADDA and T-matrix, but not strongly: the difference is throughout below 4 %. These results give additional indications on the accuracy of the database, but also indicate that the microwave extinction of ice particles primarily follows overall particle properties, such as mass and maximum size, while surface features, such as roughness, have a relatively low impact. That is, if plates and columns are treated to be hexagonal or cylindric has only a marginal impact on microwave optical properties.

For habits, where external shape definitions have been used, comparison with their source databases (Liu, 2008; Hong et al., 2009) is possible. However, these comparisons involve interpolation in temperature, frequency and size, and again a perfect agreement can not be expected. In order to minimise the impact of these interpolations, close matches in frequency and temperature between the databases have been selected. Discrepancies will also depend on differences in DDA convergence criteria selected as well as on the exact sampling methodology used when discretising the particle shapes. Of concern is also that Hong et al. (2009) set dipole sizes following our Eq. (8), while Liu (2008) used a stricter criterion resulting in a factor of 2 smaller dipole size. Both performed DDA calculations for a set of specific particle orientations and based the orientation averaging on these data, but the number of orientation angles differ. Tests to estimate the impact of these later factors are reported by both Liu (2008) and Hong et al. (2009).

As we do not have access to the discretised shape data used in other databases, it is not possible to assess applied particle shapes on that level, but reported masses can be compared. The masses obtained by us show insignificant deviations to

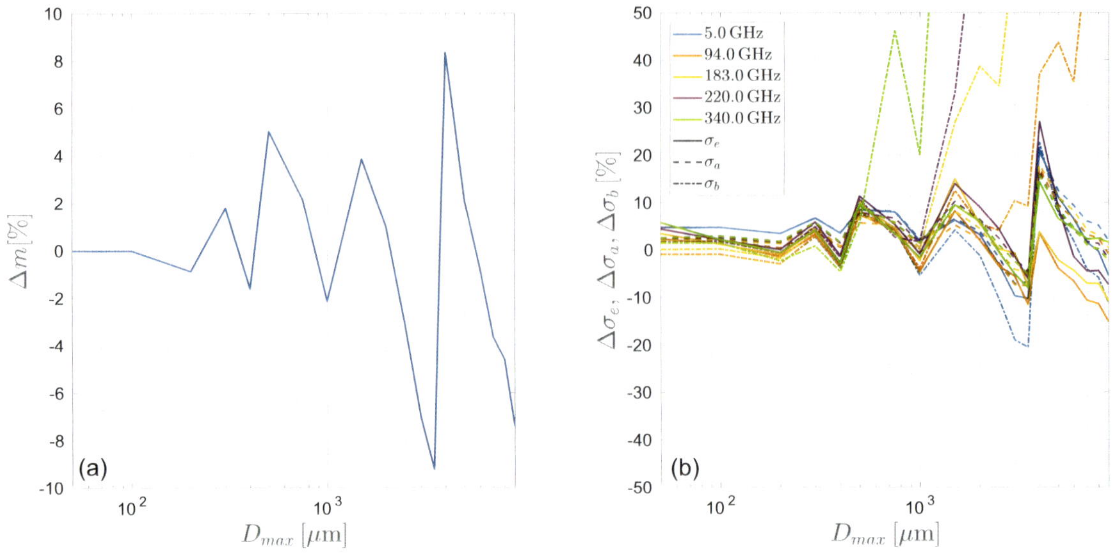

Figure 8. Comparison of sector snowflake data from Liu (2008) and this paper. **(a)** Relative differences in mass. **(b)** Relative differences in extinction (solid lines), absorption (dashed lines) and backscattering (σ_b; dot–dashed) cross sections. Shown as functions of D_{max} and using the data from Liu (2008) as a reference. Differences are reported as $\Delta m = 100\,\% \cdot (m - m_{Liu})/m_{Liu}$, etc. The comparison was made for one temperature (233.15 K) and sizes covered by Liu (2008).

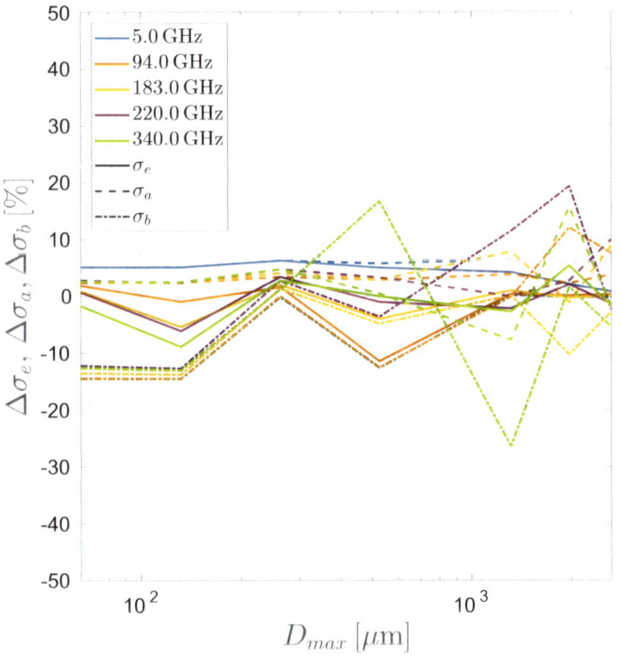

Figure 9. Comparison with block column data of Liu (2008). Otherwise as the right panel of Fig. 8.

the corresponding values reported by Liu (2008) and Hong et al. (2009), with the exception of the sector snowflakes. For this habit, discrepancies in mass of up to 9 % are found (Fig. 8). A simple discretisation of the particle shape normally results in a change in the particle mass. Our strategy has been to apply a rescaling of the discrete data to ensure that the final data correspond to the target mass. Already, Geer and Baordo (2014) noted that the sector snowflake habit in Liu (2008) shows an non-smooth variation in mass, and it seems that no similar rescaling was applied by Liu (2008). The discretisation is a particularly critical issue for the sector snowflake habit, as these particles have the highest aspect ratios and can be only a few dipoles thick. Just the decision to place dipole centres at $\ldots, \; l, 0, l, \ldots$ or $\ldots, \frac{-l}{2}, \frac{l}{2}, \ldots$ (where l is the dipole size) has here a significant impact, both for the effective mass after discretisation and the final optical properties.

The differences in the mass of sector snowflakes give a direct imprint in the extinction reported for this habit by the two databases (Fig. 8), but it is also noticeable that the difference in extinction is for some sizes considerably larger than the mass difference. However, the most striking feature is the high deviation in backscattering. Again, the sector snowflake stands out, and, as a much more representative example, a comparison for the block column habit is shown in Fig. 9. The agreement is better for the block columns in overall terms, but the backscattering still frequently deviates by 10 % or more.

A comparison with data of Hong et al. (2009) is found in Fig. 10. Clear systematic differences originating from the different refractive index models applied can be recognised. The scattering data of Hong et al. (2009) are based on the refractive index of Warren (1984). This parameterisation deviates significantly from more recent ones (Sect. 4.1), particularly regarding the imaginary part. The deviations in the refractive index are such that the data of Hong et al. (2009) exhibit

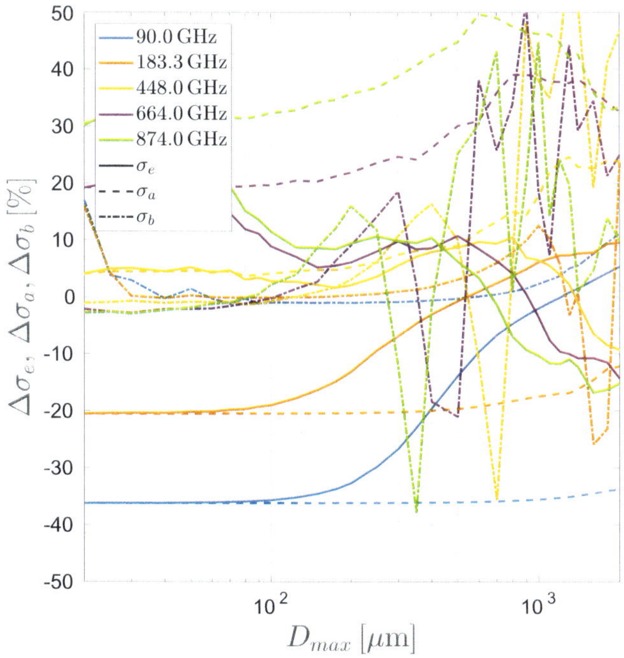

Figure 10. Comparison to Hong et al. (2009) of 8-column aggregate data: relative differences in extinction (solid lines), absorption (dashed lines) and backscattering (dot–dashed). Differences are reported as $\Delta\sigma = 100\,\% \cdot (\sigma - \sigma_{\text{Hong}})/\sigma_{\text{Hong}}$. The comparison was made for one temperature (243.15 K) and sizes covered by Hong et al. (2009).

a higher/lower absorption below/above about 400 GHz. The difference in the real part of the refractive index is much less pronounced and there is a fairly good agreement on the extinction due to scattering. The same is true for the backscattering at 90 GHz, while the backscattering at, for example, 664 GHz deviates significantly for millimetre-sized particles. Putting aside deviations that should be caused by the older refractive index data used by Hong et al. (2009), the remaining data indicate a good underlying agreement between the two sets of DDA calculations.

The difference in refractive index with respect to Hong et al. (2009) is relatively small at 448 GHz. With this in mind, the combined interpretation of Figs. 7, 9 and 10 is that the calculation of extinction disagrees by 10 % at most, but is generally inside 5 %. Larger differences are found for backscattering, particularly at higher frequencies. If only the radar frequencies of today (i.e. < 100 GHz) are considered, then the general agreement for backscattering is also of the order of 10 %, with a strong exception in the sector snowflake habit and some data points with early stop of the Romberg scheme.

The results found for backscattering give a direct indication on the accuracy of the phase matrix for the direct backward direction. This is likely the direction with the highest calculation errors. However, the finer details of the phase matrix should not be critical for passive measurements as the radiation field is relatively smooth (except around the limb direction), and the overall accuracy of the phase matrix should be similar to the one of the extinction. To study the effective accuracy of the phase matrix data provided would require full radiative transfer simulations, which are beyond the scope of this article.

5.2 Effective density

Figure 11 displays effective densities for a selection of habits. The effective density is defined as

$$\rho_e = \frac{6m}{\pi D_{\text{max}}^3}, \tag{17}$$

i.e. the mass divided by the volume of the minimum circumsphere. It is linked to the a and b coefficients of a habit as

$$\rho_e = \frac{6a D_{\text{max}}^{b-3}}{\pi}, \tag{18}$$

where the m has been replaced by Eq. (5). Accordingly, the data in Fig. 11 should ideally end up along straight lines. This is also the case for habits having an analytic description of the shape, while deviations are observed for habits having stochastically generated shape data.

The effective density is governed by two main factors, the particle's "fluffiness" and aspect ratio. Habits having throughout a high ρ_e are generally dense particles with a spherical appearance. Particles with low values either have a high aspect ratio (such as the sector snowflake) or are very porous. Habits with $b < 3$ have negative slopes in Fig. 11 and tend to grow differently in the spatial dimensions, e.g. the plate type 1 habit has an increasing aspect ratio with size. However, $b < 3$ can also correspond to particles that increase in fluffiness with increasing distance to the particles' centre, e.g. the GEM snow habit.

The legend of Fig. 11 is divided into groups of habits, with single crystals having solid lines, aggregate and snow particles dashed lines, and dense graupel or hail-type particles dot–dashed lines. The ice sphere habit serves as an upper bound, while the long column, the flat 3-bullet rosette and the sector snowflake exhibit the lowest density, depending on the D_{max} range. More generally, the graupel and hail habits occupy the upper part of the plot, being the densest habit group. The aggregates and crystals tend to share the same area in the middle to lower parts of the graph. It will be seen later on that the effective density has a noticeable impact on the observed optical properties.

5.3 Extinction and backscattering

Example data of extinction and backscattering efficiencies are displayed in Fig. 12 for some selected frequencies and habits. The habits and legends are the same as in Fig. 11, i.e. single crystals have solid lines, aggregate or snow particles dashed lines, and dense graupel or hail-type particles

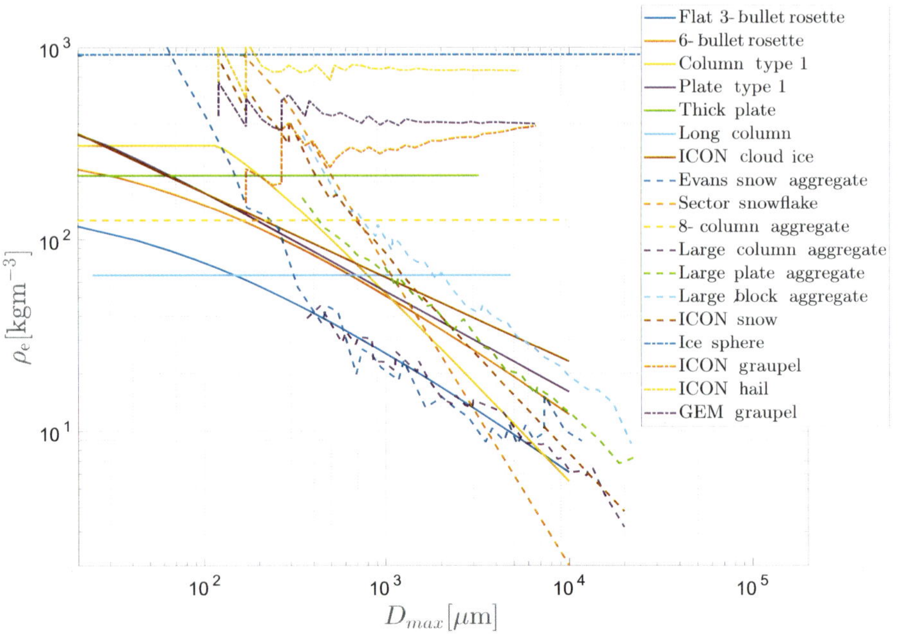

Figure 11. Effective densities ρ_e of a selection of habits as a function $D_{\max}$.

dot–dashed lines. The extinction and backscattering efficiencies are calculated as

$$Q = \frac{4\sigma}{\pi D_{\text{veq}}^2}, \tag{19}$$

where σ is the given cross section. That is, both x and Q are defined with respect to D_{veq}. A related remark is that efficiencies are thus compared in Fig. 12 between particles with the same mass (as a given x implies a specific mass).

A few characteristics can be observed in Fig. 12. Firstly, the groups tend to cluster. For example, the aggregate habits tend to have low extinction at 175 GHz and high extinction at 670 GHz. An opposite tendency can be observed for the dense particles given dot–dashed lines, which have relatively high and low extinction at 175 and 670 GHz, respectively. The crystal habits (solid lines) have efficiencies similar to heavy/dense particles at 175 GHz, but varying, intermediate, extinction values at 670 GHz. Similar observations can be made for backscattering; dense particles tend to have high backscattering, and aggregates low backscattering. Interestingly, the features observed for effective density in Fig. 11 are to a large degree transferred to the scattering properties in Fig. 12, but exceptions are also noted. For example, the spheres and ICON hail have comparatively low extinction at $x > 3$ at 175 GHz, and several crystal habits have higher backscattering at 94.1 GHz than the dense particles. Overall, the effective density can be considered an important parameter, but other aspects, such as exact particle shape, can exert influence as well. This discussion refers to how extinction and backscattering are related to particle mass. Other relationships can be found if the comparison is made with respect to $D_{\max}$.

5.4 Triple-frequency signatures

A common manner to evaluate scattering properties in the radar community is the triple-frequency signature (Kneifel et al., 2011). This is a multi-frequency bulk backscattering analysis involving three frequencies with data reported as two dual-wavelength ratios (DWR). The approach involves the effective reflectivity factor Z_e:

$$Z_e = 10^{18} \frac{\lambda^4}{\pi^5 \cdot |K|^2} \int_0^\infty \sigma_b(D_{\max}) N(D_{\max}) \, dD_{\max}, \tag{20}$$

where $\sigma_b(D_{\max})$ is the backscattering cross section, $N(D_{\max})$ the number of particles of size $D_{\max}$, and K is calculated as $K = (m^2-1)/(m^2+1)$, with m being the complex refractive index of liquid water. The DWR for two wavelengths λ is then obtained as

$$\text{DWR}_{\lambda_1,\lambda_2} = 10 \cdot \log\left(\frac{Z_e(\lambda_1)}{Z_e(\lambda_2)}\right). \tag{21}$$

Example triple-frequency results are shown in Fig. 13, using three standard radar frequency bands. Refractive index of water (m) was taken from Ellison (2007), assuming a reference temperature of 270 K. Further, an exponential PSD, defined over $D_{\max}$,

$$\frac{dN}{dD_{\max}} = N_0 \exp^{-\Lambda D_{\max}}, \tag{22}$$

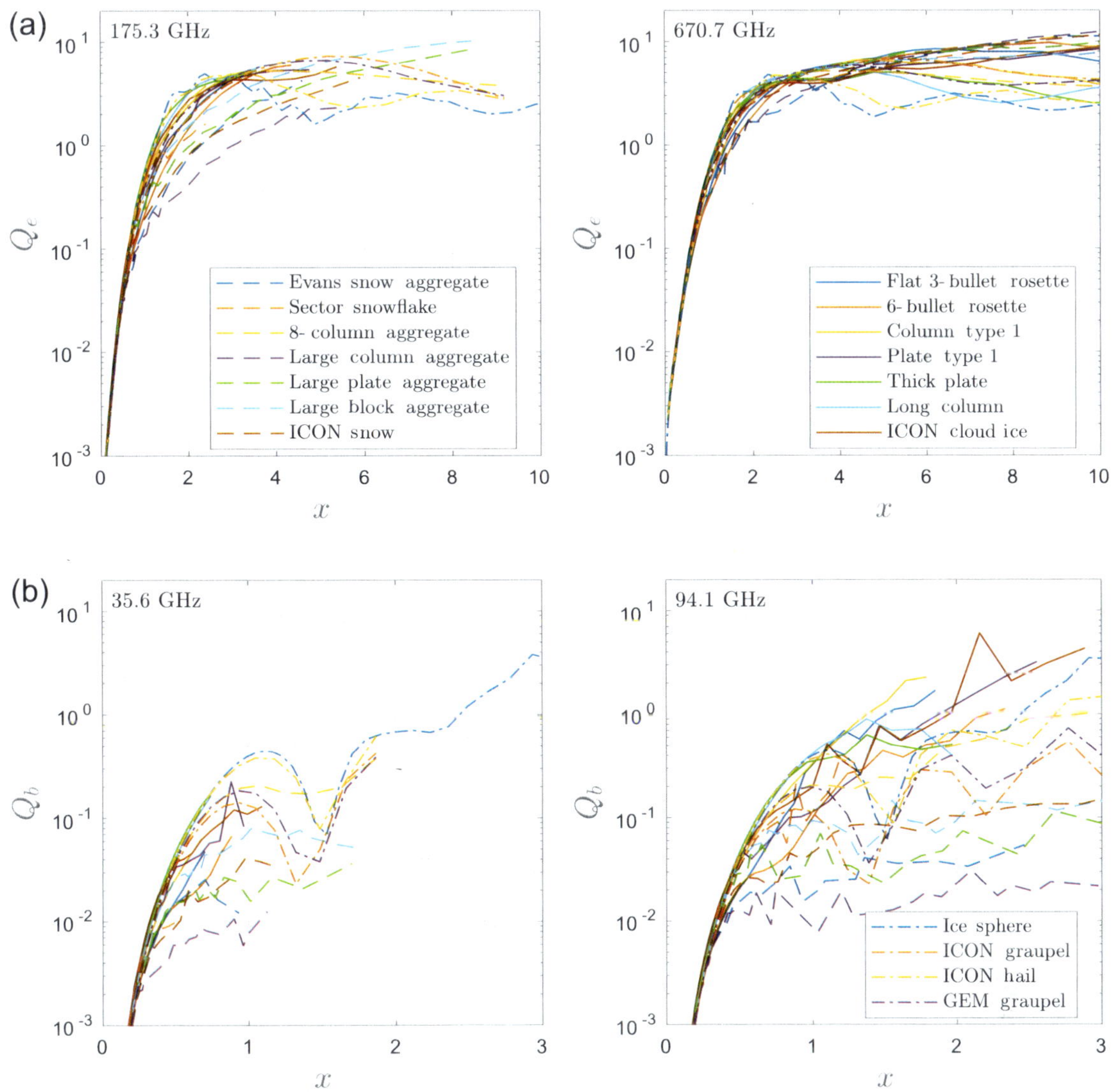

Figure 12. Example extinction **(a)** and backscattering **(b)** efficiencies (Eq. 19) as a function of the size parameter (Eq. 4). Legends are valid across all panels.

was selected. This PSD is identical to the one applied by Kneifel et al. (2011) and Kulie et al. (2014). N_0 is set to a constant value of 1, which is of no consequence since this constant is cancelled out in Eq. (21). In order to not let truncation issues influence strongly, Δ was varied for each habit separately, in such a way that the mean size of the PSD ranges from 1.5 times the smallest available $D_{\max}$ to 1/3 times the largest available $D_{\max}$.

Effective reflectivity is defined in such a way that if all particles scatter according to Rayleigh theory, Z_e is independent of frequency. This has the consequence that most high slope parameters (Δ), which correspond to low ice water contents, give rise to zero or low DWR. The other end of each triple-frequency curve corresponds to the highest Δ that could be accommodated with the size range available for the habit.

The triple-frequency signatures have been shown to act as an indication of effective density, frequently linked to the degree of riming (Kneifel et al., 2015; Yin et al., 2017). Figure 13 shows that the backscattering of the database habits matches the general expectations. First of all, the data span the same general ranges of DWR as reported by Kneifel et al. (2011) and Kulie et al. (2014). Further, the single crystal habits (solid lines), which are of comparatively low effective density (see Fig. 11), tend to occupy the left part of the graph. On the other hand, the dense graupel and hail habits (dot–dashed lines) are found in the rightmost part of the graph. The aggregate-type habits generally occupy the space in between the crystals and denser habits.

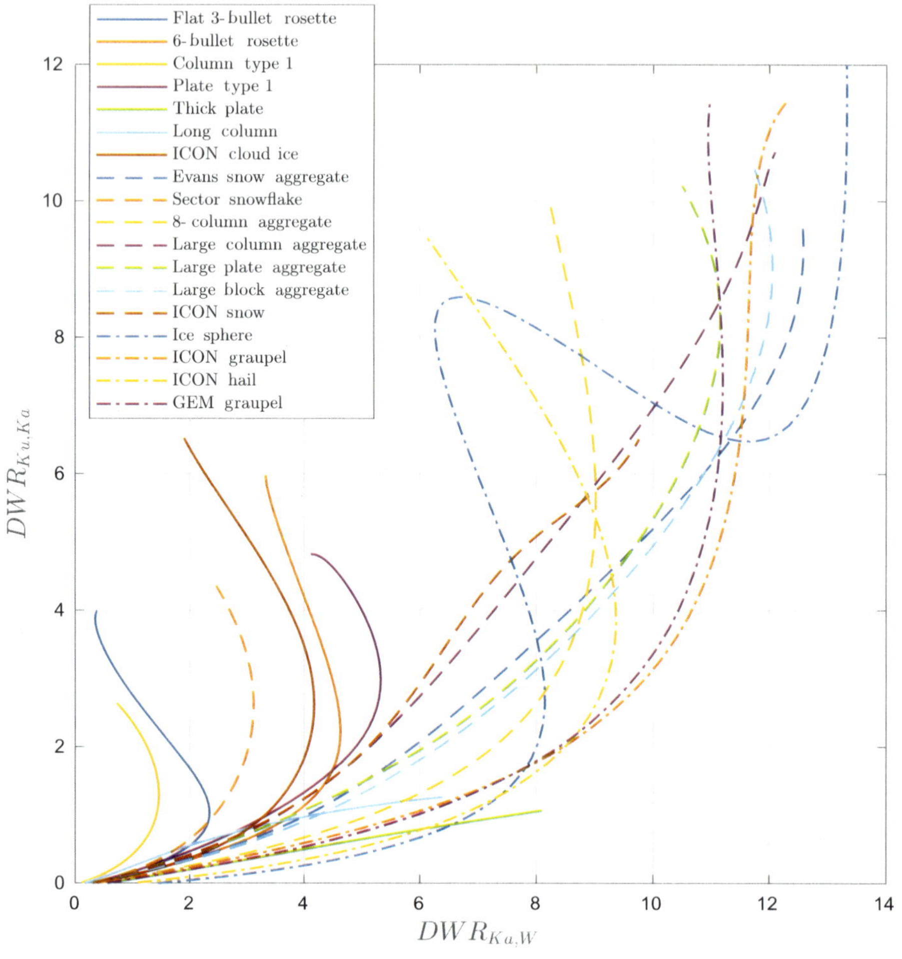

Figure 13. Triple-frequency signatures of selected habits, assuming an exponential particle size distribution (Eq. 22) and radar frequencies of 13.4 (K_u band), 35.6 (K_a band) and 94.1 (W band) GHz. The x-axis displays DWR between K_a and W bands and the y-axis shows DWR between K_u and K_a bands.

6 Database: format and access

This section outlines database format and accessibility, while the detailed documentation is provided with the database itself.

6.1 Format

The database is structured in a folder hierarchy, following the habit classification displayed in Table 2. At the bottom of the folder structure, there is one folder for each habit. These folders contain auto-generated pdf documents and text files that summarise the scattering data as well as a "habit logo" figure (Figs. 1 and 2). The actual data are also found in these folders, with one netCDF4 file for each particle size of the habit. Inside each netCDF4 file, there is an internal folder structure to organise the data according to frequency and temperature. These folders contain the scattering data, as well as particle shape and processing information. The stored variables and their attributes are summarised in Table 5. The variable names should hopefully all be self-explanatory, but for clarity they are accompanied by a description tag.

The `SingleScatteringData` group contains not only the core scattering properties (**Z**, **K**, and ***a***), but also information on frequency, temperature, and angular grids. The `ShapeData` group provides information on the particle behind the scattering properties, such as mass, D_{max}, D_{veq}, aspect ratio, aerodynamic area equivalent diameter, and refractive index. The `CalculationData` group contains ADDA settings and log data (DDA residual error used, date of completion, etc.). In particular, an identifier of the used shape file is given, and these files are part of the database. Thus, each scattering calculation is fully reproducible.

For orientation-averaged scattering properties, the optical properties exhibit features that can be used to reduce the overall data volume. For example, the scattering matrix has only six unique, non-zero, elements: Z_{11}, Z_{12}, Z_{22}, Z_{33}, Z_{34} and Z_{44}. The extinction matrix has only one such

Table 5. Summary of the core data format.

Attributes:	date, version	
Groups:		
SingleScatteringData:		
	Attributes:	orient_type
	Dimensions:	aa_scat, za_scat, aa_inc, za_inc, scatMat_row, scatMat_col, phaMatElem, extMatElem, absVecElem
	Variables:	frequency, temperature, aa_scat, za_scat, aa_inc, za_inc, phaMat_index, extMat_index, absVec_index, phaMat_data, extMat_data, absVec_data
ShapeData:		
	Attributes:	description, source, refrIndex_model, habit_file_id, habit_id, phase, refrIndex_homogenous_bool, density_homogenous_bool
	Variables:	diameter_max, diameter_vol_eq, aspect_ratio diameter_area_eq_aerodynamical, mass, dpl, N_dipoles, refrIndex_real, refrIndex_imag, alpha, beta, gamma
CalculationData:		
	Attributes:	method, software, software_version, system, n_nodes, n_cores, date_completion, ADDA_eps, ADDA_avgParam_file, ADDA_scatParam_file

element, K_{11}. Similarly, only one element of the absorption vector, a_{11}, is non-zero. Only these unique elements are stored, together with information on how full matrices and vectors will be populated. For example, the scattering matrix is stored in the database as a tensor (phaMat_data) with dimensions (aa_scat, za_scat, aa_inc, za_inc and phaMatElem), where phaMatElem corresponds to the six stored values ($Z_{11}, Z_{12}, \ldots$). Information on how to reconstruct the full matrix is provided separately, in phaMat_index that contains the position matrix

$$\begin{pmatrix} 1 & 2 & 0 & 0 \\ 2 & 3 & 0 & 0 \\ 0 & 0 & 4 & 5 \\ 0 & 0 & -5 & 6 \end{pmatrix}. \tag{23}$$

The numbers correspond to the indices of phaMatElem in phaMat_data, i.e. where to place those values in the matrix. The indexing is 1-based, where zero flags mean that the matrix is zero at that position. A negative number means the database value will be multiplied by -1 before it is inserted into the matrix. Corresponding systems are used for the extinction matrix and the absorption vector.

6.2 Interface and standard habits

The scattering data described above can easily be extracted using any netCDF or HDF5 interface. However, for enhancing the user friendliness, two data interfaces are distributed with the database. The interfaces are implemented in MATLAB and Python. The aim has been to keep the functionality as similar as possible between the two interfaces (but details of the programming languages used make some differences unavoidable). The MATLAB interface should function with any relatively new MATLAB version. The Python interface should function with both Python 2 and Python 3, but Python 3 (≥ 3.5) is recommended for full functionality. There are no extensive requirements on the interfaces. The Python interface requires the netCDF4, numpy and os packages, and the MATLAB interface the netcdf package.

The functionalities provided by the two interfaces are exploration of the database content, extraction of selected parts of the data (with respect to habit, size, frequencies and temperature), interpolation of data in temperature, frequency, size and angles, preparation of habit mixes and calculation of bulk properties.

The interfaces also provide support for converting data to the format used by the ARTS (Eriksson et al., 2011a) and RTTOV-SCATT (Bauer et al., 2006) forward models. Both interfaces can deal with the ARTS scattering format, on the condition that the Atmlab (MATLAB) and Typhon (Python) packages are installed. Only the Python interface supports RTTOV, i.e. conversion of the database scattering properties to data tables that can be digested by RTTOV-SCATT's Mie coefficient module. (This requires an update of the Mie coefficient module developed along with the ARTS database. This will be part of future RTTOV releases, but is also available as a patch file provided with the database.)

To allow a quick introduction to the database, a set of "standard habits" has been generated and provided in both the ARTS and RTTOV-SCATT formats. These habits should cover a range of conditions, such as cloud ice, snow, graupel, and hail. Most of these habits are a direct compilation of the database content, but some can be denoted as habit mixes. These combinations were created to simplify the usage of some of the aggregate and rimed habits that do not give full coverage in size. In order to also cover the lower size range, those habits were complemented with data from some crystal habit. Further details about the standard habits are provided in the accompanying `ReadMe` file.

7 Data availability

The database and its interfaces are publicly available at Zenodo, a free-of-charge research data repository hosted by CERN. The interfaces and the database itself are available under different uploads due to different licensing, with separate DOIs. The database DOI is https://doi.org/10.5281/zenodo.1175572 (Ekelund et al., 2018b) and the interfaces' DOI is https://doi.org/10.5281/zenodo.1175588 (Mendrok et al., 2018). This article refers to the data stored at Zenodo. As the database will be extended gradually, there is also a "development version" where the new data will be added until a new official version is released. This version is located on an ftp server hosted by Hamburg University, accessible through the ARTS homepage (http://www.radiativetransfer.org/tools/, last access: 11 July 2018). The database itself is provided under the CC BY-SA licence (https://creativecommons.org/licenses/by-sa/4.0/, last access: 11 July 2018), allowing the user to share and adapt the material, under the conditions that appropriate credit is given and indication of any changes made is given. Also, distribution of any modified or built-upon content must be done under the same licence. All source code falls under the GPL (http://www.gnu.org/licenses/gpl-3.0.en.html, last access: 11 July 2018) (General Public License).

8 Conclusions

The first version of a general, and fully publicly available, database of hydrometeor optical properties is presented. The database covers frequencies of 1 to 886 GHz, i.e. the full microwave region is targeted. This initial version is restricted to totally random particle orientation and focuses on particles consisting solely of ice. A variety of particle shapes, organised into "habits", are provided, in order to allow representation of cloud ice, snow, graupel and hail in radiative transfer simulations. In total, 34 ice hydrometeor habits are provided, and, as a complement, data for liquid spheres are included. Considered temperatures are 190, 230, and 270 K for ice, and 230, 250, 270, 290, and 310 K for liquid water. The data fully support polarimetric applications of both passive and active type.

The "ARTS database" presented here is the most extensive one at hand for totally random orientation. All except one of the existing databases (Lu et al., 2016) are also restricted to this particle orientation case. Only some main remarks are given here; for a review, see Sect. 1 and Ekelund et al. (2017). Only Ding et al. (2017) offer the same wide frequency coverage. In comparison to those data, the ARTS database includes a considerably higher number of habits (34 vs. 12) and a somewhat higher coverage in particle size. As all the aggregate habits in Ding et al. (2017), as well as in Hong et al. (2009), have $b = 3$ (see Eq. 5), it can be claimed that the ARTS database provides the first publicly available realistic representation of snow aggregates at microwave frequencies above 200 GHz. For lower frequencies, a broader coverage of snow is provided by the database of Kuo et al. (2016).

The aggregates of Ding et al. (2017) can potentially work as a proxy for hail and heavily rimed particles. The ARTS database contains a more diverse set of particle shapes for these classes, but also covers the 8-column aggregate part of both Hong et al. (2009) and Ding et al. (2017). A large part of the effort behind the database is the development of tools to generate realisations of these particle types. Regarding column and plate ice crystals, it was noted that such habits in Liu (2008) all have $b = 3$, and for these particle categories the data provided with the variation of aspect ratio taken from Hong et al. (2009) should be more realistic. The "update" of the data of Hong et al. (2009) using a better parameterisation of the refractive index resulted in changes of up to 50 % (Fig. 10), as well as a broader coverage in frequency, temperature and sizes.

All optical properties, except some reference data for spheres calculated by Mie code, were derived by discrete dipole approximation (DDA) handling arbitrarily shaped particles. Considering the high calculation burden of DDA (particularly at higher size parameters), it was not possible to both provide extremely accurate data and a broad database. It was decided at this stage to aim for a moderate accuracy, in order to first get a relatively broad overview. The basic idea is to explore the general properties of the data generated (manuscript in preparation), in order to better understand which particle shape properties govern the optical properties. It is not feasible, or desirable, to have data for every possible particle shape, but at this moment it is not clear to what extent for example micromorphological aspects like the shape and size of the crystals inside an aggregate need to be considered.

Comparisons to T-matrix, Liu (2008) and Hong et al. (2009) indicate that accuracy of the data is in general within 10 % (Sect. 5.1). This potential inaccuracy must not be ignored, but should today be relatively small compared to all other uncertainties involved in full simulations and retrievals. For example, Baran et al. (2017) reasoned in the same way and were also content with a 10 % accuracy. However, much higher discrepancies were found occasionally, particularly regarding backscattering. Further tests will be performed to better characterise the accuracy of the optical properties provided by the database, and, if found necessary, data will be recalculated with more demanding settings to provide more accurate backscattering.

The calculation burden of DDA also resulted in a cutoff around size parameter (x) 10 being introduced, but it is argued that this in general has an insignificant or low impact on bulk properties (Sect. 2.5). In other works, the need for calculations to much higher x is discussed, but it is here essential to notice how x is defined. We calculate x based on volume equivalent diameter (D_{veq}), while, for example, in Baran et al. (2017) it is defined with respect to maximum diameter (D_{max}), resulting in a higher x for a given particle (except for spheres).

Besides accuracy improvements, the next version of the database is planned to also incorporate data for melting and moderately rimed particles. A more major addition will be to include data for oriented particles. Such data are today very sparse and limited to a few frequencies below 200 GHz (e.g. Tyynelä and Chandrasekar, 2014). Initial calculations for "azimuthally random orientation" have been performed, with the conclusion that these calculations are much more demanding than the totally random case, with respect to both calculation burden and final data storage. Thus, practical hurdles must be overcome, and it is essential to carefully analyse the data at hand to guide the future calculations towards the most representative habits, as well as determining how many frequencies, temperatures and sizes the database must cover to be of practical use.

Author contributions. PE initiated the database, has acted as project leader, coded RimeCraft and the MATLAB interface, and written the article's text. RE has produced all data, designed the database structure and written the article's text. JM developed the RTTOV and Python interfaces, provided feedback on data and methods, and contributed parts of the article's text. MB, OL and SAB participated in planning of the database, have contributed to data and text, and started the production of the data for oriented particles that will be a main addition of the next database version.

Competing interests. The authors declare that they have no conflict of interest.

Acknowledgements. A large part of the database and associated tools were produced inside a study funded by EUMETSAT (Contract No. EUM/COS/LET/16/879389). The study manager at EUMETSAT was Christophe Accadia, who provided appreciated feedback and inspiration. Remaining funding for Patrick Eriksson, Robin Ekelund and Jana Mendrok was provided by the Swedish National Space Board. Stefan A. Buehler was partially supported through the Cluster of Excellence CliSAP (EXC177), Universität Hamburg, funded by the German Science Foundation (DFG), through BMBF project HD(CP)^2, Fkz. 01LK1502B and Fkz. 01LK1505D, and through the HALO/HAMP DFG project (BU 2253/3-1). Special thanks go to Torbjörn Rathsman, who implemented the SnowFlake Toolkit as his master thesis project. We also thank Frank Evans and Jani Tyynelä for kindly providing shape data. The RTTOV interface was designed with input from Alan Geer at ECMWF. ADDA calculations have been done at central computing facilities at both involved universities (C3SE/Chalmers and DKRZ(PID 878)/Hamburg). CERN is thanked for providing data storage.

Edited by: David Carlson

References

Adams, I. S., Gaiser, P., and Jones, W. L.: Simulation of the Stokes vector in inhomogeneous precipitation, Radio Sci., 43, RS5006, https://doi.org/10.1029/2007RS003744, 2008.

Auer Jr., A. H. and Veal, D. L.: The dimension of ice crystals in natural clouds, J. Atmos. Sci., 27, 919–926, https://doi.org/10.1175/1520-0469(1970)027<0919:TDOICI>2.0.CO;2, 1970.

Baran, A. J.: A new application of a multifrequency submillimeter radiometer in determining the microphysical and macrophysical properties of volcanic plumes: A sensitivity study, J. Geophys. Res., 117, D00U18, https://doi.org/10.1029/2011JD016781, 2012a.

Baran, A. J.: From the single-scattering properties of ice crystals to climate prediction: A way forward, Atmos. Res., 112, 45–69, 2012b.

Baran, A. J., Hesse, E., and Sourdeval, O.: The applicability of physical optics in the millimetre and sub-millimetre spectral region. Part I: The ray tracing with diffraction on facets method, J. Quant. Spectrosc. Ra., 190, 13–25, 2017.

Baran, A. J., Ishimoto, H., Sourdeval, O., Hesse, E., and Harlow, C.: The applicability of physical optics in the millimetre and sub-millimetre spectral region. Part II: Application to a three-component model of ice cloud and its evaluation against the bulk single-scattering properties of various other aggregate models, J. Quant. Spectrosc. Ra., 206, 83–100, 2018.

Bauer, P., Moreau, E., Chevallier, F., and O'keeffe, U.: Multiple-scattering microwave radiative transfer for data assimilation applications, Q. J. Roy. Meteor. Soc., 132, 1259–1281, 2006.

Birman, C., Mahfouf, J.-F., Milz, M., Mendrok, J., Buehler, S. A., and Brath, M.: Information content on hydrometeors from millimeter and sub-millimeter wavelengths, Tellus A, 69, 1271562, https://doi.org/10.1080/16000870.2016.1271562, 2017.

Boukabara, S. A., Garrett, K., Chen, W., Iturbide-Sanchez, F., Grassotti, C., Kongoli, C., Chen, R., Liu, Q., Yan, B., Weng, F., Ferraro, R., Kleespies, T. J., and Meng, H.: MiRS: An all-weather 1DVAR satellite data assimilation and retrieval system, IEETGRS, 49, 3249–3272, https://doi.org/10.1109/TGRS.2011.2158438, 2011.

Boukabara, S.-A., Garrett, K., Grassotti, C., Iturbide-Sanchez, F., Chen, W., Jiang, Z., Clough, S. A., Zhan, X., Liang, P., Liu, Q., Islam, T., Zubko, V., and Mims, A.: A physical approach for a simultaneous retrieval of sounding, surface, hydrometeor, and cryospheric parameters from SNPP/ATMS, J. Geophys. Res., 118, 12–600, 2013.

Bouniol, D., Delanoë, J., Duroure, C., Protat, A., Giraud, V., and Penide, G.: Microphysical characterisation of West African MCS anvils, Q. J. Roy. Meteor. Soc., 136, 323–344, 2010.

Brath, M., Fox, S., Eriksson, P., Harlow, R. C., Burgdorf, M., and Buehler, S. A.: Retrieval of an ice water path over the ocean from ISMAR and MARSS millimeter and submillimeter brightness temperatures, Atmos. Meas. Tech., 11, 611–632, https://doi.org/10.5194/amt-11-611-2018, 2018.

Buehler, S. A., Jiménez, C., Evans, K. F., Eriksson, P., Rydberg, B., Heymsfield, A. J., Stubenrauch, C., Lohmann, U., Emde, C., John, V. O., Sreerekha, T. R., and Davis, C. P.: A concept for a satellite mission to measure cloud ice water path and ice particle size, Q. J. Roy. Meteor. Soc., 133, 109–128, https://doi.org/10.1002/qj.143, 2007a.

Buehler, S. A., Kuvatov, M., Sreerekha, T. R., John, V. O., Rydberg, B., Eriksson, P., and Notholt, J.: A cloud filtering method for microwave upper tropospheric humidity measurements, Atmos. Chem. Phys., 7, 5531–5542, https://doi.org/10.5194/acp-7-5531-2007, 2007b.

Buehler, S. A., Defer, E., Evans, F., Eliasson, S., Mendrok, J., Eriksson, P., Lee, C., Jiménez, C., Prigent, C., Crewell, S., Kasai, Y., Bennartz, R., and Gasiewski, A. J.: Observing ice clouds in the submillimeter spectral range: the CloudIce mission proposal for ESA's Earth Explorer 8, Atmos. Meas. Tech., 5, 1529–1549, https://doi.org/10.5194/amt-5-1529-2012, 2012.

Buehler, S. A., Mendrok, J., Eriksson, P., Perrin, A., Larsson, R., and Lemke, O.: ARTS, the Atmospheric Radiative Transfer Simulator – version 2.2, the planetary toolbox edition, Geosci. Model Dev., 11, 1537–1556, https://doi.org/10.5194/gmd-11-1537-2018, 2018.

Côté, J., Gravel, S., Méthot, A., Patoine, A., Roch, M., and Staniforth, A.: The operational CMC-MRB Global Environmental Multiscale (GEM) Model. Part I: Design considerations and formulation, Mon. Weather Rev., 126, 1373–1395, https://doi.org/10.1175/1520-0493(1998)126<1373:TOCMGE>2.0.CO;2, 1998.

Davis, C., Emde, C., and Harwood, R.: A 3D polarized reversed Monte Carlo radiative transfer model for mm and sub-mm passive remote sensing in cloudy atmospheres, IEEE T. Geosci. Remote, 43, 1096–1101, https://doi.org/10.1109/TGRS.2004.837505, 2005.

Davis, P. J. and Rabinowitz, P.: Methods of numerical integration, Academic Press, New York, 1984.

Ding, J., Bi, L., Yang, P., Kattawar, G. W., Weng, F., Liu, Q., and Greenwald, T.: Single-scattering properties of ice particles in the microwave regime: Temperature effect on the ice refractive index with implications in remote sensing, J. Quant. Spectrosc. Ra., 190, 26–37, 2017.

Doviak, R. J. and Zrnic, D. S.: Doppler Radar & Weather Observations, Academic press, 2014.

Draine, B. T. and Flatau, P. J.: Discrete-Dipole Approximation For Scattering Calculations, J. Climate, 11, 1491–1499, https://doi.org/10.1364/JOSAA.11.001491, 1994.

Ekelund, R., Brath, M., Eriksson, P., and Mendrok, J.: Scattering properties of frozen hydrometeors at MW and sub-mm wavelengths, Tech. rep., Final report of EUMETSAT Contract No EUM/C0/16/4600001742/CJA, 2017.

Ekelund, R., Brath, M., Eriksson, P., and Mendrok, J.: Technical Report: ARTS Microwave Single Scattering Properties Database, Tech. rep., Dept. of Space, Earth and Environment, Chalmers University of Technology, available at: https://zenodo.org/record/1175573#.WzIXkJ9fjCI (last access: 11 July 2018), 2018a.

Ekelund, R., Brath, M., Mendrok, J., and Eriksson, P.: ARTS Microwave Single Scattering Properties Database (Version 1.0.0) [Data set], Zenodo, https://doi.org/10.5281/zenodo.1175573 (last access: 11 July 2018), 2018b.

Ellison, W. J.: Permittivity of pure water, at standard atmospheric pressure, over the frequency range 0–25 THz and the temperature range 0–100°C, J. Phys. Chem. Ref. Data, 36, 1–18, https://doi.org/10.1063/1.2360986, 2007.

Ellison, W. J., Lamkaouchi, K., and Moreau, J.-M.: Water: a dielectric reference, J. Mol. Liq., 68, 171–279, https://doi.org/10.1016/0167-7322(96)00926-9, 1996.

Eresmaa, R. and McNally, A. P.: Diverse profile datasets from the ECMWF 137-level short-range forecasts, Tech. rep., ECMWF, https://doi.org/10.13140/2.1.4476.8963, 2014.

Eriksson, P., Ekström, M., Rydberg, B., Wu, D. L., Austin, R. T., and Murtagh, D. P.: Comparison between early Odin-SMR, Aura MLS and CloudSat retrievals of cloud ice mass in the upper tropical troposphere, Atmos. Chem. Phys., 8, 1937–1948, https://doi.org/10.5194/acp-8-1937-2008, 2008.

Eriksson, P., Buehler, S. A., Davis, C. P., Emde, C., and Lemke, O.: ARTS, the atmospheric radiative transfer simulator, Version 2, J. Quant. Spectrosc. Ra., 112, 1551–1558, https://doi.org/10.1016/j.jqsrt.2011.03.001, 2011a.

Eriksson, P., Rydberg, B., and Buehler, S. A.: On cloud ice induced absorption and polarisation effects in microwave limb sounding, Atmos. Meas. Tech., 4, 1305–1318, https://doi.org/10.5194/amt-4-1305-2011, 2011b.

Eriksson, P., Rydberg, B., Sagawa, H., Johnston, M. S., and Kasai, Y.: Overview and sample applications of SMILES and Odin-SMR retrievals of upper tropospheric humidity and cloud ice mass, Atmos. Chem. Phys., 14, 12613–12629, https://doi.org/10.5194/acp-14-12613-2014, 2014.

Eriksson, P., Jamali, M., Mendrok, J., and Buehler, S. A.: On the microwave optical properties of randomly oriented ice hydrometeors, Atmos. Meas. Tech., 8, 1913–1933, https://doi.org/10.5194/amt-8-1913-2015, 2015.

Evans, K. F. and Stephens, G. L.: Microwave radiative transfer through clouds composed of realistically shaped ice crystals. Part II. Remote sensing of ice clouds, J. Atmos. Sci., 52, 2058–2072, 1995.

Evans, K. F., Wang, J. R., O'C Starr, D., Heymsfield, G., Li, L., Tian, L., Lawson, R. P., Heymsfield, A. J., and Bansemer, A.: Ice hydrometeor profile retrieval algorithm for high-frequency microwave radiometers: application to the CoSSIR instrument during TC4, Atmos. Meas. Tech., 5, 2277–2306, https://doi.org/10.5194/amt-5-2277-2012, 2012.

Field, P. R., Heymsfield, A. J., and Bansemer, A.: Snow size distribution parameterization for midlatitude and tropical ice clouds, J. Atmos. Sci., 64, 4346–4365, 2007.

Fox, S., Lee, C., Moyna, B., Philipp, M., Rule, I., Rogers, S., King, R., Oldfield, M., Rea, S., Henry, M., Wang, H., and Harlow,

R. C.: ISMAR: an airborne submillimetre radiometer, Atmos. Meas. Tech., 10, 477–490, https://doi.org/10.5194/amt-10-477-2017, 2017.

Garrett, T. J., Yuter, S. E., Fallgatter, C., Shkurko, K., Rhodes, S. R., and Endries, J. L.: Orientations and aspect ratios of falling snow, Geophys. Res. Lett., 42, 4617–4622, https://doi.org/10.1002/2015GL064040, 2015.

Geer, A. J. and Baordo, F.: Improved scattering radiative transfer for frozen hydrometeors at microwave frequencies, Atmos. Meas. Tech., 7, 1839–1860, https://doi.org/10.5194/amt-7-1839-2014, 2014.

Gong, J. and Wu, D. L.: Microphysical properties of frozen particles inferred from Global Precipitation Measurement (GPM) Microwave Imager (GMI) polarimetric measurements, Atmos. Chem. Phys., 17, 2741–2757, https://doi.org/10.5194/acp-17-2741-2017, 2017.

Guerbette, J., Mahfouf, J.-F., and Plu, M.: Towards the assimilation of all-sky microwave radiances from the SAPHIR humidity sounder in a limited area NWP model over tropical regions, Tellus A, 68, 28620, https://doi.org/10.3402/tellusa.v68.28620, 2016.

Hogan, R. J., Illingworth, A. J., and Sauvageot, H.: Measuring crystal size in cirrus using 35-and 94-GHz radars, J. Atmos. Ocean. Tech., 17, 27–37, 2000.

Holl, G., Eliasson, S., Mendrok, J., and Buehler, S.: SPARE-ICE: Synergistic ice water path from passive operational sensors, J. Geophys. Res., 119, 1504–1523, 2014.

Hong, G.: Parameterization of scattering and absorption properties of nonspherical ice crystals at microwave frequencies, J. Geophys. Res., 112, D11208, https://doi.org/10.1029/2006JD008364, 2007.

Hong, G., Yang, P., Baum, B. A., Heymsfield, A. J., Weng, F., Liu, Q., Heygster, G., and Buehler, S. A.: Scattering database in the millimeter and submillimeter wave range of 100–1000 GHz for nonspherical ice particles, J. Geophys. Res., 114, D06201, https://doi.org/10.1029/2008JD010451, 2009.

Hufford, G.: A model for the complex permittivity of ice at frequencies below 1 THz, Int. J. Infrared Milli., 12, 677–683, 1991.

Illingworth, A. J., Barker, H., Beljaars, A., Ceccaldi, M., Chepfer, H., Clerbaux, N., Cole, J., Delanoë, J., Domenech, C., Donovan, D. P., Fukuda, S., Hirakata, M.,Hogan, R. J., Huenerbein, A., Kollias, P., Kubota, T., Nakajima, T., Nakajima, T. Y., Nishizawa, T., Ohno, Y., Okamoto, H., Oki, R., Sato, K., Satoh, M., Shephard, M. W., Velázquez-Blázquez, A., Wandinger, U., Wehr, T., and van Zadelhoff, G.-J.: The EarthCARE satellite: The next step forward in global measurements of clouds, aerosols, precipitation, and radiation, B. Am. Meteorol. Soc., 96, 1311–1332, 2015.

Iwabuchi, H. and Yang, P.: Temperature dependence of ice optical constants: Implications for simulating the single-scattering properties of cold ice clouds, J. Quant. Spectrosc. Ra., 112, 2520–2525, https://doi.org/10.1016/j.jqsrt.2011.06.017, 2011.

Johnson, B. T., Olson, W. S., and Skofronick-Jackson, G.: The microwave properties of simulated melting precipitation particles: sensitivity to initial melting, Atmos. Meas. Tech., 9, 9–21, https://doi.org/10.5194/amt-9-9-2016, 2016.

Kajikawa, M.: Observation of the falling motion of early snowflakes, Pt. 1, Relationship between the free-fall pattern and the number and shape of component snow crystals, J. Meteorol. Soc. Jpn. Ser. II, 60, 797–803, 1982.

Kangas, V., D'Addio, S., Betto, M., Barre, H., Loiselet, M., and Mason, G.: Metop second generation microwave sounding and microwave imaging missions, in: 2012 EUMETSAT Meteorological Satellite Conference, 3–7, 2012.

Kneifel, S., Kulie, M. S., and Bennartz, R.: A triple-frequency approach to retrieve microphysical snowfall parameters, J. Geophys. Res., 116, D11203, https://doi.org/10.1029/2010JD015430, 2011.

Kneifel, S., Lerber, A., Tiira, J., Moisseev, D., Kollias, P., and Leinonen, J.: Observed relations between snowfall microphysics and triple-frequency radar measurements, J. Geophys. Res.-Atmos., 120, 6034–6055, 2015.

Kollias, P., Clothiaux, E., Miller, M., Albrecht, B., Stephens, G., and Ackerman, T.: Millimeter-wavelength radars: New frontier in atmospheric cloud and precipitation research, B. Am. Meterol. Soc., 88, 1608–1624, 2007.

Kulie, M. S., Hiley, M. J., Bennartz, R., Kneifel, S., and Tanelli, S.: Triple-Frequency Radar Reflectivity Signatures of Snow: Observations and Comparisons with Theoretical Ice Particle Scattering Models, J. Appl. Meteorol. Clim., 53, 1080–1098, https://doi.org/10.1175/JAMC-D-13-066.1, 2014.

Kunzi, K., Bauer, P., Eresmaa, R., Eriksson, P., Healy, S. B., Mugnai, A., Livesey, N., Prigent, C., Smith, E. A., and Stephens, G.: Microwave absorption, emission and scattering: trace gases and meteorological parameters, in: The remote sensing of tropospheric composition from space, edited by: Burrows, J. P., Platt, U., and Borrell, P., Physics of Earth and Space Environments, Springer Berlin Heidelberg, 153–230, https://doi.org/10.1007/978-3-642-14791-3_4, 2011.

Kuo, K.-S., Olson, W. S., Johnson, B. T., Grecu, M., Tian, L., Clune, T. L., van Aartsen, B. H., Heymsfield, A. J., Liao, L., and Meneghini, R.: The microwave radiative properties of falling snow derived from nonspherical ice particle models. Part I: An extensive database of simulated pristine crystals and aggregate particles, and their scattering properties, J. Appl. Meteorol. Clim., 55, 691–708, 2016.

Leinonen, J. and Szyrmer, W.: Radar signatures of snowflake riming: A modeling study, Earth Space Sci., 2, 346–358, https://doi.org/10.1002/2015EA000102, 2015.

Liou, K. N.: An Introduction to Atmospheric Radiation, Academic Press, 2002.

Liu, G.: A database of microwave single-scattering properties for nonspherical ice particles, B. Am. Meteorol. Soc., 89, 1563, https://doi.org/10.1175/2008BAMS2486.1, 2008.

Lu, Y., Jiang, Z., Aydin, K., Verlinde, J., Clothiaux, E. E., and Botta, G.: A polarimetric scattering database for non-spherical ice particles at microwave wavelengths, Atmos. Meas. Tech., 9, 5119–5134, https://doi.org/10.5194/amt-9-5119-2016, 2016.

Lynch, D. K., Sassen, K., Starr, D., and Stephens, G. (Eds.): Cirrus, Oxford University Press, 2002.

Magono, C. and Lee, C. W.: Meteorological classification of natural snow crystals, Journal of the Faculty of Science, Hokkaido University, Series 7, Geophysics, 2, 321–335, 1966.

Maruyama, K.-I. and Fujiyoshi, Y.: Monte Carlo simulation of the formation of snowflakes, J. Atmos. Sci., 62, 1529–1544, https://doi.org/10.1175/JAS3416.1, 2005.

Mätzler, C.: MATLAB functions for Mie scattering and absorption, version 2, Tech. rep., IAP Res. Rep, University of Bern, 2002.

Mätzler, C.: Thermal microwave radiation: applications for remote sensing, vol. 52 of IET electromagnetic waves series, Institution of Engineering and Technology, 2006.

McFarquhar, G. M. and Heymsfield, A. J.: Parameterization of tropical cirrus ice crystal size distribution and implications for radiative transfer: Results from CEPEX, J. Atmos. Sci., 54, 2187–2200, 1997.

Mendrok, J., Eriksson, P., Ekelund, R., and Brath, M.: ARTS Microwave Single Scattering Properties Database Interfaces (Version 1.0.0) [Software], Zenodo, available at: https://doi.org/10.5281/zenodo.1175589 (last access" 11 July 2018), 2018.

Min, M., Hovenier, J., Dominik, C., de Koter, A., and Yurkin, M.: Absorption and scattering properties of arbitrarily shaped particles in the Rayleigh domain: A rapid computational method and a theoretical foundation for the statistical approach, J. Quant. Spectrosc. Ra., 97, 161–180, https://doi.org/10.1016/j.jqsrt.2005.05.059, 2006.

Mishchenko, M. I. and Travis, L. D.: Capabilities and limitations of a current FORTRAN implementation of the T-matrix method for randomly oriented, rotationally symmetric scatterers, J. Quant. Spectrosc. Ra., 60, 309–324, 1998.

Mishchenko, M. I., Travis, L., and Lacis, A.: Scattering, absorption, and emission of light by small particles, Cambridge University Press, Cambridge, UK, 2002.

Mitchell, D. L.: Use of mass- and area-dimensional power laws for determining precipitation particle terminal velocities, J. Atmos. Sci., 53, 1710–1723, https://doi.org/10.1175/1520-0469(1996)053<1710:UOMAAD>2.0.CO;2, 1996.

Mitchell, D. L. and Arnott, W. P.: A model predicting the evolution of ice particle size spectra and radiative properties of cirrus clouds. Part II: Dependence of absorption and extinction on ice crystal morphology, J. Atmos. Sci., 51, 817–832, https://doi.org/10.1175/1520-0469(1994)051<0817:AMPTEO>2.0.CO;2, 1994.

Ori, D., Maestri, T., Rizzi, R., Cimini, D., Montopoli, M., and Marzano, F.: Scattering properties of modeled complex snowflakes and mixed-phase particles at microwave and millimeter frequencies, J. Geophys. Res., 119, 9931–9947, 2014.

O'Shea, S., Choularton, T., Lloyd, G., Crosier, J., Bower, K., Gallagher, M., Abel, S., Cotton, R., Brown, P., Fugal, J., Schlenczek, O., Borrmann, S., and Pickering, J. C.: Airborne observations of the microphysical structure of two contrasting cirrus clouds, J. Geophys. Res., 121, 13510–1353, https://doi.org/10.1002/2016JD025278, 2016.

Piyush, D. N., Goyal, J., and Srinivasan, J.: Retrieval of cloud ice water path using SAPHIR on board Megha-Tropiques over the tropical ocean, Adv. Space. Res., 59, 1895–1906, 2017.

Rathsman, T.: A software toolkit for generating ice and snow particle sharp data, Master's thesis, Msc. thesis, Earth and Space Sciences, Chalmers University of Technology, Gothenburg, Sweden, available at: (last access: 11 July 2018), 2016.

Ray, P.: Broadband complex refractive indices of ice and water, Appl. Optics, 11, 1836–1844, 1972.

Sassen, K., Matrosov, S., and Campbell, J.: CloudSat spaceborne 94 GHz radar bright bands in the melting layer: An attenuation-driven upside-down lidar analog, Geophys. Res. Lett., 34, L16818, https://doi.org/10.1029/2007GL030291, 2007.

Schmitt, C. G. and Heymsfield, A. J.: Observational quantification of the separation of simple and complex atmospheric ice particles, Geophys. Res. Lett., 41, 1301–1307, 2014.

Stephens, G. L., Vane, D. G., Tanelli, S., Im, E., Durden, S., Rokey, M., Reinke, D., Partain, P., Mace, G. G., Austin, R., L'Ecuyer, T., Haynes, J., Lebsock, M., Suzuki, K., Waliser, D., Wu, D., Kay, J., Gettelman, A., Wang, Z., and Marchand, R.: CloudSat mission: Performance and early science after the first year of operation, J. Geophys. Res., 113, D00A18, https://doi.org/10.1029/2008JD009982, 2008.

Turner, D. D., Kneifel, S., and Cadeddu, M. P.: An improved liquid water absorption model at microwave frequencies for supercooled liquid water clouds, J. Atmos. Ocean. Tech., 33, 33–44, https://doi.org/10.1175/JTECH-D-15-0074.1, 2016.

Tyynelä, J.: Radar backscattering from snowflakes: Comparison of fractal, aggregate, and soft spheroid models, J. Atmos. Ocean. Tech., 28, 1365–1372, https://doi.org/10.1175/JTECH-D-11-00004.1, 2011.

Tyynelä, J. and Chandrasekar, V.: Characterizing falling snow using multifrequency dual-polarization measurements, J. Geophys. Res., 119, 8268–8283, https://doi.org/10.1002/2013JD021369, 2014.

Warren, S.: Optical constants of ice from the ultraviolet to the microwave, Appl. Optics, 23, 1206–1225, https://doi.org/10.1364/AO.23.001206, 1984.

Warren, S. G. and Brandt, R. E.: Optical constants of ice from the ultraviolet to the microwave: A revised compilation, J. Geophys. Res., 113, D14220, https://doi.org/10.1029/2007JD009744, 2008.

Weng, F., Zou, X., Wang, X., Yang, S., and Goldberg, M.: IIntroduction to Suomi national polar-orbiting partnership advanced technology microwave sounder for numerical weather prediction and tropical cyclone applications, J. Geophys. Res., 117, D19112, https://doi.org/10.1029/2012JD018144, 2012.

Wu, D., Esper, J., Ehsan, N., Johnson, T., Mast, W., Piepmeier, J. R., and Racette, P. E.: IceCube: cubeSat 883-GHz radiometry for future ice cloud remote sensing, in: Earth Science Technology Forum, Pasadena, California, 2015.

Wu, D. L., Jiang, J. H., Read, W. G., Austin, R. T., Davis, C. P., Lambert, A., Stephens, G. L., Vane, D. G., and Waters, J. W.: Validation of the Aura MLS cloud ice water content measurements, J. Geophys. Res., 113, D15S10, https://doi.org/10.1029/2007JD008931, 2008.

Yang, P. and Liou, K. N.: Single-scattering properties of complex ice crystals in terrestrial atmosphere, Contrib. Atmos. Phys., 71, 223–248, 1998.

Yang, P., Liou, K. N., Wyser, K., and Mitchell, D. L.: Parameterization of the scattering and absorption properties of individual ice crystals, J. Geophys. Res., 105, 4699–4718, https://doi.org/10.1029/1999JD900755, 2000.

Yin, M., Liu, G., Honeyager, R., and Joseph Turk, F.: Observed differences of triple-frequency radar signatures between snowflakes in stratiform and convective clouds, J. Quant. Spectrosc. Ra., 193, 13–20, https://doi.org/10.1016/j.jqsrt.2017.02.017, 2017.

Yurkin, M. A. and Hoekstra, A. G.: The discrete-dipole-approximation code ADDA: capabilities and known limitations, J. Quant. Spectrosc. Ra., 112, 2234–2247, 2011.

8

ReOBS: a new approach to synthesize long-term multi-variable dataset and application to the SIRTA supersite

Marjolaine Chiriaco[1], Jean-Charles Dupont[2], Sophie Bastin[1], Jordi Badosa[4], Julio Lopez[2], Martial Haeffelin[2], Helene Chepfer[3], and Rodrigo Guzman[3]

[1]LATMOS/IPSL, UVSQ Université Paris-Saclay, UPMC Univ. Paris 06, CNRS, Guyancourt, France

[2]Institute Pierre Simon Laplace, Ecole Polytechnique, Université Paris-Saclay, Palaiseau, France

[3]Laboratoire de Météorologie Dynamique, Univ. Pierre and Marie Curie, Paris, France

[4]Laboratoire de Météorologie Dynamique, Ecole Polytechnique, Palaiseau, France

Correspondence: Marjolaine Chiriaco (marjolaine.chiriaco@latmos.ipsl.fr)

Abstract. A scientific approach is presented to aggregate and harmonize a set of 60 geophysical variables at hourly timescale over a decade, and to allow multiannual and multi-variable studies combining atmospheric dynamics and thermodynamics, radiation, clouds and aerosols from ground-based observations. Many datasets from ground-based observations are currently in use worldwide. They are very valuable because they contain complete and precise information due to their spatio-temporal co-localization over more than a decade. These datasets, in particular the synergy between different type of observations, are under-used because of their complexity and diversity due to calibration, quality control, treatment, format, temporal averaging, metadata, etc. Two main results are presented in this article. (1) a set of methods available for the community to robustly and reliably process ground-based data at an hourly timescale over a decade is described and (2) a single netCDF file is provided based on the SIRTA supersite observations. This file contains approximately 60 geophysical variables (atmospheric and in ground) hourly averaged over a decade for the longest variables. The netCDF file is available and easy to use for the community. In this article, observations are "re-analyzed". The prefix "re" refers to six main steps: calibration, quality control, treatment, hourly averaging, homogenization of the formats and associated metadata, as well as expertise on more than a decade of observations. In contrast, previous studies (i) took only some of these six steps into account for each variable, (ii) did not aggregate all variables together in a single file and (iii) did not offer an hourly resolution for about 60 variables over a decade (for the longest variables). The approach described in this article can be applied to different supersites and to additional variables. The main implication of this work is that complex atmospheric observations are made readily available for scientists who are non-experts in measurements. The dataset from SIRTA observations can be downloaded at http://sirta.ipsl.fr/reobs.html (last access: April 2017) (Downloads tab, no password required) under https://doi.org/10.14768/4F63BAD4-E6AF-4101-AD5A-61D4A34620DE.

1 Introduction

The Intergovernmental Panel on Climate Change (IPCC) simulations show a large spread between models when predicting future climate at a global scale, but also when representing the observed current climate. These model uncertainties are larger at the regional scale and at short timescales (e.g., seasonal scale). These scales are, however, key for the impact assessment. For example, models do not reproduce observed magnitudes of interannual and seasonal variability and extremes in temperature and precipitation (Terray and Boé, 2013). Hawkins and Sutton (2009) also show that climate's natural variability is the main source of uncertainty to predict regional climate evolution at the timescale of 10–20 years (compared to the selected scenario or model). Observations of the atmosphere must be considered in order to improve both our knowledge of the processes that create this temporal variability and the simulation uncertainties. These observations must describe atmospheric processes that involve a large number of variables in the atmospheric columns and in the ground, and at various spatial and temporal scales.

Multiannual and multi-variable datasets are therefore necessary. Many of these datasets from ground-based observations have a significant scientific value because they contain complete and precise information on one or several decades, due to their spatio-temporal co-localization. Supersite observatories such as the Site Instrumental de Recherche par Télédétection Atmosphérqiue (SIRTA, Haeffelin et al., 2005) or the different Atmospheric Radiation Measurements (ARM, Ackerman et al., 2003) are among these sets of observations. But they are under-used, in particular the observed synergy aspects, because of their complexity and diversity in terms of calibration procedures, quality control, data treatment, file format, temporal representativeness, metadata etc., and because of the weak magnitude of the signals to be highlighted (e.g., trend versus natural variability), and also because of the complex connections between local-scale processes and climatic-scale anomalies (e.g., links between ground–boundary-layer–atmosphere processes and heat waves; as in Chiriaco et al., 2014).

An important homogenization work was needed for these observations. Homogenization has been performed for ARM observatories leading to the ARMBE (ARM Best Estimate) data product (Xie et al., 2010), which is the "ARM data streams specifically tailored to climate modelers for use in the evaluation of global climate models. They contain a best estimate of several cloud, radiation and atmospheric quantities. The ARMBE dataset was created to showcase all the flagship products of ARM" (from https://www.arm.gov/capabilities/vaps/armbe, last access: April 2017). A specificity of ARMBE products is that all variables are gathered in only two files: ARMBEATM (ATM for atmosphere) for many atmospheric state profiles and surface quantities, and ARMBECLDRAD (CLDRAD for cloud and radiation) that contains a best estimate of several selected ARM and satellite-measured cloud- and radiation-relevant quantities.

In this article, additional steps are applied to the observations and precisely described in order to understand how the observations are "re-analyzed". This method is called ReOBS. The prefix "Re" refers to six main steps on more than a decade of observations: calibration, quality control, algorithmic treatment, hourly averaging, homogenization of the data formats and associated metadata, as well as scientist expertise. In contrast, previous studies (i) only take into account some of these six steps for each variable, (ii) do not aggregate together all variables in a single file and (iii) do not offer an hourly resolution for about 60 variables over a decade (for the oldest variables).

The ReOBS method was initially inspired by the ARMBE project and has been developed at SIRTA (located 20 km southwest of Paris, France). The SIRTA observatory has been collecting data for 15 years from active and passive remote sensing, in situ measurements at the surface, in the ground and in the planetary boundary layer (PBL). Early versions of the SIRTA ReOBS dataset have already been used in scientific studies that required the multi-variables and multi-temporal scales available in the SIRTA-ReOBS dataset (Cheruy et al., 2012; Chiriaco et al., 2014; Pal and Haeffelin, 2015, Bastin et al., 2016; Dione et al., 2017). The ReOBS method has also been tested for other supersites for some variables (classical meteorology, radiative fluxes, heat fluxes): Cabauw (in the Netherlands) and Chilbolton (in England) supersites in the framework of the EUCLIPSE European project (European Union Cloud Intercomparison, Process Study and Evaluation project), CO-PDD (Cézeaux – Opme – Puy De Dôme at Clermont Ferrand in France) and P2OA (Plateforme Pyrénéenne d'Observations Atmosphériques at Lannemezan in France) in Dione et al. (2016).

The objective of the current paper is to present a scientific approach (ReOBS) to aggregate and harmonize about 60 geophysical variables at hourly timescale over at least a decade, and to study atmospheric dynamics and thermodynamics, radiation, clouds and aerosols from ground-based observations. This paper presents two main results: (1) a set of methods available for the community to process ground-based data robustly and reliably at an hourly timescale over at least a decade and (2) provision of a single netCDF file containing about 60 substantial geophysical variables hourly averaged over 15 years for the oldest ones, and easily usable for the community.

The SIRTA observations used for applying the ReOBS method are described in Sect. 2. The method used for ReOBS is then detailed in Sect. 3. Section 4 presents the contents of the SIRTA-ReOBS file and its major strengths: the vertical profiles, the multi-temporal scales and the multi-parameter specificity. Discussion and conclusions are presented in Sect. 5.

2 Observations

2.1 SIRTA observatory

SIRTA is a French national observatory dedicated to the monitoring of tropospheric clouds and aerosols, the dynamics and thermodynamics of the boundary layer, and the turbulent and organized transport of water and energy near the surface. The SIRTA observatory is a mid-latitude site (48.71° N, 2.2° E) located in a semi-urban area on the Saclay plateau 20 km southwest of Paris, and hosts active and passive remote-sensing instruments since 2002 (Haeffelin et al., 2005). The SIRTA missions are (1) to monitor continuously and in the long-term the atmospheric column using a core ensemble of instruments; (2) to coordinate field campaigns in order to address specific scientific questions, such as processes related to water vapor and clouds, the ultraviolet radiation, or the aerosol physics and chemistry; and (3) to provide teaching resources and to host experimental training activities.

Figure 1 shows a selection of SIRTA routine measurements from the different on-site locations (e.g., roof, mast, plain). The measurements used in the current study are listed in Table 1. Lidars play a special role in the SIRTA instrumental park because several lidars have been deployed at the SIRTA observatory over the past 15 years, providing a unique 3-D database: (1) a dual-wavelength (532 and 1064 nm) depolarization lidar (called LNA for "cloud and aerosol lidar", used in the current study) from 2002 until 2015 (Haeffelin et al., 2005), (2) a multi-wavelength elastic (355, 532, 1064 nm) and Raman (387, 408, 607 nm) depolarization lidar (called IPRAL for "IPSL Hi-Performance multi-wavelength Raman Lidar for Cloud Aerosol Water Vapor Research") since mid-2015, (3) an automatic 355 nm backscatter and depolarization lidar (Leosphere ALS 450, used in this study) from 2008 until 2014 and (4) an automatic 1064 nm lidar ceilometer (Lufft CHM15k) since mid-2015. The different lidars differ significantly in complexity, emitted power, detection channels, signal-to-noise ratio and capacity to operate autonomously. For instance, the LNA backscattered signal provides information on the presence of clouds and aerosols in the vertical column between 0.5 and 15 km altitude, whereas the ALS 450 backscatter lidar signal is exploited between 0.2 and 10 km.

2.2 SIRTA measurements used as inputs for ReOBS

Table 1 shows the measurements used as inputs to create the SIRTA-ReOBS file. The table contains the instruments name, the physical bounds of the measurements, the native resolution of the measurements and the available period of observation. This set of variables includes in situ measurements (1–6 and 11–14 in Table 1), passive remote-sensing measurements (7–10 and 17–20) and active remote-sensing measurements (15–16 in Table 1).

These different measurements are used to create the geophysical variables listed in Table 2. Some of the geophysical variables are directly measured, and some others require advanced data processing, such as substantial quality control or algorithm application. Data processing performed independently of the ReOBS processing chain and already published is described and referenced in Table 2. The data processing developed in the framework of the ReOBS project is described in Sect. 3.

In the rest of the article, the geophysical variables are split into four groups. Group A contains the *standard meteorology variables* (first block in Table 2), such as 2 m temperature, pressure, wind speed and direction, relative humidity, etc. Group B contains the *advanced non-standard meteorology variables* (second block in Table 2), such as radiative fluxes, heat fluxes, in-ground temperature and moisture, etc. These latter variables are directly measured but are usually not available from typical weather stations because they require advanced technologies, for instance based on remote sensing. Group C contains *variables retrieved from measurements* using algorithms applied to remote-sensing measurements (third block in Table 2), such as cloud fraction, water vapor content, etc. Finally, group D contains atmospheric vertical profiles from lidar (fourth block in Table 2).

3 The ReOBS method

3.1 ReOBS general processing chain

The 15-year-long SIRTA-ReOBS dataset is contained in a single netCDF file containing hourly values of 63 physical variables listed in Table 2. The short and standard name used for each variable in the ReOBS dataset follows the Coupled Model Intercomparison Project (CMIP) and the Climate Forecast (CF) conventions, respectively, when available. For variables not included in CMIP or CF conventions, classical names are used or new ones are created.

The strength of the ReOBS dataset is that all variables are processed using the same high-level processing chain, completed by some sub-processing computations specific to each variable. Figure 2 shows the ReOBS processing chain (in blue in Fig. 2), which starts after the acquisition process (in orange in Fig. 2). Steps outside of the ReOBS processing chain are marked in green.

For each variable (except lidar profiles), the hourly mean values are calculated from the native resolution data (5 s to 1 min) by averaging all the data available within ±30 min around the full hour in order to be consistent with outputs from global circulation models (GCMs) and regional climate models (RCMs). Each hourly variable is completed by its intra-hour standard deviation. The hourly standard deviation (SD) of each variable helps in detecting non-physical spikes (i.e., successive increase and decrease) and dips (i.e., successive decrease and increase in the signal). This temporal variability information is also useful to document large changes

Figure 1. Photos of the routine instruments at the SIRTA supersite.

in the atmospheric conditions such as a cold front for air temperature, broken clouds for radiative fluxes and summer storms for precipitations or latent heat fluxes.

Variables entering the ReOBS dataset are quality-controlled at their native time resolution. The quality control test consists of verifying that the variable lies within physical bounds (Table 1). Calculations of hourly mean and standard deviation only use native resolution data that have passed quality control. A simple informative quality flag is associated with each hourly value of a variable:

- 0: quality control is OK.
- 1: there are valid data but for less than 50 % of the period (that is, for less than 30 min).
- 2: flag 2 is only used for internal control and is never used as an informative output in the ReOBS file.
- 3: data is unavailable for the entire hour (no measurements or less than 50 % of the measurements in the hour passes the quality control). In this case, the hourly value is set by convention to −999.96.

Besides the systematic quality tests described above, some additional complementary quality controls have been applied to specific variables, as described in the following subsections (Sects. 3.2 to 3.5).

3.2 Specific computations for standard meteorological variables

Classical meteorological variables collected at three different locations are included in the ReOBS dataset: (1) the first group of variables is collected at the SIRTA supersite and has the advantage of being representative of the very local meteorology since the beginning of the supersite activities, (2) the second group of variables aims at characterizing the surrounding meteorology around the SIRTA site and (3) the third group of variables is from the standardized Météo-France station, collected at Trappes, 15 km away from the SIRTA supersite. These three different datasets are identified

Table 1. List of variables measured at SIRTA and used as inputs for ReOBS.

Measured variable, unit	Instrument	Reference	Physical bounds – Sensor uncertainty	Native resolution	Period of observation
(1) 2 m air temperature, K	Platinum resistance thermometer (PT-100 sensors)	Haeffelin et al., 2005	−30/50 °C – 0.2 °C	5 s	2003–2016
(2) 2 m relative humidity, %	HMP110 hygrometer		3/103 % – 2 %	5 s	2003–2016
(3) Pressure, Pa	PTB110 barometer		850/1050 hPa – 0.2 hPa	5 s	2003–2016
(4) 2 m wind speed, $m\,s^{-1}$	A100R cup anemometer		$0/40\,m\,s^{-1}$ – $0.2\,m\,s^{-1}$	5 s	2003–2016
(5) 2 m wind direction, °	W200P wind vane		0–360°	5 s	2003–2016
(6) Precipitation at surface, $kg\,m^{-2}\,s^{-1}$	R3070 rain gauge		$0/50\,mm\,h^{-1}$ – 0.1 mm	5 s	2003–2016
(7) Surface downwelling LW radiation, $W\,m^{-2}$	CG4 or CGR4 pyrgeometers	Ohmura et al., 1998 + BSRN procedures: McArthur, 2004	$100/500\,W\,m^{-2}$ – $4\,W\,m^{-2}$	1 s	2003–2016
(8) Surface downwelling SW radiation, $W\,m^{-2}$	Diffuse: Kipp & Zonen CMP22 or CM22 pyranometers Direct: CH1 or CHP1 pyrheliometers		$-5/1200\,W\,m^{-2}$ – $5\,W\,m^{-2}$	1 s	2003–2016
(9) Surface upwelling LW radiation, $W\,m^{-2}$	cg2 30 m above ground		$250/500\,W\,m^{-2}$ – $8\,W\,m^{-2}$	10 s	2007–2016
(10) Surface upwelling SW radiation, $W\,m^{-2}$	cm21 30 m above ground		$-5/400\,W\,m^{-2}$ – $10\,W\,m^{-2}$	10 s	2007–2016
(11) Soil temperature x^{a} cm below ground[a], K	Platinum Resistance Thermometer (PT-100 sensors)	–	−30/50 °C	5 s	2007–2016
(12) Soil moisture x^{a} cm below ground[a], $g\,cm^{-3}$	Capacitive sensor (ML2x model from Delta-T Devices)	Roth et al. (1992)	0.05–0.6 $m^3\,m^{-3}$	5 s	2007–2016
(13) 3-D wind velocities and virtual air temperature, $m\,s^{-1}$	METEK (USA-1 standard model) sonic anenometer	Wieser et al. (2001)	$0–30\,m\,s^{-1}$, $0.02\,m\,s^{-1}$	10 Hz	2006–2016
(14) water vapor fluctuations, ppt	Open-Path Krypton hygrometer IRGA (Infrared Gas Analyzer)		0–60 ppt, 2 %	10 Hz	
(15) lidar backscattered signal, –	Leosphere automatic lidar (355 nm)	Haeffelin et al. (2011)	–	30 s, 15 m vertical	2008–2013
(16) lidar backscattered signal, –	LNA lidar (532 and 1064 nm)	Haeffelin et al. (2005)	–	30 s, 15 m vertical	2003–2016

Table 1. Continued.

Measured variable, unity	Instrument	Reference	Physical bounds – Sensor uncertainty	Native resolution	Period of observation
(17) 360° sky image, –	Yankee Environmental System Total Sky Imager (TSI)	Long et al. (1998)	–	1 min	2009–2016
(18) 440–870 nm spectral irradiance	Cimel sunphotometer	Dubovik et al. (2000)	–	when sun disc is visible	2008–2016
(19) zenith path delay (ZPD), s	GPS	Champolion et al. (2004)	–	15 min	2008–2016
(20) liquid water path	RPG-HATPRO microwave radiometer	Rose et al. (2005)	–	1 s	2010–2016

[a] x is 5, 10, 20, 30, 50 cm.

with the suffixes -sirta, -regional and -trps, respectively, in the following.

3.2.1 Description of surrounding meteorology around the SIRTA site

Figure 3b–d illustrate the air temperature, wind speed, and cumulated precipitations probability density functions (PDFs) in relative occurrence at three Météo-France stations within a 50 × 50 km domain around the SIRTA supersite: in Trappes (48.8° N, 2.0° W), in Paris-Montsouris (48.8° N, 2.3° W) and in Orly (48.7° N 2.4° W); these plots highlight the eventual differences from one site to another.

The PDF of the 2 m air temperature (noted tas in SIRTA-ReOBS) shows an offset of about 2 °C for the Paris-Montsouris site compared to the Orly, Trappes and SIRTA sites, which is due to the urban heat. Maximum values for the mean wind speed value (noted sfcWind in SIRTA-ReOBS) are measured at the Orly site and the mean wind speed is around $3\,\mathrm{m\,s^{-1}}$ at SIRTA. Note that measurements at the SIRTA site are performed over a roof: wind speed is thus measured at 10 m above the roof, corresponding to 25 m above ground level, whereas it is measured at 10 m above ground level for the other stations. Even if a ground level standard (Météo-France-like) meteorological station is also present at SIRTA, the rooftop measurements were preferred for the ReOBS file because they started earlier (in 2003) than the standard meteorological station (in 2006). The four stations are characterized with a cumulated annual precipitation ranging between 600 and $700\,\mathrm{mm\,yr^{-1}}$.

The data collected at the three stations around the SIRTA supersite are used to characterize the surrounding 2 m meteorology. A weight is assigned to each of the three stations based on the following method: the 50 km× 50 km domain is divided into 90×10^3 grid boxes (300 × 300), the distance between each box and each site is calculated and then each box is linked to its nearest site. Then the percentage number of boxes linked to each site gives the weight of the site within the domain. The weight of the Trappes station is then 44.4 %, the weight of the Orly station is 34.5 % and the weight of the Paris-Montsouris station is 21.1 % (Fig. 3a). The regional-scale meteorology variables v (−REG) included in ReOBS are then obtained from

$$\bar{v} = \sum_{i}^{n} x_i w_i, \tag{1}$$

where $x = \{x_1, \ldots, x_4\}$ is the set of values taken by a variable v (2 m temperature, humidity, wind, etc.) at each of the four stations, and $w = \{w_1, \ldots, w_4\}$ is the station weight.

3.2.2 Quality control of the standard meteorological variables

The quality control for meteorological variables listed in Table 2 consists of two additional tests compared to what was indicated in Sect. 3.1. The goal of the quality control is to reject unphysical values and to reject values with unrealistic temporal variability (Tables 1 and 3), e.g., non-physical jump in the data record, non-physical persistence in time of the measured values.

Non-physical jumps in the data are detected at native high temporal resolution (5 s). If the difference between two successive measurements is more than a specified limit given in Table 3 (these tests are about to be refined in a new study that will give a new version of the SIRTA-ReOBS file) the current measurement is rejected but it is used for checking the temporal consistency with the next measurement. Two examples of measurements that did not pass the quality control tests are shown in Fig. 4a and b for pressure and soil temperature jumps, respectively. In the first example, an unphysical change of 2 hPa within 1 min is observed in pressure (larger than 5 hPa during 5 min, see Table 3). In the second example,

Table 2. Variables included in SIRTA-ReOBS. First block (category A) is for classical meteorological measurements, second block (category B) is for more advanced measurements, third block (category C) is for parameters retrieved from observations and fourth block (category D) is for vertical lidar measurements.

	Variable, unit	ReOBS short name	Based on Table 1 variables	Treatment before ReOBS processing chain
A	SIRTA 2 m air temperature, K	tas_sirta	(1)	Direct measurement
	SIRTA 2 m relative humidity, %	hurs_sirta	(2)	Direct measurement
	SIRTA 2 m specific humidity, $kg\,kg^{-1}$	huss_sirta	(2)	Simply derived from (2)
	SIRTA Sea-level pressure, Pa	psl_sirta	(3)	Simply derived from (3)
	SIRTA 2 m wind speed, $m\,s^{-1}$	sfcWind_sirta	(4)	Direct measurement
	SIRTA 2 m northward wind, $m\,s^{-1}$	vas_sirta	(4) (5)	Simply derived from (4) & (5)
	SIRTA 2 m eastward wind, $m\,s^{-1}$	uas_sirta	(4) (5)	Simply derived from (4) & (5)
	SIRTA precipitation at surface, $kg\,m^{-2}\,s^{-1}$	pr_sirta	(6)	Direct measurement
	Trappes 2 m air temperature, K	tas_trps	Meteo-FR	Direct measurement
	Trappes 2 m northward wind, $m\,s^{-1}$	vas_trps	Meteo-FR	Derived from wind speed and direction
	Trappes 2 m eastward wind, $m\,s^{-1}$	uas_trps	Meteo-FR	Derived from wind speed and direction
	Trappes precipitation at surface, $kg\,m^{-2}\,s^{-1}$	pr_trps	Meteo-FR	Direct measurement
	Regional 2 m air temperature, K	tas_regional	Meteo-FR	Weighted averaging
	Regional 2 m northward wind, $m\,s^{-1}$	vas_regional	Meteo-FR	Weighted averaging
	Regional 2 m eastward wind, $m\,s^{-1}$	uas_regional	Meteo-FR	Weighted averaging
	Regional precipitation at surface, $kg\,m^{-2}\,s^{-1}$	pr_regional	Meteo-FR	Weighted averaging
B	Surface downwelling LW radiation, $W\,m^{-2}$	rlds	(7)	Direct measurement
	Surface downwelling SW radiation, $W\,m^{-2}$	rsds	(8)	Direct measurement
	Surface upwelling LW radiation, $W\,m^{-2}$	rlus	(9)	Direct measurement
	Surface upwelling SW radiation, $W\,m^{-2}$	rsus	(10)	Direct measurement
	Soil temperature x^a cm below ground, K	stx^a	(11)	Direct measurement
	Soil moisture x^a cm below ground, $g\,cm^{-3}$	smx^a	(12)	Direct measurement

Table 2. Continued.

	Variable	ReOBS short name	Based on Table 1 variables	Treatment before ReOBS processing chain
C	Lidar cloud fraction	cf_nfov	(15)	Developed for ReOBS based on Morille et al. (2007): Sect. 3.4.1
	Surface downwelling SW radiation for clear sky, $W\,m^{-2}$	rsdscs	(8)	Data parameterization fitting an equation to measured data, accounting zenithal angle, effects of Sun-Earth geometry, mean cloud-free atmospheric components, local surface albedo, subset of measurements error (Dutton et al., 2004)
	Surface downwelling LW radiation for clear sky, $W\,m^{-2}$	rldscs	(7)	Analysis of surface irradiance, air temperature, humidity measurements (Long and Turner, 2008); technique with repeatability about $3\,W\,m^{-2}$ (Durr and Philipona, 2004; Long, 2004)
	Cloud fraction from sky imager	tot_cld_tsi		Analysis of color ratio, filtering image into clear or cloudy (Long and DeLuisi, 1998; Long et al., 2006b)
	Cloud fraction from LW radiation	cflw	(7)	APCADA algorithm (Durr and Philipona, 2004)
	Cloud fraction from SW radiation	cfsw	(8)	Long et al. (2006a)
	Surface upward sensible, $W\,m^{-2}$	hfss	(13) (14)	Derived from fluctuations of heat and moisture covariances with respect to vertical wind velocity (Brutsaert, 1982; Panofsky and Dutton, 1984) Variances and covariances rotated to streamwise coordinate for flux computation (Kaimal and Finnigan, 1994) Corrections for sonic virtual temperature (Schotanus et al., 1983) and density correction for latent heat flux (Webb et al., 1980)
	Surface upward latent, $W\,m^{-2}$	hfls	(13) (14)	
	Lidar cloud base height, m	cbh$x$[b]	(17)	Developed for ReOBS based on Morille et al. (2007): Sect. 3.4.1
	Aerosol optical thickness at x nm	aot_x[d]	(18)	Holben et al. (1998)
	Regional 2 m air temperature, K	tas_REG	Meteo-FR	Developed for ReOBS: Sect. 3.2
	Regional 2 m northward wind, $m\,s^{-1}$	vas_REG	Meteo-FR	
	Regional 2 m eastward wind, $m\,s^{-1}$	uas_REG	Meteo-FR	
	Regional precipitation at surface, $kg\,m^{-2}\,s^{-1}$	pr_REG	Meteo-FR	
	Clear sky integrated water vapor, $kg\,m^{-2}$	water	(18)	Using 675, 870, 940 nm channels (Schmid et al., 2001)

Table 2. Continued.

	Variable, unit	ReOBS short name	Based on Table 1 variables	Treatment before ReOBS processing chain
	Aerosol optical thickness at x^c nm	aot_x^c	(18)	Beer–Lambert-Bouguer law (Holben et al., 1998)
	Ångström exponent[e] between x^d and y^d nm, nm	x_yangstrom[d]	(18)	Eck et al. (1999)
	Mixing layer depth, m	mld	??	Developed in the context of ReOBS: Sect. 3.4.2 (Pal and Haeffelin., 2015)
	Total GPS water vapor, kg m^{-2}	iwv	(19)	Businger et al. (1996)
	Liquid water content, g m^{-2}	lwp	(20)	Brightness temperature at 23.8 and 31.4 GHz + input from temperature and humidity sensors (Bosisio and Mallet, 1998). Accuracy about 10–20 g m^{-2}
D	Lidar scattering ratio vertical histograms	SRhisto	(16)	Developed for ReOBS following GOCCP method (Chepfer et al., 2010): Sect. 3.5
	Lidar STRAT classification vertical histograms	STRAThisto	(16)	Developed for ReOBS applying STRAT algorithm (Morille et al., 2007): Sect. 3.5
	Lidar molecular profile	Molecular	(16)	Developed for ReOBS applying STRAT algorithm (Morille et al., 2007): Sect. 3.5
	Altitude of nor malization of lidar profiles, m	Alt norm	(16)	Developed for ReOBS applying STRAT algorithm (Morille et al., 2007): Sect. 3.5

[a] x is 5, 10, 20, 30, 50 cm;
[b] x is first layer (1), second layer (2), third layer (3);
[c] x is 1020, 870, 675, 500, 440, 380, 340 nm;
[d] x and y are the interval between [c] values.
[e] negative slope (or first derivative) of aerosol optical depth with wavelength in logarithmic scale is the Angstrom parameter (Eck et al., 1999, see Fig. 4r for significance value).

several temperature spikes (0.6 °C within 1 min for ground at −5 cm) are detected and we reject the data when the increase reaches +3 °C and the decrease −4 °C within 15 min (i.e., +0.2 and −0.27 °C for 1 min resolution).

The unphysical persistence in time of the measured values are detected by verifying that the variability within 1 hour is physical, following values in Table 3. If the 1 minute val ues do not vary by more than a specified lower limit (given in Table 3) within 1 hour, the current value fails the check. Figure 4d shows an example of an unphysical wind speed measured by a cup anemometer. The value is 0 m s^{-1} after 18:00 UT because of frost deposition on the sensor (shown by low temperature and high relative humidity in Fig. 4c). The persistence test is completed by a calculation of the standard deviation of temperature, pressure, humidity and wind speed for the last 1-hour period. In combination with the persistence test, the evaluation of the standard deviation is a very good tool for the detection of a blocked sensor, as well as a 1 h sensor drift.

3.3 Specific computation for advanced meteorological variables

The data quality of in-ground temperature and permeability of the soil is checked using the tests above (Table 1 and 3).

The quality of the downwelling shortwave (SW) and longwave (LW) fluxes is tested following the recommendation of the Baseline Surface Radiation Network (BSRN; test version 2.0; Roesch et al., 2011). Additional semi-automatic controls were developed and applied to SW irradiances in order to reject data collected when the sun tracker failed (used for the direct and diffuse SW radiation measurements) and to remove values that are inconsistent between measured global SW fluxes and global SW fluxes calculated from direct and diffuse measured ones. Individual 1 min native data not passing the test are automatically removed before performing the 1 h averages. For SW fluxes, the global as well as the direct and diffuse irradiance components are included in the ReOBS dataset. A best estimate of the global SW is calculated as a combination of the global irradiance measurement and the sum of the diffuse and horizontal direct irradiance mea-

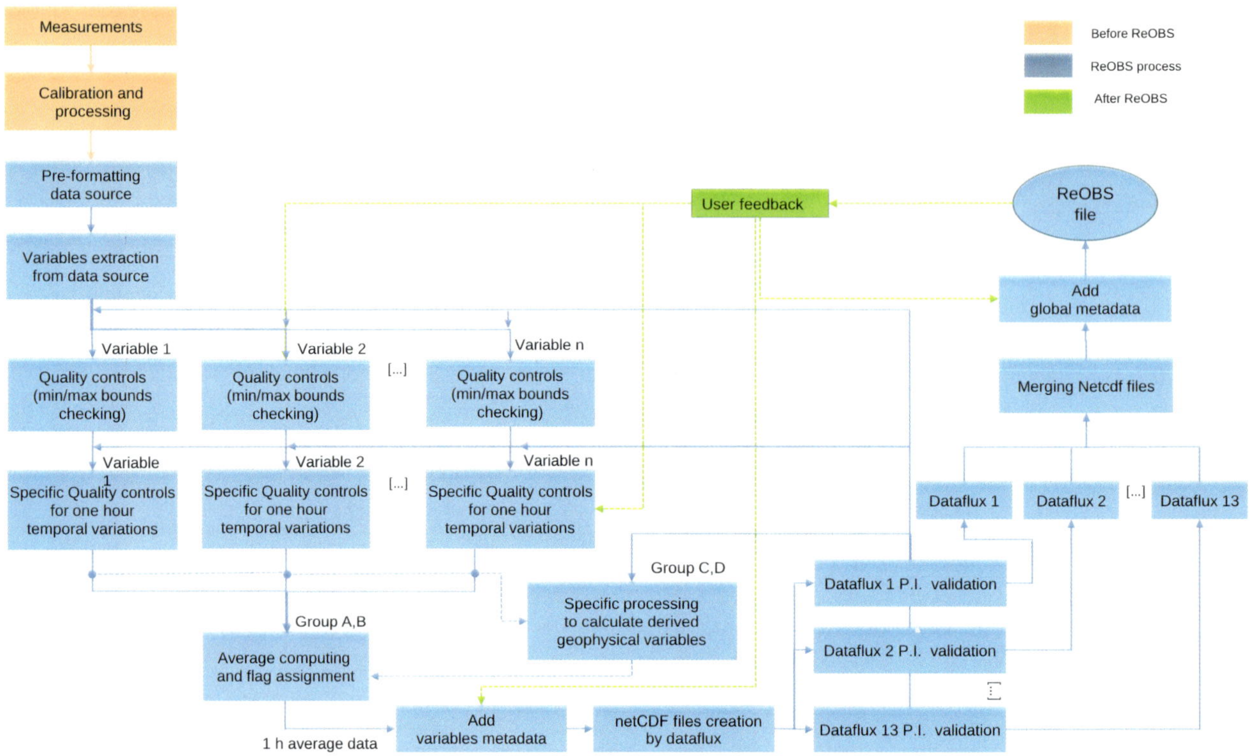

Figure 2. Schematic of the ReOBS general processing chain. Orange, blue, and green boxes and arrows are for steps before, during, and after, respectively, the ReOBS processing chain.

surements. The sum is taken as default and the blanks in observations are filled with the global irradiance measurement.

The sensible and latent heat flux data are subjected to spike detection and rejection algorithms. Sensible and latent heat fluxes are based on sonic measurements and a gas analyzer. The lag between the sonic measurements and the gas analyzer is set to the lag of maximum correlation over the averaging interval between the sonic anemometer temperature and the absolute humidity measured by the gas analyzer. At hourly intervals, sensible and latent heat fluxes are derived from eddy-covariance techniques, as well as turbulence statistics. Raw data and calculated statistics are subjected to strict data limits to reject unphysical values (13 and 14 in Table 1). For the latent heat flux, the open-path infrared gas analyzer (IRGA) used between 2005 and 2012 could be damaged by precipitations and was therefore manually switched on and off. The temporal sampling was thus relatively low and we decided to exchange the IRGA with an open-path LI-COR LI-7500 in 2012. With the open-path InfraRed Gas Analyser the molar density fluctuations are accounted for in the processing by following the classic formulation of Webb et al. (1980). Moreover, an automatic method has been applied to correct wind statistics for any misalignment of the sonic anemometer with respect to the local wind streamlines of the sonic anemometer with respect to the local wind streamlines according to Wilczak et al. (2001).

3.4 Retrievals based on remote-sensing measurements developed for ReOBS

3.4.1 Computations for the cloud fraction and cloud base height from lidar

The ReOBS dataset contains cloud base height and time series of the cloud fraction (CF), deduced from the SIRTA 355 nm lidar and processed with the STRAT algorithm (STRucture of the ATmosphere; Morille et al., 2007). The cloud fraction (noted cf_nfov, where “nfov” stands for “narrow field of view”) is defined as the number of profiles containing clouds divided by the total number of profiles collected in 1 hour. The cloud base height of the first layer (noted CBH1) corresponds to the altitude of the first cloud layer from the ground as detected by the STRAT algorithm. An hourly cloud base height is reported in ReOBS only if at least 33 % of the profiles collected during this hour are cloudy and only if less than 40 % of the profiles collected during this hour are noisy (i.e., at least 60 % of profiles are valid). Sensitivity tests based on several case studies have shown that taking less than this 33 % or more than this 40 % threshold leads to cloud base height values non-representative of what happens in the current hour. CBH2 and CBH3 are the altitudes of the base of a second and a third cloud layer, respectively, detected above CBH1 and sep-

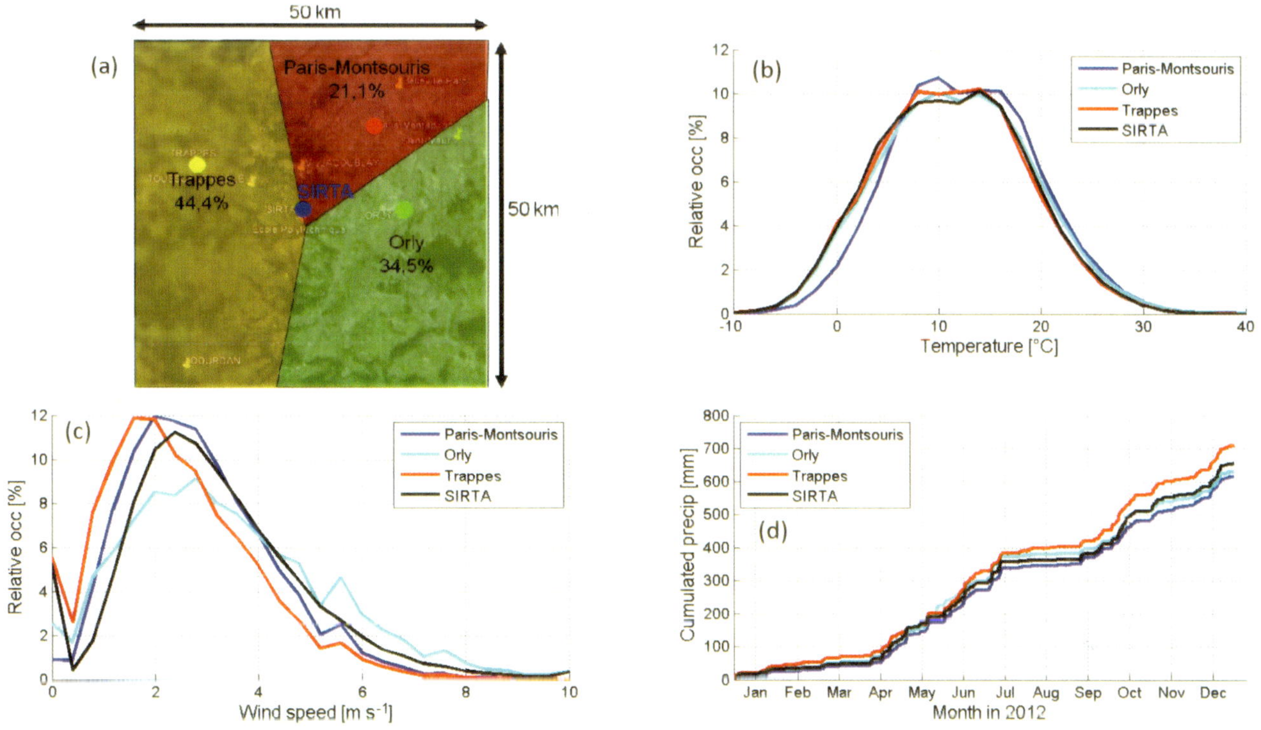

Figure 3. **(a)** Location of the SIRTA supersite and the three neighboring Météo France stations with their associated weight as defined in the text. Relative occurrence of hourly mean air temperature **(b)** and wind speed **(c)** at SIRTA and at the neighboring Météo France stations between 2005 and 2014, and cumulated precipitation at SIRTA and at the neighboring Météo France stations in 2012 **(d)**.

arated from the first cloud (the one with CBH1) with clear sky.

3.4.2 The mixing layer depth product

The mixing layer depth (MLD, noted mld in SIRTA-ReOBS) is part of the SIRTA-ReOBS database. It is retrieved from routine lidar measurements (ALS 450 from the Leosphere company) following the method described in Pal et al. (2013) and Haeffelin et al. (2012).

In this method, the intensity of the lidar-derived aerosol backscatter signal at different altitudes is used to determine the hourly-averaged vertical profiles of variance. Next, the location of maximum turbulent mixing within the mixing layer is determined and corresponds to the mean MLD. Micrometeorological measurements of the Monin–Obukhov length scale are used (effect of buoyancy on turbulent flow, Monin and Obukhov, 1954) to better determine the MLD, especially for the early morning transition and evening transition periods. For these two specific periods, a first-order approximation of the boundary layer growth rates is obtained and the variance-based results guides the attribution by searching the minimum altitude of the gradient closest to the mean MLD. Two transition periods of a day are used to distinguish the turbulent regimes during the well-mixed convective ABL (atmospheric boundary layer) and nocturnal or stable MLD.

Table 3. Range of temporal variabilities considered when performing the quality control for the variables listed in table. Upper arrows mean an increase during the time window indicated, and lower arrows mean a decrease during the time window indicated.

Variable	Temporal variability
tas	5 min: ↗ < 6 °C and ↘ < −9 °C 60 min: ↗↘ > 0.1 °C
hurs	5 min: ↗ < 22 % and ↘ < −23 % 60 min: ↗↘ > 0.05 %
psl	5 min: ↗ < 5 hPa and ↘ < −4 hPa 60 min: ↗↘ > 0.1 hPa
sfcWind	5 min: ↗ < 30 m s^{-1}
pr	5 min: ↗ < 40 mm
st$x$[a]	15 min: ↗ < 3 °C and ↘ < −4 °C at −5 cm 15 min: ↗↘ < 3 °C at −10 cm 15 min: ↗↘ < 1.5 °C at −30 cm 60 min: ↗↘ > 0.05 °C

[a] x is 5, 10, 20, 30, 50 cm.

3.5 Computations for the lidar profiles

The SIRTA ReOBS dataset contains information on the detailed vertical description of the atmosphere since 2002 from the LNA instrument. A drawback of this instrument is that it requires human intervention and does not operate when it rains, which introduces gaps in the data record.

Two different hourly variables are included in the ReOBS dataset:

- One variable called STRAThisto, which contains the number of occurrences of clear sky, aerosols, clouds, invalid data and fully attenuated laser within 1 hour for each vertical level. The vertical resolution is 15 m up to 15 km and the layer type classification is based on the STRAT algorithm.

- One variable called SRhisto is a 2-D height-intensity number of occurrences accumulated during 1 hour, as defined in Chepfer et al. (2010). We use the lidar scattering ratio $SR = ATB / ATB_{mol}$, where ATB is the total attenuated backscatter lidar signal and ATBmol is the signal in clear-sky conditions. The vertical resolution is 15 m and the intensity axis contains 18 bins; −999 / −777 / −666 / 0 / 0.01 / 1.2 / 3 / 5 / 7 / 10 / 15 / 20 / 25 / 30 / 40 / 50 / 60 / 80. The value "−999" indicates non-normalized noisy profiles, the value "−777" is for profiles that cannot be normalized due to the presence of a very low cloud, and the value "−666" is for invalid data. ATB profiles are normalized to a daily molecular profile based on radiosounding measurements launched every day 10 km away from the SIRTA supersite (at the Météo-France station in Trappes). The altitude of normalization of ATB (which must be clear sky) is determined for each profile using the STRAT algorithm.

4 Results

4.1 Description of the ReOBS database content

All data passing the quality control tests are included in the ReOBS final netCDF file. The variables included in the SIRTA-ReOBS are listed in Table 2 together with their nomenclature (Table 2, second column). There are 42 lines in Table 2, corresponding to 34 variables currently in the file. Figure 5 shows the temporal coverage of each variable. Some variables such as the classical meteorological variables or the downwelling radiative fluxes are very well sampled since 2002 when SIRTA activities started. In contrast, the record for lidar profiles, which started in 2002, contains many gaps. The sampling of the latent heat flux is much more intermittent than the sampling of the sensible heat flux due to instrumental issues (see Sect. 3.3).

There are two versions of the SIRTA-ReOBS file: a complete file, which includes all information available (1.2 Gb), and another one which contains all data except for the vertical information from the lidar (11.5 Mb): this last is significantly smaller and so it is easier to handle. Both data files are available on the following website: http://sirta.ipsl.fr/reobs.html (Downloads tab, no password required), which also includes quicklooks and a documentation.

The main advantages of ReOBS compared to classical supersite databases are (1) the vertical profile information coming from lidar measurements, which is user friendly thanks to the GOCCP (GCM Oriented CALIPSO Cloud Product) method (Chepfer et al., 2010), (2) the possibility to study the troposphere at different timescales (from daily to decadal timescales) and (3) the availability of a multi-variable synergetic view of the atmosphere. And of course a mix of these three aspects. These three main added values are detailed in the following subsections.

4.2 Vertical profile information

The lidar profiles included in SIRTA-ReOBS provide useful information on the vertical distribution of clouds and aerosols in the atmosphere. This information together with many other SIRTA-ReOBS variables have been used recently in various studies (Cheruy et al., 2012; Chiriaco et al., 2014; Bastin et al., 2016). We first show examples of the two main ReOBS variables build from lidar measurements (SRhisto and STRAThisto) and we then describe how these data are used to built cloud fraction profiles. Finally, we describe how to use these data to evaluate clouds simulated by models.

4.2.1 SRhisto and STRAThisto

Figure 6 shows SRhisto (Fig. 6a) and STRAThisto (Fig. 6b) for every hour containing measurements from 2003 to 2016. Periods without lidar measurements are not included in this figure and this happens frequently (see Fig. 5) because measurements are only performed when it is not raining and with human intervention (i.e., not during night and weekends). Using SRhisto, the repartition of clouds can be analyzed as a function of altitude and as a function of the intensity of the lidar signal (SR), which is a proxy of the particle optical thickness. Fully attenuated lidar signals are located in the bin $0 < SR < 1$, clear sky are found in the bin $SR = 1$, uncertain are in the bin $1 < SR < 5$ (it could be aerosols for instance), and $SR > 5$ is for clouds (white vertical line in Fig. 6a; bins defined in Sect. 3.5). The analysis of SRhisto shows that non-precipitating clouds observed at SIRTA (note: the LNA lidar instrument does not operate when it rains) are mostly thin, low clouds (under 4 km with $SR < 15$) or thin, high clouds (above 7 km with $SR < 20$), and there are also thicker clouds with $SR > 80$ or fully attenuated lidar signal. There are almost no mid-level clouds (between 4 and 7 km) and only few clouds with $20 < SR < 80$. The analysis of STRAThisto indicates that for these non-precipitating cases, the amount of clouds that fully attenuates the lidar signal (i.e., "noise") is

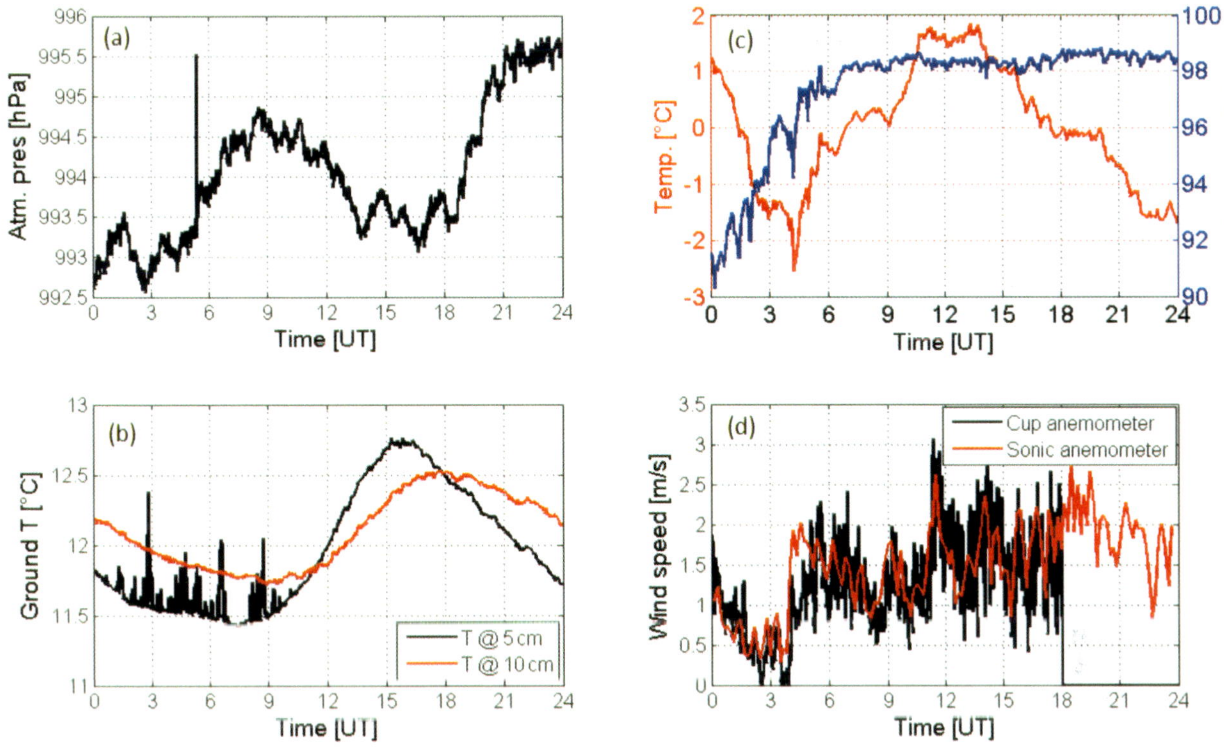

Figure 4. **(a)** Example of an unphysical jump in instantaneous values of pressure and **(b)** temperature in ground at 5 cm (black) and at 10 cm (red). **(d)** Example of unphysical persistence of high wind speed measurements using a cup anemometer due to frost in **(c)** (negative temperature in red and high humidity values in blue).

approximately on the same order of magnitude as the amount of thinner clouds.

Lidar profiles that would be measured in clear-sky conditions (so called molecular profiles) are necessary to build SRhisto and STRAThisto as it is used in the SR estimation and in the STRAT lidar profile normalization. These molecular profiles are estimated based on temperature and pressure profiles measured twice a day by METEO-FRANCE radiosounding at Trappes (10 km from SIRTA). These molecular lidar profiles are included in SIRTA-ReOBS under the Molecular variable, as well as the altitude of normalization used for STRAT under the Alt norm variable.

4.2.2 Cloud fraction profiles

Cloud fraction profiles are derived from the SRhisto or from the STRAThisto variables at a temporal scale ranging from 1 hour up to several years. At a given altitude level, the cloud fraction is the ratio between the occurrence of cloudy cases and the occurrence of all cases excluding the noisy ones. In STRAThisto the occurrence of cloudy layers is given in flag "clouds". In SRhisto, a layer is declared cloudy in a lidar profile when $SR > 5$ and $SR > 1+\varepsilon/ATB_{mol}$ with $\varepsilon = 1.3 \times 10^{-6}$ SI (ATB_{mol} is included in SIRTA-ReOBS). As expected, the cloud fraction profiles obtained from SRhisto or from STRAThisto (Fig. 6d) are different due to the differences in the definition of the cloud detection in the two algorithms. In particular, SRhisto features less low-level clouds ($z < 4$ km) than STRAThisto. The magenta curve in Fig. 6c is the SR distribution for cloudy cases during a given hour for the STRAT algorithm. This distribution shows that about 28 % of these cases correspond to cases where SR cannot be estimated because of the presence of a very low cloud (−777 in Fig. 6c) preventing the normalization of the profile (no detection of molecular signal under the cloud). This could explain the differences in low cloud fractions between STRAThisto and SRhisto in Fig. 6d. The part of the magenta curve with values of SR between 0.01 and 5 corresponds to cloudy cases for the STRAT algorithm but not based on the SR threshold method. This could explain the bias between CF SR and CF STRAT that occurs at almost all vertical levels. Red and yellow curves in Fig. 6c also highlight the fact that most of the cases that are defined as PBL or aerosols for the STRAT algorithm are actually not cloudy when based on the SR threshold method (the parts of these curves above $SR = 5$ represent less than 5 %). The differences between STRAT- and SR-based algorithms illustrate the important sensitivity of the cloud fraction profile to the cloud definition. This sensitivity needs to be taken into account when comparing the measurements to simulations from GCMs or RCMs in order to

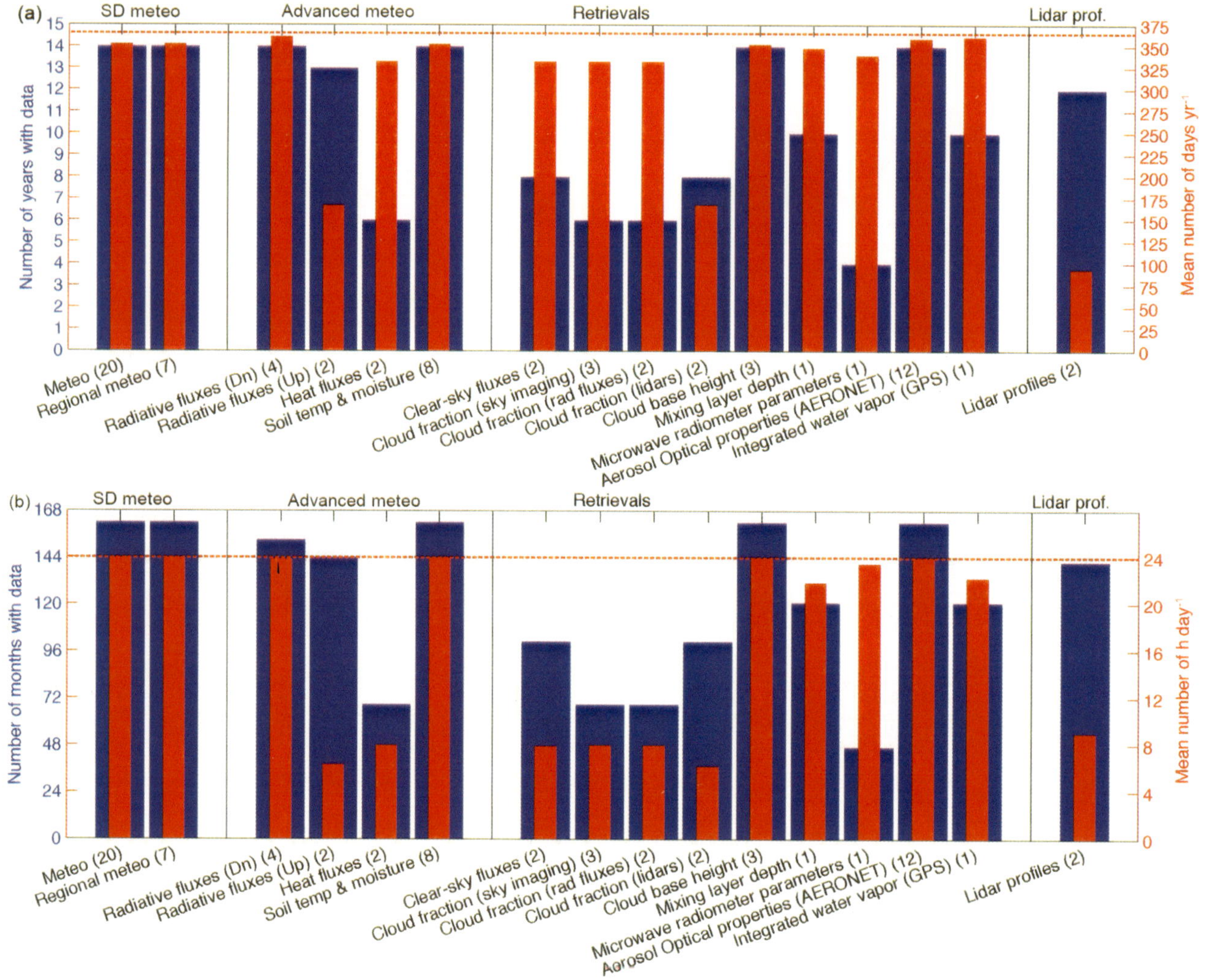

Figure 5. Temporal coverage of groups of variables in the SIRTA-ReOBS dataset. In panel **(a)**, blue bars indicate the total numbers of years with data and red bars indicate the mean numbers of days with measurements in a year. In panel **(b)**, blue bars indicate the numbers of months with data and red bars indicate the mean numbers of hours with measurement in a day. The numbers in brackets are the number of variables in each sub-group. Variables are separated in four categories: classical meteorological measurements (group A, left), more advanced measurements (group B, center-left), variables retrieved from measurements (group C, center-right) and lidar profiles (group D, right). Dn means downward, Up means upward.

reproduce the algorithm hypotheses in the simulations; it is usually done using a lidar simulator described below.

4.2.3 Lidar simulator

For comparing the SIRTA-ReOBS SRhisto variable to GCM or RCM outputs, we have developed a ground-based lidar simulator, which is an adaptation of the CALIOP (Cloud-Aerosol Lidar with Orthogonal Polarization) lidar simulator (Chepfer et al., 2008), taking into account the specification due to the ground-based position of the instrument. In particular, the vertical turnaround of the lidar equation following the very first version of lidar simulator described in Chiriaco et al. (2006). Model outputs are used as inputs for the lidar simulator to simulate what would be measured from the ground-based lidar if the atmosphere was the simulated one. First the lidar equation that gives the ATB as a function of altitude is used to simulate SR from model outputs. Then the same space and time resolutions as in observations and the SR thresholds are used for the simulated lidar profiles as in the actual data algorithm (SRhisto for SIRTA-ReOBS), making the lidar profiles directly comparable to the measured ones.

This ground-based version of the lidar simulator has been used for comparisons between the SIRTA-ReOBS lidar profiles and the WRF/MED-CORDEX (Weather Research and Forecast model; Coordinated Regional Climate Downscaling

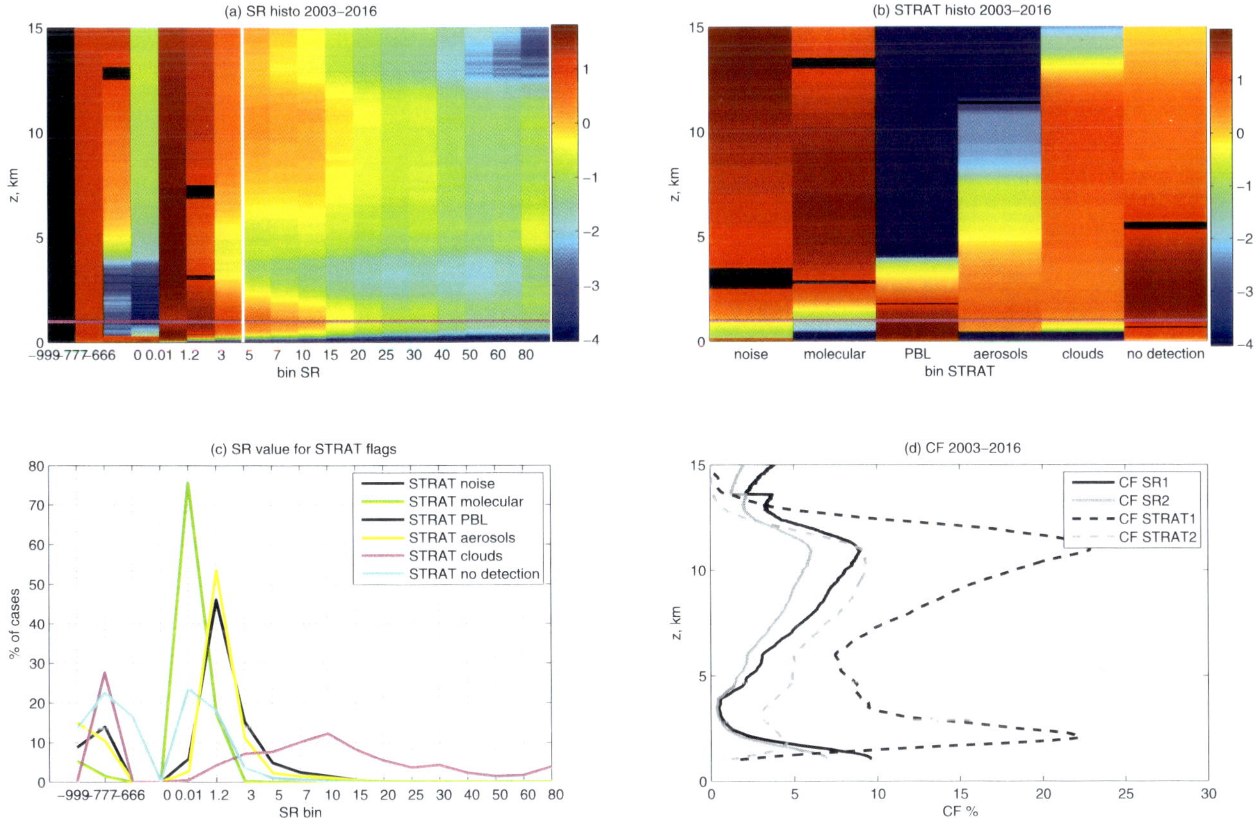

Figure 6. (a) Lidar scattering ratio (SR) histogram obtained by cumulating all SIRTA observation data from 2003 to 2016. The color bar is the logarithm of the percentage of occurrence (the sum of one line is equal to log(100 %)); the pink horizontal line corresponds to the altitude of recovery of the lidar ($z = 1$ km; below this altitude, lidar data is more complicated to use); the white vertical line corresponds to the threshold of cloud detection (SR = 5). **(b)** STRAT histogram obtained by cumulating all data from 2003 to 2016. The color bar is the logarithm of the percentage of occurrence. **(c)** Percentage of occurrence of SR values for the different STRAT flags (noise in blue – no cases actually –, molecular in green, PBL in red, aerosols in yellow, clouds in magenta and no detection in cyan), cumulating all altitudes above 1 km and only for hours containing a single STRAT flag. **(d)** Fraction of clouds (in %): CF SR1 (black solid line) is the occurrence of SR > 5 versus the occurrence of SR > 0, CF SR2 (gray solid line) is the occurrence of SR > 5 vs. the total occurrence of profiles, CF STRAT1 (black dashed line) is the occurrence of STRAT cloudy profiles versus the occurrence of STRAT molecular + PBL + aerosols + cloud profiles, CF STRAT2 (gray dashed line) is the occurrence of STRAT cloudy profiles vs. the total occurrence of profiles.

Experiment for Mediterranean area) simulation in Bastin et al. (2016).

This lidar simulator is currently implemented in the new COSP2 simulator package (version 2 of COSP CFMIP; Cloud Feedback Model Intercomparison Project, Observation Simulator Package; currently developed for CMIP6 simulations), following these steps: (1) computation of the molecular optical thickness of each layer (i.e., the atmosphere clear of any particles), (2) computation of the particles optical thickness of each layer, (3) computation of the total optical thickness of each layer by adding the molecular and the particles optical thicknesses, (4) computation of the total backscatter lidar signal as it would have been measured by a ground-based lidar by integrating progressively these optical thicknesses from the lowest atmospheric layer to the top of the atmosphere and (5) computation of the SR profile by dividing the attenuated total backscatter lidar profile by the clear-sky profile.

4.3 From the daily timescale to the decadal timescale

The temporal variability in the variables included in SIRTA-ReOBS is synthesized in a single figure, as shown in Fig. 7a for the 2 m temperature. Each row represents a year and in each row, the x axis indicates the day of the year and the y axis indicates the hour of the day. This figure allows for the visualization of the presence of gaps in the record and the different temporal scales of variability: diurnal, seasonal and interannual. A first visual inspection leads to the identification of significant anomalies in terms of amplitude and in terms of persistence. Figure 7b shows the mean temperature diurnal cycle split by seasons. Solid lines indicate the local SIRTA temperature (–SIR) and dashed lines indicate the sur-

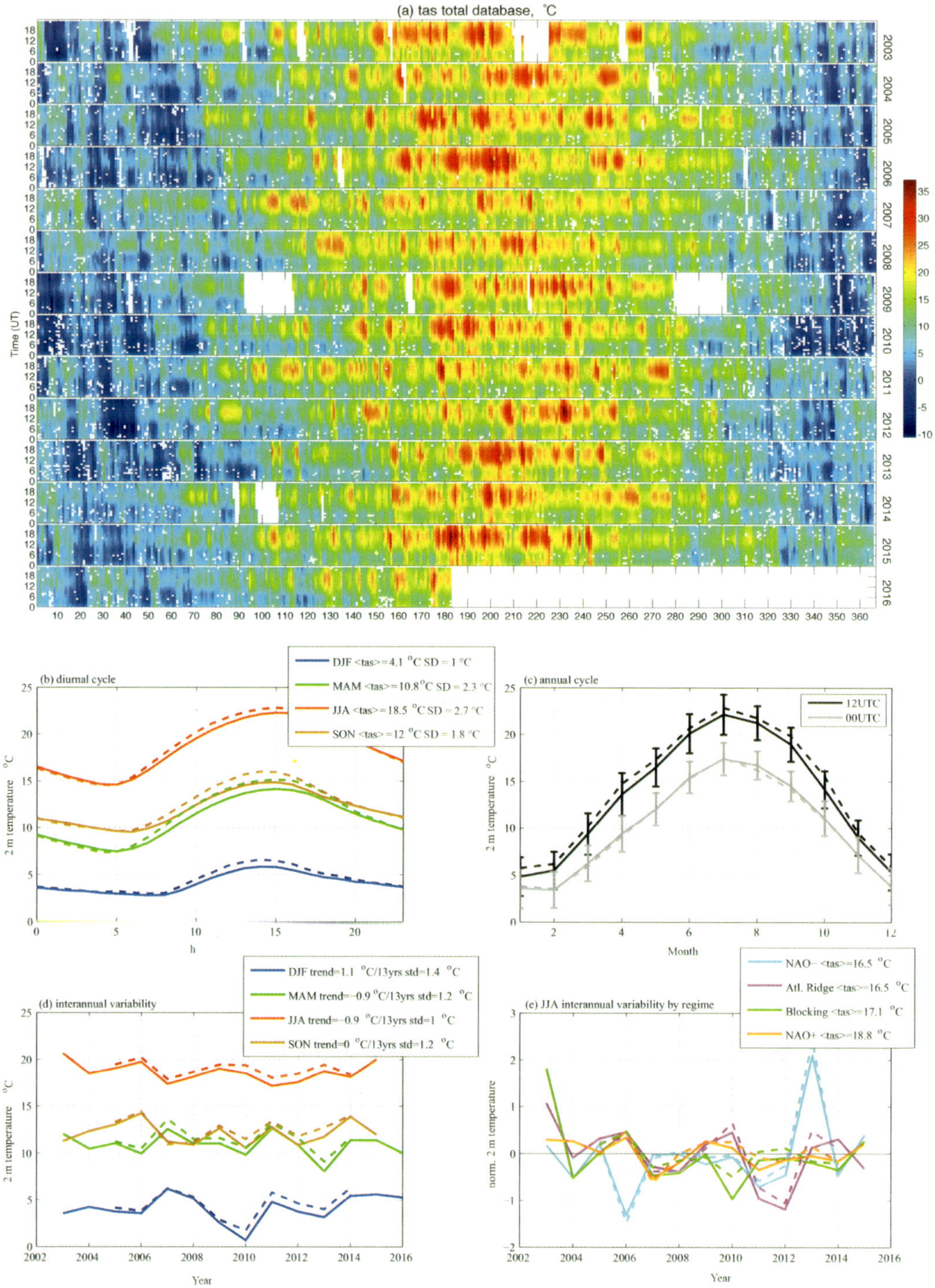

Figure 7. Contribution of the multi-temporal scale for the 2 m temperature (in °C). **(a)** Hourly values, each row corresponds to a year with the day of the year in x axis, and the hour of the day in y axis. **(b)** Mean diurnal cycle averaged from 2003 to 2016 split into seasons (DJF in blue, MAM in green, JJA in red, SON in brown). The mean values and the standard deviation of the 2 m temperature in each season are indicated. **(c)** Mean annual cycle averaged monthly from 2003 to 2016 at 12.00 UTC (black line) and at 00:00 UTC (gray line) with interannual SD in error bars. **(d)** Interannual evolution from 2003 to 2016, averaged by season (same colors as **b**). The trends of the curve (i.e., the slope of the curve linear regression multiplied by the number of years – 13) and its standard deviation are indicated. **(e)** Same curve as in **(d)** for JJA only, split per weather regimes (NAO− in cyan, Atlantic Ridge in magenta, blocking in green, NAO+ in orange) and plotted in anomaly (i.e., the mean value of all years is subtracted), where "norm. T2" is calculated following the Eq. (2). In all panels, solid lines are for the SIRTA local 2 m temperature, and dashed lines are for the regional 2 m temperature (around the SIRTA supersite).

rounding temperature (–REG, Sect. 3.2.). Since air temperature is at first order controlled by radiation, the coldest season is winter (mean value 4.1 °C) followed by spring (10.8 °C), fall (12 °C), and summer (18.5 °C), as expected. The amplitude of the diurnal cycle is greater in summer (standard deviation of 2.7 °C), than spring (SD =2.3 °C), fall (SD =1.7 °C) and winter (SD =1 °C). The specificity of SIRTA seems to lead to an attenuation of this diurnal cycle, as it is less pronounced than in the surrounding areas (note that temperatures during daytime are lower at SIRTA than in the surroundings, whereas they are equivalent during the night), likely due to urban, vegetation or soil moisture effects. Figure 7c shows the mean annual cycle of the 2 m temperature at 12:00 UTC (noon; black lines) and at 00:00 UTC (midnight; gray lines). As for the diurnal cycle, differences between local SIRTA measurements (solid lines) and regional 2 m temperature (dashed lines) are more pronounced at noon than at midnight. Figure 7d shows the interannual variability in the 2 m temperature split into seasons. There is no significant trend in the four seasons (the linear regression of each of the four curves multiplied by the number of years is weaker than 1σ, where σ is the SD, of the curve). Nevertheless, significant temperature anomalies are detected such as the cold winter 2010, the cold spring 2013, the warm fall 2006, the warm winter 2007 or the hot summer 2003. Summer mean values are split into weather regimes following the classification of Yiou et al. (2008) from the A2C2 (Atmospheric flow Analogues for Climate Change) project. In summer at SIRTA, the daily temperature is maximal when the weather regime is NAO+ (North Atlantic Oscillation+), it is weaker when the weather regime is blocking or NAO– and it is minimal when the weather regime is "Atlantic Ridge", as expected based on literature (numbers in the box in Fig. 7e). The anomaly (i.e., the mean value of all years is subtracted from each year value) of $V(y, r_i)$ for June–July–August in a given year y and a given regime r_i (where r_i is one of the four weather regimes mentioned above) plotted in Fig. 7e is calculated as follows:

$$V(y, r_i) = \langle \mathrm{tas}(y, r_i) \rangle / \mathrm{SD}(\mathrm{tas}(y, r_i)), \quad (2)$$

where $\langle \mathrm{tas}(y, r_i) \rangle$ is the mean value of the 2 m air temperature in year y and for days in regime r_i, and $\mathrm{SD}(\mathrm{tas}(y, r_i))$ is its standard deviation. Hence $V(y, r_i)$ is a mean temperature normalized by its variability and is unitless. Using this estimation, strong anomalies (i.e., anomalies that have a strong standard deviation) due to only a few numbers of days are minimized. This representation shows that summers that are not particularly warm or cold could actually contain significant anomalies. During summer 2013 for instance, NAO– days have been significantly warmer than NAO– days of the other summers, meaning that during these particular days, the temperature anomaly was due to processes and not only due to the large-scale circulation condition.

Figure 8 is the same as Fig. 7 but illustrates the cloud radiative effect (CRE) in the longwave by the following equation:

$$\mathrm{CRE_{LW}} = \mathrm{rlds} - \mathrm{rldscs}, \quad (3)$$

where rlds and rldscs are the downward all-sky and clear-sky LW flux as defined in Table 2. Figure 8a highlights the fact that the database only has few gaps for these variables. It also shows that the diurnal cycle does not seem to be very intense (about $5\,\mathrm{W\,m^{-2}}$ amplitude in DJF: December–January–February and SON: September–October–November; and about $10\,\mathrm{W\,m^{-2}}$ in JJA: June–July–August and MAM: March–April–May), whereas the annual cycle is significant (about $25\,\mathrm{W\,m^{-2}}$ difference between summer and winter, in particular during the night). Figure 8a–d show that clouds have a stronger radiative effect in the longwave during winter than during the other seasons regardless of the hour of the day, for every year. It could simply be due to the amount of clouds that occur more often during winter, or due to cloud radiative properties that are different between the seasons. This variable does not have a significant trend from 2003 to 2015 for all season (i.e., the trend is smaller than the standard deviation). Nevertheless, the mean seasonal values are significantly anti-correlated to the temperature values in spring (–0.7) and in summer (–0.9). At first order, the $\mathrm{CRE_{LW}}$ is driven by the amount of clouds, and the more clouds the cooler the temperature. This anti-correlation is less pronounced in winter and fall (–0.5). This is explained by the fact that (1) the temperature variability must be driven by the air mass circulation more than by clouds and that (2) in winter there is less solar radiation even if there are no clouds so the difference between a clear-sky day and a cloudy day is not as pronounced as in summer. Particular anomalies of $\mathrm{CRE_{LW}}$ can be related to the temperature ones: for instance winter 2007 was particularly mild (Fig. 7d) and was associated with weak longwave cloud radiative effect (Fig. 8d) that could be due to a deficit of clouds. On the contrary, winter 2010 was colder than other winters in the period of study and is associated with strong $\mathrm{CRE_{LW}}$. This correlation is also observed in summer (e.g., summers 2007 and 2011 are cold and have strong $\mathrm{CRE_{LW}}$). The distinction of $\mathrm{CRE_{LW}}$ for each of the four weather regimes in summer (Fig. 8e) shows the part of the $\mathrm{CRE_{LW}}$ anomaly that is not due to the large-scale dynamical conditions, which is the first order driver. The 2013 positive temperature anomaly for NAO– cases is associated with an important deficit of $\mathrm{CRE_{LW}}$ in this weather regime.

4.4 Multi-variables synergetic view of the atmosphere

One of the main advantages of ReOBS is that all variables are synthesized in a single file at the same temporal resolution, facilitating studies with multi-variable synergy, particularly useful for the understanding of atmospheric processes. This synergy aspect has been exploited in previous studies

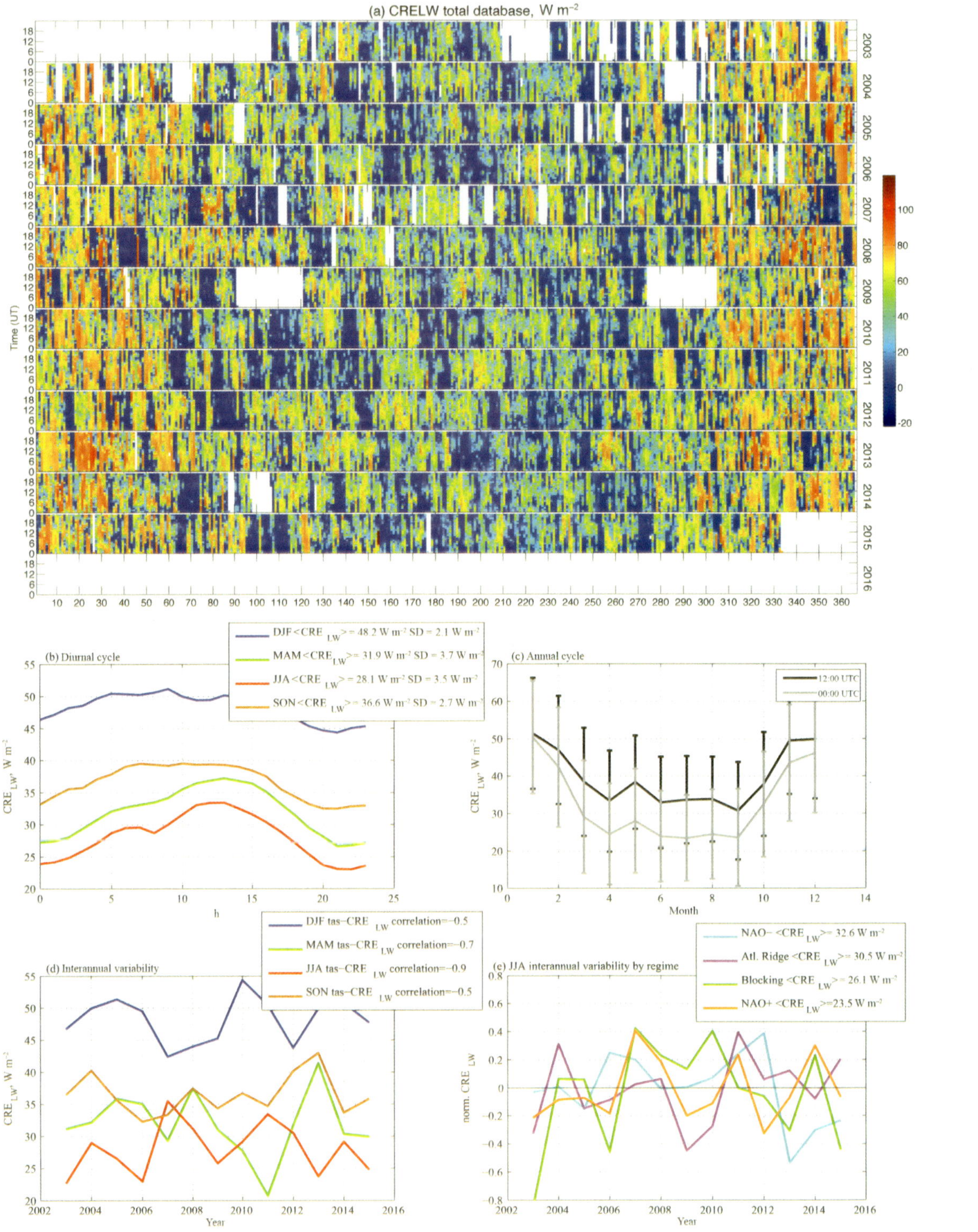

Figure 8. Same as Fig. 7 but for the longwave cloud radiative effect (in W m^{-2}). In panel **(d)**, the correlation between values in Fig. 7d (tas) and values in Fig. 8d (CRE_{LW}) are indicated.

using the SIRTA-ReOBS data; for instance to study the diurnal cycle, the annual cycle and the interannual variability but for multiple variables, (Cheruy et al., 2012 and Bastin et al., 2016), to study the different components and scales of the mixing layer depth variability (Pal and Haeffelin, 2015),

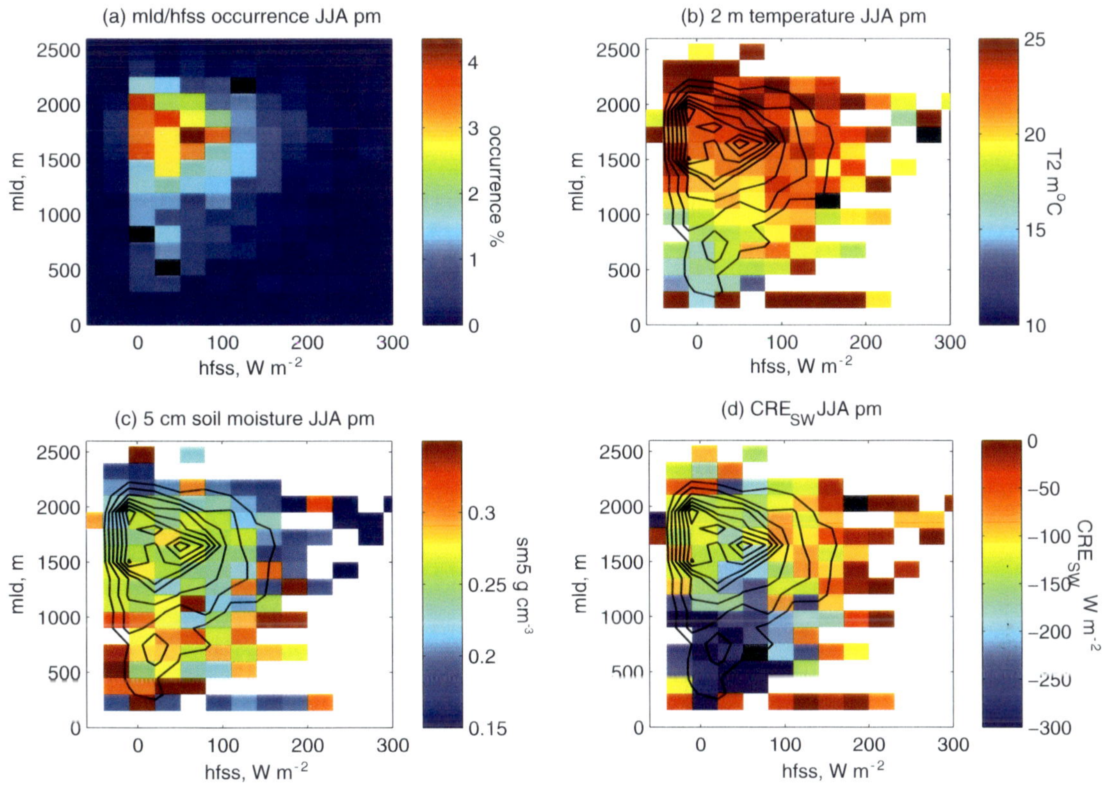

Figure 9. **(a)** Occurrence distribution (in percent) of the mixing layer depth (y axis) and the sensible heat flux (x axis) variables in summer (JJA) for the afternoon (02:00 to 06:00 LST). Averaged 2 m temperature **(b)**, averaged soil moisture at 5 cm depth **(c)**, and averaged shortwave cloud radiative effect **(d)**. Data are cumulated before the averaging. Black contours in **(b)**, **(c)** and **(d)** are isolines of **(a)**: the outermost isoline indicates 0.5 % of occurrence and each curve is then incremented of 0.5 %, and the innermost curve corresponds to 4 %.

and to perform in addition a dynamical analysis (Dione et al., 2016 and Chiriaco et al., 2014).

Figure 9 illustrates a possible synergy of multi-variables. The distribution of three variables affecting boundary layer processes in summer (JJA) is plotted (colors) as a function of mixing layer depth (y axis) and sensible heat flux (x axis) in the afternoon (between 02:00 and 06:00 LST, local solar time). The occurrence distribution of mixing layer depth versus sensible heat flux is reported in Fig. 9a and then as black contours in each other subplot: each isoline represents an increment of 0.5 %; pixels outside the most external isoline represent less than 0.5 % of the cases (per pixel). The 2 m temperature distribution is shown in Fig. 9b soil moisture at 5 cm depth on the middle one, and the cloud radiative effect on shortwave fluxes (CRE_{SW}) on the bottom one.

Figure 9a shows that shallow boundary layers (altitude of 500–1000 m) in summertime afternoon are mostly associated with low values of sensible heat flux (0–50 W m^{-2}). They are associated with strong values of shortwave cloud radiative forcing (< -200 W m^{-2}) due to the presence of clouds, high soil moisture (> 0.25 g m^{-2}) and low air temperatures (< 17 °C). Deeper boundary layers (altitude of 1500–2000 m) are associated with a wide range of sensible heat fluxes (50–150 W m^{-2}) and generally higher air temperatures (> 22 °C). For these deeper boundary layer cases, soil moisture and shortwave cloud radiative forcing are found to vary significantly.

The role of clouds in the link between mld and hfss can be easily identified in Fig. 9. In absence of clouds (CRE_{SW} close to zero), mld and hfss both have a high amplitude, while they both have a weak amplitude in the presence of clouds with strong albedo effect ($CRE_{SW} < -200$ W m^{-2}). The occurrence of clouds with strong albedo effect correlates well with low temperatures and high soil moisture values.

However, most occurrences (black contours) correspond to low hfss, relatively high mld and intermediate values of CRE_{SW}. Temperatures are generally quite high also, and sm5 also presents intermediate values. Very clear-sky and dry-soil conditions ($CRE_{SW} > -50$ W m^{-2} and sm5 < 0.2 g m^{-2}) generally lead to strong sensible heat fluxes and high temperatures, which do not necessarily translate into higher mixing layer depths than under cloudier conditions.

In summary, low mld are induced by strong cloud albedo effect and thus by low temperature and weak sensible heat flux due to weak energy reaching the surface. On the contrary, at hourly timescale, a mld higher than 1500 m is associated with a temperature higher than 20 °C and a wide range of CRE_{SW} (although greater than $-200\,W\,m^{-2}$). But this mld can be associated with a weak sensible heat flux. One reason for this is that the dominant timescale of variability for the boundary layer depth is the daily timescale, the maximum value being reached generally near 16:00 UTC in summer above SIRTA (Pal and Haeffelin, 2015), while the timescale of variability of the boundary layer forcers is hourly or less (radiative and heat fluxes). The temporal variability around the mld maximal value is often weak during this time lapse because it reacts with a delay. The energy dissipation rate in the boundary layer is slow and then the boundary layer stays deep even after the solar energy starts to decrease. So there is a delay between the decrease in mld and the decrease in the sensible heat flux. When considering the hourly timescale, many cases have high mld and low hfss. Investigating this issue in detail using the ReOBS database is beyond the scope of this paper.

5 Data availability

The ReOBS processing chain has been applied to SIRTA ground-based measurements and leads to the production of a single netCDF file containing about 60 substantial geophysical variables hourly averaged over a period of up to a decade. The netCDF file is available at http://sirta.ipsl.fr/reobs.html under https://doi.org/10.14768/4F63BAD4-E6AF-4101-AD5A-61D4A34620DE.

6 Summary and perspectives

We have presented a set of methods available for the community to robustly process ground-based data at an hourly timescale over more than a decade. The ReOBS processing chain has been applied to SIRTA ground-based measurements and leads to the production of a single netCDF file containing about 60 substantial geophysical variables hourly averaged over up to a decade. The netCDF file is available at http://sirta.ipsl.fr/reobs.html under https://doi.org/10.14768/4F63BAD4-E6AF-4101-AD5A-61D4A34620DE.

The main implication of this work is that complex observations are made available for the scientific community and allow for multiannual and multi-variable studies combining atmospheric dynamics and thermodynamics, radiation, clouds and aerosols. For example, the variability in 2 m temperature and LW cloud radiative effect can be jointly studied on the diurnal up to the interannual timescales. The multi-variable synergy is also illustrated with a focus on the boundary layer processes. As mentioned before, SIRTA-ReOBS has been already used in previous published studies: Cheruy et al. (2012) and Bastin et al. (2016) used SIRTA-ReOBS to evaluate simulations from GCMs and from RCMs, respectively. In these studies, using SIRTA-ReOBS has led to identifying the processes responsible of the model biases. Still in term of processes, Pal and Haeffelin (2015) used SIRTA-ReOBS to study the different components and scales of the mixing layer depth variability. Dione et al. (2016) and Chiriaco et al. (2014) have benefited from the SIRTA-ReOBS to study specific season anomalies. Datasets from ReOBS method are also useful tools for teaching and outreach activities such as the European KIC-Climate summer Journeys of the LABEX L-IPSL (Laboratory of Excellence Institut Pierre Simon Laplace) or the CLE-workshop (CLimate and Environment).

The ReOBS processing chain is now complete but the produced files such as SIRTA-ReOBS are continuously being improved, e.g., by adding new periods of data, by treating new variables and by improving the quality control. The SIRTA-ReOBS file presented in this paper is at a precise time. Future development for SIRTA-ReOBS include (1) improving the quality control of classical meteorological variables based on a comparative study of different methods, (2) adding vertical profiles from radiosounding launched twice a day 10 km away from the SIRTA supersite since the 90s and (3) adding new variables such as cloud radar data, gases and wind profiles from radar and lidar.

The ReOBS approach described in this paper will be applied to other supersites. Applying this approach to data from supersites of the ACTRIS-FR (Aerosol Cloud and Trace Gases Researche Infrastructure – France) infrastructure, in particular to the P2OA site located in the south of France is currently being tested. Applying ReOBS to ACTRIS-EU supersites is also under discussion. Another ongoing project is to integrate the ReOBS dataset to the OBS4MIP (Observations for Model Intercomparisons Project) database, which contains the data collected from observations developed specially for comparisons to CMIP simulations. This requires only few adaptations to fit the OBS4MIP standards.

Competing interests. The authors declare that they have no conflict of interest.

Acknowledgements. The authors would like to thank the following for financial support: Ecole Polytechnique, IPSL, FX-Conseil and the European project EUCLIPSE (European Union Cloud Intercomparison, Process Study and Evaluation project) for ReOBS since the beginning of this project. This study also benefited from the support of the Labex L-IPSL, which is funded by the Agence Nationale pour la Recherche (grant no. ANR-10-LABX-0018).

The authors would like to thank IPSL mesocenter and ESPRI teams from IPSL for providing computing and storage resources, and the SIRTA for providing measurements and data.

This work is also a contribution to the EECLAT project through LEFE-INSU and CNES supports, as well as to ACTRIS-FR, which is a national distributed research infrastructure and identified on the French road map for Research Infrastructures, published by the Ministry of Research. ACTRIS-FR is coordinated by the CONSOR-

TIUM ACTRIS-FR and comprises a large number of French research organizations and institutions.

Marjolaine Chiriaco was partly supported by Centre National d'Etudes Spatiales (CNES) until 2016.

The authors would like to thank the A2C2 European project for providing the weather regimes classification (ERC advanced grant no. 338965 – A2C2).

Edited by: David Carlson

References

Ackerman, T. and Stokes, G.: The Atmospheric Radiation Measurement Program, Phys. Today, 56, 38–45, 2003.

Bastin, S., Chiriaco, M., and Drobinski, P.: Control of radiation and evaporation on temperature variability in a WRF/MED-CORDEX simulation: comparison with collocated long term ground based observations near Paris, Clim. Dynam., https://doi.org/10.1007/s00382-016-2974-1, 2016.

Bosisio, A. V. and Mallet, C.: Influence of cloud temperature on brightness temperature and consequences for water retrieval, Radio Sci., 33, 929–939, 1998.

Brutsaert, W.: Evaporation into the atmosphere: theory, history and applications., D. Reidel Publishing Co., Norwell, MA, USA, 299 pp., 1982.

Businger, S., Chiswell, S. R., Ulmer, W. C., and Johnson, R.: Balloons as a Lagrangian measurement platform for atmospheric research, J. Geophys. Res., 101, 4363–4371, https://doi.org/10.1029/95JD00559, 1996.

Chepfer, H., Bony, S., Winker, D., Chiriaco, M., Dufresne J.-L., and Sèze, G.: Use of CALIPSO lidar observations to evaluate the cloudiness simulated by a climate model, Geoph. Res. Lett., 35, L15704, 2008.

Chepfer, H., Bony, S., Winker, D., Cesana, G., Dufresne, J. L., Minnis, P., Stubenrauch, C. J., and Zeng, S.: The GCM Oriented Calipso Cloud Product (CALIPSO-GOCCP), J. Geophys. Res., 115, D00H16, https://doi.org/10.1029/2009JD012251, 2010.

Cheruy, F., Dupont, J. C., Campoy, A., Ducharne, A., Hourdin, F., Haeffelin, M., and Chiriaco, M.: Combined influence of atmospheric physics and soil hydrology on the realism of the LMDz model compared to SIRTA measurements, Clim. Dynam., 40, 2251–2269, https://doi.org/10.1007/s00382-012-1469-y, 2012.

Chiriaco M., Vautard, R., Chepfer, H., Haeffelin, M., Dudhia, J., Wanherdrick, Y., Morille Y., and Protat, A.: The Ability of MM5 to Simulate Ice Clouds: Systematic Comparison between Simulated and Measured Fluxes and Lidar/Radar Profiles at SIRTA Atmospheric Observatory, Mon. Weather Rev., 134, 897–918, 2006.

Chiriaco, M., Bastin, S., Yiou, P., Haeffelin, M., Dupont, J. C., Klenov, L., and Stéfanon, M.: European heat-wave in July 2006: observations and modelling showing how local processes amplify conducive large-scale conditions, Geophys. Res. Lett., 41, 5644–5652, 2014.

Dione, C., Lohou, F., Chiriaco, M., Lothon, M., Bastin, S., Baray, J.-L., Yiou, P., and Colomb, A.: The influence of synoptic circulations and local processes on temperature anomalies over three French observatories, JAMC, 58, 141–158, https://doi.org/10.1175/JAMC-D-16-0113.1, 2017.

Dubovik, O. and King, M. D.: A flexible inversion algorithm for retrieval of aerosol optical properties from sun and sky radiance measurements, J. Geophys. Res., 105, 20673–20696, 2000.

Durr, B. and Philipona, R.: Automatic cloud amount detection by surface longwave downward radiation measurements, J. Geophys. Res., 109, D05201, https://doi.org/10.1029/2003JD004182, 2004.

Dutton, E. G., Farhadi, A., Stone, R. S., Long, C. N., and Nelson, D. W.: Long-term variations in the occurrence and effective solar transmission of clouds as determined from surface-based total irradiance observations, J. Geophys. Res., 109, D03204, https://doi.org/10.1029/2003JD003568, 2004.

Eck, T. F., Holben, B. N., Reid, J. S., Dubovik, O., Smirnov, A., O'Neill, N. T., Slutsker, I., and Kinne, S.: Wavelength dependence of the optical depth of biomass burning, urban and desert dust aerosols, J. Geophys. Res., 104, 31333–31350, 1999.

Haeffelin, M., Barthès, L., Bock, O., Boitel, C., Bony, S., Bouniol, D., Chepfer, H., Chiriaco, M., Cuesta, J., Delanoë, J., Drobinski, P., Dufresne, J.-L., Flamant, C., Grall, M., Hodzic, A., Hourdin, F., Lapouge, F., Lemaître, Y., Mathieu, A., Morille, Y., Naud, C., Noël, V., O'Hirok, W., Pelon, J., Pietras, C., Protat, A., Romand, B., Scialom, G., and Vautard, R.: SIRTA, a ground based atmospheric observatory for cloud and aerosol research, Ann. Geophys., 23, 253–275, https://doi.org/10.5194/angeo-23-253-2005, 2005.

Haeffelin, M., Angelini, F., Morille, Y., Martucci, G., Frey, S., Gobbi, G. P., Lolli, S., O'Dowd, C. D., Sauvage, L., Xueref-Rémy, I., Wastine, B., and Feist, D. G.: Evaluation of Mixing-Height Retrievals from Automatic Profiling Lidars and Ceilometers in View of Future Integrated Networks in Europe, Bound.-Lay. Meteorol., 143, 49–75, doi10.1007/s10546-011-9643-z, 2012.

Hawkins, E. and Stutton, R.: The potential to narrow uncertainty in regional climate predictions, American Meteorological Society, 90, 1095–1107, 2009.

Holben, B. N., Eck, T. F., Slutsker, I., Tanre, D., Buis, J. P., Setzer, A., and Smirnov, A.: AERONET – A federated instrument network and data archive for aerosol characterization, Remote Sens. Environ., 66, 1–16, 1998.

Kaimal, J. C. and Finnigan, J. J.: Atmospheric Boundary Layer Flows: Their Structure and Measurement, Oxford Univ. Press, New York, USA, 289 pp., 1994.

Long, C. N.: The Next Generation Flux Analysis: Adding Clear-sky LW and LW Cloud Effects, Cloud Optical Depths, and Improved Sky Cover Estimates, 14th ARM Science Team Meeting Proceedings, 22–26 March 2004, Albuquerque, NM, USA, 2004.

Long, C. N. and DeLuisi, J. J.: Development of an Automated Hemispheric Sky Imager for Cloud Fraction Retrievals, In Proceedings 10th Symposium on Meteorological Observations and Instrumentation, 11–16 January 1998, Phoenix, AZ, USA, 1998.

Long, C. N. and Turner, D. D.: A method for continuous estimation of clear-sky downwelling longwave radiative flux developed using ARM surface measurements, J. Geophys. Res., 113, D18206, https://doi.org/10.1029/2008JD009936, 2008.

Long, C. N., Ackerman, T. P., Gaustad, K. L., and Cole, J. N. S.: Estimation of fractional sky cover from broadband shortwave radiometer measurements, J. Geophys. Res., 111, D11204, https://doi.org/10.1029/2005JD006475, 2006a.

Long, C. N., Sabburg, J., Calbó, J., and Pagès, D.: Retrieving cloud characteristics from ground-based daytime colorall-sky images, J. Atmos. Ocean. Tech., 23, 633–652, 2006b.

McArthur, L. J. B.: Baseline Surface Radiation Network (BSRN), Operations Manual, WMO/TD-No. 1274, WCRP/WMO, 2004.

Monin, A. S. and Obukhov, A. M.: Basic laws of turbulent mixing in the surface layer of the atmosphere, Tr. Akad. Nauk SSSR Geofiz. Inst., 24, 163–187, 1954.

Morille, Y., Haeffelin, M., Drobinski, P., and Pelon, J.: STRAT: An automated algorithm to retrieve the vertical structure of the atmosphere from single channel lidar data, J. Atmos. Ocean. Tech., 24, 761–775, https://doi.org/10.1175/JTECH2008.1, 2007.

Ohmura, A., Duttonb, E. G., Forganc, B., Fröhlichd, C., Gilgena, H., Hegnera, H., Heimoe, A., König-Langlof, G., McArthurg, B., Müllera, G., Philiponad, R., Pinkerh, R., Whitlocki, C. H., Dehnej, K., and Wilda, M.: Baseline surface radiation network (BSRN/WCRP): New precision radiometry for climate research, B. Am. Meteorol. Soc., 79, 2115–2136, 1998.

Pal, S. and Haeffelin, M.: Forcing mechanisms governing diurnal, seasonal, and interannual variability in the boundary layer depths: Five years of continuous lidar observations over a suburban site near Paris, J. Geophys. Res.-Atmos., 120, 11936–11956, https://doi.org/10.1002/2015JD023268, 2015.

Pal, S., Haeffelin, M., and Batchvarova, E.: Exploring a geophysical process-based attribution technique for the determination of the atmospheric boundary layer depth using aerosol lidar and near surface meteorological measurements, J. Geophys. Res., 118, 1–19, 2013.

Panofsky, H. A. and Dutton, J. A.: Atmospheric turbulences, John Wiley, New York, USA, 397 pp., 1984.

Roesch, A., Wild, M., Ohmura, A., Dutton, E. G., Long, C. N., and Zhang, T.: Assessment of BSRN radiation records for the computation of monthly means, Atmos. Meas. Tech., 4, 339–354, https://doi.org/10.5194/amt-4-339-2011, 2011.

Rose, T., Crewell, S., Loehnert, U., and Simmer, C.: A network suitable microwave radiometer for operational monitoring of the cloudy atmosphere, Atmos. Res., 75, 183–200, 2005.

Roth, C. H., Malicki, M. A., and Plagge, R.: Empirical evaluation of the relationship between soil dielectric constant and volumetric water content as the basis for calibrating soil moisture measurements, J. Soil Sci. 43, 1–13, 1992.

Schmid, B., Michalsky, J. J., Slater, D. W., Barnard, J. C., Halthore, R. N., Liljegren, J. C., Holben, B. N., Eck, T. F., Livingston, J. M., Russell, P. B., Ingold, T., and Slutsker, I.: Comparison of Columnar Water-Vapor Measurements from Solar Transmittance Methods, Appl. Opt., 40, 1886–1896, 2001.

Schotanus, P., Nieuwstadt, F. T. M., and Debruin, H. A. R.: Temperature-measurement with a sonic anemometer and its application to heat and moisture fluxes, Bound.-Lay. Meteorol., 26, 81–93, 1983.

Terray, L. and Boé, J.: Quantifying 21st-century France climate change and related uncertainties, C. R. Geosci., 345, 136–149, 2013.

Webb, E. K., Pearman, G. I., and Leuning, R.: Correction of flux measurements for density effects due to heat and water-vapor transfer, Q. J. Roy. Meteorol. Soc., 106, 85–100, 1980.

Wieser, A., Fiedler, F., and Corsmeier, U.: The Influence of the Sensor Design on Wind Measurements with Sonic Anemometer Systems, JAOT, 18, 1585–1608, 2001.

Wilczak, J. M., Oncley, S. P., and Stage, S. A.: Sonic Anemometer Tilt Correction Algo- rithms, Bound.-Lay. Meteorol., 99, 127–150, https://doi.org/10.1023/A:1018966204465, 2001.

Xie, S., McCoy, R. B., Klein, S.,A., Cederwall, R.,T., Wiscombe, W. J., Clothiaux, E. E., Gaustad, K. L., Golaz, J. C., Hall, S. D., Jensen, M. P., Johnson, K. L., Lin, Y., Long, C. N., Mather, J. H., McCord, R. A., McFarlane, S. A., Palanisamy, G., Shi, Y., and Turner, D. D. CLOUDS AND MORE: ARM Climate Modeling Best Estimate Data, B. Am. Meteorol. Soc., 91, 13–20, https://doi.org/10.1175/2009BAMS2891.1, 2010.

Yiou, P., Goubanova, K., Li, Z. X., and Nogaj, M.: Weather regime dependence of extreme value statistics for summer temperature and precipitation, Nonlin. Processes Geophys., 15, 365 378, https://doi.org/10.5194/npg-15-365-2008, 2008.

9

Combining data from the distributed GRUAN site Lauder–Invercargill, New Zealand, to provide a site atmospheric state best estimate of temperature

Jordis S. Tradowsky[1,2,3], **Gregory E. Bodeker**[1], **Richard R. Querel**[2], **Peter J. H. Builtjes**[3], **and Jürgen Fischer**[4]

[1]Bodeker Scientific, Alexandra, New Zealand

[2]National Institute of Water and Atmospheric Research (NIWA), Lauder, New Zealand

[3]Institute for Meteorology, Freie Universität Berlin, Berlin, Germany

[4]Institute for Space Sciences, Freie Universität Berlin, Berlin, Germany

Correspondence: Jordis S. Tradowsky (jordis@bodekerscientific.com)

Abstract. A site atmospheric state best estimate (SASBE) of the temperature profile above the GCOS (Global Climate Observing System) Reference Upper-Air Network (GRUAN) site at Lauder, New Zealand, has been developed. Data from multiple sources are combined within the SASBE to generate a high temporal resolution data set that includes an estimate of the uncertainty on every value.

The SASBE has been developed to enhance the value of measurements made at the distributed GRUAN site at Lauder and Invercargill (about 180 km apart), and to demonstrate a methodology which can be adapted to other distributed sites. Within GRUAN, a distributed site consists of a cluster of instruments at different locations.

The temperature SASBE combines measurements from radiosondes and automatic weather stations at Lauder and Invercargill, and ERA5 reanalysis, which is used to calculate a diurnal temperature cycle to which the SASBE converges in the absence of any measurements.

The SASBE provides hourly temperature profiles at 16 pressure levels between the surface and 10 hPa for the years 1997 to 2012. Every temperature value has an associated uncertainty which is calculated by propagating the measurement uncertainties, the ERA5 ensemble standard deviations, and the ERA5 representativeness uncertainty through the retrieval chain. The SASBE has been long-term archived and is identified using the digital object identifier https://doi.org/10.5281/zenodo.1195779.

The study demonstrates a method to combine data collected at distributed sites. The resulting best-estimate temperature data product for Lauder is expected to be valuable for satellite and model validation as measurements of atmospheric essential climate variables are sparse in the Southern Hemisphere. The SASBE could, for example, be used to constrain a radiative transfer model to provide top-of-the-atmosphere radiances with traceable uncertainty estimates.

1 Introduction

Measurements of the upper air are essential for atmospheric research and weather forecasts. While high vertical resolution temperature profiles can be retrieved from space-based instruments, these retrievals require validation, which is typically done by comparison with ground-based or in situ measurements. Ground-based techniques to observe upper-air temperatures include lidar and microwave radiometer, while in situ measurements are typically made using balloon-borne radiosondes. While radiosondes provide vertically highly resolved profiles of temperature, pressure and humidity, they are only used at about 800 upper-air sites worldwide (Ingleby, 2017), which typically launch two sondes per day. Given the limited spatio-temporal sampling of the radiosonde measurements, their use for satellite validation can be challenging as the number of collocations within a given time interval and distance is small (see e.g. Calbet, 2016). However, if measurements from different instruments, or from collocated sites, can be combined in a best-estimate data product, the value of those measurements can be enhanced.

Site atmospheric state best estimates (SASBEs; Tobin et al., 2006) combine measurements from multiple instruments to create a vertically resolved, high temporal resolution time series of an atmospheric essential climate variable (ECV; GCOS-200, 2016; Bojinski et al., 2014) above a site. SASBEs aim to encompass all suitable knowledge of the state of the target variable at the specific site and include an estimate of the uncertainty on each value, therewith satisfying the requirements of GCOS-170 (2013). Tobin et al. (2006) developed a SASBE for the Atmospheric Radiation Measurement (ARM) site Southern Great Plains[1], which was used to validate retrievals of temperature and water vapour from the Atmospheric Infrared Sounder (AIRS; Aumann et al., 2003). Maillard Barras et al. (2015) present a methodology to combine in situ ozonesonde and space-based microwave radiometer measurements from SOMORA into an ozone SASBE. The authors found improved agreement between the ozone SASBE and the Microwave Limb Sounder (AURA/MLS) in comparison to the agreement with the operational SOMORA retrieval which does not include ozonesonde profiles.

As upper-air measurements in the Southern Hemisphere are especially sparse, it is essential to exploit all available observations. A well-equipped atmospheric measurements site, operated by the New Zealand National Institute of Water and Atmospheric Research (NIWA), is based at Lauder on the South Island of New Zealand. This site is part of the GCOS[2] Reference Upper-Air Network (GRUAN; GCOS-112, 2007), which was established to fill the need for reference-quality measurements of upper-air ECVs. Bodeker et al. (2016) document the development of GRUAN over the first 10 years after its establishment; a governance structure is in place, 24 sites have joined the network as of 2016, and a first reference-quality data product, accounting for all known biases, and including a traceable uncertainty estimate, is publicly available (see Dirksen et al., 2014).

Within GRUAN, a cluster of instruments at different (co-located) locations can be operated as a distributed site. If, at such a distributed site, the measurements made at one location are used to estimate measurements made elsewhere, the uncertainty on the estimated measurements must include the original measurement uncertainty plus the additional uncertainty introduced by the transfer algorithm (GCOS-170, 2013). A method to enhance the value of a distributed GRUAN site by combining measurements from distributed instruments is presented here. Measurements made at Invercargill (46.413° S, 168.3173° E, 1 m) are used as predictors of the temperatures above Lauder (45.0383° S, 169.6843° E, 370 m) and, as required, the uncertainty added by the transfer algorithm is calculated and included in the best estimate of the temperature.

GRUAN is also setting new standards regarding the careful documentation of data sets, i.e. every GRUAN data product has an accompanying peer-reviewed publication describing the bias correction, retrieval steps, and uncertainty components. Following GRUAN's tenets, this paper documents a method to create a temperature SASBE for Lauder, New Zealand. This paper first presents the observational data sets used herein, followed by a detailed description of the SASBE methodology used. The SASBE is presented in Sect. 4, followed by a discussion of method, results, and possible applications. A short conclusion completes the paper.

2 Observational Data and Reanalysis

The temperature SASBE combines radiosonde and automatic weather station data collected at Lauder and Invercargill. These measurements are supplemented with ERA5 reanalysis (Hersbach and Dee, 2016) which are used to provide an estimate of the diurnal temperature cycle above the sites. The atmospheric research facility at Lauder is operated by NIWA and the operational measurement site at Invercargill is run by the New Zealand MetService. Lauder is a certified GRUAN site which also includes radiosonde profiles from Invercargill in its data stream, based on the concept of a distributed site. The radiosonde data used in the SASBE described here were not GRUAN processed as the time frame for which the SASBE is calculated is prior to the availability of GRUAN profiles for Lauder. Nonetheless, the purpose if this paper is to demonstrate the construction of a SASBE for a GRUAN site for implementation across GRUAN. Lack of GRUAN data for this demonstration in no way compromises the validity of the method. Within this study, basic estimates of the 1σ uncertainty for radiosonde temperatures are used. If available, traceable uncertainty estimates such as those pro-

[1]The site has since joined GRUAN.

[2]Global Climate Observing System.

vided with the GRUAN radiosonde data product (Dirksen et al., 2014) should be used instead.

The ERA5 reanalysis is the fifth generation of atmospheric reanalysis produced by the European Centre for Medium-Range Weather Forecasts (ECMWF). ERA5 data for the years 2010 to 2017 are used in this study to provide diurnal temperature cycles above Lauder and Invercargill from which anomalies can be calculated. ERA5 reanalysis data for the years 2010 to 2017 are used here. Eventually, ERA5 will be available from 1950 onwards and monthly updates with a maximum latency of 3 months are planned, however at the beginning of 2018 only 8 years of data were available. ERA5 has an hourly temporal resolution, a 31 km horizontal grid worldwide with 137 vertical levels up to 0.01 hPa. It is the first reanalysis which includes uncertainties calculated by running a 10-member ensemble of data assimilations at 62 km resolution with 3-hourly output. The temperature and its uncertainty are interpolated to the location of the Lauder and Invercargill upper-air sites. Additional temporal interpolation is required to obtain uncertainty estimates on the hourly ERA5 temperatures.

To clarify the sources of the data and the various identification numbers for the stations, Table 1 summarises the data sources, while Table 2 provides the station details.

3 Methodology

A method to combine all available knowledge about the temperature profile at the lower South Island of New Zealand in a temperature SASBE for Lauder, located centrally on the lower South Island, is presented here. The SASBE is calculated at 16 vertical levels, i.e. surface, 925, 850, 700, 500, 400, 300, 250, 200, 150, 100, 70, 50, 30, 20, and 10 hPa, which are the so-called standard pressure level of radiosondes transmitted in the alphanumeric TEMP format. Currently, the global operational upper-air network is migrating towards high-resolution radiosonde reports in the BUFR format (Ingleby and Edwards, 2015). In the interest of clarity, the method is described as applied to one selected pressure level below.

One source of knowledge about the temperatures above Lauder is the ERA5 reanalysis which is used to calculate a diurnal temperature cycle. This diurnal cycle builds the foundation for the SASBE presented here, i.e. the SASBE converges towards the ERA5 diurnal cycle in the absence of any other sources of data. However, the upper air site at Lauder, New Zealand, performs weekly radiosondes launches. The temperature anomaly of the radiosonde measurement with respect to the ERA5 diurnal cycle is calculated and, around the launch time of the radiosonde, the temperature SASBE diverts from the diurnal cycle towards the observed temperature. The persistence (autocorrelation) of temperature anomalies can be used to determine how fast the SASBE converges towards the diurnal cycle. At the exact time of the radiosonde measurement, the autocorrelation function is 1, i.e. the measurement is perfectly correlated with itself, which implies that the SASBE adapts the measured value. With further time difference, the autocorrelation decreases and the SASBE converges towards the diurnal cycle.

While the upper-air research site in Lauder has a wide range of instruments, the temperature profile is only measured once weekly with radiosondes. Fortunately, about 180 km southwest of Lauder, the operational upper-air site at Invercargill measures the temperature profile twice daily. While the absolute temperatures above Lauder and Invercargill might differ, the temperature anomalies are well-correlated (see Sect. 3.1.2). Therefore, the Invercargill temperature anomalies with respect to the ERA5 diurnal cycle may be used to infer temperature anomalies above Lauder. Because these inferred anomalies provide knowledge about the temperature above Lauder they can be integrated into the SASBE to induce excursions from the diurnal cycle. As for the Lauder radiosonde anomalies, their influence should decrease with further time distance from the measurement time.

Within the SASBE, the inferred temperature anomalies should have less weight than the measurements made at Lauder, i.e. they should be attenuated as they originate from a different location, and should furthermore not be granted any weight at the time of a Lauder measurement. This is achieved by including the weight ϕ which is again based on the autocorrelation of temperature anomalies at Lauder. At any given time t, ϕ is the value of the autocorrelation function for the time difference to the closest Lauder launch time (see Fig. 1).

The concept of the SASBE is schematically shown in Fig. 1. The details of the preparation of the individual components and their combination is described in the following sections.

3.1 Preparing the individual components of the SASBE

3.1.1 Calculating the upper-air diurnal temperature cycle above Lauder and Invercargill

The diurnal temperature cycle above Lauder and Invercargill is calculated from the hourly ERA5 reanalysis. Bilinear interpolation in space is used to retrieve ERA5 reanalysis temperature at the exact location of the Lauder and Invercargill upper-air sites, respectively. Then, four Fourier pairs for the hour of the day (h), and four Fourier pairs for the day of the year (d) are fitted to the data (see Eqs. 1 and 2). Inclusion of the higher harmonics of the diurnal cycle, which represent the temporal variability due to atmospheric tides (Haurwitz, 1964), is possible due to the hourly resolution of ERA5. The ability to resolve atmospheric tides is typically limited by the temporal resolution of, e.g. satellite-based instruments (Huang et al., 2010).

Table 1. The data sets used. For those data sets available in New Zealand's National Climate Database, Table 2 gives the agent number and network number which are required to identify the site. The Invercargill radiosonde measurements are obtained from two sources, (i) New Zealand's Climate Database (low resolution), and (ii) as high-resolution profiles in the original Vaisala output format (MetService). The high resolution Vaisala files, which also include a range of metadata, have been processed into netCDF files (Bruno Kinoshita, personal communication, 2017) and are used for the years they are available. AWS abbreviates automatic weather station.

Data set	Source	Available at (last access: 5 December 2018)
Lauder radiosonde	Network for the Detection of Atmospheric Composition Change	ftp://ftp.cpc.ncep.noaa.gov/ndacc/station/lauder/
Lauder AWS	New Zealand's National Climate Database	https://cliflo.niwa.co.nz/
Invercargill radiosonde	MetService; New Zealand's National Climate Database	https://cliflo.niwa.co.nz/
Invercargill AWS	New Zealand's National Climate Database	https://cliflo.niwa.co.nz/
ERA5 reanalysis	ECMWF Copernicus Climate Change Service	https://climate.copernicus.eu/products/climate-reanalysis

Table 2. Station details for Lauder and Invercargill, including the World Meteorological Organization (WMO) station identifier (ID). The Agent and Network number are required to identify the station in New Zealand's National Climate Database. The Observing Systems Capability Analysis and Review (OSCAR) tool is WMO's official repository of metadata on surface-based observations.

Station	WMO ID	Cliflo Identifiers	OSCAR entry (last access: 5 December 2018)
Lauder	93817		https://oscar.wmo.int/surface/index.html#/search/station/stationReportDetails/12986
Lauder AWS		Agent 5535, Network I59065	https://oscar.wmo.int/surface/index.html#/search/station/stationReportDetails/12986
Invercargill	93844	Agent: 5814, Network: I68433	https://oscar.wmo.int/surface/index.html#/search/station/stationReportDetails/12991
Invercargill AWS	93845	Agent: 12444, Network: I68437	https://oscar.wmo.int/surface/index.html#/search/station/stationReportDetails/12992

$$T_{\text{Diur}} = \zeta_0 + \sum_{i=1}^{4} \zeta_{i\cdot 2-1} \cdot \sin\left(\frac{i \cdot 2\pi h}{24}\right) + \zeta_{i\cdot 2} \cdot \cos\left(\frac{i \cdot 2\pi h}{24}\right), \tag{1}$$

where all ζs are expanded in a Fourier series with four pairs to account for the annual cycle, i.e.

$$\zeta_i = \zeta_{i_0} + \sum_{j=1}^{4} \zeta_{i_{j\cdot 2-1}} \cdot \sin\left(\frac{j \cdot 2\pi d}{365.25}\right) + \zeta_{i_{j\cdot 2}} \cdot \cos\left(\frac{j \cdot 2\pi d}{365.25}\right). \tag{2}$$

This leads to $9 \cdot 9 = 81$ fit coefficient which are estimated for Lauder and for Invercargill and can be used to calculate a climatological mean diurnal temperature cycle for every day of the year, with a smooth transition from day to day.

The 1σ uncertainty on the regression coefficients is calculated based on the method described in Bodeker and Kremser (2015), using the 3-hourly 10-member ensemble standard deviations provided within ERA5, interpolated to an hourly resolution. The fitting uncertainty on the diurnal cycle is calculated as follows:

$$\sigma_{\text{fit}} = \sqrt{\sum_{i=1}^{81} \sigma_{\zeta_i}^2 \left(\frac{\partial T_{\text{Diur}}}{\partial \zeta_i}\right)^2}. \tag{3}$$

This uncertainty indicates how well the regression model fits the 8 years of hourly reanalysis data. However, this uncertainty does not indicate how representative the ERA5 diurnal cycle is for the temperatures measured above Lauder or Invercargill. Therefore, another uncertainty component, the representativeness uncertainty ($\sigma_{\text{representativeness}}$), is included and the uncertainty on the diurnal cycle is estimated as follows:

$$\sigma_{T_{\text{Diur}}} = \sqrt{\sigma_{\text{fit}}^2 + \sigma_{\text{representativeness}}^2}. \tag{4}$$

The representativeness uncertainty is estimated as the standard deviation of the differences between the radiosonde temperature and T_{Diur} for every available radiosonde launch at Lauder and Invercargill, respectively.

The fitted diurnal cycle with its associated uncertainty and the ERA5 temperatures at the respective day of the year are shown in Fig. 2 at 925 hPa for 2 selected days of the year. Depending on the site and the pressure level, the representativeness uncertainty reaches values from approximately 2.8 to 5.6 K and dominates the uncertainty on the diurnal cycle.

3.1.2 Inferring temperature anomalies above Lauder from temperature measurements above Invercargill

Although the absolute temperature values at Lauder (inland) and Invercargill (coastal) differ, the temperature anomalies with respect to their respective diurnal cycles are correlated. For a maximum time difference of 1.5 h between the Lauder and Invercargill radiosonde launch, the correlation of the anomalies is between $R = 0.69$ and $R = 0.92$ depending on the pressure level, with a mean correlation of $R = 0.85$. This correlation of anomalies is exploited to estimate temperature

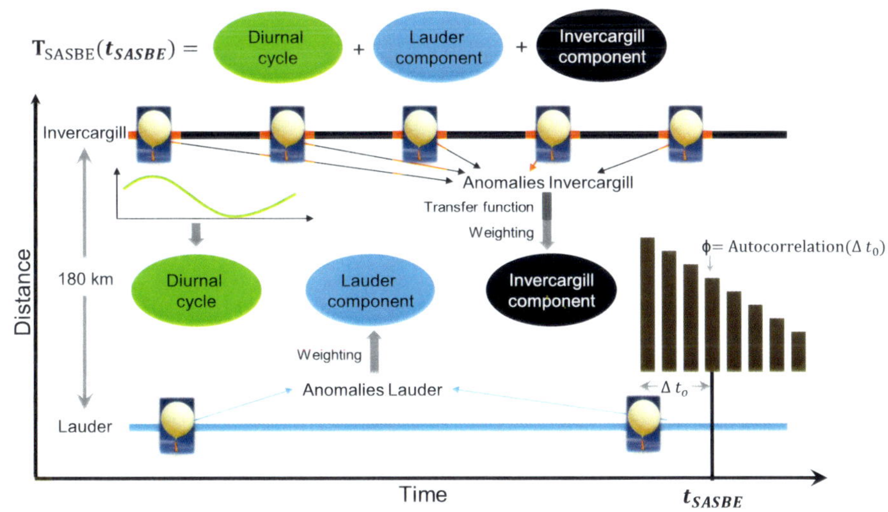

Figure 1. Schematic explanation of the temperature SASBE and its components. The diurnal cycle is determined from ERA5 as described in Sect. 3.1.1. To calculate the Invercargill component, a transfer function (see Sect. 3.1.2) and some weighting is applied to the temperature anomalies at Invercargill. To calculate the Lauder component, the temperature anomalies are weighted (see Sect. 3.2). Also shown is the weight ϕ, which is determined by the value of the autocorrelation function for the time difference between t and the closest Lauder radiosonde launch.

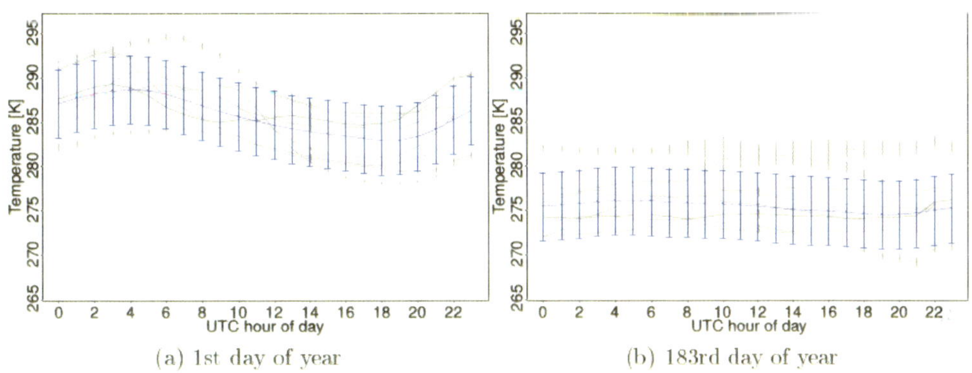

Figure 2. Fitted diurnal cycle (blue) and ERA5 temperatures for 2 example days of the year from 2010 to 2016 (grey), at 925 hPa above Lauder. The ERA5 ensemble standard deviation is shown as vertical bars around the grey lines and the uncertainty on the fitted diurnal cycle is shown in the blue error bars.

anomalies above Lauder based on temperature anomalies of radiosondes launched at Invercargill with respect to the diurnal cycle.

The temperature anomaly (T') at a given pressure level with respect to the diurnal cycle is calculated from the radiosonde measurement and the diurnal cycle calculated at the respective site as $T'(t) = T_{\mathrm{RS}}(t) - T_{\mathrm{Diur}}(t)$. The 1σ uncertainty on the temperature anomalies is calculated as $\sigma_{T'(t)} = \sqrt{\sigma^2_{T_{\mathrm{RS}}(t)} + \sigma^2_{T_{\mathrm{Diur}}(t)}}$. When the separation in launch time between Lauder and Invercargill is less than 1.5 h, the data are used to train a regression model (Eq. 5) that is applied to "translate" temperature anomalies from Invercargill to Lauder.

Four basis functions are used in the regression, i.e. the offset term that accounts for the constant bias between Lauder and Invercargill (Term 1), the temperature anomaly above Invercargill (Term 2), the difference (always Invercargill minus Lauder) in the surface pressure (ΔSP, Term 3), and the difference in the surface temperature anomaly ($\Delta\mathrm{ST}'$, Term 4).

$$\widehat{T}'_{\mathrm{Lau}} = \underbrace{\gamma}_{\text{Term 1}} + \underbrace{\beta \cdot T'_{\mathrm{Inv}}}_{\text{Term 2}} + \underbrace{\eta \cdot \Delta\mathrm{SP}}_{\text{Term 3}} + \underbrace{\kappa \cdot \Delta\mathrm{ST}'}_{\text{Term 4}} + \underbrace{\epsilon}_{\text{Term 5}}, \quad (5)$$

where γ, β, η, and κ are the regression coefficients and ϵ (Term 5) is the residual which cannot be explained with the regression model. Term 2 is expanded with a Fourier series for the wind direction θ to account for the dependence of the correlation between the Lauder and Invercargill temperature anomaly on the meteorological conditions, such as a lag/lead in the temperature anomaly based on the wind direction.

$$\beta = \beta_0 + \beta_1 \sin(2\pi\theta) + \beta_2 \cos(2\pi\theta) + \beta_3 \sin(4\pi\theta) + \beta_4 \cos(4\pi\theta) \tag{6}$$

Including the Fourier pairs and normalising all basis functions (except γ) by subtracting their mean value leads to

$$\begin{aligned}\widehat{T}'_{\text{Lau}} =\ & \gamma + \beta_0 \cdot \left(T'_{\text{Inv}} - \overline{T'_{\text{Inv}}}\right) \\ & + \beta_1 \cdot \left(T'_{\text{Inv}} \sin(2\pi\theta) - \overline{T'_{\text{Inv}} \sin(2\pi\theta)}\right) \\ & + \beta_2 \cdot \left(T'_{\text{Inv}} \cos(2\pi\theta) - \overline{T'_{\text{Inv}} \cos(2\pi\theta)}\right) \\ & + \beta_3 \cdot \left(T'_{\text{Inv}} \sin(4\pi\theta) - \overline{T'_{\text{Inv}} \sin(4\pi\theta)}\right) \\ & + \beta_4 \cdot \left(T'_{\text{Inv}} \cos(4\pi\theta) - \overline{T'_{\text{Inv}} \cos(4\pi\theta)}\right) \\ & + \eta \cdot \left(\Delta\text{SP} - \overline{\Delta\text{SP}}\right) + \kappa \cdot \left(\Delta\text{ST}' - \overline{\Delta\text{ST}'}\right) + \epsilon. \end{aligned} \tag{7}$$

The overbar denotes the mean of the basis function. At 925 hPa the wind direction is not recorded in the input data set and therefore a simplified regression model (Eq. 8), excluding the Fourier expansion is applied.

$$\widehat{T}'_{\text{Lau}} = \gamma + \beta \cdot \left(T'_{\text{Inv}} - \overline{T'_{\text{Inv}}}\right) + \eta \cdot \left(\Delta\text{SP} - \overline{\Delta\text{SP}}\right) + \kappa \cdot \left(\Delta\text{ST}' - \overline{\Delta\text{ST}'}\right) + \epsilon \tag{8}$$

The residuals (ϵ) are plotted in Fig. 3 for varying maximum time differences $\Delta t_{\text{regression}}$ between the Lauder and Invercargill launch times. Decreasing the time difference, in general, decreases the residuals slightly at most levels. However, at 30 hPa the regression model trained on collocations from within 6 h leads to the smallest residuals. Further analysis shows that the correlation between Lauder and Invercargill temperature anomalies is larger for a maximum time difference of 3 h ($R = 0.78$) and 6 h ($R = 0.76$) than for 1.5 h ($R = 0.69$) at the 30 hPa level. The sample size for collocations within 1.5 h is small and contains some large differences between the Lauder and Invercargill anomalies. The increased persistence in the stratosphere, combined with the small sample size, leads to a larger correlation when expanding the time difference. However, since the maximum time difference of 1.5 h minimises the residuals at most pressure levels, it is used as $\Delta t_{\text{regression}}$. This choice is ultimately an expert decision and further options could be tested to ensure an optimal choice is made, e.g. by testing different maximum time differences for each pressure level. For the purpose of demonstrating this methodology, a constant value has been selected with the expectation that the results would not change significantly for reasonable variation of $\Delta t_{\text{regression}}$. At the surface, plotted at 965 hPa in Fig. 3 for convenience, the residuals are zero based on the availability of hourly surface data. The calculation of the residuals at the surface level provides a test of the regression model. Within the SASBE,

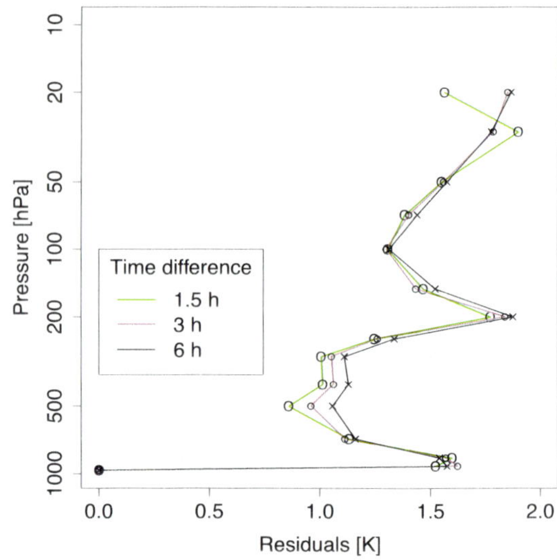

Figure 3. Temperature residuals for the regression model described in Eq. (5). The regression model is trained with three different data sets containing Lauder and Invercargill radiosonde measurements with launch times having a maximum time difference of 1.5 h (green, 324 collocations), 3 h (pink, 646 collocations), and 6 h (blue, 879 collocations).

the measurements of the automatic weather station are used at the surface level.

The uncertainties on the Invercargill and Lauder measurements and anomalies, and on the regression coefficients, are propagated through Eqs. (7) and (8), following standard error propagation rules (e.g. Bevington and Robinson, 2003). Uncertainties of different variables are assumed to be uncorrelated since the error covariance matrices are unknown. The uncertainty on the estimated Lauder temperature anomaly is then calculated including the partial derivatives with respect to all variables. The uncertainty for the full regression model is shown in Eq. (9) and a simplified version is applied at the 925 hPa level.

$$\begin{aligned}\sigma_{\widehat{T}'_{\text{Lau}}} = \Bigg(& \sigma_\gamma^2 \left(\frac{\partial \widehat{T}'_{\text{Lau}}}{\partial \gamma}\right)^2 + \sigma_{\beta_0}^2 \left(\frac{\partial \widehat{T}'_{\text{Lau}}}{\partial \beta_0}\right)^2 + \sigma_{\beta_1}^2 \left(\frac{\partial \widehat{T}'_{\text{Lau}}}{\partial \beta_1}\right)^2 \\ & + \sigma_{\beta_2}^2 \left(\frac{\partial \widehat{T}'_{\text{Lau}}}{\partial \beta_2}\right)^2 + \sigma_{\beta_3}^2 \left(\frac{\partial \widehat{T}'_{\text{Lau}}}{\partial \beta_3}\right)^2 + \sigma_{\beta_4}^2 \left(\frac{\partial \widehat{T}'_{\text{Lau}}}{\partial \beta_4}\right)^2 \\ & + \sigma_{T'_{\text{Inv}}}^2 \left(\frac{\partial \widehat{T}'_{\text{Lau}}}{\partial T'_{\text{Inv}}}\right)^2 + \sigma_\theta^2 \left(\frac{\partial \widehat{T}'_{\text{Lau}}}{\partial \theta}\right)^2 + \sigma_\eta^2 \left(\frac{\partial \widehat{T}'_{\text{Lau}}}{\partial \eta}\right)^2 \\ & + \sigma_{\Delta\text{SP}}^2 \left(\frac{\partial \widehat{T}'_{\text{Lau}}}{\partial \Delta\text{SP}}\right)^2 + \sigma_\kappa^2 \left(\frac{\partial \widehat{T}'_{\text{Lau}}}{\partial \kappa}\right)^2 + \sigma_{\Delta\text{ST}'}^2 \left(\frac{\partial \widehat{T}'_{\text{Lau}}}{\partial \Delta\text{ST}'}\right)^2 \Bigg)^{1/2} \end{aligned} \tag{9}$$

As a result, every estimated Lauder temperature anomaly has an individual uncertainty estimate which includes the uncertainty on the measurements and Invercargill anomaly, as well as the uncertainty introduced by the regression model.

Unfortunately, there appears to be considerable uncertainty about uncertainties associated with radiosonde measurements. With the exception of the GRUAN radiosonde data product, which is corrected for all known biases, and provides an uncertainty best estimate on every value, the different terminology used by different authors of publications using radiosonde measurements leads to confusion. Furthermore, the uncertainty estimates are not traceable and are commonly provided as a general estimate, rather than as an individual value for each measurement. For the purpose of this temperature SASBE, rather than unravelling the different estimates of uncertainties for the same instruments, we set the 1σ uncertainty to the values given in Table 3. The exact values of the uncertainties might differ from these estimates, but this is irrelevant to the mechanics of the methodology and to the temperature estimate itself, but could inflate or deflate the uncertainty bars. The 1σ temperature uncertainty for the Vaisala RS92 is set to 0.25 K based on Vaisala (2013), which is in agreement with Steinbrecht et al. (2008). This uncertainty is doubled for the Vaisala RS80, which is the older radiosonde model manufactured by Vaisala.

3.2 Combining the individual components of the SASBE

The temperature SASBE for Lauder is calculated from the different components presented in the Sect. 3.1.1 and 3.1.2. The diurnal temperature T_{Diur} builds the foundation for the SASBE. Estimates of the temperature anomaly above Lauder ($\widehat{T'}_{\text{Lau}}$) inferred from radiosonde launches at Invercargill are available 12-hourly (the $\widehat{\ }$ symbol is used for inferred quantities). The Lauder temperature anomalies (T'_{Lau}) are available approximately weekly.

At any time t the SASBE temperature $T_{\text{SASBE}}(t)$ may be influenced by several radiosonde measurements made at Lauder and Invercargill (the maximum time difference of measurements influencing the temperature SASBE is set to 30 days). Weighting determines how much weight is given to a certain temperature anomaly. The weight ϕ determines how much influence the Lauder and Invercargill measurements have. At the time of a radiosonde measurement at Lauder $\phi = 1$ which implies that the entire weight is given to the Lauder temperature anomalies. With further time difference from the closest launch at Lauder (Δt_0), ϕ decreases based on the autocorrelation function (acf) and Invercargill measurements gain weight. Autocorrelation is the correlation of a time series with a lagged (delayed) version of itself (Wilks, 2011). Here ϕ is calculated from ERA5 temperature anomalies with respect to the ERA5 diurnal cycle as follows:

$$\phi = \text{acf}(\Delta t_0), \quad \text{if} \quad \phi < 0.5 \quad \Rightarrow \quad \phi = 0.5. \tag{10}$$

The lowest value of ϕ is restricted to 0.5 to limit the maximum weight given to the Invercargill component. The choice of this threshold is ultimately an expert decision. A quantitative comparison of different values for ϕ could be made by analysing the effects of different lower thresholds for ϕ during the period of an extensive measurement campaign.

Table 3. Estimates of uncertainties associated with the measurements used in the SASBE.

Instrument and variable	Uncertainty
Automatic weather station temperature	0.2 K
Automatic weather station pressure sensor	0.5 hPa
Vaisala RS92 temperature	0.25 K
Vaisala RS80 temperature	0.5 K
Vaisala RS80/RS92 wind direction	3°

Furthermore, another weight is required to determine how much influence any individual temperature anomaly from Lauder or Invercargill has. These weights are taking into account the time difference (Δt_i, Δt_j) of an individual launch to the time for which the SASBE is calculated. For Lauder temperature anomalies these weights, w_i, are calculated as follows:

$$w_i = \frac{1}{\Delta t_i^2} \cdot \frac{1}{\sum_{i=1}^{N} \Delta t_i^2}, \tag{11}$$

from the N Lauder launch times t_i. For the estimated anomalies inferred from Invercargill measurement the weights, w_j, are calculated from the M Invercargill launch times t_j as follows:

$$w_j = \frac{1}{\Delta t_j^2} \cdot \frac{1}{\sum_{i=1}^{N} \frac{1}{\Delta t_i^2} + \sum_{j=1}^{M} \frac{1}{\Delta t_j^2}}. \tag{12}$$

The weights for Lauder anomalies are normalised with the sum over all Lauder weights and the anomalies implied from measurements in Invercargill are normalised with the sum of their own weight plus the sum of the Lauder weights. This insures that the inferred temperatures have limited influence.

If a positive temperature anomaly was determined from a radiosonde measurement at a given time, there is some confidence that the anomaly would still be positive in the hours before and afterwards. However, as a best estimate, the temperature anomaly a few hours after the measurement may conservatively be assumed to be lower than at the time of the measurement. This effect is included into the SASBE by attenuating the temperature anomaly with the autocorrelation function. This attenuated anomaly T^* is calculated as follows:

$$T^*_{i\text{Lau}}(t) = T'_{\text{Lau}}(t_i) \cdot \text{acf}(\Delta t_i), \tag{13}$$

with $\Delta t_i = |t - t_i|$ for Lauder. The attenuated estimate of anomalies above Lauder inferred from Invercargill radiosonde measurements is calculated as follows:

$$\widehat{T}^*_{j\text{Lau}}(t) = \widehat{T}'_{\text{Lau}}(t_j) \cdot \text{acf}(\Delta t_j), \tag{14}$$

where $\Delta t_j = |t - t_j|$.

Taking the different weights and the decaying temperature anomalies into account, the SASBE temperature is calculated as follows:

$$T_{\text{SASBE}}(t) = \underbrace{T_{\text{Diur}}(t)}_{\text{Diurnal cycle}} + \underbrace{\sum_{i=1}^{N} \phi \cdot w_i \cdot T^*_{i\text{Lau}}(t)}_{\text{Lauder component}} + \underbrace{\sum_{j=1}^{M} (1-\phi) \cdot w_j \cdot \widehat{T}^*_{j\text{Lau}}(t)}_{\text{Invercargill component}}. \quad (15)$$

As the SASBE requires an uncertainty estimate on every temperature value, the uncertainties are propagated through Eq. (15). In a first step, Eq. (15) is rewritten, by expanding the terms for the weighted temperature anomaly, as follows:

$$T_{\text{SASBE}}(t) = T_{\text{Diur}}(t) + \sum_{i=1}^{N} \phi \cdot w_i \cdot \text{acf}(\Delta t_i) \cdot \left[T_{\text{RS}_{\text{Lau}}}(t_i) - T_{\text{Diur}}(t_i)\right] + \sum_{j=1}^{M} (1-\phi) \cdot w_j \cdot \text{acf}(\Delta t_j) \cdot \left[\widehat{T}_{\text{RS}_{\text{Lau}}}(t_j) - T_{\text{Diur}}(t_j)\right], \quad (16)$$

where $T_{\text{RS}_{\text{Lau}}}$ is the temperature measured with a radiosonde above Lauder and $\widehat{T}_{\text{RS}_{\text{Lau}}}$ is a regression model estimated (see Sect. 3.1.2) temperature above Lauder based on the Invercargill radiosonde flight and the diurnal temperature cycle above Lauder. The uncertainty on the temperature SASBE is calculated from the following components:

$$\sigma_{T_{\text{SASBE}}(t)} = \sqrt{\sigma^2_{T_{\text{Diur}}} \left(\frac{\partial T_{\text{SASBE}}}{\partial T_{\text{Diur}}}\right)^2 + \sigma^2_{T_{\text{RS}_{\text{Lau}}}} \left(\frac{\partial T_{\text{SASBE}}}{\partial T_{\text{RS}_{\text{Lau}}}}\right)^2 + \sigma^2_{\widehat{T}_{\text{RS}_{\text{Lau}}}} \left(\frac{\partial T_{\text{SASBE}}}{\partial \widehat{T}_{\text{RS}_{\text{Lau}}}}\right)^2}, \quad (17)$$

with the partial derivatives:

$$\frac{\partial T_{\text{SASBE}}}{\partial T_{\text{Diur}}} = 1 - \sum_{i=1}^{N} \phi \cdot w_i \cdot \text{acf}(\Delta t_i) - \sum_{j=1}^{M} (1-\phi) \cdot w_j \cdot \text{acf}(\Delta t_j), \quad (18)$$

$$\frac{\partial T_{\text{SASBE}}}{\partial T_{\text{RS}_{\text{Lau}}}} = \sum_{i=1}^{N} \phi \cdot w_i \cdot \text{acf}(\Delta t_i), \quad (19)$$

and

$$\frac{\partial T_{\text{SASBE}}}{\partial \widehat{T}_{\text{RS}_{\text{Lau}}}} = \sum_{j=1}^{M} (1-\phi) \cdot w_j \cdot \text{acf}(\Delta t_j), \quad (20)$$

respectively, and therefore:

$$\sigma_{T_{\text{SASBE}}(t)} = \sqrt{\sigma^2_{T_{\text{Diur}}} \left(1 - \sum_{i=1}^{N} \phi \cdot w_i \cdot \text{acf}(\Delta t_i) - \sum_{j=1}^{M} (1-\phi) \cdot w_j \cdot \text{acf}(\Delta t_j)\right)^2 + \sigma^2_{T_{\text{RS}_{\text{Lau}}}} \left(\sum_{i=1}^{N} \phi \cdot w_{i_{\text{Lau}}} \cdot \text{acf}(\Delta t_i)\right)^2 + \sigma^2_{\widehat{T}_{\text{RS}_{\text{Lau}}}} \left(\sum_{j=1}^{M} (1-\phi) \cdot w_j \cdot \text{acf}(\Delta t_j)\right)^2}. \quad (21)$$

The uncertainty on inferred RS temperatures ($\widehat{T}_{\text{RS}_{\text{Lau}}}$) is estimated as follows:

$$\sigma_{\widehat{T}_{\text{RS}_{\text{Lau}}}} = \sqrt{\sigma^2_{T_{\text{RS}_{\text{Inv}}}} \left(\frac{\partial \widehat{T}_{\text{RS}_{\text{Lau}}}}{\partial T_{\text{RS}_{\text{Inv}}}}\right)^2 + \sigma^2_{T_{\text{Diur}_{\text{Lau}}}} \left(\frac{\partial \widehat{T}_{\text{RS}_{\text{Lau}}}}{\partial T_{\text{Diur}_{\text{Lau}}}}\right)^2 + \sigma^2_{T_{\text{Diur}_{\text{Inv}}}} \left(\frac{\partial \widehat{T}_{\text{RS}_{\text{Lau}}}}{\partial T_{\text{Diur}_{\text{Inv}}}}\right)^2}. \quad (22)$$

In Eq. (22), $\sigma_{T_{\text{Diur}_{\text{Lau}}}}$ and $\sigma_{T_{\text{Diur}_{\text{Inv}}}}$ include the uncertainty based on the fit only (see Eq. 3), to avoid including the representativeness uncertainty multiple times in the SASBE.

The uncertainty on the best-estimate temperature ($\sigma_{T_{\text{SASBE}}(t)}$), at the 1σ level, is calculated for every temperature value in the SASBE and is shown as black uncertainty bars in the top panels of Figs. 4–9.

4 Results

The temperature SASBE for Lauder is available in hourly resolution for 1997 to 2012 at 16 vertical levels. At the surface, the temperature SASBE consists of the measurements from the automatic weather station in Lauder. If no temperature or pressure measurement is available at a given time, the values are linearly interpolated if measurements are available within ± 1 h (temperature) or ± 12 h (pressure). In this study, the uncertainty added by the interpolation is not taken into account.

As an example, Fig. 4 show the hourly temperature SASBE at 925 hPa, and its individual components, for 6 to 9 December 2010. Panel 4a shows the SASBE temperature (black line) and its associated 1σ uncertainty bars. Also shown is the diurnal cycle and its uncertainty (green), as well as the results of a denial study (red), which calculates a best estimate of the temperature assuming that no radiosonde measurements are available at Lauder. At times when no radiosonde measurement above Lauder influences the temperature SASBE, the temperature of the denial study equals the

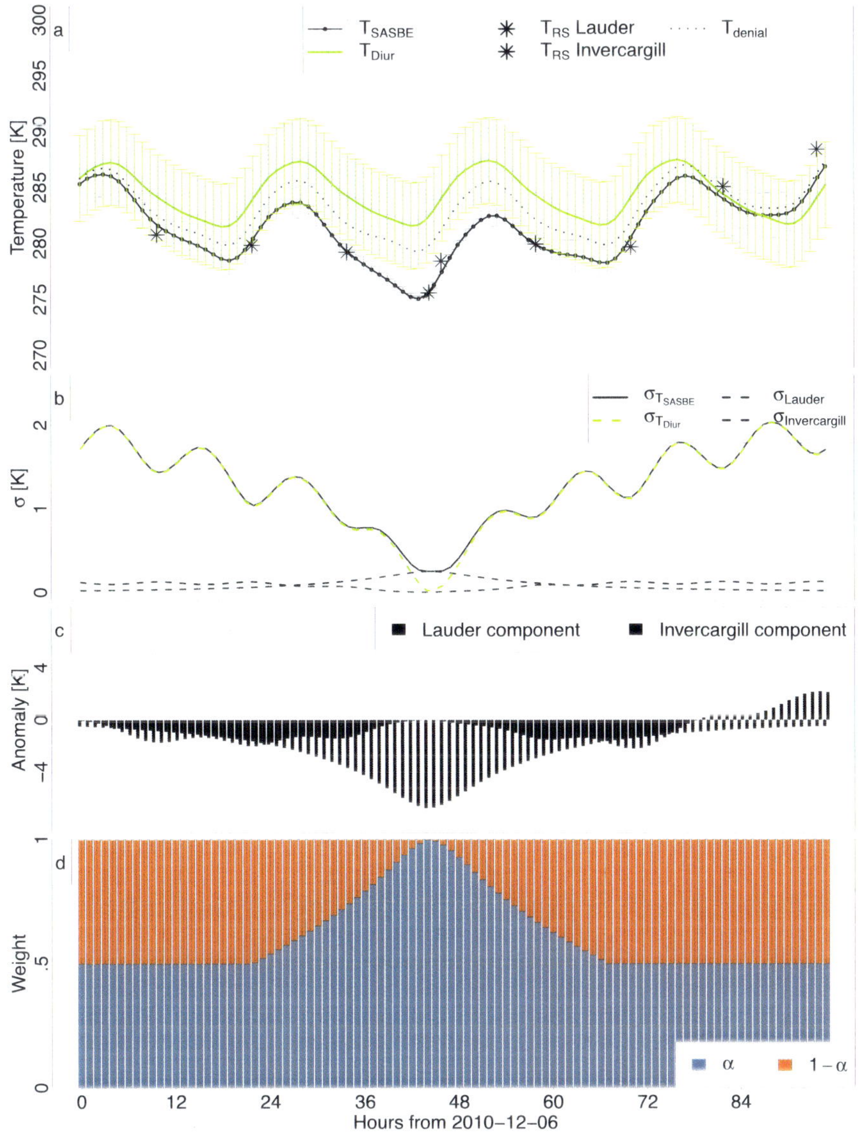

Figure 4. Temperature SASBE at 925 hPa and its individual components for 6 to 9 December 2010. **(a)** Temperature SASBE (black line), diurnal cycle (green line), denial study (i.e. pretending no measurements exist at Lauder, dotted red line), and radiosonde temperature measurement if available (Invercargill: red star; Lauder: blue star). **(b)** Uncertainty components from Eq. (17), i.e. uncertainty added by: the diurnal cycle (red), the Lauder radiosonde anomaly (blue), and the estimated Lauder anomaly (red). The combined uncertainty ($\sigma_{T_{\text{SASBE}}(t)}$) is shown as black line. **(c)** Weighted Lauder radiosonde temperature anomaly ($\sum_{i=1}^{N}\phi \cdot w \cdot T^{*}_{i_{\text{Lau}}}(t)$, blue bars) and weighted estimated Lauder temperature anomaly ($\sum_{j=1}^{M}\phi \cdot w_j \cdot \widehat{T}^{*}_{j_{\text{Lau}}}(t)$, red bars). **(d)** Weights ϕ and $(1-\phi)$ which determine how much weight is given to the Lauder radiosonde temperature anomalies and estimated Lauder anomalies based on radiosonde measurements taken above Invercargill.

temperature SASBE. Blue stars show the temperature measured with a radiosonde above Lauder, if available, and red stars show the temperature that is inferred from the Invercargill radiosondes by transferring the anomalies to Lauder using a regression model as described in Sect. 3.1.2. At times when a measurement is available above Lauder, the SASBE temperature is identical to the radiosonde measurement, as it receives the full weight, i.e. $\phi = 1$ (see panel d). Typically two radiosondes are launched daily from Invercargill at about 10:00 and 22:00 UTC (i.e. eight launches during the 4 days plotted), while typically one radiosonde is launched from Lauder each week.

Panel 4b shows the uncertainty added by the individual components and the total uncertainty on the SASBE, i.e. the terms of Eq. (17). At the exact time of a Lauder radiosonde measurement, the uncertainty on the SASBE equals the radiosonde uncertainty, and increases with time away from a measurement at Lauder, as the other terms in-

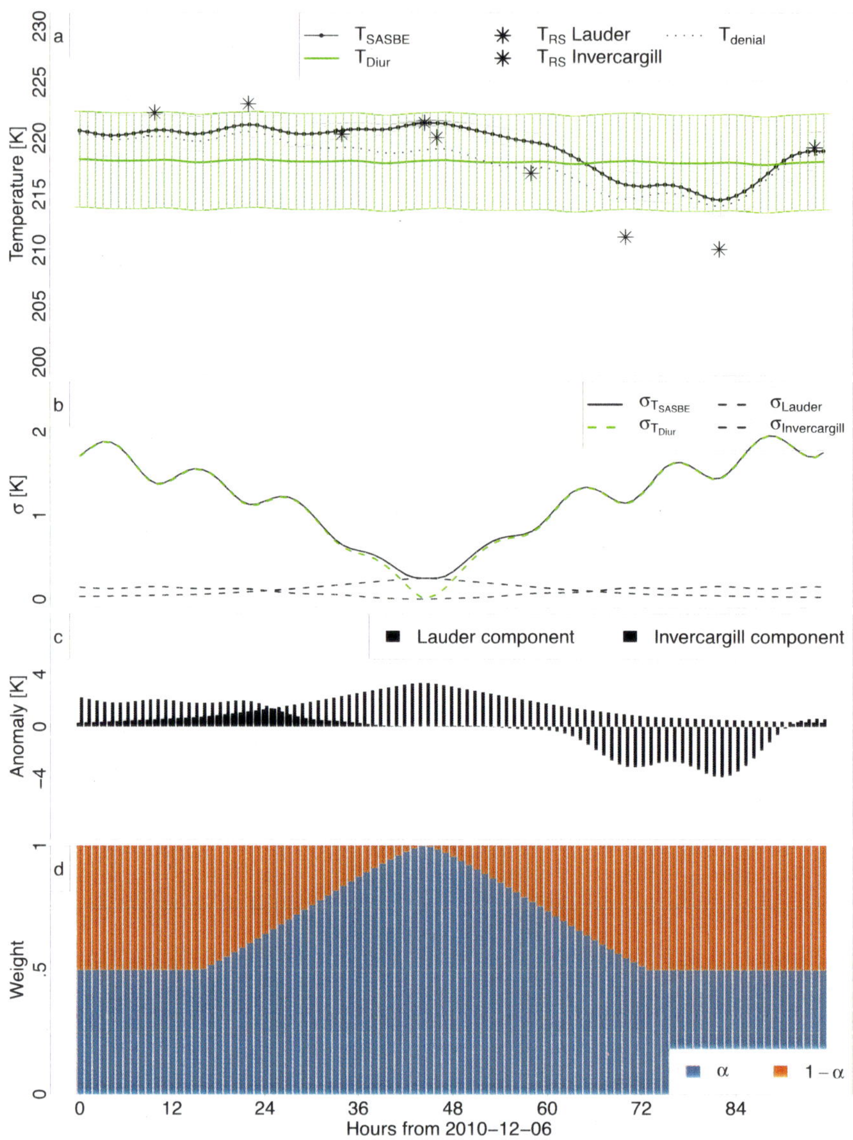

Figure 5. As in Fig. 4, but at 150 hPa.

fluence the SASBE and therefore increase the uncertainty. Panel 4c shows the weighted contribution from the Lauder radiosonde anomalies $\left(\sum_{i=1}^{N}\phi \cdot w \cdot T^{*}_{i_{\mathrm{Lau}}}(t)\right)$ in blue and the weighted contribution from the estimated Lauder anomalies $\left(\sum_{j=1}^{M}\phi \cdot w_j \cdot \widehat{T}^{*}_{j_{\mathrm{Lau}}}(t)\right)$ in red. The bottom panel shows the weights ϕ and $(1-\phi)$ which are given to the Lauder anomalies and estimated Lauder anomalies, respectively. As the autocorrelation of temperature anomalies decays slower in the stratosphere, the influence of the temperature measured in Lauder decreases more slowly at 70 hPa level (see Fig. 6) than at the 925 hPa level.

Figures 5 and 6 show the temperature SASBE for the same time frame at the 150 and 70 hPa level, respectively.

It can be seen in Figs. 4 and 5 that using the Invercargill measurements as a proxy for the temperature above Lauder is an improvement compared to using climatology, i.e. the diurnal cycle, only. This becomes obvious when comparing the differences between the SASBE and (i) denial study, and (ii) the diurnal cycle at the time of a Lauder radiosonde measurement. As the temperature of the denial study is closer to the SASBE than the diurnal cycle, including information from Invercargill would be beneficial at those instances. However at the 70 hPa level the fit to the radiosonde measurement in Lauder is slightly worse using Invercargill measurements as a proxy.

Figures 7 to 9 show the SASBE at 4 days during the austral winter at 925, 150, and 70 hPa, respectively. The diurnal temperature cycle is less pronounced in austral winter as can

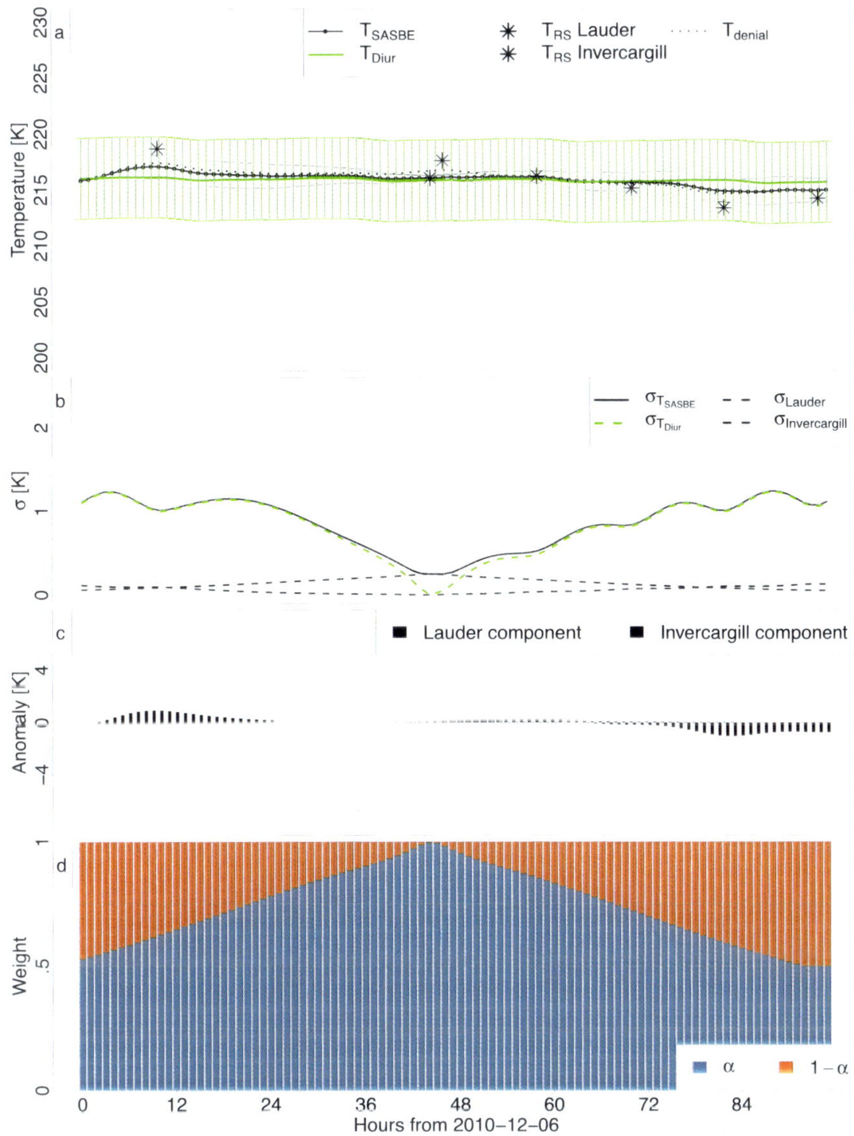

Figure 6. As in Fig. 4, but at 70 hPa.

be seen by comparing Fig. 7 with Fig. 4. For the Lauder radiosonde launch at 16 July 2003, the fit to the radiosonde is improved by using Invercargill measurements as a proxy at the 925 hPa level and at the 150 hPa level, however, the fit is slightly diminished at 70 hPa.

To evaluate if using Invercargill measurements as part of the SASBE is valuable, in general, it is tested whether using estimates of temperature anomalies above Lauder which are based on measurements from Invercargill improves the SASBE in comparison to using merely the ERA5 diurnal cycle. This is tested by analysing the yearly mean absolute residuals between the Lauder radiosonde measurements and (i) the diurnal cycle (green) and (ii) the diurnal cycle plus inferred Lauder temperature anomalies (denial study, red); see Fig. 10. At most pressure levels, and for all years, the residuals are distinctly smaller when using the estimated temperature anomalies. This means that using the Invercargill radiosondes plus a transfer algorithm improves the SASBE compared to using the diurnal cycle alone. At the 20 and 10 hPa level, improvements are only visible during certain years as most radiosondes from Invercargill do not reach these pressure levels.

5 Summary and discussion

The new data product presented here provides an hourly resolved best estimate of the temperature above Lauder, New Zealand, at 16 vertical levels from the surface to 10 hPa. This SASBE combines radiosonde and automatic weather station measurements from Lauder and Invercargill with a diurnal

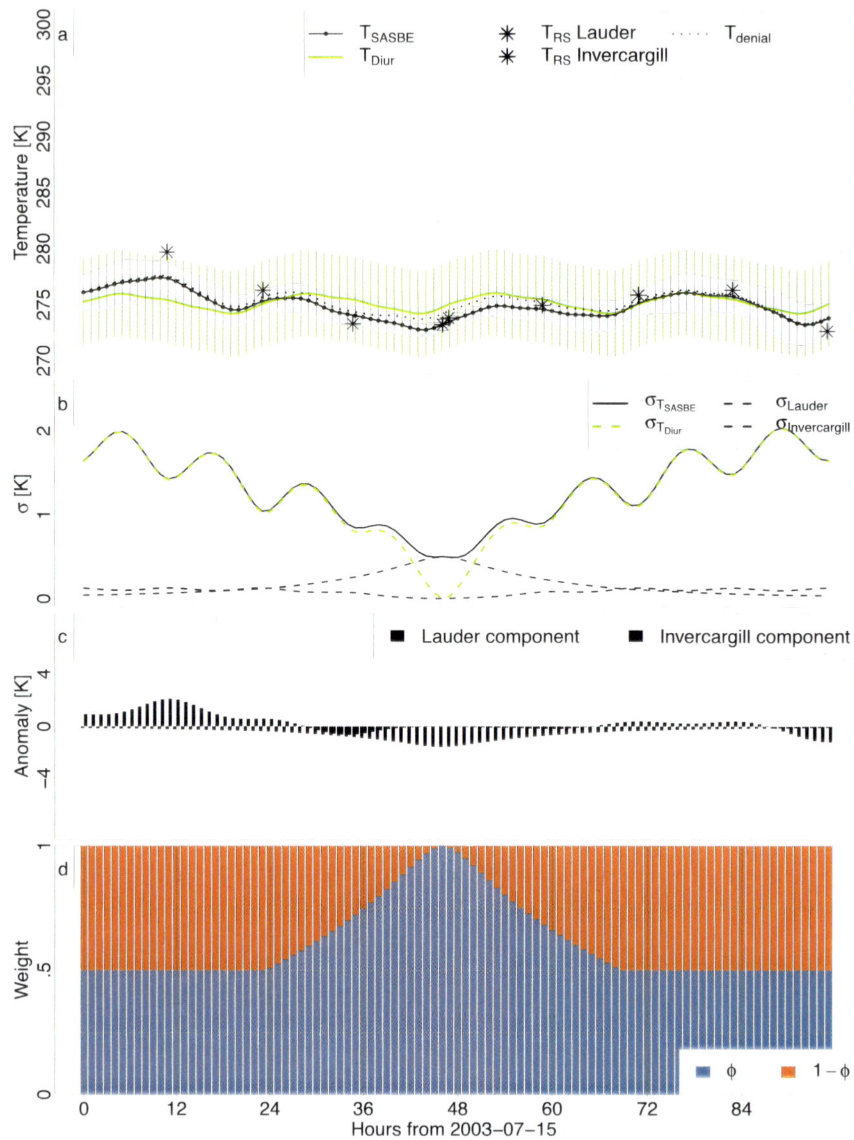

Figure 7. As in Fig. 4, but for 4 days in July 2003 at 925 hPa.

cycle calculated from the new ERA5 reanalysis by ECMWF. The information from radiosonde measurements made at Invercargill is transferred to Lauder using a regression model. While it would be preferable to have high temporal resolution (e.g daily) measurements available at Lauder, using the radiosonde measurements from Invercargill improves the estimate of the temperature compared to merely using the ERA5 diurnal cycle. Using the temperature anomalies of radiosondes launched from Invercargill with respect to the diurnal cycle, introduces uncertainty, which is calculated by error propagation through the regression model. This meets the requirements outlined by GCOS-170 (2013) regarding the handling of uncertainties when transferring information from the location of a measurement to another location. Thus, this study presents a methodology for how to estimate temperatures at a given site based on measurements made elsewhere. The described method is likely to be applicable to other ECVs that have a large spatio-temporal scale, such as ozone above the boundary layer. However, for ECVs that vary rapidly over distance and time, such as water vapour, the described method is unlikely to provide useful results. Therefore, prior to applying the method, the correlation between the time series of a chosen variable at both locations should be evaluated.

Within the development of this data product, some pragmatic choices have been made. While other researchers might choose some parameters differently, the methodology itself remains unaffected. Parameters to be selected include the maximum separation in time between data that are used to train the regression model, and the maximum weight given to

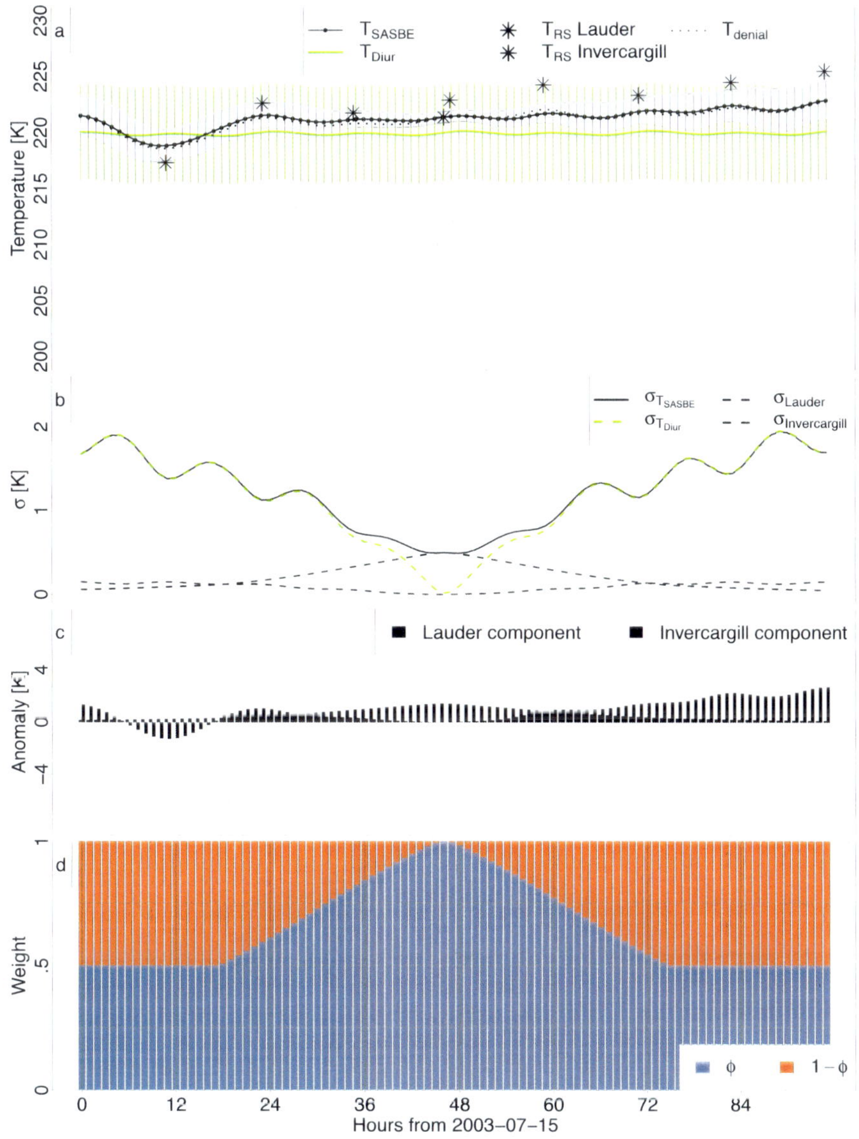

Figure 8. As in Fig. 4, but for 4 days in July 2003 at 150 hPa.

the estimated anomalies. Furthermore, different options for the calculation of the weights might be possible.

Due to its high temporal resolution, and the included estimate of the uncertainty, the SASBE is well-qualified for the validation of space-based sensors. Most overpasses of a satellite during the years 1997–2012 will lead to a close collocation in time, given the hourly resolution of the SASBE. Using individual radiosonde measurements rather than a SASBE can minimise the number of collocations significantly, as was the case in Calbet et al. (2017), who analysed differences between the Infrared Atmospheric Sounding Interferometer (IASI, Siméoni et al., 1997) and GRUAN radiosonde profiles, but was limited to using data from one GRUAN site only. If the satellite data product, i.e. the retrieved temperature profile, is provided with an uncertainty estimate, a quantitative comparison between the SASBE and the temperature retrieval is possible. The comparisons can either be made using direct collocations or by applying a numerical weather prediction model as a transfer standard (Tradowsky et al., 2017) to eliminate effects caused by imperfect spatial and, to a smaller degree, temporal collocation. However, most space-based sensors do not measure ECVs directly. For example, radiometers such as AIRS and IASI measure the top-of-the-atmosphere radiance and a radiative transfer model is required to retrieve atmospheric ECVs. As the retrieval of atmospheric temperatures from radiance is an ill-posed problem, it can be preferable to validate space-based radiometers in their native measurement space (see e.g. Calbet et al., 2017). With a radiative transfer model and additional water vapour profiles, the temperature SASBE can be used to pro-

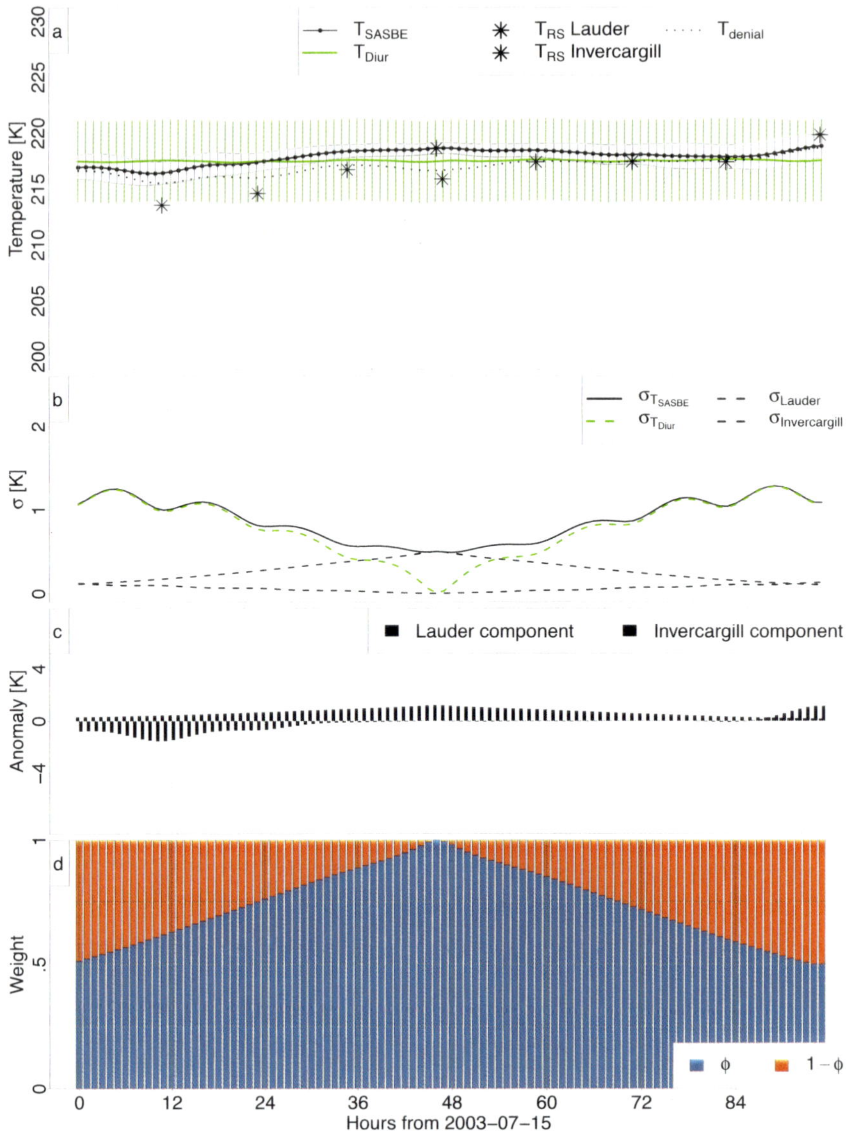

Figure 9. As in Fig. 4, but for 4 days in July 2003 at 70 hPa.

vide such a radiance space validation as it has been described in Tradowsky et al. (2016) and Tradowsky (2018). The uncertainties provided in the SASBEs can be propagated into uncertainties in the top-of-the-atmosphere radiances, by using a Monte Carlo approach. For this purpose, the top-of-the-atmosphere radiances are calculated many times (e.g. hundreds of model runs) with slightly changed input variables (i.e. temperature, ozone and water vapour). Every input variable is varied between its upper and lower uncertainty bound and different combinations of the variables are used to calculate the top-of-the-atmosphere radiances. The uncertainty of the modelled radiances is then obtained from the spread of the calculated top-of-the-atmosphere radiances for different variations of the input variables. The modelled radiances can then be compared with radiances obtained from satellite-based radiometers above the specific location. This method has the advantage that it validates the space-based radiometers in radiance space without requiring a retrieval and the a priori information required within the retrieval.

The temperature SASBE may also be used as a priori for the retrieval of other ECVs at Lauder, i.e. a temperature profile is required as a priori for the retrieval of an ozone profile from the stratospheric ozone lidar based at Lauder (Brinksma et al., 1997). As the radiosonde measurements made at Lauder have not been assimilated into numerical weather prediction models prior to 2016, the SASBE can be used as an approximately independent validation data set for model calculations (not entirely independent as the Invercargill measurements have been assimilated). Its hourly resolution and the availability of uncertainties make the SASBE

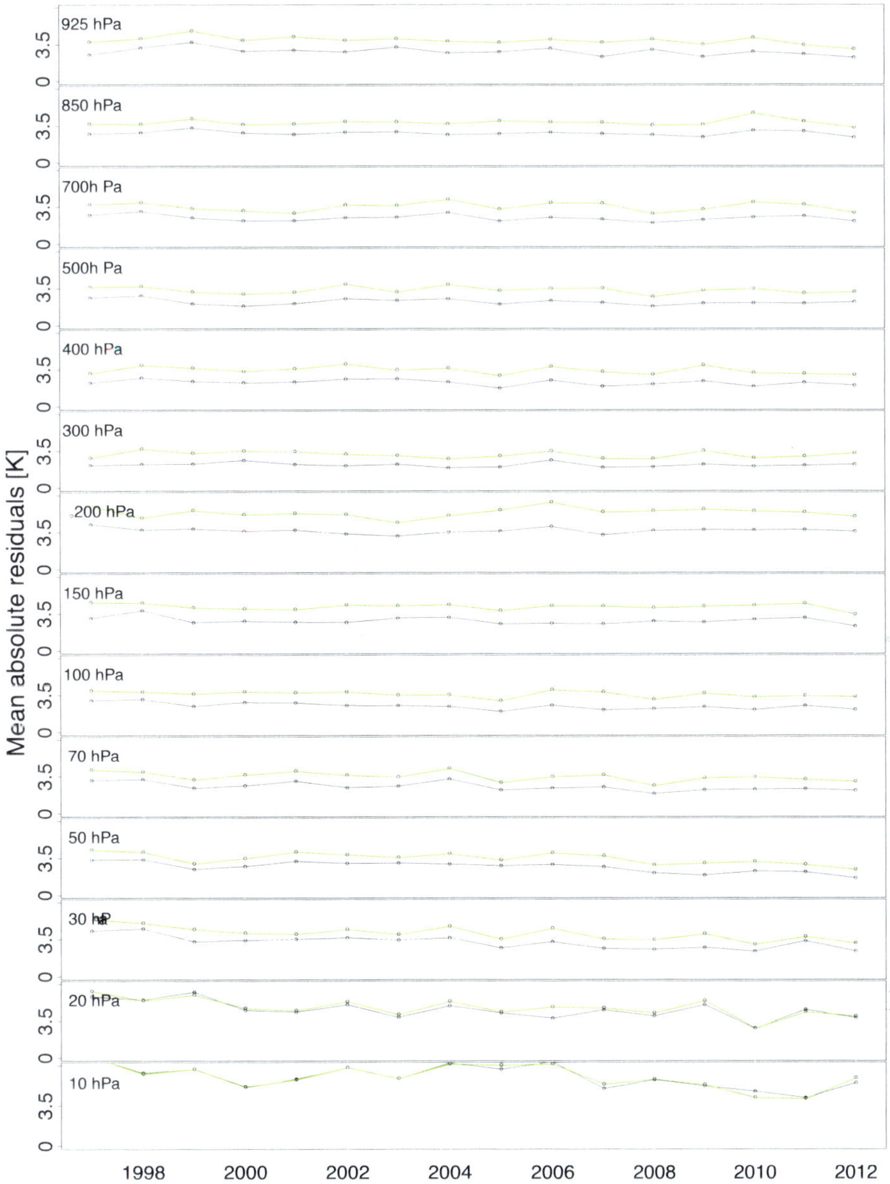

Figure 10. Yearly mean of absolute residuals between the Lauder radiosonde measurements and both (i) the diurnal cycle (green) and (ii) the denial study (red).

especially valuable. The ERA5 reanalysis is used to calculate the diurnal cycle and the anomalies with regards to this diurnal cycle above Lauder and Invercargill. However, to the best of our knowledge, Lauder radiosonde measurements have not been assimilated into ERA5, at least not prior to 2017, as they are not part of the Integrated Global Radiosonde Archive (IGRA) and have not been assimilated by ECMWF's operational forecast model prior to 2017 (Bruce Ingleby, personal communication, 2018). At times close to a Lauder radiosonde measurements, the SASBE is primarily determined by the Lauder radiosonde temperature. Thus, the SASBE may be valuable for assessing the accuracy of the ERA5 reanalysis when applying close collocation criteria for the time (i.e. around the radiosonde measurement time). While this is not different to using the Lauder radiosondes directly, the SASBE is available in an easier accessible and consistent format.

6 Code and data availability

The temperature SASBE for Lauder, New Zealand, has been long-term archived and is indexed with a doi, see https://doi.org/10.5281/zenodo.1195779 (Tradowsky et al., 2018). Users should contact the first author of the paper to inform her about the use of the data product. The code used to produce the SASBE has been developed in R and is available on request from the first author.

7 Conclusions

This publication describes the development of a temperature SASBE for the GRUAN site at Lauder, New Zealand. A method to combine measurements from distributed sites is presented which accounts for the additional uncertainty that is introduced by using the measurement of an ECV made at one location as a proxy for the ECV at another location nearby. This is an essential first step to enhance the value of distributed GRUAN sites. GRUAN operates several distributed sites, i.e. Cabauw and De Bilt, The Netherlands, Beltsville and Sterling, US, and Lauder and Invercargill, New Zealand. While the instrumentation might differ at every upper-air sites, the methodology described here can be adapted for other distributed sites. The results of this study show that radiosonde measurements from Invercargill, together with a transfer algorithm, add value to the temperature SASBE for Lauder. This is also reflected in the uncertainties on the SASBE temperature which decrease in the vicinity of a radiosonde measurement from Invercargill, which has been transferred to Lauder and is included into the SASBE. The method presented enhances the value of upper-air observations by (i) considerable improvement of the temporal resolution, (ii) the availability of an uncertainty estimate with every temperature value, and (iii) better documentation of the data set. While the SASBE presented here will not give an accurate estimate of the true temperature at every time step, it is a best estimate of this unknown true value, which includes an estimate of the uncertainty to indicate the confidence that users of the SASBE should have in the temperature estimate at a given time. The uncertainty is an essential part of the SASBE, and users of the data product are urged to be cognisant of the uncertainties when using the temperature SASBE.

Author contributions. JST developed the methodology for the SASBE, wrote the software, and drafted the paper. GEB supervised and supported the development of the methodology and provided detailed comments on the draft paper. RRQ, PJHB, and JF supervised the work and provided comments to the draft paper.

Competing interests. The authors declare that they have no conflict of interest.

Acknowledgements. The authors would like to thank the New Zealand MetService for the radiosonde data from Invercargill and Bruno Kinoshita (NIWA) for converting a subset of radiosondes profiles from PCCORA binary into netCDF format. The measurements and activities at Lauder are supported by the Strategic Science Investment Fund (SSIF) through the New Zealand Ministry of Business, Innovation and Employment (MBIE). Jordis S. Tradowsky acknowledges support from the German Academic Exchange Service. The SASBE contains modified Copernicus Climate Change Service information (2017). The authors would further like to thank two anonymous reviewers for their helpful comments that improved the paper.

Edited by: David Carlson

References

Aumann, H., Chahine, M., Gautier, C., Goldberg, M., Kalnay, E., McMillin, L., Revercomb, H., Rosenkranz, P., Smith, W., Staelin, D., and Strow, L. L.: AIRS/AMSU/HSB on the Aqua mission: Design, science objectives, data products, and processing systems, IEEE T. Geosci. Remote, 41, 253–264, 2003.

Bevington, P. and Robinson, D.: Data Reduction and Error Analysis for the Physical Sciences, 3rd edn., McGraw Hill, New York, USA, 2003.

Bodeker, G. E. and Kremser, S.: Techniques for analyses of trends in GRUAN data, Atmos. Meas. Tech., 8, 1673–1684, https://doi.org/10.5194/amt-8-1673-2015, 2015.

Bodeker, G. E., Bojinski, S., Cimini, D., Dirksen, R., Haeffelin, M., Hannigan, J., Hurst, D., Leblanc, T., Madonna, F., Maturilli, M., Mikalsen, A., Philipona, R., Reale, T., Seidel, D., Tan, D., Thorne, P., Vömel, H., and Wang, J.: Reference Upper-Air Observations for Climate: From Concept to Reality, B. Am. Meteorol. Soc., 97, 123–135, https://doi.org/10.1175/bams-d-14-00072.1, 2016.

Bojinski, S., Verstraete, M., Peterson, T., Richter, C., Simmons, A., and Zemp, M.: The Concept of Essential Climate Variables in Support of Climate Research, Applications, and Policy, B. Am. Meteorol. Soc., 95, 1431–1443, https://doi.org/10.1175/BAMS-D-13-00047.1, 2014.

Brinksma, E., Swart, D., Bergwerff, J., Meijer, Y., and Ormel, F.: Advances in Atmospheric Remote Sensing with Lidar, chap. Stratospheric and Mesospheric Profiling, RIVM Stratospheric Ozone Lidar at NDSC Station Lauder: Routine Measurements and Validation During the OPAL Campaign, 529–532, Springer, Berlin, Heidelberg, https://doi.org/10.1007/978-3-642-60612-0_128, 1997.

Calbet, X.: Assessment of adequate quality and collocation of reference measurements with space-borne hyperspectral infrared instruments to validate retrievals of temperature and water vapour, Atmos. Meas. Tech., 9, 1–8, https://doi.org/10.5194/amt-9-1-2016, 2016.

Calbet, X., Peinado-Galan, N., Rípodas, P., Trent, T., Dirksen, R., and Sommer, M.: Consistency between GRUAN sondes, LBLRTM and IASI, Atmos. Meas. Tech., 10, 2323–2335, https://doi.org/10.5194/amt-10-2323-2017, 2017.

Dirksen, R. J., Sommer, M., Immler, F. J., Hurst, D. F., Kivi, R., and Vömel, H.: Reference quality upper-air measurements: GRUAN data processing for the Vaisala RS92 radiosonde, Atmos. Meas. Tech., 7, 4463–4490, https://doi.org/10.5194/amt-7-4463-2014, 2014.

GCOS-112: GCOS REFERENCE UPPER-AIR NETWORK (GRUAN): Justification, requirements, siting and instrumentation options, Tech. Rep. GCOS-112 (WMO/TD No. 1379), World Meteorological Organization, available at: https://www.gruan.org/gruan/editor/documents/gcos/gcos-112.pdf (last access: 5 December 2018), 2007.

GCOS-170: The GCOS Reference Upper-Air Network (GRUAN), WIGOS Technical Report 2013-2, World Meteorological Organization, Geneva, Switzerland, 2013.

GCOS-200: The Global Observing System for Climate: Implementation Needs, Tech. rep., World Meteorological Organization, Geneva, Switzerland, 2016.

Haurwitz, B.: Atmospheric Tides, Science, 144, 1415–1422, 1964.

Hersbach, H. and Dee, D.: ERA5 reanalysis is in production, ECMWF Newsletter No. 147, available at: https://www.ecmwf.int/en/newsletter/147/news/era5-reanalysis-production (last access: 5 December 2018), 2016.

Huang, F., McPeters, R., Bhartia, P., Mayr, H., Frith, S., Russell, J., and Mlynczak, M.: Temperature diurnal variations (migrating tides) in the stratosphere and lower mesosphere based on measurements from SABER on TIMED, J. Geophys. Res.-Atmos., 115, D16121, https://doi.org/10.1029/2009JD013698, 2010.

Ingleby, B.: An assessment of different radiosonde types 2015/2016, Technical Memorandum 807, European Centre for Medium-Range Weather Forecasts, Reading, UK, 2017.

Ingleby, B. and Edwards, D.: Changes to radiosonde reports and their processing for numerical weather prediction, Atmos. Sci. Lett., 16, 44–49, https://doi.org/10.1002/asl2.518, 2015.

Maillard Barras, E., Haefele, A., Stübi, R., and Ruffieux, D.: A method to derive the Site Atmospheric State Best Estimate (SASBE) of ozone profiles from radiosonde and passive microwave data, Atmos. Meas. Tech. Discuss., 8, 3399–3422, https://doi.org/10.5194/amtd-8-3399-2015, 2015.

Siméoni, D., Singer, C., and Chalon, G.: Infrared atmospheric sounding interferometer, Acta Astronaut., 40, 113–118, 1997.

Steinbrecht, W., Claude, H., Schönenborn, F., Leiterer, U., Dier, H., and Lanzinger, E.: Pressure and Temperature Differences between Vaisala RS80 and RS92 Radiosonde Systems, J. Atmos. Ocean. Tech., 25, 909–927, https://doi.org/10.1175/2007JTECHA999.1, 2008.

Tobin, D., Revercomb, H., Knuteson, R., Lesht, B., Strow, L., Hannon, S., Feltz, W., Moy, L., Fetzer, E., and Cress, T.: Atmospheric Radiation Measurement site atmospheric state best estimates for Atmospheric Infrared Sounder temperature and water vapour retrieval, J. Geophys. Res., 111, D09S14, https://doi.org/10.1029/2005JD006103, 2006.

Tradowsky, J.: Enhancing the Upper-Air Observational Temperature Record to Improve Satellite Validation and Weather Forecasts, PhD thesis, Freie Universität Berlin, available at: https://refubium.fu-berlin.de/handle/fub188/22207, last access: 5 December 2018.

Tradowsky, J., Bodeker, G., Thorne, P., Carminati, F., and Bell, W.: GRUAN in the service of GSICS: Using reference ground-based profile measurements to provide traceable radiance calibration for space-based radiometers, GSICS Quarterly Newsletter, 10, 5–6, https://doi.org/10.7289/V5GT5K7S, 2016.

Tradowsky, J., Burrows, C., Healy, S., and Eyre, J.: A New Method to Correct Radiosonde Temperature Biases Using Radio Occultation Data, J. Appl. Meteorol. Clim., 56, 1643–1661, https://doi.org/10.1175/JAMC-D-16-0136.1, 2017.

Tradowsky, J. S., Bodeker, G. E., Querel, R. R., Builtjes, P. J. H., and Fischer, J.: A Site Atmospheric State Best Estimate of Temperature for Lauder, New Zealand (1997–2012), Version 2.3.0 [Data set], Zenodo, https://doi.org/10.5281/zenodo.1195779, 2018.

Vaisala: Vaisala Radiosonde RS92 Performance in the WMO Intercomparison of High Quality Radiosonde Systems/Yangjiang, China 2010, Vaisala White Paper, Vantaa, Finland, 2013.

Wilks, D.: Statistical methods in the atmospheric sciences, 3rd edn., Academic Press, San Diego, USA, 2011.

10

The Total Carbon Column Observing Network site description for Lauder, New Zealand

David F. Pollard[1], Vanessa Sherlock[1], John Robinson[1], Nicholas M. Deutscher[2], Brian Connor[3], and Hisako Shiona[1]

[1]National Institute of Water and Atmospheric Research Ltd (NIWA), Lauder, New Zealand
[2]School of Chemistry, University of Wollongong, Northfields Ave, Wollongong, NSW, 2522, Australia
[3]BC Consulting Limited, Martinborough, New Zealand

Correspondence to: David F. Pollard (dave.pollard@niwa.co.nz)

Abstract. In this paper we describe the retrievals of atmospheric trace gases from near-infrared, high-resolution solar absorption spectroscopy measurements at the Lauder atmospheric research station in New Zealand and submitted to the Total Carbon Column Observing Network (TCCON) archive.

The Lauder site (45.034° S, 169.68° E, 370 m a.s.l.) is located within a sparsely populated region of the South Island of New Zealand and is sheltered from the prevailing wind direction by the Southern Alps, which gives the site a high number of clear-sky days and an air mass that is largely unmodified by regional anthropogenic sources. The Lauder TCCON archive consists of data from two instruments: a Bruker IFS 120HR from June 2004 to February 2010 and a Bruker IFS 125HR from February 2010 to present. The bias between the two instruments is assessed to be 0.068 % for CO_2. Since measurements using the IFS 125HR began, the SD about the hourly mean has been better than 0.1 % for 96.81 % of CO_2 column retrievals.

The retrievals have been calibrated against in situ airborne measurements to correct for biases and provide traceability to the World Meteorological Organization (WMO) scales with an accuracy of 0.1 % for CO_2.

The Lauder TCCON time series of retrieved dry-air mole fractions of CO_2, CH_4, N_2O, HF, H_2O, HDO and CO are available from the TCCON data archive.

1 Introduction

Atmospheric trace gases that absorb infrared radiation have an important influence on the Earth's climate due to the greenhouse effect. Accumulation of some of these species, most notably carbon dioxide (CO_2) and methane (CH_4), are causing changes to the global radiation budget, which in turn is modifying the climate system. In order to understand the direct and indirect influences this process is having on the climate, it is important to understand the global distribution, sources and sinks of these gases.

One method for measuring the total column abundance of these gases is high-resolution solar absorption spectroscopy using Fourier transform spectrometers (FTSs) in the near-infrared spectral region. The Total Carbon Column Observing Network (TCCON, Wunch et al., 2011a) consists of a geographically dispersed network of such instruments designed to provide consistent and precise retrievals of a number of the atmospheric species. For CO_2 column measurements, Rayner and O'Brien (2001) demonstrated that a precision of better than 0.25 % is necessary to improve understanding of the carbon cycle while Olsen and Randerson (2004) showed that a precision of better than 0.1 % would allow the strength of the Northern Hemisphere carbon sink to be assessed. The TCCON has adopted these values as measurement precision

Table 1. Instrument information.

Date	Instrument	Detectors	Spectral range	LSE correction method	Data processing version	TCCON archive name
20 Jun 2004–28 Feb 2011	120HR	Extended InGaAs	3900–9000 cm^{-1}	None	GGG2014R01	lauder01
02 Feb 2010–present	125HR	Extended InGaAs + Si	3900–15 000 cm^{-1}	Si-based	GGG2014R01	lauder02

targets for the network with an expected site-to-site bias of less than 0.2 % (Wunch et al., 2015).

In this paper we describe the TCCON site at Lauder, New Zealand (45.034° S, 169.68° E, 370 m a.s.l.), and the time series of column-averaged dry-air mole fractions (DMFs, denoted by X_{gas}) of CO_2, CH_4, nitrous oxide (N_2O), hydrogen fluoride (HF), water vapour (H_2O and HDO) and carbon monoxide (CO), which have been retrieved using the 2014 version of the GGG processing software (Wunch et al., 2015) and are archived on the TCCON data archive (Sherlock et al., 2014a, b). This time series spans from 2004 to the present and represents the longest TCCON dataset in the Southern Hemisphere.

TCCON data from Lauder have been used extensively to support satellite missions, such as the Greenhouse Gases Observing Satellite (GOSAT, Yokota et al., 2009; Wunch et al., 2011b; Crisp et al., 2012) and the Orbiting Carbon Observatory 2 (OCO-2, Crisp et al., 2008), by providing data for retrieval validation and algorithm development (e.g. Schepers et al., 2016; Dupuy et al., 2016; Connor et al., 2016; Wunch et al., 2016), as well as for carbon cycle studies and model validation (e.g. Chevallier et al., 2011; Deutscher et al., 2014).

2 Site details

The Lauder atmospheric research station, operated by New Zealand's National Institute of Water and Atmospheric Research (NIWA), is located in a sparsely populated rural area on the South Island of New Zealand. The station is 2 km north of the settlement of Lauder (population of approximately 50) in a broad valley with mountain ranges on three sides. Prevailing westerly winds have to cross the Southern Alps before reaching the site, and this orographic drying leads to annual rainfall of around 400 mm evenly distributed throughout the year, resulting in a semi-arid climate with a high number of clear-sky days.

Land around the site is predominantly used for irrigated livestock grazing or seasonal cropping or it is dry scrub. Farming practices in the area have shifted in recent years from low-intensity sheep and cattle grazing to more intensive dairy farming with a corresponding change from flood to spray irrigation. Some upland farming takes place in the region, which sometimes requires austral springtime tussock burn off. Plumes from this biomass burning may affect in situ measurements at the site on a small number of days each year.

In the region, most domestic heating is by wood burning. The nearest major population centre is the township of Alexandra, with a population of approximately 5000, located 40 km southwest of the site. Although air quality in this and other settlements in the region suffers from the inefficient burning of wood, in most cases these plumes are capped by an inversion layer and drain away from the research station, rarely affecting measurements. There are no other major anthropogenic sources of greenhouse gases or pollutants in the region.

Southern New Zealand is exposed to strong westerly flows that have had little or no interaction with significant land masses. Consequently, the atmospheric composition observed at the site can be considered representative of the atmospheric background state (Steinkamp et al., 2017).

As well as being part of TCCON, the Lauder station is a member of the Network for the Detection of Atmospheric Composition Change (NDACC), the Baseline Surface Radiation Network (BSRN), Global Atmosphere Watch (GAW) and the GCOS (Global Climate Observing System) Reference Upper-Air Network (GRUAN).

Other greenhouse gas measurements are also made at the site; these are summarised in Sect. 3.4.

3 Instrumentation

Two FTS instruments have been used to make TCCON measurements at Lauder (Table 1); the chronology of these instruments and major changes to them are listed in Table 2.

Both instruments are accommodated in a dedicated, air conditioned and temperature-stabilised laboratory. Both have dedicated solar trackers, protected by autonomous covers, which close automatically in the event of rain or high winds.

3.1 Bruker IFS 120HR

From June 2004, following the installation of an indium gallium arsenide (InGaAs) detector, an existing Bruker 120HR FTS (referred to as lauder01 in the TCCON archive) was used in a time-sharing mode to make routine measurements of solar absorption in the mid-infrared (MIR, for the

Table 2. Major events.

Date	Instrument	Details	Impact
30 Mar 2004	120HR	First NIR measurements	
20 Jun 2004	120HR	Start of TCCON archived data	
30 May 2006	120HR	Change to using single scans instead of co-added interferograms and from 20 to 40 kHz scan rate	
1 Apr 2009	120HR	Laser change	
1 Aug 2009	125HR	Instrument installed	
2 Feb 2010	125HR	Start of TCCON archived data	
28 Feb 2010	120HR	End of TCCON archived data	
17–25 Mar 2010	125HR	Shear alignment	
7–14 Apr 2010	125HR	IR beam steer alignment	
1–7 Jul 2010	125HR	Alignment using Hase and Blumenstock method	
5 Oct 2010–20 Jan 2011	125HR	Timing errors	X_{air} uncertainty increased
20 Jan 2011	125HR	Laser sampling board replaced	
18 Mar 2011	125HR	New CaF_2 beam splitter installed	
4–11 May 2011	125HR	Alignment optimised	
20 Apr 2012	125HR	Laser replaced	Ghosts minimised 24 Apr 2012
12 Sep–24 Oct 2012	125HR	Timing errors	X_{air} uncertainty increased
21 and 24 Oct 2013	125HR	Measurements affected by smoke from Australian bushfires.	Data excluded from archive
12 Sep–24 Oct 2012	125HR	Timing errors	X_{air} uncertainty increased
09 Mar–30 Jun 2014	125HR	Timing errors	X_{air} uncertainty increased
28 Sep–1 Oct 2014	125HR	Timing errors	X_{air} uncertainty increased
5 Jan 2015	125HR	Laser replaced	Ghosts minimised
28 Jan 2015	125HR	Laser replaced	Ghosts minimised
20 Oct–16 Nov 2015	125HR	Control electronics upgrade and various firmware updates	Some data lost
30 Mar 2016	125HR	Switch from scan rate of 20 to 10 kHz	

NDACC) and near-infrared (NIR, for the TCCON). Acquisition of spectra in the two spectral regions (MIR and NIR) requires a beam splitter exchange, necessitating daily venting and re-evacuation of the instrument (and loss of observation time for the original MIR measurement programme). This was the operational TCCON instrument from 20 June 2004 until 28 February 2010. From 20 June 2004 to 30 May 2006, each retrieval consisted of 10 co-added interferograms measured at a scanner velocity of 20 kHz. After this period retrievals were made from individual interferograms measured at 40 kHz.

A prototype active tracking system was installed in the IFS 120HR solar tracker telescope in 2006, which adjusted the position of a 45° mirror immediately before the instrument input port based on feedback from quadrant diodes. The in-house-developed NIWA360 tracker as used by the IFS 125HR instrument was not installed for the IFS 120HR until 2011, after the end of the lauder01 dataset.

After installation of the IFS 125HR instrument (see below) 1 month of measurements using both instruments have been released to the TCCON archive to allow users to assess any biases between instruments. An assessment of bias between instruments is also made in Sect. 7.1.

Measurements in TCCON mode are still made periodically (approximately monthly) with this instrument to provide an independent check of the primary TCCON instrument. These data are not released to the TCCON archive.

3.2 Bruker IFS 125HR

In 2009 NIWA purchased a Bruker IFS 125HR spectrometer dedicated to TCCON measurements (lauder02 in the TCCON data archive). This instrument began routine TCCON measurements in February 2010.

Active tracking for this instrument is achieved using the in-house-developed NIWA360 system, which directs a small amount of the incoming solar radiation through a telescope and onto quadrant photo diodes, which steer the tracker to keep the solar image centred on the FTS input aperture to within 10 % of the radial extent of the solar disc, an accuracy of better than 0.025°. The NIWA360 system has been used for the IFS 125HR since installation.

As with most TCCON sites, this instrument was originally supplied with a faulty ECL03 laser sampling board, resulting

Table 3. Airborne calibration flights at the Lauder TCCON site.

Campaign	Flight	Date	Alt. range	TCCON instrument	Species compared
HIPPO 1	RF 7	21 Jan 2009	0.7–14.6 km	120HR	CO_2, CH_4, CO and N_2O
HIPPO 2	RF 5	10 Nov 2009	0.8–13.1 km	120HR	CO_2, CH_4, CO and N_2O
HIPPO 3	RF 6	6 Apr 2010	0.6–12.9 km	125HR	CO_2, CH_4, CO and N_2O
HIPPO 5	RF 8	28 Aug 2011	2.7–13.7 km	125HR	CO_2, CH_4, CO and N_2O
HIPPO 5	RF 9	30 Aug 2011	2.8–14.1 km	125HR	CO_2, CH_4, CO and N_2O
AirCore	1	9 Sep 2015	0.7–16.0 km	125HR	CO_2 and CH_4
AirCore	2	11 Sep 2015	1.2–21.0 km	125HR	CO_2 and CH_4

in spurious "ghost" signals, which interfere with the spectral fitting. The electronics board in the Lauder IFS 125HR was upgraded to an ECL05 board in January 2011 and the ghost-to-parent ratio (GPR) has since been monitored regularly and minimised whenever necessary. A full discussion on the treatment of this issue is given in Sect. 4.1.

Apart from some early testing between February and October 2010 when some measurements were taken at scanner velocities of 10 and 40 kHz, the majority of the IFS 125HR time series have been taken at 20 kHz. Following the installation of new control electronics in October 2015, measurements were transitioned to 10 kHz at the end of March 2016.

3.3 Auxiliary measurements

In order to satisfy the requirements of a TCCON station, and to carry out retrievals, a number of auxiliary measurements of meteorological parameters are required. A NIWA climate monitoring station is sited at Lauder, approximately 220 m from the FTS instruments. This includes measurements of

- pressure: Vaisala PTB100A ±0.3 hPa.
- temperature and humidity: Vaisala HMP155D ±0.3 °C, ±3.5 %.
- wind speed and direction: Vector A101M anemometer ($\pm 0.1 < 10\,m\,s^{-1}$, ±2 % rest of range) and Vector W200P vane (±2°).

3.4 Other instruments present at the site

The atmospheric research station at Lauder hosts a wide range of atmospheric remote sensing and in situ instrumentation. With respect to atmospheric greenhouse gas measurements, these include in situ FTS measurements of CO_2, CO, CH_4 and N_2O (Griffith et al., 2012), non-dispersive infrared (NDIR) measurements of CO_2 using a LI-COR LI-7000 instrument and flask samples for subsequent measurements of CO_2, CO, CH_4, N_2O and stable isotopologues of CO_2, which are collected between 15:00 and 16:00 NZST when the wind speed is above $5\,m\,s^{-1}$ in order to sample a well mixed boundary layer (Brailsford et al., 2012). The in situ measurements at Lauder are described more fully in Appendix A of Steinkamp et al. (2017). The site also hosts lidar measurements of ozone and aerosols.

4 Data processing

The TCCON uses a common processing and retrieval system known as GGG (currently version GGG2014 described by Wunch et al., 2015) in order to minimise biases across the network. This section will describe the data processing as it applies to the IFS 125HR data. The processing is broadly similar for IFS 120HR except that for the IFS 120HR the Silicon (Si) detector is not used, only forward scan interferograms are recorded and in principle there is no requirement to correct for laser sampling errors because the IFS 120HRs did not contain the defective laser sampling boards.

The IFS 125HR records direct-current-corrected, dual channel (InGaAs and Si), forward and reverse scan interferograms throughout the day whenever there are clear skies. After each day of measurements, the interferograms are assessed visually by the operator to check for data quality to remove any periods of total cloud contamination. The raw interferograms are then processed by the I2S (interferogram-to-spectrum) software package, which corrects for solar intensity fluctuations (Keppel-Aleks et al., 2007), applies a phase correction (Mertz, 1967) and a laser sampling error (LSE) correction (Sect. 4.1 below), and then computes the spectra using a fast Fourier transform (Bergland, 1969). The causes and mitigation of the LSE are described later in this section.

The resulting spectra are analysed using the GFIT non-linear, least squares fitting algorithm, which uses molecular absorption coefficients, a priori estimates of gas mole fraction profiles, profiles of temperature, pressure, and humidity and the calculated solar viewing geometry to calculate theoretical absorption spectra. The atm.101 spectral line list (Toon, 2014a) used is based on the HITRAN2012 database (Rothman et al., 2013) with exceptions as described in Wunch et al. (2011a, 2015). The solar line list used is that of Toon (2014b).

The profiles of temperature, pressure and humidity are interpolated to the site location from the National Centers for

Environmental Prediction (NCEP) reanalysis data (Kalnay et al., 1996) for local noon.

A priori profiles of all retrieved species other than water vapour are generated using a set of empirical functions that are optimised to fit observations from a number of different in situ, aircraft and balloon, and satellite sources as a function of latitude, longitude and season, as well as taking account of long-term variability. The GGG2014 a priori profiles can be generated using a stand-alone Fortran program (Toon and Wunch, 2014).

GFIT then compares the theoretical and measured spectra in a number of independent micro-windows and minimises the RMS difference between them by iteratively scaling the assumed gas profiles and fitting spectroscopic characteristics such as continuum level, tilt and some degree of curvature, and frequency shifts.

In order to reduce the sensitivity of the measurement to variations in surface pressure and atmospheric water vapour, TCCON reports the column-averaged dry-air mole fractions (DMFs or X_{gas}), calculated by dividing the column of gas by the column of dry air inferred from the co-retrieved O_2 column multiplied by the assumed dry-air mole fraction of O_2:

$$X_{gas} = \frac{VC_{gas}}{VC_{O_2}} \cdot 0.2095. \quad (1)$$

This scaling by the retrieved O_2 column also reduces the effect of systematic instrumental biases such as timing or pointing errors.

The DMF of dry air, X_{air}, is a special case given by

$$X_{air} = \frac{VC_{air}}{VC_{O_2}} \cdot 0.2095 - X_{H_2O} \cdot \frac{m_{H_2O}}{m_{air}^{dry}}, \quad (2)$$

where m_{H_2O} and m_{air}^{dry} are the mean molecular masses of water ($18.02\,g\,mol^{-1}$) and dry air ($28.964\,g\,mol^{-1}$), X_{H_2O} is the retrieved DMF of water vapour and VC_{air} is calculated from the surface pressure, P_s.

$$VC_{air} = \frac{P_s}{\{g\} \cdot \frac{m_{air}^{dry}}{N_a}}, \quad (3)$$

where $\{g\}$ is the column-averaged acceleration due to gravity and N_a is Avogadro's constant.

In the case of a perfectly accurate measurement, the value of X_{air} would be unity. Due to spectroscopic limitations the actual value is approximately 0.98 for all TCCON sites and varies by approximately 1 % with solar zenith angle. Deviations from the characteristic values for X_{air} are generally indicative of erroneous behaviour in the measurement and retrieval system such as tracking errors or fitting to an incorrect air mass as described in Sect. 5.2. However, these errors will generally cancel out in the calculation of DMFs of individual gases as the error will be present for both the gas and O_2.

As detailed elsewhere (Deutscher et al., 2010; Wunch et al., 2011a, 2015), the TCCON X_{CO_2} data exhibit non-physical symmetric variability with changing air mass. The data presented here have the TCCON standard air mass correction of -0.0068 applied; however, this value is derived from a limited time series of data from three sites only. Calculating the corresponding value from the uncorrected Lauder X_{CO_2} data yields an air mass correction of -0.0095 ± 0.0021; applying this value to the Lauder X_{CO_2} time series generates a seasonal cycle amplitude that is $\pm 0.6\,\mu mol\,mol^{-1}$ smaller. Users for whom the seasonal variability is critical can circumvent this issue by only selecting X_{CO_2} values corresponding to a restricted range of solar zenith angles (SZAs) that is consistent throughout the entire year, e.g. only using measurements between 35 and 55° SZA, at which the air mass correction is smallest.

A network-wide bias correction based on airborne measurements to provide traceability to WMO trace gas scales as described in Wunch et al. (2010) with updated values from Wunch et al. (2015), listed in the second column of Table 4, is also applied. The Lauder-specific calibration campaigns are described in Sect. 6 of this article.

4.1 Laser sampling error correction

The IFS 125HR FTS makes use of a helium–neon (HeNe) metrology laser running parallel to the measured solar beam in the interferometer to allow accurate measurement and control of the optical path difference by counting interference fringes. A fault with the original ECL03 laser amplifier board supplied with this instrument meant that the metrology laser was not sampled correctly at the zero crossing point between the peak and trough of the resulting sine wave (Messerschmidt et al., 2010). As a result, some of the spectral information above the Nyquist frequency of $7899\,cm^{-1}$ was folded below it and vice versa, causing features known as ghosts.

During installation the LSE was minimised for a scanner velocity of 40 kHz, resulting in a small LSE at a rate of 20 kHz used for the majority of operational measurements.

Following the discovery of this fault, Bruker supplied all TCCON sites with replacement ECL05 laser amplifier boards, which allowed the zero level (i.e. the level at which the electronics identify the zero crossing point) to be adjusted. The rectified electronics board was installed in the Lauder IFS 125HR in January 2011. Since this time, regular (approx. every 3 months) checks have been conducted to check for the presence of ghosts and to minimise them by adjusting the zero level if necessary. This is achieved by taking measurements of a tungsten lamp, placing a band pass filter (5500–$6600\,cm^{-1}$) between the interferometer and detector, and measuring the signal aliased at $9800\,cm^{-1}$. If any signal is detected, the zero level is iteratively adjusted in order to minimise it.

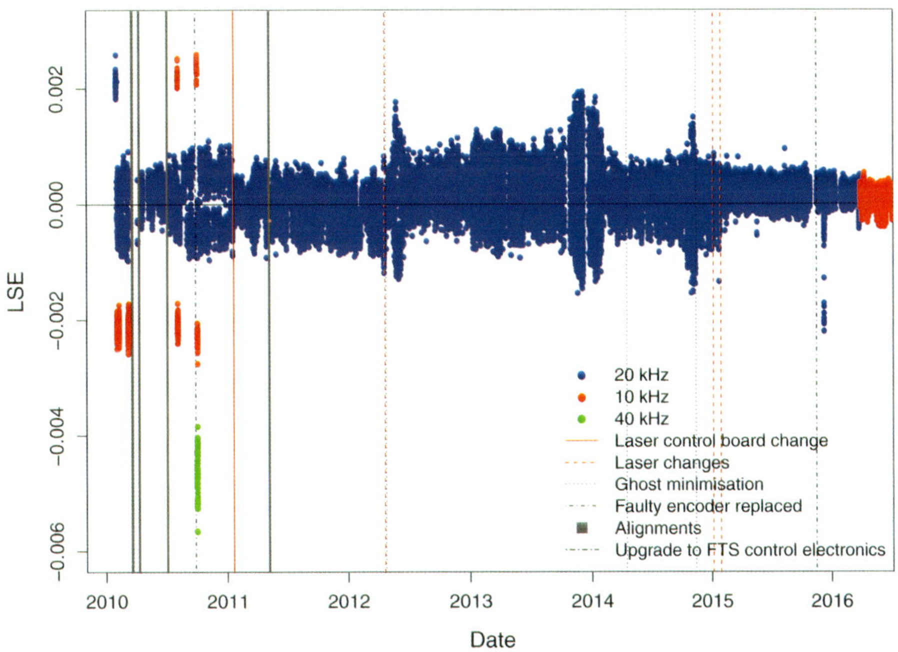

Figure 1. Time series of LSE calculated using the I2S algorithm for the IFS 125HR spectra. For 20 kHz data only every 10th data point is plotted for clarity. Vertical lines show significant changes or adjustments to the instrument

The 2014 release of the GGG processing software I2S algorithm makes use of the spectra recorded using the Si detector, which is wholly contained in the upper half of the alias, to derive the magnitude of the LSE and allow the spectra of both detectors to be resampled, removing the ghost signal. This method is described fully in Wunch et al. (2015).

Figure 1 shows the time series of LSE determined using the I2S algorithm over the period of measurements made using the IFS 125HR. Periods of instrument alignment are denoted by the shaded vertical bands whilst other notable instrument changes are indicated by the various vertical lines. The solid vertical line in early 2011 represents the change of laser amplifier board and subsequent improvement in LSE. Also of note is the change in LSE performance with individual metrology lasers, with the performance of the unit installed between April 2012 and January 2015 being poorer and less stable than other devices as indicated by the relative scatter in the derived LSEs.

5 Instrument stability

5.1 Instrumental line shape

The performance and alignment of both instruments is monitored through monthly retrievals of the instrumental line shape (ILS) from spectra obtained from a lamp source with a reference hydrogen chloride (HCl) cell, using the LINEFIT 14 software (Hase et al., 2013). The retrieved ILS describes the deviation from a theoretical ideal ILS, which is assumed by the GFIT retrieval code, in terms of the modulation efficiency (ME) and phase error, both of which vary as a function of optical path difference (OPD) of the interferometer. In the following discussion, ME will refer to the modulation efficiency at the maximum OPD of 45 cm used for TCCON measurements.

Figure 2 shows a time series of the retrieved ME and maximum phase error for both instruments. For the IFS 125HR, the mean of the ME is 0.9995 with a SD of 0.0106 indicating the stability of the instrument optical alignment over time. This value is significantly better than 4 % deviation in ME, which Hase et al. (2013) demonstrated was necessary in order to achieve the X_{CO_2} retrieval accuracy required by the TCCON. The results for the IFS 120HR show more scatter and less stability, with a mean of 1.0124 and SD of 0.0256. Regular ILS retrievals from lamp spectra have been carried out since August 2010.

5.2 IFS 125HR timing errors

Due to the instability of the IFS 125HR instrument's internal clock, there have been a number of occasions when timing errors have been noted in the retrieved data. These are listed in Table 2. The errors have generally occurred when updates to the operating system or experimental set-up have prevented the instrument control computer from periodically resetting the time in the instrument's firmware. Such errors cause the retrieval scheme to attempt to fit the spectra to an incorrect air mass due to small differences in the calculated and true SZA. At the magnitude at which these errors occurred, the effect on the retrieved X_{gas} values is negligible

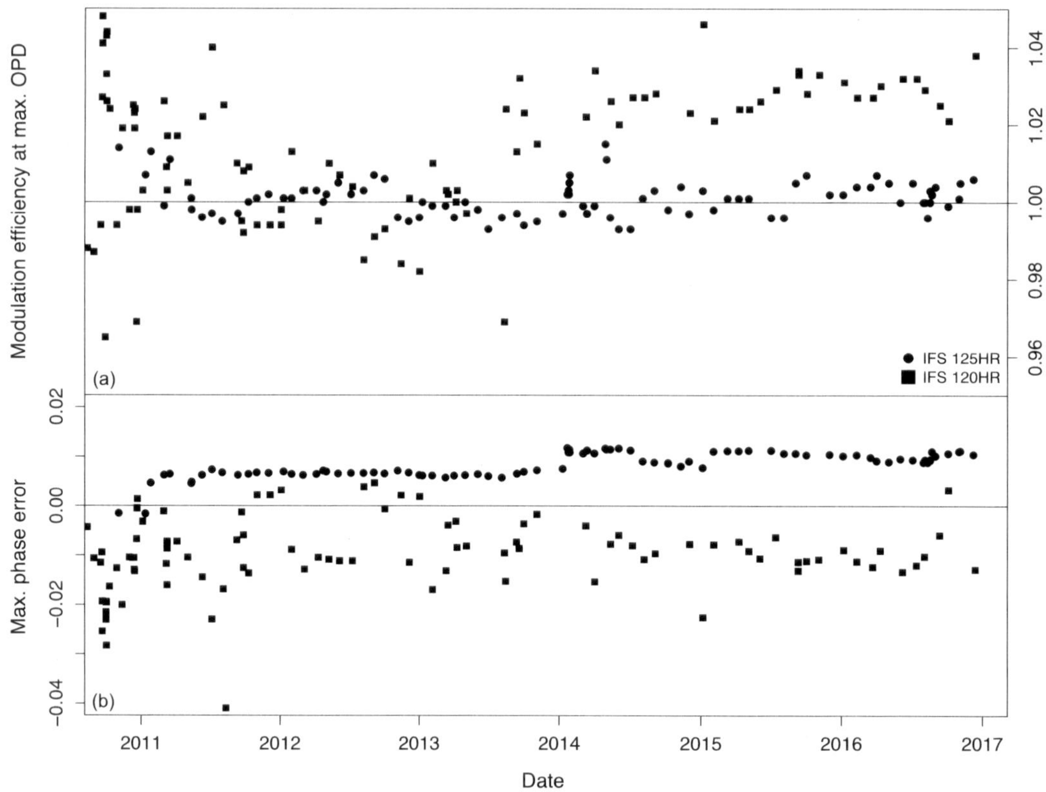

Figure 2. The time series of retrieved modulation efficiencies at 45 cm optical path difference **(a)** and maximum phase error **(b)** for the Bruker IFS120HR and IFS125HR instruments at Lauder, based on lamp measurements of a 10 cm HCl cell, indicating the improved stability of the IFS125HR instrument line shape over time.

as these are scaled by the retrieved column of O_2 in order to remove such instrumental artefacts. However, the values of X_{air} that are retrieved and used as a diagnostic and quality indicator within TCCON are perturbed, as in this case it is the theoretical column of dry air that is scaled by the retrieved O_2 column. By artificially introducing a timing error into retrievals that are known to have accurate timing, it has been shown that for a timing error of 60 s and a SZA less than 85°, the perturbation of X_{CO_2} is less than 0.02 %, compared to 1 % for X_{air}. Therefore, for the 2014 TCCON data release, data subject to these errors have been included without any attempt to correct them.

5.3 Time series of X_{air}

As noted previously, for an ideal retrieval, X_{air} would be unity, but due to spectroscopic limitations there is a TCCON wide bias and SZA dependence. These are, however, stable both temporally and across the network. Thus, X_{air} is a useful diagnostic of the quality of measurements, with retrievals deviating more than 1 % from the nominal value of 0.98 indicating systematic errors, e.g. the timing errors mentioned above, or deviations from an ideal ILS.

Figure 3 shows the time series of X_{air} retrievals from the IFS 125HR instrument that have satisfied the data quality criteria for release to the TCCON archive. It can be seen that most of these values fall within 1 % of 0.98 (indicated by the dashed lines); however, there are some periods with values outside of these limits, which have been identified as having timing errors. These periods are highlighted in green and defined in Table 2.

In Fig. 4 we show the hourly mean and SD (upper panel) of X_{air} values for the entire Lauder TCCON dataset covering both instruments.

There is a discontinuity in the SD of the hourly mean time series corresponding to a change from performing retrievals using 10 co-added interferograms sampled at a scanner velocity of 20 kHz to single scans at 40 kHz in May 2006. The discontinuity in both the mean and SD that occurs in early 2010 indicates the change from the IFS 120HR to the IFS 125HR and demonstrates the improvement in performance with the newer instrument.

6 Airborne calibration

The accuracy of TCCON measurements is limited by the available spectral line parameters to around 2–3 %. However,

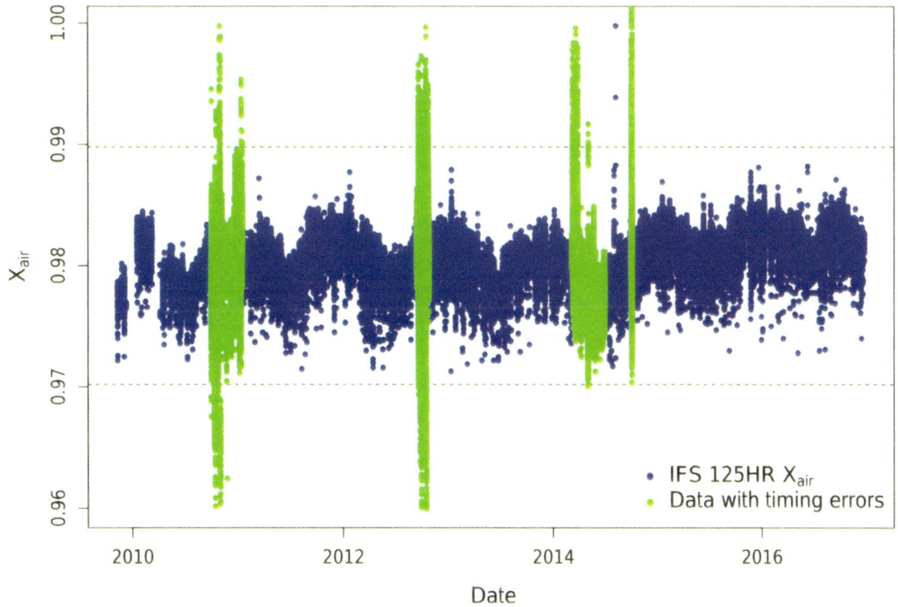

Figure 3. Time series of IFS 125HR X_{air}. Data affected by timing errors are shown in green and defined in Table 2. For values not affected by timing errors, only every 10th data point is plotted for clarity.

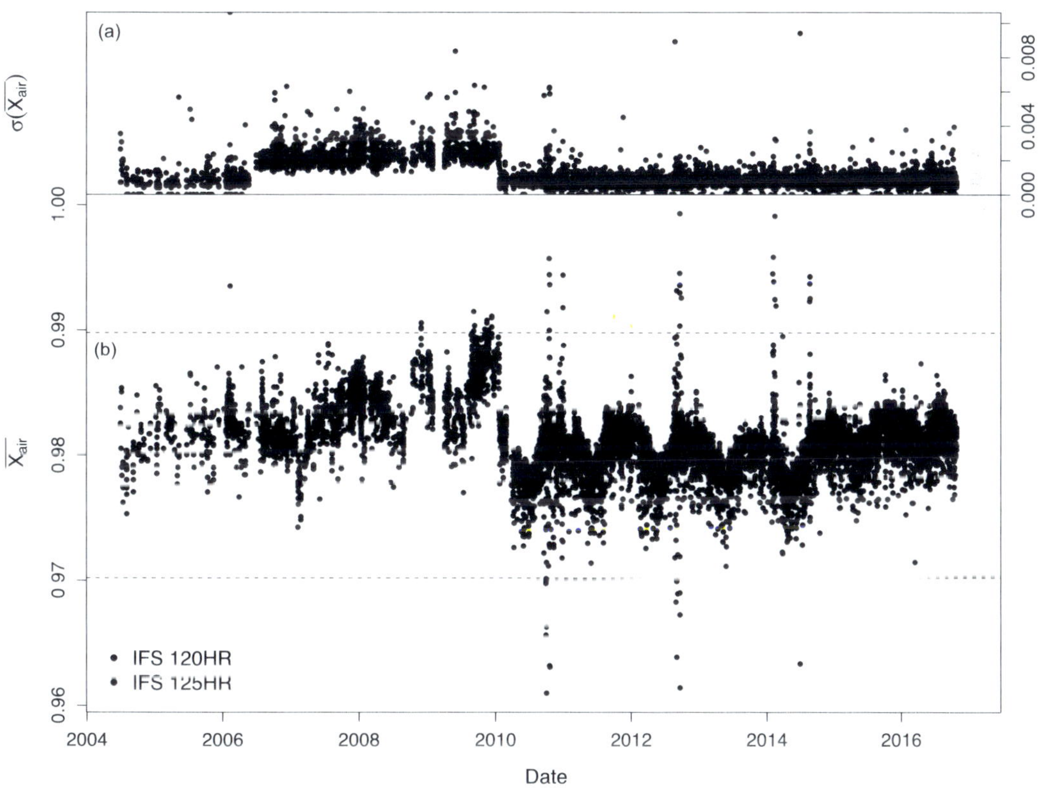

Figure 4. Time series of hourly means of X_{air} **(b)** and SDs about them **(a)** retrieved from measurements made with both FTS instruments at Lauder.

Table 4. Comparison of airborne calibration scale factors, derived from Lauder flights, with TCCON network-wide values.

Species	TCCON scale factor	120HR	125HR
X_{CO_2}	0.9898 ± 0.0010	0.9880 ± 0.0041	0.9911 ± 0.0014
X_{CH_4}	0.9765 ± 0.0020	0.9879 ± 0.0052	0.9867 ± 0.0016
X_{CO}	1.0672 ± 0.0200	1.0124 ± 0.1419	0.9655 ± 0.1116
X_{N_2O}	0.9638 ± 0.0100	0.9833 ± 0.0130	0.9809 ± 0.0081

because of the use of a common retrieval system across the network, the resulting bias is consistent for all retrievals and sites.

In order to align TCCON measurements with the accepted WMO gas standard scale, a number of in situ, airborne profile measurements have been made at many of the TCCON sites, including Lauder, to derive a network-wide scaling factor for a subset of retrieved species (Wunch et al., 2010). In this section we compare the network wide scaling factors with those derived for the two Lauder instruments.

The Lauder site was included in the flight schedule of each of the HIAPER Pole-to-Pole Observations (HIPPO) campaigns (Wofsy, 2011), and in 2015 it was the base for an AirCore (Karion et al., 2010) campaign. Details of successful calibration flights are listed in Table 3. Each flight consists of one or more altitude profiles that have been averaged together to give measured profiles of the mole fractions of each target species. In some cases operational requirements meant that these profiles could not be performed directly over Lauder and were instead carried out during the approach into, or climb out of, Christchurch International Airport approximately 285 km to the northeast. It is considered that for the target species, the spatial variability is sufficiently small that these profiles can be considered representative of the air mass at the TCCON site if the boundary layer is discarded. In order to extend the profiles from their lowest level to the surface, they are interpolated using mole fractions measured from surface flask samples collected at the time of overflight at Lauder. Above the maximum altitude attained during the flight, these profiles have been supplemented using the TCCON a priori profiles.

In order to derive DMFs that can be directly compared to the TCCON measurements, the airborne profiles are integrated following the method of Rodgers and Connor (2003), as it was applied by Wunch et al. (2010), weighting with the FTS retrieval averaging kernels to give a smoothed column-average DMF x_s:

$$x_s = x_a + A(x_h - x_a), \tag{4}$$

where x_a is the a priori column DMF, x_h is the DMF of the merged aircraft profile and A is the column-averaging kernel.

Figure 5 shows the resulting calibration curves for the smoothed airborne DMFs and the mean of the corresponding day's FTS retrievals for the four species compared. Uncertainties in the airborne measurements are assigned as in Wunch et al. (2010), and those for the TCCON measurements are the SD of the day's retrievals. Calibration scaling factors are then calculated assuming a linear relationship, forced through the origin and taking account of the uncertainties in both the smoothed aircraft and TCCON values. The resulting scaling factors for the Lauder instruments and those derived for the entire TCCON network in Wunch et al. (2010), updated in Wunch et al. (2015), are listed in Table 4. These show good agreement and the accuracy, based on the uncertainty in the fitted linear relationship, following this correction is 0.1 % for X_{CO_2}.

Table 5. Comparison of retrievals from both instruments. Mean difference (IFS 120HR minus IFS 125HR) of daily mean values, their SD and standard error (SD of the sample mean) expressed as a percentage, and the number of daily means compared.

Species	Mean difference %	SD	SE	N
X_{CO_2}	0.068	0.113	0.015	55
X_{CH_4}	−0.069	0.141	0.019	55
X_{CO}	1.143	1.498	0.202	55
X_{N_2O}	−0.005	0.237	0.032	55
X_{H_2O}	0.684	1.555	0.210	55
X_{HDO}	0.666	1.917	0.258	55
X_{HF}	−0.867	0.784	0.106	55

7 Results

7.1 Instrument intercomparison

In order to assess any biases between the two instruments used to collect TCCON data at Lauder, we have compared days on which measurements were taken with both instruments between October 2009 and December 2012. Figures 6 to 8 show in the upper panels the daily mean retrieved values for X_{CO_2}, X_{CH_4} and X_{CO} respectively. Error bars represent the SD about the daily means. The lower panels show the difference (IFS 120HR – IFS 125HR) between the daily means, and the error bars are the sum in quadrature of the SDs about the daily means. Also plotted are the mean of the daily mean differences and the associated SD and standard error. These values are listed as percentages for all compared species in Table 5. For X_{CO_2} the mean bias is 0.068 % with a standard error of 0.015 %.

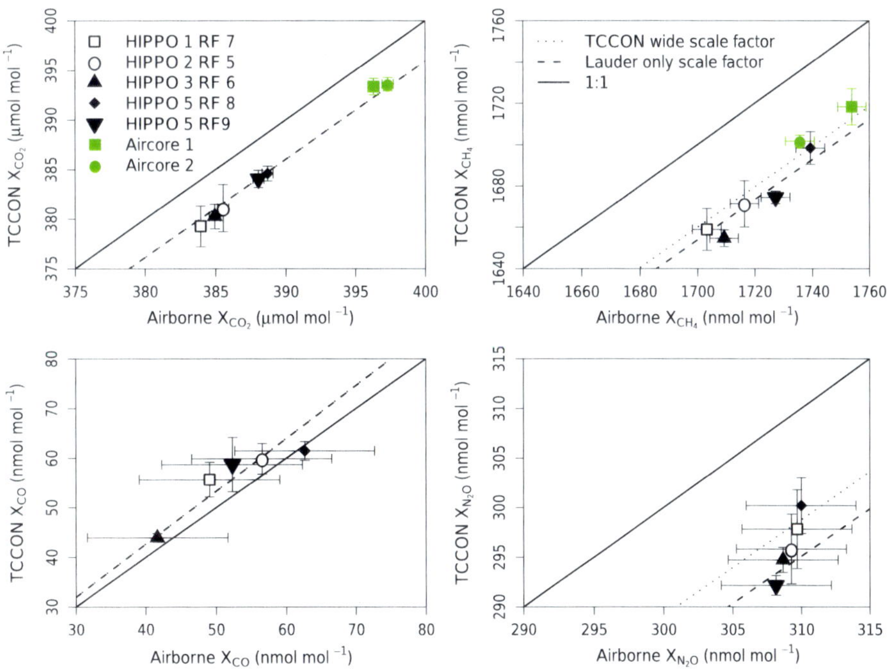

Figure 5. Airborne calibration curves for X_{CO_2}, X_{CH_4}, X_{CO} and X_{N_2O} with fitted linear relationships derived for the TCCON network and Lauder-only intercomparisons (forced through origin). The HIPPO 1 and 2 flights were compared to the IFS 120HR (open symbols); all others were compared to the IFS 125HR (filled symbols).

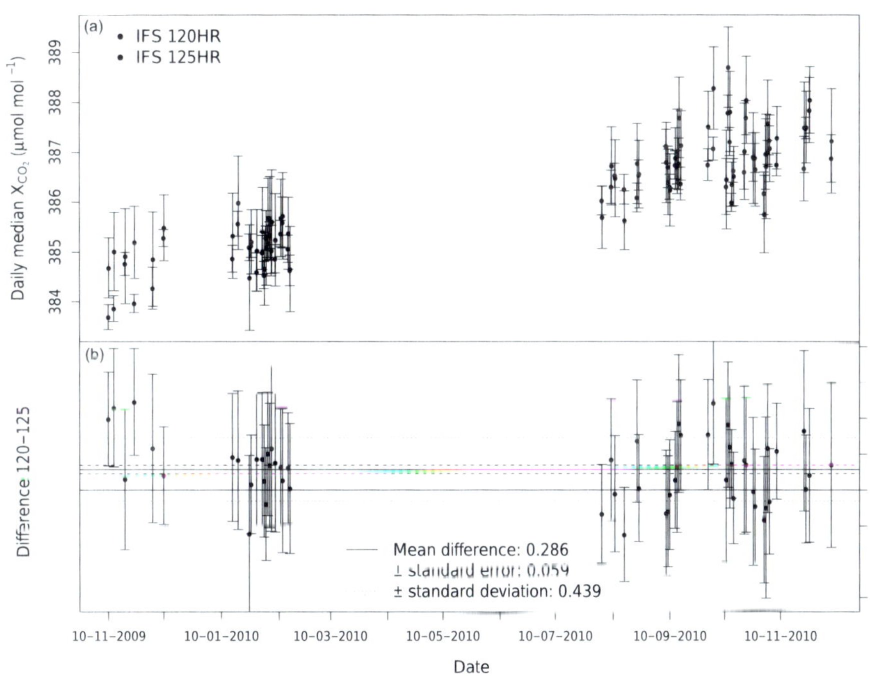

Figure 6. Time series of daily means of retrieved X_{CO_2} from measurements made with both FTS instruments at Lauder **(a)** and the difference between them **(b)**. Error bars in the upper panel indicate the SD about the daily mean, while those in the lower panel are the sum in quadrature of the SD about the daily means. The horizontal lines in the lower panel show the mean of the daily mean differences and the SD and standard error (SD of the sample mean) about it.

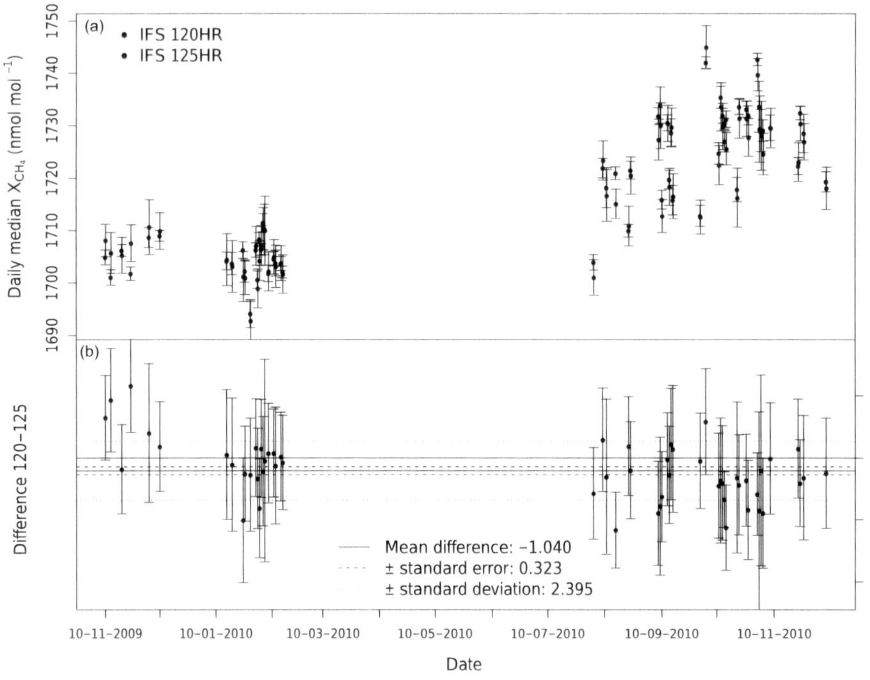

Figure 7. As Fig. 6 but for X_{CH_4}.

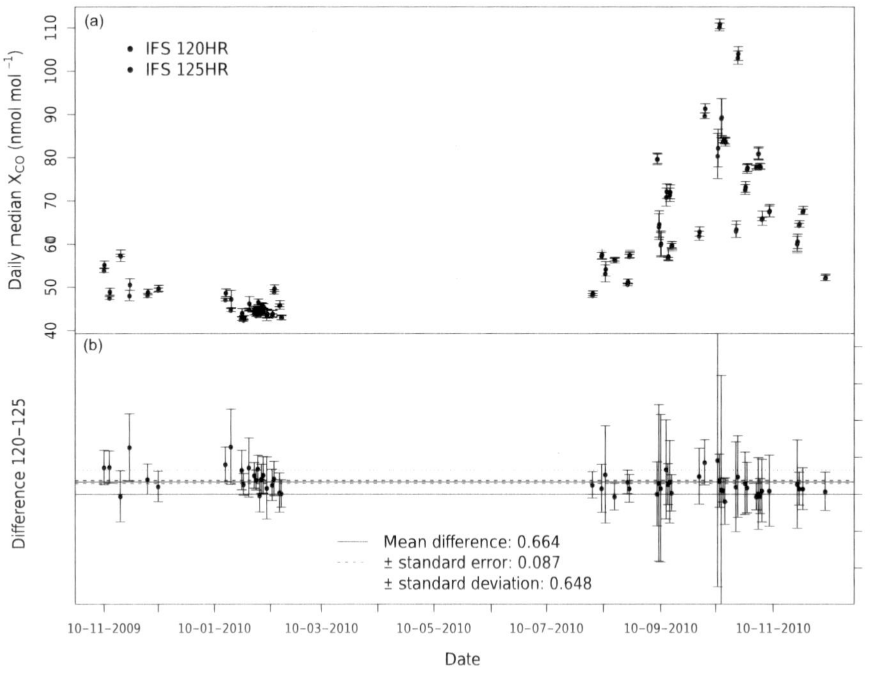

Figure 8. As Fig. 6 but for X_{CO}.

7.2 Time series of retrieved species

Here we present a subset of retrieval time series available in the Lauder TCCON data archive. Figures 9 to 11 show the hourly means of the retrieved X_{gas}, along with their SD in the upper panels. These figures are made up of data from both of the Lauder instruments with the change-over occurring on 1 February 2010.

The time series of X_{CO_2} values is shown in Fig. 9. Over the length of this time series X_{CO_2} has been increasing by approximately 2 µmol mol^{-1} yr^{-1}. There is a seasonal cy-

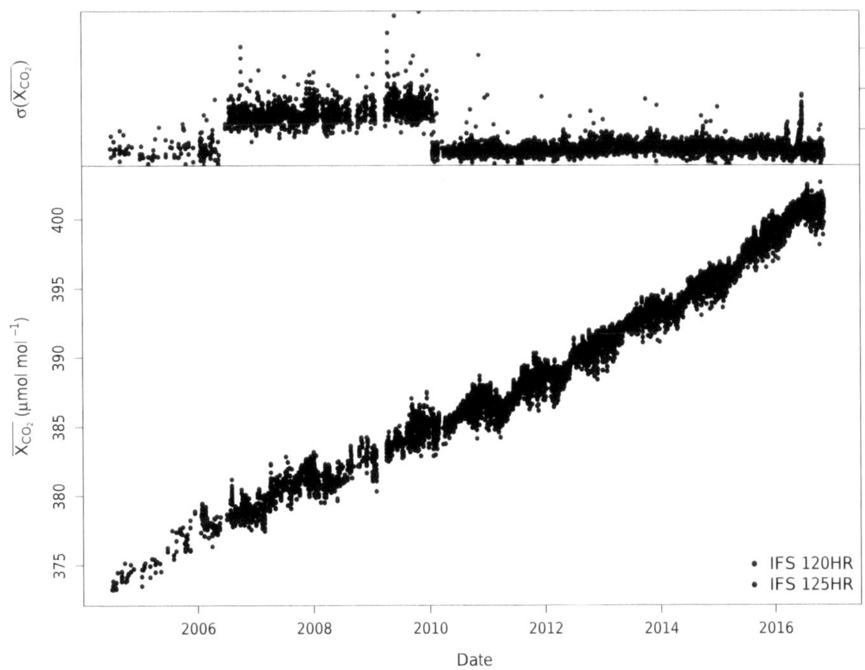

Figure 9. Time series of hourly means of X_{CO_2} retrieved from measurements made with both FTS instruments at Lauder.

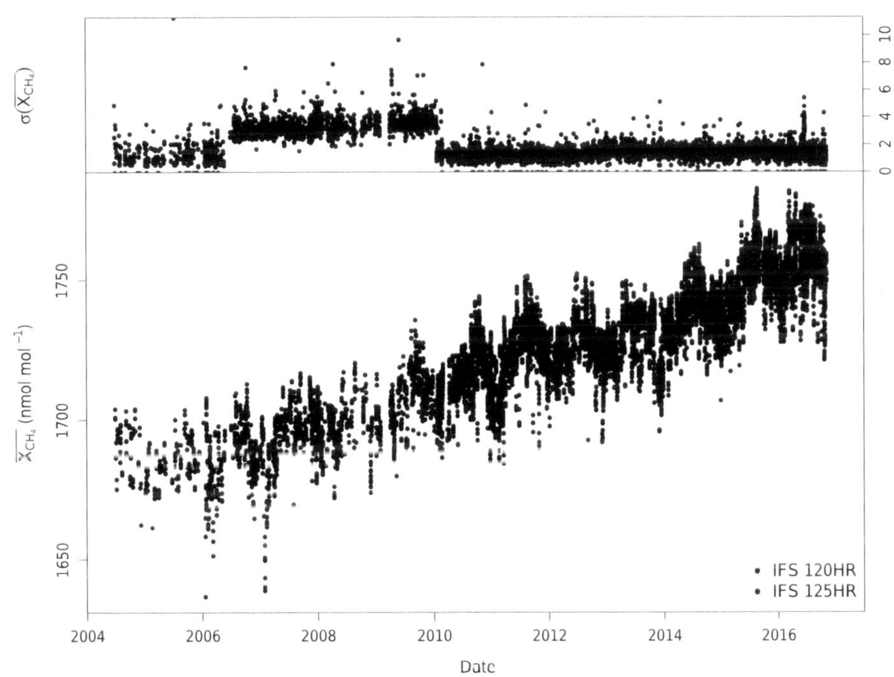

Figure 10. Time series of hourly means of X_{CH_4} retrieved from measurements made with both FTS instruments at Lauder.

cle due to summer drawdown, but it is smaller than in the Northern Hemisphere (Deutscher et al., 2014). Since measurements using the IFS 125HR began, the SD about the hourly mean has been better than 0.1 % for 96.81 % of retrievals and better than 0.25 % for 99.96 % of retrievals. The precision of the IFS 120HR was poorer, with only 14.95 % of hourly SDs better than 0.1 and 95.93 % better than 0.25 %.

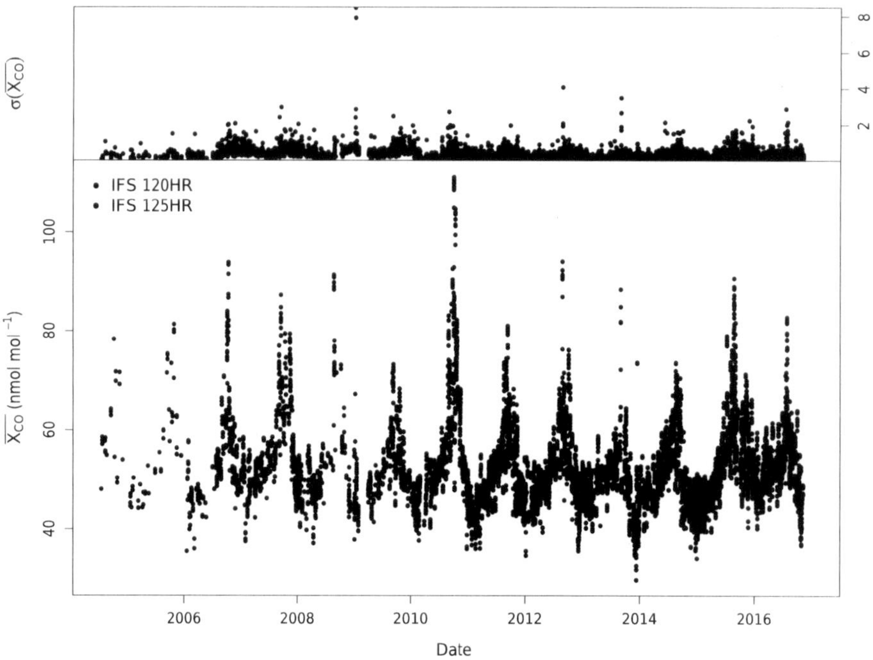

Figure 11. Time series of hourly means of X_{CO} retrieved from measurements made with both FTS instruments at Lauder.

Table 6. Precision of Lauder X_{CO_2} datasets compared to TCCON targets.

Instrument	IFS 120HR	IFS 125HR
Mean SD about hourly means	0.17 %	0.07 %
% < 0.25 %	95.93	99.96
% < 0.1 %	14.95	96.81

The precision of both instruments compared to the TCCON targets are shown in Table 6.

Figure 10 shows the X_{CH_4} time series. X_{CH_4} has a smaller positive trend of approximately $5.6\,\text{nmol}\,\text{mol}^{-1}\,\text{yr}^{-1}$ and a seasonal cycle of similar magnitude with a summer minimum due to oxidation by photochemically produced OH radicals. The Lauder TCCON X_{CH_4} time series starts around 2 years before the end of a period of stable atmospheric CH_4 concentrations that has been observed in a number of different time series (Schaefer et al., 2016, and references therein) and this is also apparent in the data shown here.

The time series of X_{CO} shown in Fig. 11 has a small, negative trend and a stronger seasonal cycle with a summer minimum due to OH oxidation and a springtime peak attributed to CO transported from Southern Hemisphere biomass burning sources (Buchholz et al., 2016).

Figure 12 shows the time series of the remaining retrieved species.

8 Data availability

The GGG2014 versions of the processed Lauder TCCON data are available from the TCCON data archive (Sherlock et al., 2014a, b). The data DOIs are https://doi.org/10.14291/tccon.ggg2014.lauder01.R0/1149293 for the IFS 120HR data and https://doi.org/10.14291/tccon.ggg2014.lauder02.R0/1149298 for the IFS 125HR data.

The IFS 120HR data are available from 20 June 2004 to 28 February 2010, while the IFS 125HR data cover the range 2 February 2010 to the present day, with data becoming publicly available 180 days after collection.

9 Conclusions

The Lauder TCCON dataset contains retrievals of X_{CO_2}, X_{CH_4}, X_{N_2O}, X_{HF}, X_{H_2O}, X_{HDO} and X_{CO} from June 2004 to the present date using two FTS instruments. Overlapping measurements were made using both instruments and from these, the bias between them was assessed to be 0.068 % for X_{CO_2}. Regular monitoring of the ILS (monthly, both instruments) and LSE (quarterly, IFS 125HR only) is carried out to ensure the consistent performance of the instruments and the quality of the retrievals and to achieve the TCCON target precision of 0.1 % for X_{CO_2} as well as consistency with the rest of the network.

Airborne calibration campaigns have shown that the scale factors required to align the Lauder retrievals of X_{CO_2}, X_{CH_4},

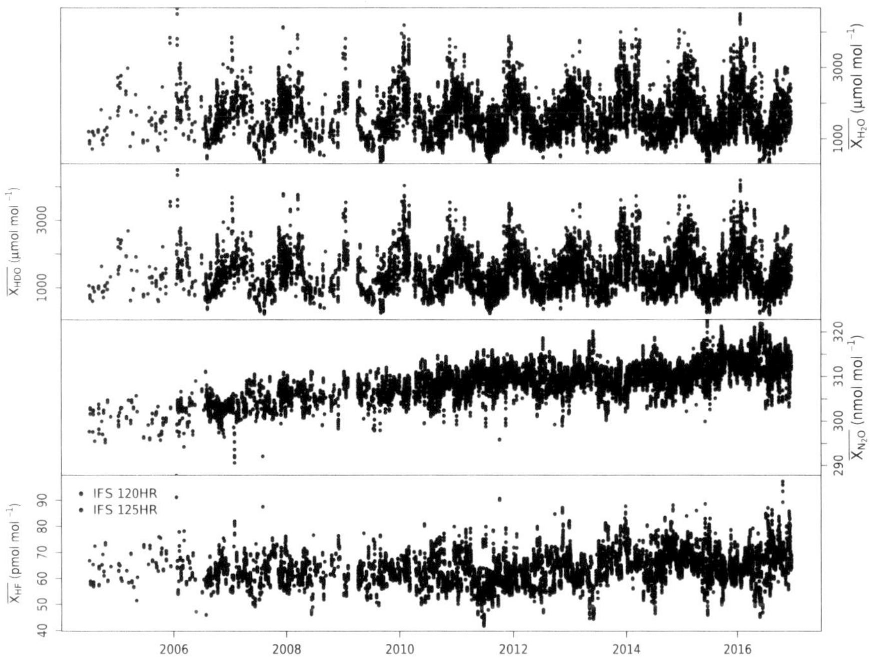

Figure 12. Time series of hourly means of X_{H_2O} **(a)**, X_{HDO} **(b)**, X_{N_2O} **(c)** and X_{HF} **(d)** retrieved from measurements made with both FTS instruments at Lauder.

X_{CO} and X_{N_2O} with WMO scales are consistent with those used across the TCCON.

Timing errors have sporadically affected measurements taken with the IFS 125HR instrument, which have perturbed the retrieved values for X_{air} by up to 2 %. However, due to the scaling by the co-retrieved column of O_2, the effect on retrievals of other gases has been minimal, less than 0.02 % in the case of CO_2 at SZAs less than 85°. In a future data release it is expected that these timing errors will be retrospectively corrected.

Author contributions. BC was the instigator of the Lauder TCCON programme and was the Lauder TCCON principal investigator (PI) until 2008. VS was involved in the early development of the Lauder TCCON project and was the site PI from 2008 to 2014. VS was responsible for the retrieval of the TCCON products from the Lauder spectra. JR is responsible for the operation and alignment of the FTS instruments and developed the NIWA solar trackers. ND analysed the air mass dependence and provided advice on data processing and methods for carrying out airborne calibrations. HS maintains the GGG processing system (developed by VS) and quality control of the TCCON data products. DP is the current TCCON site PI for Lauder and wrote this paper. All authors have read and provided feedback on the paper.

Competing interests. The authors declare that they have no conflict of interest.

Acknowledgements. The Lauder TCCON programme is core-funded by NIWA through New Zealand's Ministry of Business, Innovation and Employment with additional support from the National Institute for Environmental Studies, Japan, GOSAT project.

Nicholas Deutscher is supported by an Australian Research Council – Discovery Early Career Researcher Award DE140100178.

The authors would like to acknowledge the following: Colm Sweeney, Sonja Wolter, Tim Newberger and Jack Higgs of NOAA in organising the AirCore campaign at Lauder and the logistical support of Richard Querel, Wills Dobson, Jocelyn Turnbull and Joel Wilson during the AirCore flights; Steve Wofsy and Britt Stevens for providing calibrated HIPPO data; Paul Wennberg, Debra Wunch, and Coleen Roehl of Caltech and Geoff Toon of JPL for technical and hardware assistance over the years; and Dan Smale and Mike Harvey for the useful and informative discussions and advice.

Edited by: David Carlson

References

Bergland, G. D.: A radix-eight fast Fourier transform subroutine for real-valued series, IEEE T. Acoust. Speech, 17, 138–144, https://doi.org/10.1109/tau.1969.1162043, 1969.

Brailsford, G. W., Stephens, B. B., Gomez, A. J., Riedel, K., Mikaloff Fletcher, S. E., Nichol, S. E., and Manning, M. R.: Long-term continuous atmospheric CO_2 measurements at Bar-ing Head, New Zealand, Atmos. Meas. Tech., 5, 3109–3117, https://doi.org/10.5194/amt-5-3109-2012, 2012.

Buchholz, R. R., Paton-Walsh, C., Griffith, D. W. T., Kubistin, D., Caldow, C., Fisher, J. A., Deutscher, N. M., Kettlewell, G., Riggenbach, M., Macatangay, R., Krummel, P. B., and Langenfelds, R. L.: Source and meteorological influences on air quality (CO, CH_4 & CO_2) at a Southern Hemisphere urban site, Atmos. Environ., 126, 274–289, https://doi.org/10.1016/j.atmosenv.2015.11.041, 2016.

Chevallier, F., Deutscher, N. M., Conway, T. J., Ciais, P., Ciattaglia, L., Dohe, S., Fröhlich, M., Gomez-Pelaez, A. J., Griffith, D., Hase, F., Haszpra, L., Krummel, P., Kyrö, E., Labuschagne, C., Langenfelds, R., Machida, T., Maignan, F., Matsueda, H., Morino, I., Notholt, J., Ramonet, M., Sawa, Y., Schmidt, M., Sherlock, V., Steele, P., Strong, K., Sussmann, R., Wennberg, P., Wofsy, S., Worthy, D., Wunch, D., and Zimnoch, M.: Global CO_2 fluxes inferred from surface air-sample measurements and from TCCON retrievals of the CO_2 total column, Geophys. Res. Lett., 38, L24810, https://doi.org/10.1029/2011GL049899, 2011.

Connor, B. J., Sherlock, V., Toon, G., Wunch, D., and Wennberg, P. O.: GFIT2: an experimental algorithm for vertical profile retrieval from near-IR spectra, Atmos. Meas. Tech., 9, 3513–3525, https://doi.org/10.5194/amt-9-3513-2016, 2016.

Crisp, D., Miller, C. E., and DeCola, P. L.: NASA Orbiting Carbon Observatory: measuring the column averaged carbon dioxide mole fraction from space, J. Appl. Remote Sens., 2, 023508, https://doi.org/10.1117/1.2898457, 2008.

Crisp, D., Fisher, B. M., O'Dell, C., Frankenberg, C., Basilio, R., Bösch, H., Brown, L. R., Castano, R., Connor, B., Deutscher, N. M., Eldering, A., Griffith, D., Gunson, M., Kuze, A., Mandrake, L., McDuffie, J., Messerschmidt, J., Miller, C. E., Morino, I., Natraj, V., Notholt, J., O'Brien, D. M., Oyafuso, F., Polonsky, I., Robinson, J., Salawitch, R., Sherlock, V., Smyth, M., Suto, H., Taylor, T. E., Thompson, D. R., Wennberg, P. O., Wunch, D., and Yung, Y. L.: The ACOS CO_2 retrieval algorithm – Part II: Global X_{CO_2} data characterization, Atmos. Meas. Tech., 5, 687–707, https://doi.org/10.5194/amt-5-687-2012, 2012.

Deutscher, N. M., Griffith, D. W. T., Bryant, G. W., Wennberg, P. O., Toon, G. C., Washenfelder, R. A., Keppel-Aleks, G., Wunch, D., Yavin, Y., Allen, N. T., Blavier, J.-F., Jiménez, R., Daube, B. C., Bright, A. V., Matross, D. M., Wofsy, S. C., and Park, S.: Total column CO_2 measurements at Darwin, Australia – site description and calibration against in situ aircraft profiles, Atmos. Meas. Tech., 3, 947–958, https://doi.org/10.5194/amt-3-947-2010, 2010.

Deutscher, N. M., Sherlock, V., Mikaloff Fletcher, S. E., Griffith, D. W. T., Notholt, J., Macatangay, R., Connor, B. J., Robinson, J., Shiona, H., Velazco, V. A., Wang, Y., Wennberg, P. O., and Wunch, D.: Drivers of column-average CO_2 variability at Southern Hemispheric Total Carbon Column Observing Network sites, Atmos. Chem. Phys., 14, 9883–9901, https://doi.org/10.5194/acp-14-9883-2014, 2014.

Dupuy, E., Morino, I., Deutscher, N., Yoshida, Y., Uchino, O., Connor, B., De Mazière, M., Griffith, D., Hase, F., Heikkinen, P., Hillyard, P., Iraci, L., Kawakami, S., Kivi, R., Matsunaga, T., Notholt, J., Petri, C., Podolske, J., Pollard, D., Rettinger, M., Roehl, C., Sherlock, V., Sussmann, R., Toon, G., Velazco, V., Warneke, T., Wennberg, P., Wunch, D., and Yokota, T.: Comparison of X_{H_2O} retrieved from GOSAT short-wavelength infrared spectra with observations from the TCCON Network, Remote Sens., 8, 414, https://doi.org/10.3390/rs8050414, 2016.

Griffith, D. W. T., Deutscher, N. M., Caldow, C., Kettlewell, G., Riggenbach, M., and Hammer, S.: A Fourier transform infrared trace gas and isotope analyser for atmospheric applications, Atmos. Meas. Tech., 5, 2481–2498, https://doi.org/10.5194/amt-5-2481-2012, 2012.

Hase, F., Drouin, B. J., Roehl, C. M., Toon, G. C., Wennberg, P. O., Wunch, D., Blumenstock, T., Desmet, F., Feist, D. G., Heikkinen, P., De Mazière, M., Rettinger, M., Robinson, J., Schneider, M., Sherlock, V., Sussmann, R., Té, Y., Warneke, T., and Weinzierl, C.: Calibration of sealed HCl cells used for TCCON instrumental line shape monitoring, Atmos. Meas. Tech., 6, 3527–3537, https://doi.org/10.5194/amt-6-3527-2013, 2013.

Kalnay, E., Kanamitsu, M., Kistler, R., Collins, W., Deaven, D., Gandin, L., Iredell, M., Saha, S., White, G., Woollen, J., Zhu, Y., Chelliah, M., Ebisuzaki, W., Higgins, W., Janowiak, J., Mo, K. C., Ropelewski, C., Wang, J., Leetmaa, A., Reynolds, R., Jenne, R., and Joseph, D.: The NCEP/NCAR 40-year reanalysis project, B. Am. Meteorol. Soc. , 77, 437–471, https://doi.org/10.1175/1520-0477(1996)077<0437:tnyrp>2.0.co;2, 1996.

Karion, A., Sweeney, C., Tans, P., and Newberger, T.: AirCore: an innovative atmospheric sampling system, J. Atmos. Ocean. Tech., 27, 1839–1853, https://doi.org/10.1175/2010JTECHA1448.1, 2010.

Keppel-Aleks, G., Toon, G. C., Wennberg, P. O., and Deutscher, N. M.: Reducing the impact of source brightness fluctuations on spectra obtained by Fourier-transform spectrometry, Appl. Optics, 46, 4774–4779, https://doi.org/10.1364/AO.46.004774, 2007.

Mertz, L.: Auxiliary computation for Fourier spectrometry, Infrared Phys., 7, 17–23, https://doi.org/10.1016/0020-0891(67)90026-7, 1967.

Messerschmidt, J., Macatangay, R., Notholt, J., Petri, C., Warneke, T., and Weinzierl, C.: Side by side measurements of CO_2 by ground-based Fourier transform spectrometry (FTS), Tellus B, 62, 749–758, https://doi.org/10.1111/j.1600-0889.2010.00491.x, 2010.

Olsen, S. C. and Randerson, J. T.: Differences between surface and column atmospheric CO_2 and implications for carbon cycle research, J. Geophys. Res.-Atmos., 109, D02301, https://doi.org/10.1029/2003JD003968, 2004.

Rayner, P. J. and O'Brien, D. M.: The utility of remotely sensed CO_2 concentration data in surface source inversions, Geophys. Res. Lett., 28, 175–178, https://doi.org/10.1029/2000GL011912, 2001.

Rodgers, C. D. and Connor, B. J.: Intercomparison of remote sounding instruments, J. Geophys. Res.-Atmos., 108, 4116–4229, https://doi.org/10.1029/2002jd002299, 2003.

Rothman, L. S., Gordon, I. E., Babikov, Y., Barbe, A., Benner, D. C., Bernath, P. F., Birk, M., Bizzocchi, L., Boudon, V., Brown, L. R., Campargue, A., Chance, K., Cohen, E. A., Coudert, L. H., Devi, V. M., Drouin, B. J., Fayt, A., Flaud, J. M., Gamache, R. R., Harrison, J. J., Hartmann, J. M., Hill, C., Hodges, J. T., Jacquemart, D., Jolly, A., Lamouroux, J., Le Roy, R. J., Li, G., Long, D. A., Lyulin, O. M., Mackie, C. J., Massie, S. T., Mikhailenko, S., Muller, H. S. P., Naumenko, O. V., Nikitin, A. V., Orphal, J., Perevalov, V., Perrin, A., Polovtseva, E. R., Richard, C., Smith, M. A. H., Starikova, E., Sung, K., Tashkun, S., Tennyson, J., Toon, G. C., Tyuterev, V. G., and Wagner, G.: The HITRAN2012 molecular spectroscopic database, J. Quant. Spectrosc. Ra., 130, 4–50, https://doi.org/10.1016/j.jqsrt.2013.07.002, 2013.

Schaefer, H., Fletcher, S. E. M., Veidt, C., Lassey, K. R., Brailsford, G. W., Bromley, T. M., Dlugokencky, E. J., Michel, S. E., Miller, J. B., Levin, I., Lowe, D. C., Martin, R. J., Vaughn, B. H., and White, J. W. C.: A 21st-century shift from fossil-fuel to biogenic methane emissions indicated by $^{13}CH_4$, Science, 352, 80–84, https://doi.org/10.1126/science.aad2705, 2016.

Schepers, D., Butz, A., Hu, H., Hasekamp, O. P., Arnold, S. G., Schneider, M., Feist, D. G., Morino, I., Pollard, D., Aben, I., and Landgraf, J.: Methane and carbon dioxide total column retrievals from cloudy GOSAT soundings over the oceans, J. Geophys. Res.-Atmos., 121, 5031–5050, https://doi.org/10.1002/2015JD023389, 2016.

Sherlock, V., Connor, B., Robinson, J., Shiona, H., Smale, D., and Pollard, D.: TCCON data from Lauder, New Zealand, 120HR, Release GGG2014R0, https://doi.org/10.14291/tccon.ggg2014.lauder01.R0/1149293, TCCON data archive, hosted by the Carbon Dioxide Information Analysis Center, Oak Ridge National Laboratory, Oak Ridge, Tennessee, USA, 2014a.

Sherlock, V., Connor, B., Robinson, J., Shiona, H., Smale, D., and Pollard, D.: TCCON data from Lauder, New Zealand, 125HR, Release GGG2014R0, https://doi.org/10.14291/tccon.ggg2014.lauder02.R0/1149298, TCCON data archive, hosted by the Carbon Dioxide Information Analysis Center, Oak Ridge National Laboratory, Oak Ridge, Tennessee, USA, 2014b.

Steinkamp, K., Mikaloff Fletcher, S. E., Brailsford, G., Smale, D., Moore, S., Keller, E. D., Baisden, W. T., Mukai, H., and Stephens, B. B.: Atmospheric CO_2 observations and models suggest strong carbon uptake by forests in New Zealand, Atmos. Chem. Phys., 17, 47–76, https://doi.org/10.5194/acp-17-47-2017, 2017.

Toon, G. C.: Telluric line list for GGG2014, available at: https://doi.org/10.14291/tccon.ggg2014.atm.R0/1221656, last access: 2 May 2014a.

Toon, G. C.: Solar line list for GGG2014, available at: https://doi.org/10.14291/tccon.ggg2014.solar.R0/1221658, last access: 2 May 2014b.

Toon, G. C. and Wunch, D.: A stand-alone a priori profile generation tool for GGG2014, available at: https://doi.org/10.14291/tccon.ggg2014.priors.R0/1221661 (last access: 15 June 2016), 2014.

Wofsy, S. C.: HIAPER Pole to Pole Observations (HIPPO): fine-grained, global-scale measurements of climatically important atmospheric gases and aerosols, Philos. T. R. Soc. A, 369, 2073–2086, https://doi.org/10.1098/rsta.2010.0313, 2011.

Wunch, D., Toon, G. C., Wennberg, P. O., Wofsy, S. C., Stephens, B. B., Fischer, M. L., Uchino, O., Abshire, J. B., Bernath, P., Biraud, S. C., Blavier, J.-F. L., Boone, C., Bowman, K. P., Browell, E. V., Campos, T., Connor, B. J., Daube, B. C., Deutscher, N. M., Diao, M., Elkins, J. W., Gerbig, C., Gottlieb, E., Griffith, D. W. T., Hurst, D. F., Jiménez, R., Keppel-Aleks, G., Kort, E. A., Macatangay, R., Machida, T., Matsueda, H., Moore, F., Morino, I., Park, S., Robinson, J., Roehl, C. M., Sawa, Y., Sherlock, V., Sweeney, C., Tanaka, T., and Zondlo, M. A.: Calibration of the Total Carbon Column Observing Network using aircraft profile data, Atmos. Meas. Tech., 3, 1351–1362, https://doi.org/10.5194/amt-3-1351-2010, 2010.

Wunch, D., Toon, G. C., Blavier, J. F. L., Washenfelder, R. A., Notholt, J., Connor, B. J., Griffith, D. W. T., Sherlock, V., and Wennberg, P. O.: The Total Carbon Column Observing Network, Philos. T. R. Soc. A, 369, 2087–2112, https://doi.org/10.1098/rsta.2010.0240, 2011a.

Wunch, D., Wennberg, P. O., Toon, G. C., Connor, B. J., Fisher, B., Osterman, G. B., Frankenberg, C., Mandrake, L., O'Dell, C., Ahonen, P., Biraud, S. C., Castano, R., Cressie, N., Crisp, D., Deutscher, N. M., Eldering, A., Fisher, M. L., Griffith, D. W. T., Gunson, M., Heikkinen, P., Keppel-Aleks, G., Kyro, E., Lindenmaier, R., Macatangay, R., Mendonca, J., Messerschmidt, J., Miller, C. E., Morino, I., Notholt, J., Oyafuso, F. A., Rettinger, M., Robinson, J., Roehl, C. M., Salawitch, R. J., Sherlock, V., Strong, K., Sussmann, R., Tanaka, T., Thompson, D. R., Uchino, O., Warneke, T., and Wofsy, S. C.: A method for evaluating bias in global measurements of CO_2 total columns from space, Atmos. Chem. Phys., 11, 12317–12337, https://doi.org/10.5194/acp-11-12317-2011, 2011b.

Wunch, D., Toon, G., Sherlock, V., Deutscher, N. M., Liu, C., Feist, D. G., and Wennberg, P. O.: The Total Carbon Column Observing Network's GGG2014 Data Version, available at: https://doi.org/10.14291/tccon.ggg2014.documentation.R0/1221662, last access: 16 December 2015.

Wunch, D., Wennberg, P. O., Osterman, G., Fisher, B., Naylor, B., Roehl, C. M., O'Dell, C., Mandrake, L., Viatte, C., Kiel, M., Griffith, D. W. T., Deutscher, N. M., Velazco, V. A., Notholt, J., Warneke, T., Petri, C., De Maziere, M., Sha, M. K., Sussmann, R., Rettinger, M., Pollard, D., Robinson, J., Morino, I., Uchino, O., Hase, F., Blumenstock, T., Feist, D. G., Arnold, S. G., Strong, K., Mendonca, J., Kivi, R., Heikkinen, P., Iraci, L., Podolske, J., Hillyard, P. W., Kawakami, S., Dubey, M. K., Parker, H. A., Sepulveda, E., García, O. E., Te, Y., Jeseck, P., Gunson, M. R., Crisp, D., and Eldering, A.: Comparisons of the Orbiting Carbon Observatory-2 (OCO-2) X_{CO_2} measurements with TCCON, Atmos. Meas. Tech., 10, 2209–2238, https://doi.org/10.5194/amt-10-2209-2017, 2017.

Yokota, T., Yoshida, Y., Eguchi, N., Ota, Y., Tanaka, T., Watanabe, H., and Maksyutov, S.: Global concentrations of CO_2 and CH_4 retrieved from GOSAT: first preliminary results, Sola, 5, 160–163, https://doi.org/10.2151/sola.2009-041, 2009.

Permissions

All chapters in this book were first published in ESSD, by Copernicus Publications; hereby published with permission under the Creative Commons Attribution License or equivalent. Every chapter published in this book has been scrutinized by our experts. Their significance has been extensively debated. The topics covered herein carry significant findings which will fuel the growth of the discipline. They may even be implemented as practical applications or may be referred to as a beginning point for another development.

The contributors of this book come from diverse backgrounds, making this book a truly international effort. This book will bring forth new frontiers with its revolutionizing research information and detailed analysis of the nascent developments around the world.

We would like to thank all the contributing authors for lending their expertise to make the book truly unique. They have played a crucial role in the development of this book. Without their invaluable contributions this book wouldn't have been possible. They have made vital efforts to compile up to date information on the varied aspects of this subject to make this book a valuable addition to the collection of many professionals and students.

This book was conceptualized with the vision of imparting up-to-date information and advanced data in this field. To ensure the same, a matchless editorial board was set up. Every individual on the board went through rigorous rounds of assessment to prove their worth. After which they invested a large part of their time researching and compiling the most relevant data for our readers.

The editorial board has been involved in producing this book since its inception. They have spent rigorous hours researching and exploring the diverse topics which have resulted in the successful publishing of this book. They have passed on their knowledge of decades through this book. To expedite this challenging task, the publisher supported the team at every step. A small team of assistant editors was also appointed to further simplify the editing procedure and attain best results for the readers.

Apart from the editorial board, the designing team has also invested a significant amount of their time in understanding the subject and creating the most relevant covers. They scrutinized every image to scout for the most suitable representation of the subject and create an appropriate cover for the book.

The publishing team has been an ardent support to the editorial, designing and production team. Their endless efforts to recruit the best for this project, has resulted in the accomplishment of this book. They are a veteran in the field of academics and their pool of knowledge is as vast as their experience in printing. Their expertise and guidance has proved useful at every step. Their uncompromising quality standards have made this book an exceptional effort. Their encouragement from time to time has been an inspiration for everyone.

The publisher and the editorial board hope that this book will prove to be a valuable piece of knowledge for researchers, students, practitioners and scholars across the globe.

List of Contributors

Amelie Driemel, Bernd Loose, Hannes Grobe, Rainer Sieger and Gert König-Langlo
Alfred-Wegener-Institut Helmholtz-Zentrum für Polar- und Meeresforschung, Bremerhaven, Germany

Leslie M. Hartten, Christopher J. Cox, Paul E. Johnston and Daniel E. Wolfe
Cooperative Institute for Research in Environmental Science (CIRES), University of Colorado, Boulder, CO 80309-0216, USA
NOAA/Earth System Research Laboratory, Physical Sciences Division, Boulder, Colorado, CO 80305, USA

H. Alex McColl
Cooperative Institute for Research in Environmental Science (CIRES), University of Colorado, Boulder, CO 80309-0216, USA
NOAA/Earth System Research Laboratory, Physical Sciences Division, Boulder, Colorado, CO 80305, USA Berthoud, CO 80513, USA

Scott Abbott
NOAA/Earth System Research Laboratory, Physical Sciences Division, Boulder, Colorado, CO 80305, USA

Miguel Ángel Saz, Luis Alberto Longares and Martín de Luis
Department of Geography and Regional Planning, University of Zaragoza, Zaragoza, 50009, Spain
Environmental Sciences Institute (IUCA), University of Zaragoza, Zaragoza, 50009, Spain

Santiago Beguería
Estación Experimental de Aula Dei, Consejo Superior de Investigaciones Científicas (EEAD-CSIC), Zaragoza, 50059, Spain

Roberto Serrano-Notivoli
Department of Geography and Regional Planning, University of Zaragoza, Zaragoza, 50009, Spain
Environmental Sciences Institute (IUCA), University of Zaragoza, Zaragoza, 50009, Spain
Estación Experimental de Aula Dei, Consejo Superior de Investigaciones Científicas (EEAD-CSIC), Zaragoza, 50059, Spain

Wendy H. Wood, Shawn J. Marshall and Terri L. Whitehead
Department of Geography, University of Calgary, Calgary AB, T2N 1N4, Canada

Shannon E. Fargey
Department of Geography, University of Victoria, Victoria BC, Canada

Ronald G. Prinn and H. Langley DeWitt
Center for Global Change Science, Massachusetts Institute of Technology, Cambridge, MA, USA

Ray F. Weiss, Christina M. Harth, Jooil Kim, Jens Mühle, Peter K. Salameh and Roland Schmidt
Scripps Institution of Oceanography, University of California San Diego, La Jolla, CA, USA

Jgor Arduini and Michela Maione
Department of Pure and Applied Sciences, University of Urbino, Urbino, Italy

Tim Arnold
National Physical Laboratory, Teddington, Middlesex, UK and School of GeoSciences, University of Edinburgh, Edinburgh, UK

Paul J. Fraser, Paul B. Krummel, Zoë M. Loh, Blagoj Mitrevski and L. Paul Steele
Climate Science Centre, Oceans and Atmosphere, Commonwealth Scientific and Industrial Research Organization (CSIRO), Aspendale, Victoria, Australia

Anita L. Ganesan and Peter G. Simmonds
School of Geographical Sciences, University of Bristol, Bristol, UK

Jimmy Gasore
Rwanda Climate Observatory Secretariat, Ministry of Education of Rwanda, Kigali, Rwanda

Ove Hermansen and Chris R. Lunder
Norwegian Institute for Air Research (NILU), Kjeller, Norway

Shanlan Li and Sunyoung Park
Department of Oceanography, Kyungpook National University, Daegu, Republic of Korea

Alistair J. Manning
Hadley Centre, The Met Office, Exeter, UK
School of Chemistry, University of Bristol, Bristol, UK

Simon O'Doherty, Matt Rigby and Dickon Young
School of Chemistry, University of Bristol, Bristol, UK

Ben R. Miller
National Oceanic and Atmospheric Administration (NOAA), Earth System Research Laboratory, Boulder, CO, USA

Stefan Reimann and Martin K. Vollmer
Laboratory for Air Pollution and Environmental Technology (Empa), Swiss Federal Laboratories for Materials Science and Technology, Dübendorf, Switzerland

Takuya Saito and Yoko Yokouchi
National Institute for Environmental Studies (NIES), Tsukuba, Japan

Ray H. Wang
Georgia Institute of Technology, Atlanta, GA, USA

Bo Yao and Lingxi Zhou
China Meteorological Administration (CMA), Beijing, China

Wolfram Birmili
Leibniz Institute for Tropospheric Research (TROPOS), Leipzig, Germany
German Federal Environment Agency (UBA), Berlin, Germany

Kay Weinhold, Fabian Rasch, André Sonntag, Jia Sun, Maik Merkel and Alfred Wiedensohler
Leibniz Institute for Tropospheric Research (TROPOS), Leipzig, Germany

Susanne Bastian and Gunter Löschau
Saxon State Office for Environment, Agriculture and Geology (LfULG), Dresden, Germany

Alexander Schladitz
Saxon State Office for Environment, Agriculture and Geology (LfULG), Dresden, Germany
SICK Engineering GmbH, Ottendorf-Okrilla, Germany

Josef Cyrys, Jianwei Gu and Thomas Kusch
Helmholtz Zentrum Munich (HMGU), Institute of Epidemiology II, Neuherberg, Germany
University of Augsburg (UA), Wissenschaftszentrum Umwelt, Augsburg, Germany

Mike Pitz
Helmholtz Zentrum Munich (HMGU), Institute of Epidemiology II, Neuherberg, Germany
Bavarian Environment Agency (LUA), Augsburg, Germany

Harald Flentje
Deutscher Wetterdienst (DWD), Meteorologisches Observatorium Hohenpeißenberg, Hohenpeißenberg, Germany

Ulrich Quass and Heinz Kaminski
Institute of Energy and Environmental Technology (IUTA), Duisburg, Germany

Thomas A. J. Kuhlbusch
Institute of Energy and Environmental Technology (IUTA), Duisburg, Germany
Federal Institute for Occupational Safety and Health (BAuA), Dortmund, Germany

Frank Meinhardt, Andreas Schwerin, Olaf Bath, Ludwig Ries, Holger Gerwig and Klaus Wirtz
German Federal Environment Agency (UBA), Dessau-Rosslau, Germany

Markus Fiebig
Norwegian Institute for Air Research (NILU), Kjeller, Norway

Patrick Eriksson, Robin Ekelund and Jana Mendrok
Department of Space, Earth and Environment, Chalmers University of Technology, Gothenburg, Sweden

Manfred Brath, Oliver Lemke and Stefan A. Buehler
Meteorological Institute, Department of Earth Sciences, Center for Earth System Research and Sustainability, Universität Hamburg, Hamburg, Germany

Marjolaine Chiriaco and Sophie Bastin
LATMOS/IPSL, UVSQ Université Paris-Saclay, UPMC Univ. Paris 06, CNRS, Guyancourt, France

Jean-Charles Dupont, Julio Lopez and Martial Haeffelin
Institute Pierre Simon Laplace, Ecole Polytechnique, Université Paris-Saclay, Palaiseau, France

Helene Chepfer and Rodrigo Guzman
Laboratoire de Météorologie Dynamique, Univ. Pierre and Marie Curie, Paris, France

Jordi Badosa
Laboratoire de Météorologie Dynamique, Ecole Polytechnique, Palaiseau, France

Jordis S. Tradowsky
Bodeker Scientific, Alexandra, New Zealand
National Institute of Water and Atmospheric Research (NIWA), Lauder, New Zealand
Institute for Meteorology, Freie Universität Berlin, Berlin, Germany

Gregory E. Bodeker
Bodeker Scientific, Alexandra, New Zealand

Richard R. Querel
National Institute of Water and Atmospheric Research (NIWA), Lauder, New Zealand

Peter J. H. Builtjes
Institute for Meteorology, Freie Universität Berlin, Berlin, Germany

Jürgen Fischer
Institute for Space Sciences, Freie Universität Berlin, Berlin, Germany

David F. Pollard, Vanessa Sherlock, John Robinson and Hisako Shiona
National Institute of Water and Atmospheric Research Ltd (NIWA), Lauder, New Zealand

Nicholas M. Deutscher
School of Chemistry, University of Wollongong, Northfields Ave, Wollongong, NSW, 2522, Australia

Brian Connor
BC Consulting Limited, Martinborough, New Zealand

Index